The Codes of Life

BIOSEMIOTICS

VOLUME 1

Series Editors   **Marcello Barbieri**
Professor of Embryology
University of Ferrara, Italy

President
Italian Association for Theoretical Biology

Editor-in-Chief
Biosemiotics

Jesper Hoffmeyer
Associate Professor in Biochemistry
University of Copenhagen

President
International Society for Biosemiotic Studies

Aims and Scope of the Series
Combining research approaches from biology, philosophy and linguistics, the emerging field of biosemiotics proposes that animals, plants and single cells all engage in semiosis – the conversion of physical signals into conventional signs. This has important implications and applications for issues ranging from natural selection to animal behaviour and human psychology, leaving biosemiotics at the cutting edge of the research on the fundamentals of life.

The Springer book series *Biosemiotics* draws together contributions from leading players in international biosemiotics, producing an unparalleled series that will appeal to all those interested in the origins and evolution of life, including molecular and evolutionary biologists, ecologists, anthropologists, psychologists, philosophers and historians of science, linguists, semioticians and researchers in artificial life, information theory and communication technology.

The Codes of Life
The Rules of Macroevolution

Edited by

Marcello Barbieri
University of Ferrara
Italy

 Springer

Marcello Barbieri
University of Ferrara
Italy

ISBN 978-1-4020-6339-8   e-ISBN 978-1-4020-6340-4

Library of Congress Control Number: 2007939395

Front Cover: The illustrations represent cell samples that have been utilized in signal transduction studies
of the kind described in Chapter 9 by Nadir Maraldi. The micrographs show a hippocampal neuron
labelled with biocytin (the yellow signal indicates synaptic contacts) at the left, and mouse myoblasts
stained for emerin and f-actin, at the right. (By kind permission of the authors of the micrographs,
S. Santi and G. Lattanzi, who obtained them in the laboratory directed by Nadir Maraldi).

Printed on acid-free paper.

9 8 7 6 5 4 3 2 1

springer.com

Editorial

The New Frontier of Biology

Marcello Barbieri

Codes and conventions are the basis of all cultural phenomena and from time immemorial have divided the world of culture from the world of nature. The rules of grammar, the laws of government, the precepts of religion, the value of money, the cooking recipes, the fairy tales, and the rules of chess are all human conventions that are profoundly different from the laws of physics and chemistry, and this has led to the conclusion that there is an unbridgeable gap between nature and culture. Nature is *governed* by objective immutable laws, whereas culture is *produced* by the mutable conventions of the human mind.

In this century-old framework, the discovery of the genetic code, in the early 1960s, came as a bolt from the blue, but strangely enough it did not bring down the barrier between nature and culture. On the contrary, a "protective belt" was quickly built around the old divide with an argument that effectively emptied the discovery of the genetic code of all its revolutionary potential. The argument is that the genetic code is fundamentally a *metaphor* because it must be reducible, in principle, to physical quantities. It is a secondary structure, like those computer programs that allow us to write our instructions in English, thus saving us the trouble to write them in binary digits. Ultimately, however, there are only binary digits in the machine language of the computer, and in the same way, it is argued, there are only physical quantities at the most fundamental level of Nature.

This conclusion, known as *physicalism*, is based on one fact and one assumption. The fact is that all spontaneous reactions are completely accounted for by the laws of physics and chemistry. The assumption is that it was spontaneous reactions that gave origin to the first cells on the primitive Earth. According to physicalism, in short, genes and proteins are spontaneous molecules that evolved into the first cells by spontaneous processes.

This, however, is precisely the point that molecular biology has proved wrong. Genes and proteins are *not* produced by spontaneous processes in living systems. They are produced by molecular machines which physically stick their subunits together in the order provided by *external* templates. They are assembled by molecular robots on the basis of outside instructions, and this makes them as different from ordinary molecules as *artificial* objects are from *natural* ones. Indeed, if we agree that objects are natural when their structure is determined from within and artificial when it is determined from without, we can truly say that genes and proteins are *artificial molecules*, that they are *artifacts made by molecular machines*. This in turn implies that all biological objects are artifacts, i.e. that the whole of life is artifact-making.

Spontaneous genes and spontaneous proteins did appear on the primitive Earth but they did not evolve into the first cells, because spontaneous processes do not have biological specificity. They gave origin to *molecular machines* and it was these machines and their products that evolved into the first cells. The simplest molecular machines we can think of are molecules that can join other molecules together by chemical bonds, and for this reason we may call them *bondmakers*. Some could form bonds between amino acids, some between nucleotides, others between sugars, and so on. Among the various types of bondmakers, some developed the ability to join nucleotides together in the order provided by a *template*. Those bondmakers started *making copies* of nucleic acids, so we can call them *copymakers*. The first Major Transition of the history of life (Maynard Smith and Szathmáry, 1995) is generally described as the origin of genes, but it seems more accurate to say that it was the origin of molecular *copying*, or the origin of *copymakers*, the first molecular machines that started multiplying nucleic acids by making copies of them.

Proteins, on the other hand, cannot be made by copying, and yet the information to make them must come from molecules that can be copied, because only those molecules can be inherited. The information for manufacturing proteins, therefore, had to come from genes, so it was necessary to bring together a carrier of genetic information (a messenger RNA), a peptide bondmaker (a piece of ribosomal RNA), and molecules that could carry both nucleotides and amino acids (the transfer RNAs). The first protein-makers, in short, had to bring together three different types of molecules (messenger, ribosomal, and transfer RNAs), and were therefore much more complex than copymakers. The outstanding feature of the protein-makers, however, was not the number of components. It was the ability to ensure a one-to-one correspondence between genes and proteins, because without it there would be no biological specificity, and without specificity there would be no heredity and no reproduction. Life, as we know, it simply would not exist without a one-to-one correspondence between genes and proteins.

Such a correspondence would be automatically ensured if the bridge between genes and proteins could have been determined by *stereochemistry*, as one of the earliest models suggested, but that is not what happened. The bridge was provided by molecules called *adaptors* (transfer RNAs) that have two recognition sites: one for a group of three nucleotides (a *codon*) and another for an amino acid. The crucial point is that the two recognition sites are physically separated and chemically independent. There is no deterministic link between codons and amino acids, and a one-to-one correspondence between them could only be the result of conventional rules. Only a real code, in short, could guarantee biological specificity, and this means that the evolution of the translation apparatus had to be coupled with the evolution of the genetic code.

Protein synthesis arose, therefore, from the integration of two different processes, and the final machine was a *code-and-template-dependent-peptide-maker*, or, more simply, a *codemaker*. The second Major Transition of the history of life is generally described as the origin of proteins, but it would be more accurate to say that it was the origin of *codemaking*, or the origin of *codemakers*, the first molecular machines that discovered molecular coding and started populating the Earth with codified proteins.

But what happened afterwards? According to modern biology, the genetic code is the only organic code that exists in the living world, whereas the world of culture has a virtually unlimited number of codes. We know, furthermore, that the genetic code came into being at the origin of life, whereas the cultural codes arrived almost 4 billion years later. This appears to suggest that evolution went on for almost the entire history of life on Earth without producing any other organic code after the first one. According to modern biology, in short, the genetic code was a single extraordinary exception, and if nature has only one exceptional code whereas culture contains an unlimited number of them, the real world of codes is culture and the barrier between the two worlds remains intact.

At a closer inspection, however, we realize that the existence of other organic codes cannot be ruled out, and that we can actually test it. Any organic code is a set of rules of correspondence between two independent worlds, and this requires molecular structures that act like *adaptors*, i.e. that perform two independent recognition processes. The adaptors are required because there is no necessary link between the two worlds, and a set of rules is required in order to guarantee the specificity of the correspondence. The adaptors, in short, are necessary in all organic codes. They are the molecular *fingerprints* of the codes, and their presence in a biological process is a sure sign that the process is based on a code. In splicing and signal transduction, for example, it has been shown that there are true adaptors at work, and that allows us to conclude that those processes are based on *splicing codes* and on *signal transduction codes* (Barbieri, 1998, 2003). In a similar way, the presence of adaptors has suggested the existence of *cytoskeleton codes* and of *compartment codes* (Barbieri, 2003).

Many other organic codes have been discovered with different theoretical and experimental criteria. Among them, are the *Sequence Codes* (Trifonov, 1989, 1996, 1999), the *Adhesive Code* (Redies and Takeichi, 1996; Shapiro and Colman, 1999), the *Sugar Code* (Gabius, 2000; Gabius et al., 2002), and the *Histone Code* (Strahl and Allis, 2000; Jenuwein and Allis, 2001; Turner, 2000, 2002; Gamble and Freedman, 2002; Richards and Elgin, 2002). Even if the definition of code has been somewhat different from case to case, these findings tell us that the living world is teeming with organic codes.

What is most important is the realization that organic codes appeared throughout the history of life and that their appearance marked the origin of great biological innovations. When the splicing codes appeared, for example, the *temporal* separation between transcription and translation could be transformed into a *spatial* separation between nucleus and cytoplasm. That set the stage for the origin of the nucleus, and we can say, therefore, that the splicing codes created the conditions for the origin of the eukaryotes. In a similar way, we can associate the origin of multicellular organisms with membrane codes (*cell adhesion codes*), and the origin of animals with embryonic codes (in particular, with the *codes of pattern*).

We realize in this way that there is a deep link between codes and macroevolution, and that opens up an entirely new scenario. The appearance of new organic codes went on throughout the history of life and was responsible for most of its major transitions, from the origin of protein synthesis, with the genetic code, all the way up to the origin of culture with the codes of language. That is the new frontier of biology.

This book is addressed to students, researchers, and academics who are interested in *all* codes of life, from the genetic code to the codes of language. The aim of the book is to show not only that a *plurality* of organic codes exists in nature, but also that codes exist at all levels of life, from the molecular world to the world of language. They exist in cells, embryos, nervous systems, minds, and cultures, and are fundamental components of those systems. In order to illustrate this plurality of codes at a plurality of levels, the book has been divided into four parts: (1) codes and evolution; (2) the genetic code; (3) protein, lipid, and sugar codes; and (4) neural, mental, and cultural codes.

Part 1 describes the first organic codes that have been recognized at the molecular level in addition to the classical triplet code of protein synthesis, and discusses the evolutionary consequences of that extraordinary fact. They range from the idea that the Major Transitions of the history of life were associated with the origin of new organic codes, to the concept that natural selection and natural conventions are two distinct mechanisms of evolution.

Part 2 is dedicated to the genetic code and describes the results of five distinct lines of investigation. More precisely, three chapters illustrate three different biological models on the origin of the genetic code and two chapters argue for the existence of mathematical features in it. This makes us realize that the study of the genetic code is still a very active field of research and, above all, that we have barely scratched its surface. Contrary to a widespread impression, the genetic code has not been explained by modern biology, and this should convince us that the problem of coding is a very central issue in biology.

Part 3 presents a collection of independent research projects that have revealed the existence of organic codes in the worlds of proteins, lipids, and sugars. Each of these cases has individual characteristics and represents a field of research in its own right. This may give an impression of fragmentation, at first, but in reality it is precisely such a *diversity* that makes us realize that codes are *normal* components of living systems. Diversity is the quintessence of the living world and the presence of codes at all levels of diversity is clearly a sign that coding is a basic tool of life.

Part 4 is dedicated to the codes that are associated with the nervous system, the animal mind, and finally language and culture. Here we have the impression to be in a more familiar territory, but in reality the mystery is possibly even deeper. The origin of language does not look any simpler than the origin of life, and some believe that it is probably the hardest problem of all. The point made here is that there has been a long stream of biological codes before language and that it was those codes that created the conditions for the origin of language. The great obstacle to our understanding of language, in short, is the present paradigm where biological codes are regarded as unlikely accidents or extraordinary exceptions rather than normal components of nature.

Acknowledgements

This is the first collective book on the codes of life and it has become a reality because its authors have accepted to take part in a project that goes beyond their individual fields. Each of them has contributed a vital piece to the new mosaic of life that we now have before us, and I wish I could adequately thank them for joining the enterprise. I am also profoundly grateful to the publishing manager, Catherine Cotton, who has created the Springer Book Series in Biosemiotics with the specific purpose to promote the study of the codes of life and to turn it into a new academic discipline. This book is the beginning of that project and I wish to acknowledge, with thanks, that Catherine has been the driving force behind it.

References

Barbieri M (1998) The organic codes. The basic mechanism of macroevolution. Rivista di Biologia-Biology Forum 91:481–514

Barbieri M (2003) The Organic Codes. An Introduction to Semantic Biology. Cambridge University Press, Cambridge

Gabius H-J (2000) Biological information transfer beyond the genetic code: the sugar code. Naturwissenschaften 87:108–121

Gabius H-J, André S, Kaltner H, Siebert H-C (2002) The sugar code: functional lectinomics. Biochimica et Biophysica Acta 1572:165–177

Gamble MJ, Freedman LP (2002) A coactivator code for transcription. Trends Biochem Sci 27(4):165–167

Jenuwein T, Allis D (2001) Translating the histone code. Science 293:1074–1080

Maynard Smith J, Szathmáry E (1995) The Major Transitions in Evolution. Oxford University Press, Oxford

Readies C, Takeichi M (1996) Cadherine in the developing central nervous system: an adhesive code for segmental and functional subdivisions. Develop Biol 180:413–423

Richards EJ, Elgin SCR (2002) Epigenetic codes for heterochromatin formation and silencing: rounding up the usual suspects. Cell 108:489–500

Shapiro L, Colman DR (1999) The diversity of cadherins and implications for a synaptic adhesive code in the CNS. Neuron 23:427–430

Strahl BD, Allis D (2000) The language of covalent histone modifications. Nature 403:41–45

Trifonov EN (1989) The multiple codes of nucleotide sequences. Bull Math Biol 51:417–432

Trifonov EN (1996) Interfering contexts of regulatory sequence elements. Cabios 12:423–429

Trifonov EN (1999) Elucidating sequence codes: three codes for evolution. Annal NY Acad Sci 870:330–338

Turner BM (2000) Histone acetylation and an epigenetic code. BioEssay 22:836–845

Turner BM (2002) Cellular memory and the histone code. Cell 111:285–291

Contents

Part 2   The Genetic Code

Chapter 3   Catalytic Propensity of Amino Acids and the Origins of the Genetic Code and Proteins **39**
Ádám Kun, Sándor Pongor, Ferenc Jordán,
and Eörs Szathmáry

Chapter 4   Why the Genetic Code Originated: Implications for the Origin of Protein Synthesis **59**
Massimo Di Giulio

Part 1
Codes and Evolution

Chapter 1

Codes of Biosequences

Edward N. Trifonov

Abstract Contrary to common belief that the nucleotide sequences only encode proteins, there are numerous additional codes, each of a different nature. The codes, at DNA, RNA, and protein sequence levels, are superposed, i.e. the same nucleotide in a given sequence may be simultaneously involved in several different encoded functions, at different levels. Such coexistence is possible due to degeneracy of the messages present in the sequence. Protein sequences are degenerate as well: involved not only in the functions related to the protein, but also adjusting to sequence requirements at the DNA level.

1 Introduction

All manifestations of life, from elementary biomolecular interactions to human behavior, are tightly associated with, if not in full command of, sequence-specific interactions. Nucleic acid or protein sequence patterns involved in the molecular or higher-level functions stand for the sequence codes of the functions. The genome that carries or encodes all these sequence patterns is, thus, a compact, intricately organized, informational depot. To single out all major sequence codes and trace them in action may be viewed as the major challenge of modern molecular biology, sequence biology.

The nucleotide sequences, thus, not only encode proteins, as an inexperienced reader would think. Various sequence instructions are read from the DNA, RNA, or protein molecule each in its own way, via one or another specific molecular interaction or a whole network of interactions. In the triplet code the reading device is the ribosome. In gene splicing the sequence signals are recognized by the spliceosome. There are also numerous relatively simple sequence-specific DNA–protein and RNA–protein interactions, where the respective sequences are read by a single protein.

After the triplet code was spectacularly cracked (Ochoa et al. 1963; Khorana et al., 1966; Nirenberg et al., 1966), the impact of this event was such that

Genome Diversity Center, Institute of Evolution, University of Haifa, Haifa 31905, Israel,
e-mail: trifonov@research.haifa.ac.il

nobody could even think of other possible codes. The triplet code was even called "genetic code," in other words *the only* code, not leaving any room for doubts. All early history of bioinformatics revolved around this single code (Trifonov, 2000a). Yet, already in 1968, R. Holliday noted almost en passant that, perhaps, recombination signals in yeast might reside on the same sequence that encodes proteins. This remark not only introduced the notion of other possible codes, but also the overlapping of different codes on the same sequence. The existence of codes, other than the classical translation triplet code, is already suggested by degeneracy of the triplet code (Schaap, 1971). Freedom in the choice of codons allows significant changes in the nucleotide sequence without changing the encoded protein sequence. This makes it possible, in principle, to utilize the interchangeable bases of the mRNA sequence for some additional, different codes. In this case, the codes would coexist in interspersed form as mosaics of two or more "colors." It is known today that a more general and widespread case is when the codes literally overlap so that some letters in specific positions of a given sequence (nucleotides or amino acids) are simultaneously involved in two or more different codes (sequence patterns). Such is the case with the coexisting triplet code and chromatin code – sequence instructions for nucleosome positioning (Trifonov, 1980; Mengeritsky and Trifonov, 1983). This was the first demonstration of the actual existence (Trifonov, 1981) of the hypothetical overlapping codes. Sequences that do not encode proteins, despite their traditional classification as noncoding, carry some important messages (codes) as well. Especially striking are the cases of sequence conservation in the noncoding regions (Koop and Hood, 1994), suggesting that the so-called noncoding sequences are associated with some function.

Amongst known general sequence codes, other than the triplet code, are transcription signals (*transcription code*) in promoters such as TATAAA box in eukaryotes, and TATAAT and TTGACA boxes in bacteria coding for initiation of transcription. Another broadly known sequence code is the *gene splicing code*, the GT–AG rule (Breathnach and Chambon, 1981) and some sequence preferences around the intron–exon junctions. A complex set of sequence rules describes details of DNA shape important for DNA–protein interactions and DNA folding in the cell.

At the level of amino acid sequences, the most important is the *protein folding code*, which is not yet described as a sequence pattern. One can single out the modular component of the folding code – organization of the globular proteins as linear succession of the modules in the form of loops of 25–30 residues closed at the ends by interactions between hydrophobic residues (Berezovsky et al., 2000; Berezovsky and Trifonov, 2002). The 3D structure of proteins appears to be encoded largely by a *binary code* (Trifonov et al., 2001; Trifonov, 2006; Gabdank et al., 2006) that, essentially, reduces the 20-letter alphabet to only two letters, for nonpolar and polar residues (more accurately, residues encoded by codons with pyrimidine or purine in the middle). The binary code also suggests the ancestral form for any given sequence.

As the carriers of instructions, biological sequences may be considered a language. Indeed, according to an appealing definition of Russian philosopher V. Nalimov (1981), language is a communication tool to carry instructions to the operator at the receiving end. Such languages as computer programs (frequently called "codes" as well) and written (spoken) human languages convey instructions expressed in the form of one code, for one reading device that takes consecutively letter by letter, word by word, until the transmitted command is fully uttered. As mentioned above, a unique property of the biological sequences is the super-position of the codes they carry. That is, the same sequence is meant to be read by several reading devices, each geared to its own specific code. Many cases of such overlapping are known (Trifonov, 1981; Normark et al., 1983). The overlapping is possible due to degeneracy of the codes. There is, of course, an informational limit for such superposition, when the freedom of degeneracy becomes insufficient to accommodate additional messages without loss of quality of many or all other messages present.

2 Hierarchy of the Codes

The commonly considered information flow from DNA to RNA and to protein is accompanied by massive loss of the sequences involved. Indeed, neither all DNA is transcribed, nor is the whole mass of RNA transcripts translated. This is espe-cially obvious in eukaryotic genomes that contain large intergenic regions, and large intervening sequences that are passed from DNA to pre-mRNA. Is that loss of sequences also a loss of information? The multiplicity of the codes and their superposition suggest that some information is lost, indeed, together with those sequences that are not transcribed and not translated. In other words, DNA carries the sequence codes, serving at the DNA level, of which some are transferred to pre-mRNA. The sequences of the transcripts carry codes serving at RNA level, of which some are passed to the protein sequences, via mRNA. One, thus, has to consider the codes characteristic for the three sequence levels, hierarchically.

One could think of yet higher-level codes, beyond the purely molecular level. Among them would be organ/tissue-specific codes, i.e. genomic sequence features characteristic for one or another physiological function. These could be specifically placed tandem repeats, dispersed repeats, amplified genes, or whole groups of genes. One could also imagine "personal code(s)" – various sequence details responsible for individual traits, such as distinct facial features (Fondon and Garner, 2004) and mimic (Peleg et al., 2006) body set, favorite postures and gestures, and, perhaps, personal behavioral traits. Well-documented existence of population-specific genetic diseases and disorders indicates that there are also sequence features responsible for ethnicity traits. These may include specific sequence polymorphisms and, perhaps, some "guest" sequences present in one ethnical group and absent in others. The higher-level codes are likely to become a major focus of molecular medicine in coming decades. In the mean time the sequence codes of molecular levels are still struggling to make it from singular to plural.

2.1 DNA Level Codes

The DNA structure is not monotonously uniform. It is modulated by the sequence-dependent local deviations from standard geometry, which may accumulate, for example, to a net DNA curvature (Trifonov and Sussman, 1980). Geometry of every base-pair step in the simple wedge model is described by three angles – wedge roll, wedge tilt, and twist. By following the sequence and deflecting the DNA axis at every step, according to the wedge and twist angles from the table of the dinucleotide codons (Bolshoy et al., 1991; Trifonov, 1991), one can calculate the predicted path of DNA axis – its local shape for any given sequence (Shpigelman et al., 1993). Hence, *DNA shape code*.

The *chromatin code* is a set of rules directing sequence-specific positioning of the nucleosomes. Sequence-dependent deformational anisotropy (bendability) of DNA appears to be an underlying principle of the nucleosome sequence specificity (Trifonov, 1980). As the strands of the nucleosome DNA follow the path of the deformed DNA duplex, they pass through inner contact points with histones (inter-face positions) and outward points (exposed to nucleoplasm). Various sequence elements that prefer the inner or outward positions would thus, ideally, reappear in the sequence at the distances that are multiples of nucleosome DNA. Indeed, the sequence periodicity is the most conspicuous feature of the nucleosome DNA sequences (Trifonov and Sussman, 1980). According to the latest updates (Cohanim et al., 2005, 2006a; Kogan et al., 2006; Trifonov et al., 2006a), there are at least three major periodical patterns in the nucleosome DNA: counter-phase AA/TT pattern, counter-phase GG/CC pattern (both combined in RR/YY pattern), and in-phase AA/TT pattern. Several other possible patterns are discussed in literature (reviewed in Kiyama and Trifonov, 2002; Segal et al., 2006).

An important issue in the elucidation of the chromatin sequence code is manda-tory weakness of the nucleosome positioning sequence signal. This is required by the necessity of unfolding the nucleosomes during template processes. That is, the DNA complexes with the histone cores in the nucleosomes should be of marginal stability only. Accordingly, the sequence elements associated with the DNA bend-ability should be rather scarce in the nucleosome DNA sequence, especially those elements that are strong contributors to the bendability. Regrettably, it makes the deciphering of the nucleosome positioning code quite a challenge.

One of the factors influencing the nucleosome positioning is sterical exclusion of the nucleosomes by other nucleosomes, neighbors in 3D space (Ulanovsky and Trifonov, 1986). The most obvious sterical rule is the rule of linkers, first formu-lated and experimentally observed by Noll et al. (1980). Since every extra base pair in the linker causes rotation of the nucleosome around the axis of the linker by ~34°, the rotation may result in a sterical clash between the nucleosomes connected by the common linker. This effect, indeed, is observed at short linkers. It is expressed in preferential appearance of the linkers of lengths about 5–11, 16–21, and 26–31 bases (Noll et al., 1980; Mengeritsky and Trifonov, 1983; Ulanovsky and Trifonov, 1986; Cohanim et al., 2006a). Intermediate linker lengths are forbidden due to the sterical clashes ("interpenetration" of the nucleosomes). The rule of linkers, thus, is an important part of the chromatin code.

2.2 RNA Level Codes

Those messages contained in the transcribed DNA are passed to RNA. The transcribed DNA, thus, contains overlapping messages of both DNA and RNA levels. The major mRNA level message is the classical triplet code – *RNA-to-protein translation code*. The chapters about this code appear in every textbook on molecular biology, and it will not be described here.

Eukaryotic transcripts also carry the *RNA splicing code*. This code is only poorly described (Breathnach and Chambon, 1981; Mount, 1982), so that existing sequence-based algorithms are not sufficient for detection of the splice sites in the sequences with as high a precision as in natural splicing process.

Overlapping with the protein-coding message, sequence of codons-triplets, is the universal 3-base periodicity with the consensus $(G\text{-}nonG\text{-}N)_n$ (Trifonov, 1987) or, more accurately, $(GCU)_n$ (Lagunez-Otero and Trifonov, 1992). Since the mRNA binding sites in the ribosome possess a complementary periodicity $(xxC)_n$, with obligatory cytosines complementary to the frequent guanines of the first codon positions in mRNA, these 3-base periodicities have been interpreted as a device to maintain correct reading frame during translation of mRNA – the *framing code* (Trifonov, 1987). As described below, the periodical pattern $(GCU)_n$ in mRNA appears to be a fossil of very ancient organization of codons (Trifonov and Bettecken, 1997).

The usage of codons corresponding to the same amino acid is known to be different for different organisms and even different genes. Among the alternative codons, the rare codons are of special interest. Their occurrence along the mRNA sequence is not random. It is shown, for example, that clusters of infrequently used codons in prokaryotic mRNA often follow at a distance about 150 triplets from one another. This is interpreted as *translation pausing code*, to slow down the translation after a protein domain (fold) is synthesized: to give the newly synthesized chain sufficient time for its proper folding (Makhoul and Trifonov, 2002).

2.3 Codes of Protein Sequences

According to common belief, the protein sequence carries instructions on how the polypeptide chain folds, for the reliable performance of respective function of the protein, encoded in the sequence as well. At the same time, it is well known that proteins with the same fold and the same function may have rather different sequences. As in the case of the triplet code, this degeneracy of the protein sequence may allow incorporation in the same sequence of some additional messages.

The *protein folding code* is a major challenge for the protein structure community. There are plenty of sophisticated approaches offering partial solutions of the problem, but the conclusive sequence rules for protein folding are still to be found.

An apparent major obstacle is estimated colossal time required for the unfolded polypeptide chain to go through all intermediate states until the final native fold structure is reached – the so-called Levinthal paradox. By some trick of nature, a special sequence organization should be there, in the protein sequences, to ensure the folding in realistic time of milliseconds to seconds. One possible way out is suggested by modular organization of the protein folds (Berezovsky and Trifonov, 2002). Indeed, if the chain length of the module is 20–30 amino acid residues, the time required for its folding fits well to the realistic limits. And, as numerous recent studies demonstrate, globular proteins *are* built of such modules of standard size 25–30 residues in form of closed loops (Berezovsky et al., 2000; Trifonov and Berezovsky, 2003; Berezovsky et al., 2003a, b; Aharonovsky and Trifonov, 2005; Sobolevsky and Trifonov, 2006).

The modular structure of proteins suggests a principally new, compressed way of presentation of amino acid sequences rather as, sequences of the modules, descendants of the early sequence/structure/function prototypes (Berezovsky et al., 2003a, b), in a new alphabet of the prototypes. This would represent the *proteomic code* contained in the amino acid sequences. The prototype modules, then, would appear as the codons of the proteomic code.

2.4   Fast Adaptation Code

This code resides and functions in all three types of genetic sequences. It is believed to be responsible for special type of quick, significant changes in the sequences, apparently, in response to environmental changes. It involves the most variable sequences – simple tandem repeats of the structure $(AB...MN)_n$. Remarkably, the information carried in the sequences resides not as much in the sequence $AB...MN$ of the repeating unit, as rather in the copy number n of the repeats (Trifonov, 1989, 2004). Indeed, after the spontaneous change in the repeating sequence, its extension or shortening, the sequence in brackets stays intact while the copy number n becomes larger or smaller, respectively. Since the repeats are involved in gene expression in one or another way, the change of n results in the modulation of gene activities, as a response to environmental challenges, and thus in fast adaptation (Trifonov, 1989, 1990, 1999, 2004; Holliday, 1991; King, 1994; Künzler et al., 1995, King et al., 1997). An important faculty of this mechanism is an apparent directionality of the mutational changes of this type (Trifonov, 2004). Indeed, small variations in the n values corresponding to repeats serving genes *irrelevant* to a given environmental stress do not change the expression patterns of these genes. On the contrary, if *relevant* responsive genes are involved, the copy numbers of the respective repeats become subject of systematic selection towards better repeat copy number (better gene expression) patterns. The relevant genes (but only relevant ones) become, thus, retuned (King et al., 1997; Trifonov, 1999).

2.5   The Codes of Evolutionary Past

Every sequence has its evolutionary history, and those sequences or sequence fragments, that have been successful in the earliest times of molecular evolution, are, perhaps, still around in hidden form or even unchanged since those times. The proteomic code described above is an example of such code of evolutionary record. The modern sequence modules are not the same as their ancestral prototypes, but a certain degree of resemblance to the ancestors is conserved allowing classification of present-day modules.

The earliest traced sequence elements go back to the very first codons, which are described as the triplets GGU, GCC, and their point mutational versions (Trifonov and Bettecken, 1997). More detailed reconstruction confirmed this conclusion (Trifonov, 2000b, 2004). According to the reconstruction of the earliest stages of molecular evolution, the very first "genes" had a duplex structure with complementary sequences $(GGC)_n$ and $(GCC)_n$, encoding, Gly_n and Ala_n, respectively. Thus, the mRNA consensus $(GCU)_n$ and the consensus $(xxC)_n$ of the mRNA binding sites in the ribosome are both fossils of the earliest mRNA sequences (Trifonov, 1987; Lagunez-Otero and Trifonov, 1992; Trifonov and Bettecken, 1997).

The size of the earliest minigenes, as it turns out, can be estimated by distance analysis of modern mRNA sequences (Trifonov et al., 2001). For this purpose the sequences were first rewritten in binary form, in an alphabet of two letters, G and A, for Gly series of amino acids and codons and Ala series (see above). Respective codons contain in their middle positions either purines (in G) or pyrimidines (in A). From the reconstructed chart of evolution of the codons (Trifonov, 2000b, 2004), it follows that all codons of G-series are descendants of the GGC codon, with purine in the middle, while codons of A-series originate from GCC codon, with pyrimidine in the middle. If the products of very first genes had the structures either G_n or A_n, of a certain size n, then after fusion of the minigenes the alternating patterns $G_nA_nG_nA_n$ ... may have been formed. Later mutations could, of course, have completely destroyed this pattern, but they did not. Analysis of large ensembles of the mRNA sequences showed that the pattern did survive, though in rather hidden form (Berezovsky and Trifonov, 2001; Trifonov et al., 2001) so that the estimation of the very first gene size became possible, 6–7 codons encoding hexa- and heptapeptides. This estimate is strongly supported by independent calculation of the sizes of the most ancient mRNA hairpins that arrived at the same minigene size (Gabdank et al., 2006; Trifonov et al., 2006b). Moreover, most conserved oligopeptide sequences, present in every prokaryotic proteome, also have the size of 6–9 amino acids (Sobolevsky and Trifonov, 2005, Sobolevsky et al., submitted).

The ancient conservation of the middle purines and pyrimidines in the codons during the evolution of the codon table, actually, has very much survived till now. This is confirmed by an analysis of amino acid substitutions in modern proteins (Trifonov, 2006; Gabdank et al., 2006). Every modern protein sequence, thus, can be written in the *A* and *G* alphabet. Such presentations of modern sequences in the *binary code* would suggest the most ancient version of the sequences.

The binary code, the mosaic of *A*- and *G*-minigenes, and the proteomic code describe various stages of protein evolution, from simple to more complex. Today one can also detect the next stage – combining the closed loop modules in the protein folds, domains.

First, the next level is seen already in protein sizes, which appear to be multiples of 120–150 amino acid units (Berman et al., 1994; Kolker et al., 2002). This size is a good match to the optimal DNA ring closure size, about 400 base pairs (Shore et al., 1981). This attractive numerology may well reflect original formation of modern genes and genomes by fusion of individual DNA circles (genome units) of this standard size (Trifonov, 1995, 2002). This would constitute the *genome segmentation code*. How this code is expressed in the sequence form is not yet specified, except for preferential appearance of methionines (former translation starts) at genome unit size distances (Kolker and Trifonov, 1995).

3   Superposition of the Codes and Interactions Between Them

As most of the codes described above are degenerate, allowing alternative or sometimes even wrong letters here and there, they may coexist as a superposition of several codes, on the same sequence (reviewed in Normark et al., 1983; Trifonov, 1981, 1989, 1996, 1997). The most spectacular case is the overlapping of the chromatin code (nucleosome positioning) with protein coding and gene splicing. Indeed, the alternating AA/TT nucleosome pattern is demonstrated to be located largely, if not fully, on those sections of the protein-coding regions that correspond to amphipathic α-helices (Cohanim et al., 2006a,b). The third positions of the codons within the region occupied by the nucleosome are responsible as well for the creation of the periodical AA/TT pattern. Moreover, even the encoded amino acid sequence is also biased to a certain degree to contribute to the nucleosome sequence pattern (Cohanim et al., 2006b). In addition, the nucleosomes are preferentially centered at the splice junctions, apparently for their protection (Denisov et al., 1997; Kogan and Trifonov, 2005). Since the coding sequences also carry at least one more message – translation framing, the nucleosome sequences display superposition of at least four different codes, on the same sequence.

The adjustment of the protein sequence, to contribute to the DNA sequence periodicity, both in prokaryotes and in eukaryotes (Cohanim et al., 2006b), is an interesting case. Apparently, on one hand, the 10–11 base DNA sequence periodicity is of no less importance for the cell than the proteins encoded in the DNA sequence. On the other hand, this example of interactions between the codes shows that the DNA sequence level message is projected all the way through mRNA to the protein sequence level. The latter one, thus, carries (reflects) the sequence patterns of the whole hierarchy – of DNA, RNA, and protein levels.

A neat example of the overlapping at the level of protein sequences is the "moonlighting" of intrinsically unfolded proteins (IUPs) (Tompa et al., 2005). That is, the same molecule of the IUP, the same sequence, can be involved in more than one function, thus, carrying different superimposed messages. Structural and functional promiscuity of the IUPs is carried through, perhaps, since the earliest times of molecular evolution. Highly structured functionally specialized proteins were not yet around, and the multi-functionality of simpler IUP molecules was of an obvious advantage for survival.

4   Is That All?

There are still many nondeciphered codes around. Nature would utilize every useful combination of letters. This is because of eternal molecular opportunism ' (Doolittle, 1988) that drives the molecules of life towards better and more diverse performance in the challenging conditions of the changing environment. In this struggle for survival (natural selection) and for better well-being (opportunism), living matter developed intricate levels of complexity, including sequence complexity. It would be naive to say that all the codes are already known, as it was, indeed, naive to content oneself with the single "genetic code" 30 years ago.

On the one hand, there are sequence biases and patterns that are still not fully explained, such as species-specific G + C content of genomes – genomic code (D'Onofrio and Bernardi, 1992), and general avoidance of the CG dinucleotides. On the other hand, many of the known molecular functions still do not have explicit sequence descriptions, such as RNA interference (Fire et al., 1998) or RNA editing (Gott and Emeson, 2000). The so-called noncoding sequences have the provocative property of being rather dispensable, though they do carry some of the codes described in the review (chromatin code, fast adaptation code). The famous case of the Fugu-fish genome, with the reduced amount of noncoding sequences in it (Aparicio et al., 2002), is often taken as an example of a seemingly insignificant role the noncoding sequences play. Yet, it is known that the noncoding sequences harbor various repeats, of dispersed type (transposons), and tandem repeats. It is also known that transposable elements play an important role in evolution and adaptation (Reanney, 1976). The tandem repeats serve as tuners of gene expression (Trifonov, 1989, 2004; King et al., 1997; Fondon and Garner, 2004) (see *Fast adaptation code*, above). Could it be that the Fugu-fish is in an evolutionary steady state, with virtually no need for adaptive sequence changes? That could be only if there are no environmental challenges for this species. Indeed, the small-genome Fugu-fish has a narrow habitat (Hinegardner, 1976), living only in coral reefs with well-defined fauna, around the islands of Japan. Thus, even dispensable sequences deserve respect, as they seem to code for the vital ability for adaptation.

The conspicuously primitive simple tandem repeats are the best advocates in favor of all sequences, no matter how nonsensical, primitive, or even dispensable they appear. In a recent study (Bacolla et al., 2006), the pure purine or pyrimidine repeats are shown to be the only difference between human and chimpanzee sequences (over 800 large segments studied). The repeats are also the same, but the copy numbers of the repeat units (total lengths of the repeat regions) are different in these two species. Referring to the fast adaptation code (above), one would think that humans and chimpanzees are nearly the same species, only well adapted to completely different living conditions. So much for even the primitive sequences.

The answer to the question in the title of this section, thus, is a firm "No."

References

Aharonovsky E, Trifonov EN (2005) Protein sequence modules. J Biomol Str Dyn 23:237–242

Aparicio S, *et al.* (2002) Whole-genome shotgun assembly and analysis of the genome of *Fugu rubripes*. Science 297:1301–1310

Bacolla A, Collins JR, Gold B et al. (2006) Long homopurine* homopyrimidine sequences are characteristic of genes expressed in brain and the pseudoautosomal region. Nucl Acids Res 34:2663–2675

Berezovsky IN, Trifonov EN (2001) Evolutionary aspects of protein structure and folding. Mol Biol 35:233–239

Berezovsky IN, Trifonov EN (2002) Loop fold structure of proteins: resolution of Levinthal's paradox. J Biomolec Str Dyn 20:5–6

Berezovsky IN, Grosberg AY, Trifonov EN (2000) Closed loops of nearly standard size: common basic element of protein structure. FEBS Lett 466:283–286

Berezovsky IN, Kirzhner VM, Kirzhner A et al. (2003a) Protein sequences yield a proteomic code. J Biomol Struct Dyn 21:317–325

Berezovsky IN, Kirzhner A, Kirzhner VM, Trifonov EN (2003b) Spelling protein structure. J Biomol Struct Dyn 21:327–339

Berman AL, Kolker E, Trifonov EN (1994) Underlying order in protein sequence organization. Proc Natl Acad Sci USA 91:4044–4047

Bolshoy A, McNamara P, Harrington RE, Trifonov EN (1991) Curved DNA without AA: experimental estimation of all 16 wedge angles. Proc Natl Acad Sci USA 88:2312–2316

Breathnach R, Chambon P (1981) Organization and expression of eukaryotic split genes coding for proteins. Ann Rev Bioch 50:349–383

Cohanim AB, Kashi Y, Trifonov EN (2005) Yeast nucleosome DNA pattern: deconvolution from genome sequences *of S. cerevisiae*. J Biomol Str Dyn 22:687–694

Cohanim AB, Kashi Y, Trifonov EN (2006a) Three sequence rules for chromatin. J Biomol Struct Dyn 23:559–566

Cohanim AB, Trifonov EN, Kashi Y (2006b) Specific selection pressure on the third codon positions: contribution to 10 - 11 base periodicity in prokaryotic genomes. J Molec Evol (in press)

Denisov DA, Shpigelman ES, Trifonov EN (1997) Protective nucleosome centering at splice sites as suggested by sequence-directed mapping of the nucleosomes. Gene 205:145–149

D'Onofrio G, Bernardi G (1992) A universal compositional correlation among codon positions. Gene 110:81–88

Doolittle RF (1988) More molecular opportunism. Nature 336:18

Fire A, Xu S, Montgomery MK et al. (1998) Potent and specific genetic interference by double-stranded RNA in *Caenorhabditis elegans*. Nature 391:806–811

Fondon JW, Garner HR (2004) Molecular origin of rapid and continuous morphological evolution. Proc Natl Acad Sci USA 101:18058–18063

Gabdank I, Barash D, Trifonov EN (2006) Tracing ancient mRNA hairpins. J Biomol Str Dyn 24:163–170

Gott JM, Emeson RB (2000) Functions and mechanisms of RNA editing. Ann Rev Genet 34:499–531

Hinegardner R (1976) Evolution of genome size. In: Ayala FJ (ed) Molecular Evolution. Sinauer Association, Sunderland

Holliday R (1968) Genetic recombination in fungi. In: Peacock WJ, Brock RD (eds) Replication and Recombination of Genetic Material. Australian Academy of Science, Canberra, Australia

Holliday R (1991) Quantitative genetic variation and developmental clocks. J Theor Biol 151:351–358

Khorana HG, Büchi H, Ghosh H et al. (1966) Polynucleotide synthesis and the genetic code. Cold Spring Harb Symp Quant Biol 31:39–49

King DG (1994) Triple repeat DNA as a highly mutable regulatory mechanism. Science 263:595–596

King DG, Soller M, Kashi Y (1997) Evolutionary tuning knobs. Endeavor 21:36–40

Kiyama R, Trifonov EN (2002) What positiones nucleosomes? A model. FEBS Lett 523:7–11

Kogan S, Trifonov EN (2005) Gene splice sites correlate with nucleosome positions. Gene 352:57–62

Kogan SB, Kato M, Kiyama R, Trifonov EN (2006) Sequence structure of human nucleosome DNA. J Biomol Struct Dyn 24:43–48

Kolker E, Trifonov EN (1995) Periodic recurrence of methionines: Fossil of gene fusion? Proc Natl Acad Sci USA 92:557–560

Kolker E, Tjaden BC, Hubley R et al. (2002) Spectral analysis of distributions: finding periodic components in eukaryotic enzyme length data. OMICS: J Integr Biol 6:123–130

Koop BF, Hood L (1994) Striking sequence similarity over almost 100 kilobases of human and mouse T-cell receptor DNA. Nature Genet 7:48–53

Künzler P, Matsuo K, Schaffner W (1995) Pathological, physiological, and evolutionary aspects of short unstable DNA repeats in the human genome. Biol Chem Hoppe-Seyler 376:201–211

Lagunez-Otero J, Trifonov EN (1992) mRNA periodical infrastructure complementary to the proof-reading site in the ribosome. J Biomol Struct Dyn 10:455–464

Makhoul CH, Trifonov EN (2002) Distribution of rare triplets along mRNA and their relation to protein folding. J Biomol Struct Dyn 20:413–420

Mengeritsky G, Trifonov EN (1983) Nucleotide sequence-directed mapping of the nucleosomes. Nucl Acids Res 11:3833–3851

Mount SM (1982) A catalogue of splice junction sequences. Nucl Acids Res 10:459–472

Nalimov VV (1981) In the labyrinths of language: A Mathematician's Journey. ISI Press, Philadelphia, USA

Nirenberg M, Caskey T, Marshall R et al. (1966) The RNA code and protein synthesis. Cold Spring Harb Symp Quant Biol 31:11–24

Noll M, Zimmer S, Engel A, Dubochet J (1980) Self-assembly of single and closely spaced nucleosome core particles. Nucl Acids Res 8:21–42

Normark S, Bergstrom S, Edlund T et al (1983) Overlapping genes. Ann Rev Genet 17:499-525

Ochoa S (1963) Synthetic polynucleotides and the amino acid code. Cold Spring Harb Symp Quant Biol 28:559–567

Peleg G, Katzir G, Peleg O et al. (2006) Hereditary family signature of facial expression. Proc Natl Acad Sci USA 103:15921–15926

Reanney DC (1976) Extrachromosomal elements as possible agents of adaption and development. Bact Rev 40:552–590

Schaap T (1971) Dual information in DNA and the evolution of the genetic code. J Theor Biol 32:293–298

Segal E, Fondufe-Mittendorf Y, Chen L, Thastrom A, Field Y, Moore IK, Wang JP, Widom J (2006) A genome code for nucleosome positioning. Nature 442:772–778

Shore D, Langowski J, Baldwin RL (1981) DNA flexibility studied by covalent closure of short fragments into circles. Proc Natl Acad Sci USA 78:4833–4838

Shpigelman ES, Trifonov EN, Bolshoy A (1993) CURVATURE: software for the analysis of curved DNA. CABIOS 9:435–440

Sobolevsky Y, Trifonov EN (2005) Conserved sequences of prokaryotic proteomes and their compositional age. J Mol Evol 61:591–596

Sobolevsky Y, Trifonov EN (2006) Protein modules conserved since LUCA. J Mol Evol 63:622–634

Tompa P, Scasz C, Buday L (2005) Structural disorder throws new light on moonlighting. Trends Bioch Sci 30:484–489

Trifonov EN (1980) Sequence-dependent deformational anisotropy of chromatin DNA. Nucl Acids Res 8:4041–4053

Trifonov EN (1981) Structure of DNA in chromatin. In: Schweiger H (ed) International Cell Biology 1980–1981. Springer-Verlag, Berlin

Trifonov EN (1987) Translation framing code and frame-monitoring mechanism as suggested by the analysis of mRNA and 16S rRNA nucleotide sequences. J Mol Biol 194:643–652

Trifonov EN (1989) The multiple codes of nucleotide sequences. Bull Math Biol 51:417–432

Trifonov EN (1991) DNA in profile. Trends Biochem Sci 16:467–470

Trifonov EN (1990) Making sense of the human genome. In: Sarma RH, Sarma MH (eds) Structure and Methods, vol. 1, Human Genome Initiative and DNA Recombination. Adenine Press, New York

Trifonov EN (1995) Segmented structure of protein sequences and early evolution of genome by combinatorial fusion of DNA elements. J Mol Evol 40:337–342

Trifonov EN (1996) Interfering contexts of regulatory sequence elements. CABIOS 12:423–429

Trifonov EN (1997) Genetic sequences as product of compression by inclusive superposition of many codes. Mol Biol 31:759–767

Trifonov EN (1999) Elucidating sequence codes: three codes for evolution. Annals NY Acad Sci 870:330–338

Trifonov EN (2000a) Earliest pages of bioinformatics. Bioinformatics 16:5–9

Trifonov EN (2000b) Consensus temporal order of amino acids and evolution of the triplet code. Gene 261:139–151

Trifonov EN (2002) Segmented genome: elementary units of genome structure. Russian J Genet. 38:659–663

Trifonov EN (2004) The triplet code from first principles. J Biomol Struct Dyn 22:1–11

Trifonov EN (2006) Theory of early molecular evolution: predictions and confirmations. In: Eisenhaber F (ed) Discovering Biomolecular Mechanisms with Computational Biology. Landes Bioscience, Georgetown

Trifonov EN, Berezovsky IN (2003) Evolutionary aspects of protein structure and folding, Curr Opinion Struct Biol 13:110–114

Trifonov EN, Bettecken T (1997) Sequence fossils, triplet expansion, and reconstruction of earliest codons. Gene 205:1–6

Trifonov EN, Sussman JL (1980) The pitch of chromatin DNA is reflected in its nucleotide sequence. Proc Natl Acad Sci USA 77:3816–3820

Trifonov EN, Kirzhner A, Kirzhner VM, Berezovsky IN (2001) Distinct stages of protein evolution as suggested by protein sequence analysis. J Mol Evol 53:394–401

Trifonov EN, Kogan S, Cohanim AB (2006a) Latest on the nucleosome positioning sequence patterns. In: Kiyama R, Shimizu M (eds) DNA Structure, Chromatin and Gene Expression. Transworld Research Network. Trivandrum, India

Trifonov EN, Gabdank I, Barash D, Sobolevsky Y (2006b) *Primordia vita*. Deconvolution from modern sequences. Origin Life Evol Biosph 36(5–6):559–565

Ulanovsky LE, Trifonov EN (1986) A different view point on the chromatin higher order structure: steric exclusion effects. In: Sarma RH, Sarma MH (eds) Biomolecular stereodynamics III. Adenine Press, New York

Chapter 2

The Mechanisms of Evolution

Natural Selection and Natural Conventions

Marcello Barbieri

Abstract The mechanisms of evolution have been one of the most controversial issues in Biology and the great debate about them has culminated, in the 1930s and 1940s, in the Modern Synthesis, the theoretical framework where natural selection is regarded as the sole mechanism of evolutionary change. Here it is shown that a new approach to these great problems is provided by two concepts that are firmly based on the evidence of molecular biology. The first is the idea that all biological objects are artifacts, in the sense that they are manufactured structures. Genes and proteins, for example, are produced by molecular machines that physically stick their subunits together in an order provided by external templates. The second concept is the idea that there are two different ways of producing biological artifacts, two distinct mechanisms that here are referred to as copying and coding. The copying of templates is the process that accounts for heredity and, in the long run, for natural selection. Coding is the process that establishes rules of correspondence between two independent worlds, thus giving origin to natural conventions. Copying and coding account, respectively, for the origins of genes and proteins, i.e. for the first two Major Transitions of the history of life. As for the other transitions, it is generally assumed that natural selection has been their sole mechanism, but here it is proposed that a key role was also played by natural conventions, i.e. by the origin of new organic codes. More precisely, it is proposed that natural selection and natural conventions are complementary mechanisms of evolution, the first accounting for the gradual transformation of existing objects and the second for the origin of absolute novelties. This conclusion is a direct consequence of the difference between copying and coding and of the existence of many organic codes in Nature, a fact for which an increasing amount of evidence has been accumulating in recent years.

Dipartimento di Morfologia ed Embriologia, Via Fossato di Mortara 64,
44100 Ferrara, Italy, e-mail: brr@unife.it

M. Barbieri (ed.), *The Codes of Life: The Rules of Macroevolution.*   15
© Springer 2008

Introduction

The history of life has been shaped by great events that today are referred to as Major Transitions (Maynard Smith and Szathmáry, 1995), or steps of macroevolution, or, more simply, origins. The most important are the origins of: (1) genes; (2) proteins; (3) first cells; (4) eukaryotes; (5) embryos; (6) mind; and (7) language.

The feature that defines these great events is the appearance of new biological objects, and this raises immediately a fundamental question: "What is the mechanism that generates novelties in life?" It is no secret that today most biologists accept the conclusion of the Modern Synthesis, i.e. the idea that natural selection *alone* has the power to create evolutionary novelty. A different answer, however, has consistently come from embryology where a long-standing tradition has repeatedly claimed that the major biological novelties are generated by development and that natural selection can act on them only when they have been brought into existence.

This classical claim has not been been invalidated by the discoveries of Evolutionary Developmental Biology (EvoDevo), and recently it has been reproposed by Scott Gilbert (2006a,b) who has stressed that "the generation of novelty is the province of developmental biology." Gilbert's conclusion is based on two groups of arguments. The first is that we finally have the essential lines of a new theory of evolutionary variation (Arthur 2004; Callebaut and Rasskin-Gutman, 2005; Gilbert 2006a), a theory based on two preconditions (gene duplication and modularity) and four mechanisms of *bricolage* (heterotopy, heterochrony, heterotypy, and heterometry). The second group of arguments comes from the discoveries of developmental genetics that have falsified two major expectations of the Modern Synthesis (the alleged impossibility of *deep genetic homology*, and the claim that *parallel evolution* is proof of the creative power of natural selection).

These unexpected discoveries have been the starting point of EvoDevo, but Gilbert's interpretation of them is still a minority view in this field. Most supporters of EvoDevo regard them as proof that developmental biology has a very substantial contribution to make to evolutionary theory, but do not see any need to abandon the paradigm of the centrality of natural selection (Holland, 1999; Davidson, 2001; Gould, 2002; Wilkins, 2002; Carroll, 2005; Ruse, 2006).

Here I will try to show that Gilbert is right in challenging this paradigm and in claiming that there are two distinct mechanisms of evolution, but that he is wrong in saying that the second mechanism is embryonic development. If this were true, we would have to conclude that natural selection has been the sole mechanism of evolution for the entire period that preceded the origin of embryonic development, i.e. for about 3000 million years. More than that, if natural selection was the sole novelty-generating mechanism before the origin of embryos, it must have been natural selection that brought the first embryos into existence. And if the first steps toward development were taken by natural selection, why should the other steps have been different? When did the break occur between natural selection and embryonic development? Arguments like these make it difficult to accept the long-standing claim of embryology, and that is why the centrality of natural selection appears to be the only reasonable option that we are left with, even when we learn that recent discoveries have falsified some predictions of the Modern Synthesis.

There is, however, another solution to the problem of evolutionary novelty. It is the idea that there have been two distinct mechanisms of evolution throughout the whole history of life. The idea that evolution took place *by natural selection and by natural conventions* from the very first cells onwards (Barbieri, 1985, 2003). The second mechanism of evolution, in short, is based on natural conventions, i.e. on organic codes, and has been present on Earth since the origin of protein life because specific proteins cannot exist without a genetic code. In order to back up this new theory, however, we need to prove: (1) that many organic codes have appeared throughout the history of life; and (2) that organic coding is a mechanism of evolution that is distinct from natural selection. These are the key points that will be addressed here and to that purpose the paper has been divided into two parts: (1) the organic codes; and (2) the mechanisms of evolution.

PART 1 – THE ORGANIC CODES[*]

1   The First Major Transition: The Origin of Genes

The discovery that genes and proteins are "manufactured" by molecular machines in all cells, has direct implications for the history of life because it tells us that primitive molecular machines came into existence long before the origin of the first cells. We can divide, therefore, the history of precellular evolution into two great stages: a first period in which the Earth became populated by "spontaneous" molecules formed by spontaneous assemblies, and a second period in which molecular machines appeared and started manufacturing "artificial" molecules. The first period is what is normally known as *Chemical Evolution*, whereas the second can be referred to as *Postchemical Evolution*, in order to underline that the appearance of molecular machines introduced a qualitative change and evolution ceased to be a purely chemical process even if it had not yet become a fully biological one.

The simplest molecular machines we can think of are molecules that could join other molecules together by chemical bonds, and for this reason we may call them *bondmakers*. Some could form bonds between amino acids, some between nucleotides, others between sugars, and so on. It has been shown, for example, that short pieces of ribosomal RNA have the ability to form peptide bonds, so it is possible that the first bondmakers were RNA molecules of small- or medium-size molecular weights. Among the various types of bondmakers, furthermore,

---

[*]The sections of this Part have been reprinted, with permission, from the paper "Is the Cell a Semiotic System?" in Barbieri (ed.), Introduction to Biosemiotics, 2007, Springer, Berlin, pp. 179–207.

some developed the ability to join nucleotides together in the order provided by a *template*. Those bondmakers were making *copies* of nucleic acids, so we can call them *copymakers*. We do not know when they appeared on the primitive Earth, but at some stage they did and that was a real turning point because it set in motion an extraordinary sequence of events.

The copying of a template is the elementary act of gene duplication, the very first step toward the phenomenon of *heredity*. When a process of copying is repeated indefinitely, furthermore, another phenomenon comes into being. Copying mistakes become inevitable, and in a world of limited resources not all changes can be implemented, which means that a process of selection is bound to take place. Molecular copying, in short, leads to *heredity*, and the indefinite repetition of molecular copying leads to *natural selection*. That is how natural selection came into existence. Molecular copying started it and molecular copying has perpetuated it ever since.

In the history of life, molecular copying came into being when the first copymakers appeared on the primitive Earth and started making copies of nucleic acids. This implies that *spontaneous* nucleic acids had already been formed by spontaneous reactions on our planet, but that was no guarantee of evolution. Only the copying of genes could ensure their survival and have long-term effects, so it was really the arrival of copymaking that set in motion the extraordinary chain of processes that we call evolution. The first Major Transition of the history of life (Maynard Smith and Szathmáry, 1995) is generally described as the origin of genes, but it seems more accurate to say that it was the origin of molecular *copying*, or the origin of *copymakers*, the first molecular machines that started multiplying nucleic acids by making copies of them.

2   The Second Major Transition: The Origin of Proteins

Proteins are the key building blocks of all living structures, as well as the engines of countless reactions that go on within those structures. For all their extraordinary versatility, however, there is one thing they cannot do. Unlike genes, they cannot be their own templates. It is simply not possible to make proteins by copying other proteins. The transition from spontaneous to manufactured molecules, therefore, was relatively simple for genes but much more complex for proteins. Manufactured genes could be made simply by copying natural genes, and all that was required to that purpose were molecules which had a polymerase-like activity. Manufactured proteins, instead, could not be made by copying, and yet the information to make them had to come from molecules that can be copied, because only those molecules can be inherited. The information for manufacturing proteins, therefore, had to come from genes, so it was necessary to bring together a carrier of genetic information (a messenger RNA), a peptide-bondmaker (a piece of ribosomal RNA) and molecules that could carry both nucleotides and amino acids (the transfer RNAs). The first protein-makers, in short, had to bring together three different types of molecules (messenger, ribosomal and transfer RNAs), and were therefore much more complex than copymakers.

The outstanding feature of the protein-makers, however, was not the number of components. It was the ability to ensure a one-to-one correspondence between genes and proteins, because without it there would be no biological specificity, and without specificity there would be no heredity and no reproduction. Life as we know it simply would not exist without a one-to-one correspondence between genes and proteins.

Such a correspondence would be automatically ensured if the bridge between genes and proteins could be determined by *stereochemistry*, as one of the earliest models suggested, but that is not what happens in Nature. The bridge is always provided by molecules of transfer RNA, first called *adaptors*, that have two recognition sites: one for a group of three nucleotides (a *codon*) and another for an amino acid. In this case, a one-to-one correspondence could still be guaranteed automatically if one recognition site could determine the other, but again that is not what happens. The two recognition sites of the adaptors are physically separated in space and are chemically independent. There simply is no necessary link between codons and amino acids, and a one-to-one correspondence between them can only be the result of conventional rules. Only a real code, in short, could guarantee biological specificity, and this means that the evolution of the translation apparatus had to go hand in hand with the evolution of the genetic code.

Protein synthesis arose, therefore, from the integration of two different processes, and the final machine was a *code-and-template-dependent-peptide-maker*, or, more simply, a *codemaker*. The second Major Transition of the history of life is generally described as the origin of proteins, but it would be more accurate to say that it was the origin of *codemaking*, or the origin of *codemakers*, the first molecular machines that discovered molecular coding and started populating the Earth with codified proteins.

3   The Fingerprints of the Organic Codes

The first Major Transition gave origin to genes by the mechanism of copying, while the second Major Transition gave origin to proteins by the mechanism of coding. At the very beginning of the history of life we find two different mechanisms: copying and coding. But what about the rest of evolution? Do we find copying and coding even in the other Major Transitions? Did other organic codes appear after the genetic code? Luckily, this is a problem that we can deal with, because if other organic codes exist in Nature we should be able to find them by experiments, just as we have found the genetic code.

A code is a set of rules that establish *a correspondence between two independent worlds*. The Morse code, for example, is a correspondence between dots and dashes and the letters of the alphabet. A language is a correspondence between words and mental objects. The genetic code is a correspondence between the world of nucleic acids and the world of proteins. The genetic code is the only organic code that is officially recognized by modern biology, but it is also a model where we find characteristics that all organic codes must have. To start with, it allows us to appreciate the difference that exists between *copying* and *coding* by comparing the seminal examples of these processes, i.e. *transcription* and *translation*.

In transcription, an RNA chain is assembled from the linear information of a DNA chain, and in this case a normal biological catalyst (an RNA polymerase) is sufficient, because each step requires a single recognition process (a DNA-RNA coupling). In translation, instead, two independent recognition processes must be performed at each step, and the catalyst of the reaction (the ribosome) needs special molecules, first called *adaptors* and then *transfer RNAs*, in order to link the two processes. The codon recognition site and the amino acid recognition site are independent from a physico-chemical point of view, and it is precisely this independence that makes a genetic code absolutely essential. Without a code, in fact, a codon could be associated with different amino acids and *biological specificity*, the most precious of life's properties, would be lost.

These concepts can easily be generalized. We are used to thinking that all biochemical processes are made of *catalyzed* reactions, but in reality we must distinguish between *catalyzed* and *codified* reactions. The catalyzed reactions are processes (like transcription) that require only one recognition process at each step. The codified reactions, instead, require two independent recognition processes at each step and a set of coding rules. The catalyzed reactions, in other words, require only *catalysts*, while the codified reactions require *adaptors*, i.e. catalysts plus a code.

Any organic code is a set of rules that establish a correspondence between two independent worlds, and this necessarily requires molecular structures that act like *adaptors*, i.e. that perform two independent recognition processes. The adaptors are required because the two worlds would no longer be independent if there is a necessary link between them, and a set of rules is required in order to guarantee the specificity of the correspondence. Any organic code, in short, must have three major features: (1) a correspondence between two independent worlds; (2) a system of molecular adaptors; and (3) a set of rules that guarantee biological specificity. We conclude that the adaptors are the key molecules in all organic codes. They are the molecular *fingerprints* of the codes, and their presence in a biological process is a sure sign that the process is based on a code. This gives us an *objective criterion* for the search of organic codes, and their existence in Nature becomes therefore, first and foremost, an experimental problem.

4 The Splicing Codes

One of the greatest surprises of molecular biology was the discovery that the primary transcripts of the genes are often transformed into messenger RNAs by removing some RNA strings (called *introns*) and by joining together the remaining pieces (the *exons*). The result is a true assembly, because exons are assembled into messengers, and we need, therefore, to find out if it is a *catalyzed* assembly (like transcription) or a *codified* assembly (like translation). In the first case the cutting and sealing operations, collectively known as *splicing*, would require only a *catalyst* (comparable to RNA polymerase), whereas in the second case they would need a catalyst and a set of *adaptors* (comparable to ribosome and tRNAs).

This suggests immediately that splicing is a codified process because it is implemented by structures that are very similar to those of protein synthesis. The splicing systems, known as *spliceosomes*, are huge molecular machines like ribosomes, and employ small molecular structures, known as *snRNAs* or *snurps*, which are very much comparable to tRNAs. The similarity, however, goes much deeper than that, because the snRNAs have properties that fully qualify them as *adaptors*. They bring together, in a single molecule, two independent recognition processes, one for the beginning and one for the end of each intron, thus creating a specific correspondence between the world of the primary transcripts and the world of messengers.

The two recognition steps are independent not only because there is a physical distance between them, but above all because the first step could be associated with different types of the second one, as demonstrated by the cases of *alternative splicing*. The choice of the beginning and of the end of an intron, furthermore, is the operation that actually defines the introns and gives them a *meaning*. Without a complete set of such operations, primary transcripts could be transformed arbitrarily into messenger RNAs, and there would be no biological specificity whatsoever.

In RNA splicing, in conclusion, we find the three basic characteristics of all codes: (1) a correspondence between two independent worlds; (2) the presence of molecular adaptors; and (3) a set of rules that guarantee biological specificity. We conclude therefore that the processing of RNA transcripts into messengers is truly a codified process based on adaptors, and takes place with rules that can rightly be given the name of *splicing codes* (Barbieri, 1998; 2003).

5 The Signal Transduction Codes

Cells react to a wide variety of physical and chemical stimuli from the environment, and in general their reactions consist in the expression of specific genes. We need therefore to understand how the environment interacts with the genes, and the turning point, in this field, came from the discovery that the external signals (known as *first messengers*) never reach the genes. They are invariably transformed into a different world of internal signals (called *second messengers*) and only these, or their derivatives, reach the genes. In most cases, the molecules of the external signals do not even enter the cell and are captured by specific receptors of the cell membrane, but even those that do enter (some hormones) must interact with intracellular receptors in order to influence the genes (Sutherland, 1972).

The transfer of information from environment to genes takes place therefore in two distinct steps: one from first to second messengers, which is called *signal transduction*, and a second path from second messengers to genes which is known as *signal integration*. The surprising thing about signal transduction is that there are hundreds of first messengers (hormones, growth factors, neurotransmitters, etc.), whereas the known second messengers are only four (cyclic AMP, calcium ions, inositol trisphosphate, and diacylglycerol) (Alberts et al., 1994).

First and second messengers, in other words, belong to two very different worlds, and this suggests immediately that signal transduction may be based on organic codes. This is reinforced by the discovery that there is no necessary connection between first and second messengers, because it has been proved that the same first messengers can activate different types of second messengers, and that different first messengers can act on the same type of second messengers.

The experimental data, in brief, prove that external signals do not have any instructive effect. Cells use them to *interpret* the world, not to yield to it. Such a conclusion amounts to saying that signal transduction is based on organic codes, and this is in fact the only plausible explanation of the data, but of course we would also like a direct proof. As we have seen, the signature of an organic code is the presence of adaptors, and the molecules of signal transduction have indeed the typical characteristics of the adaptors. The transduction system consists of at least three types of molecules: a *receptor* for the first messengers, an *amplifier* for the second messengers, and a *mediator* in between (Berridge, 1985). The system performs two independent recognition processes, one for the first and the other for the second messenger, and the two steps are connected by the bridge of the mediator. The connection, however, could be implemented in countless different ways since any first messenger can be coupled with any second messenger, and this makes it imperative to have a code in order to guarantee biological specificity.

In signal transduction, in short, we find all three characteristics of the codes: (1) a correspondence between two independent worlds; (2) a system of adaptors that give meanings to molecular structures; and (3) a collective set of rules that guarantee biological specificity. The effects that external signals have on cells, in conclusion, do not depend on the energy or the information that they carry, but on the *meaning* that cells give them with rules that we can rightly refer to as *signal transduction codes* (Barbieri, 1998; 2003).

6   The Cytoskeleton Codes

A cytoskeleton is absolutely essential for typical eukaryotic processes such as phagocytosis, mitosis, meiosis, ameboid movement, organelle assembly, and three-dimensional organization of the cell, i.e. for all those features that make eukaryotic cells so radically different from bacteria. The actual cytoskeleton, in reality, is an integrated system of three different cytoskeletons made of filaments (*microfilaments, microtubules,* and *intermediate filaments*) each of which gives a specific contribution to the three-dimensional form of the cell and to its mobility.

The driving force of the cytosleton is a very unusual mechanism that biologists have decided to call *dynamic instability*. The cytoskeletal filaments, especially microtubules and microfilaments, are in a state of continuous flux where monomers are added to one end and taken away at the other, and the filament is growing or shortening according to which end is having the fastest run. But what is really most surprising is that all this requires *energy*, which means that the cell is investing enormous amounts of energy not in building a structure but *in making it unstable*!

In order to understand the logic of dynamic instability, we need to keep in mind that cytoskeletal filaments are unstable only when their ends are not attached to special molecules that have the ability to anchor them. Every microtubule, for example, starts from an organizing center (the *centrosome*), and the extremity which is attached to this structure is perfectly stable, whereas the other extremity can grow longer or shorter, and becomes stable only when it encounters an anchoring molecule in the cytoplasm. If such an anchor is not found, the whole microtubule is rapidly dismantled and another is launched in another direction, thus allowing the cytoskeleton to explore all cytoplasm's space in a short time.

Dynamic instability, in other words, is a mechanism that allows the cytoskeleton to build structures with an *exploratory strategy*, and the power of this strategy can be evaluated by considering how many different forms it can give rise to. The answer is astonishing: the number of different structures that cytoskeletons can create is *potentially unlimited*. It is the anchoring molecules (that strangely enough biologists call *accessory proteins*) that ultimately determine the three-dimensional forms of the cells and the movements that they can perform, and there could be endless varieties of anchoring molecules. The best proof of this enormous versatility is the fact that the cytoskeleton was invented by unicellular eukaryotes but was later exploited by metazoa to build completely new structures such as the axons of neurons, the myofibrils of muscles, the mobile mouths of macrophages, the tentacles of killer lymphocytes, and countless other specializations.

Dynamic instability, in conclusion, is a means of creating an endless stream of cell types with only one common structure and with the choice of a few anchoring molecules. But this is possible only because there is *no necessary relationship* between the common structure of the cytoskeleton and the cellular structures that the cytoskeleton is working on. The anchoring molecules (or accessory proteins) are true *adaptors* that perform two independent recognition processes: microtubules on one side and different cellular structures on the other side. The resulting correspondence is based therefore on *arbitrary* rules, on true natural conventions that we can refer to as *the cytoskeleton codes* (Barbieri, 2003).

7   The Compartment Codes

Eukaryotic cells not only produce molecules of countless different types but manage to deliver them to different destinations with astonishing precision, and this gives us the problem of understanding how they manage to cope with such an immensely intricate traffic. The first step in the solution of this mystery came with the discovery that the Golgi apparatus is involved not only in the biochemical modification of innumerable molecules but also in the choice of their geographical destination. But the truly remarkable thing is that all this is achieved with an extremely simple mechanism. More precisely, the Golgi apparatus delivers an astonishing number of molecules to their destinations with only three types of vesicles. One type has labels for the transport of proteins outside the cell and another for their delivery to the cell interior, whereas the vesicles of the third type carry no destination label, and are programmed, *by default*, to reach the plasma membrane. As we can see, the solution is extraordinarily efficient. With a single mechanism and only two types of labels, the cell delivers a great amount of proteins to their destinations, and also manages to continually renew its plasma membrane.

The Golgi apparatus, however, is a transit place only for a fraction of the cell proteins. The synthesis of all eukaryotic proteins begins in the soluble part of the cytoplasm (the *cytosol*), together with that of a signal that specifies their geographical destination. The piece of the amino acid chain that emerges first from the ribosome (the *peptide leader*) can contain a sequence that the cell interprets as an *export signal to the endoplasmic reticulum*. If such a signal is present, the ribosome binds itself to the reticulum and delivers the protein into its *lumen*. If not, the synthesis continues on free ribosomes, and the proteins are shed into the cytosol. Of these, however, only a fraction remains there, because the amino acid chain can carry, in its interior, one or more signals which specify other destinations, such as the *nucleus*, the *mitochondria*, and other cell compartments. Proteins, in conclusion, carry with them the signals of their geographical destination, and even the absence of such signals has a meaning, because it implies that the protein is destined to remain in the cytosol.

The crucial point is that there is *no necessary correspondence* between protein signals and geographical destinations. The export-to-the-nucleus signals, for example, could have been used for other compartments, or could have been totally different. They and all the other geographical signals are purely conventional labels, like the names that we give to streets, cities, airports, and holiday resorts. The existence of eukaryotic compartments, in other words, is based on natural conventions, and to their rules of correspondence we can legitimately give the name of *compartment codes* (Barbieri, 2003).

8   The Sequence Codes

In the 1980s and 1990s, Edward Trifonov started a lifelong campaign in favor of the idea that the nucleotide sequences of the genomes carry several messages simultaneously, and not just the message revealed by the classic triplet code. According to Trifonov, in other words, the genetic code is not alone, and there are many other codes in the nucleotide sequences of living organisms. This conclusion rests upon Trifonov's definition that "a code is any sequence pattern that can have a biological function" or "codes are messages carried by sequences" or "a code is any pattern in a sequence which corresponds to one or another specific biological function" (Trifonov, 1989; 1996; 1999).

The plurality of codes described by Trifonov is a result of his particular definition of a code, but it is not necessarily limited by that, and could well be compatible with different definitions. The splicing code, for example, is a code not only according to his criterion, but also according to the operative definition that a code is a set of rules of correspondence implemented by adaptors. This suggests that Trifonov's conclusions may have a general validity, and at least some of his sequence codes may turn out to be true organic codes. For the time being, however, let us acknowledge the fact that according to Trifonov's definition there are at least eight sequence codes in the genomes of living creatures, in addition to the classic triplet code (Trifonov, 1996):

1. The *transcription codes* include promoters and terminators, and are rather universal, though different in prokaryotes and in eukaryotes.
2. The *gene splicing code* for the processing of nuclear pre-mRNA is largely unde-ciphered. Its main components are obligatory GU- and AG- ends of introns, as well as rather conserved consensus sequence features around the ends.
3. The *translation pausing code*, for the regulation of translation, is encoded by clusters of rare triplets for which the aminoacyl-tRNAs are in limited supply.
4. The *DNA structure code*, or *DNA shape code*, is a sequence-dependent local shape of DNA which is a crucial component of the protein-DNA recognition.
5. The *chromatin code* describes those sequence features that direct the histone octamer's binding to DNA and the formation of nucleosomes.
6. The *translation framing code* is overlapping with the triplet code (Trifonov, 1987), and ensures the correct reading frame during translation.
7. The *modulation code* is about the repeating sequences and regulates the number of repeats as an adjustable variable to modulate expression of the nearby gene.
8. The *genome segmentation code* is one of the emerging new codes, and is due to fact that the genomes appear to be built of rather standard size units.

9   A Stream of Codes

Most of the papers which have been published on biological codes do not make any reference to their definition. This is the case, for example, of the reports which have described and discussed a truly remarkable wealth of experimental data on the *Adhesive Code* (Readies and Takeichi, 1996; Shapiro and Colman, 1999), on the *Sugar Code* (Gabius, 2000; Gabius et al., 2002), and on the *Histone Code* (Strahl and Allis, 2000; Jenuwein and Allis, 2001; Turner, 2000; 2002; Gamble and Freedman, 2002; Richards and Elgin, 2002).

The practice of studying something without precisely defining it is fairly common in many sciences, and biology is no exception. The paradigmatic example is life itself, a phenomenon that we keep studying even if nobody seems to agree on its definition. Another instructive case is the concept of species, for which there is no definition that is universally valid and yet this does not prevent biologists from doing experiments, obtaining results and making sensible predictions on countless species of living creatures. Precise definitions, in short, are not always essential, but in some cases they are, and this is one of them. More precisely, we should be aware that an operative definition of organic codes in terms of adaptors would have provided a crucial guideline in at least two important cases:

1. One is the research on new biological codes such as the Adhesive Code, the Sugar Code, and the Histone Code. The problem here is that the experimental data *suggest* the existence of organic codes but do not *prove* it. And yet the results could have been conclusive because they are all compatible with the existence of true adaptors. On the face of the evidence, for example, it is most likely that lectins are the adaptors of the Sugar Code and that cadherins are the adaptors of the Adhesive Code. If that had been proved, there would be no doubt that we are in the presence of true organic codes. But people did not use a definition of codes based on adaptors, so they did not look for adaptors.

2. The second case is that of the classical research on signal transduction. Here the amount of experimental data is so enormous to be beyond description, and yet there is a remarkable paradox in this field. The only logical explanation of the facts is that signal transduction is based on organic codes and yet the word "code" has never been mentioned, so people have never looked for coding rules. The evidence has actually proved that signal transducers have the experimental characteristics of true adaptors, and yet the word "adaptors" has never been mentioned with reference to a code. This habit could well go on indefinitely by inertia, and only a precise definition of organic codes can convince people that an alternative (and much more convincing) explanation of the facts already exists.

It has been the existence of adaptors which has proved the reality of the Genetic Code, and the same is going to be true for the Signal Transduction Codes, for the Adhesive Code, for the Sugar Code, and for the Histone Code. An operative definition based on adaptors, furthermore, is the only scientific instrument that can allow us to prove the existence of other organic codes in Nature. And when we really start looking for them, we may well discover that so far we have only scratched the surface. That there is a long golden stream of organic codes out there.

PART 2 – THE MECHANISMS OF EVOLUTION

1 The Molecular Mechanisms

The mechanisms of evolution have been one of the most controversial issues in Biology and the great debate about them culminated, in the 1930s and 1940s, in the Modern Synthesis, the theoretical framework where natural selection and neutral drift are regarded as the sole mechanisms of evolutionary change.

Natural selection and neutral drift are both due to chance variations in the transmission of the hereditary characters, and are based therefore on the mechanism of molecular copying. More precisely, on the indefinite repetition of a process of molecular copying in a world of limited resources. In these circumstances, copying mistakes are bound to take place and that inevitably leads either to selection or to drift. The two evolutionary mechanisms of the Modern Synthesis, in short, are both due to molecular copying, and would account for the whole of evolution if all basic processes of life could be reduced to molecular copying.

The discovery of the genetic code, however, has proved that there are *two* distinct molecular mechanisms at the basis of life, transcription and translation, or copying and coding. The discovery of other organic codes, furthermore, allows us to generalize this conclusion because it proves that coding is not limited to translation. Copying and coding, in other words, are distinct molecular mechanisms that operate in all living systems and throughout the whole history of life. This suggests that they are also two distinct mechanisms of evolution because an evolutionary mechanism is but the long-term result of a molecular mechanism. More precisely, copying leads, in the long run, to natural selection and coding leads to natural conventions. This conclusion, however, is not as straightforward as it may appear at first sight, because it rests on the assumption that *coding cannot be reduced to copying*. In order to prove that natural conventions are a distinct mechanism of evolution, therefore, we need to prove that copying and coding are fundamentally different mechanisms of molecular change.

2 Copying and Coding

Copying and coding are both capable of bringing novelties into the world, but they do it in very different ways. By its very nature, the copying mechanism produces either exact copies or slightly different versions of the same molecules. This means that natural selection produces new objects only by modifying previous ones, i.e. by making objects that are only relatively different from their predecessors. Natural selection, in short, creates *relative* novelties, not absolute ones.

In the case of coding, the situation is totally different. The rules of a code are not dictated by physical necessity, and this means that a new code can establish relationships that have never existed before in the Universe. The objects that are assembled with the rules of a new code can have no relationship whatsoever to previous objects. Natural conventions, in short, create *absolute* novelties, not relative ones.

Another difference between the two mechanisms is that copying operates on *individual* molecules, whereas coding involves a *collective* set of rules. The difference between natural selection and natural conventions, in other words, is the difference that exists between individual change and collective change. An example of this difference can be seen in any language, whose evolution is due to variations that take place not only at the level of the individual words but also at the collective level of the rules of grammar.

A third difference between copying and coding is that they involve two different types of entities. A variation in the copying of a gene changes the linear sequence, i.e. the *information* of that gene. A variation in a coding rule, instead, changes the *meaning* of that rule. The main difference between copying and coding, and therefore between natural selection and natural conventions, comes from the difference that exists between "information" and "meaning".

There are, in conclusion, three major differences between copying and coding: (1) copying modifies existing objects whereas coding brings new objects into existence; (2) copying acts on individual objects whereas coding acts by collective rules; and (3) copying is about biological information whereas coding is about biological meaning. Copying and coding, in short, are profoundly different mechanisms of molecular change, and this tells us that natural selection and natural conventions are two distinct mechanisms of evolutionary change.

3   Different Mechanisms at Different Levels

The idea that natural selection can work at different levels of organization (genes, organisms, species) has been at the center of countless debates in evolutionary biology. Less attention has been given to the alternative possibility that at different levels of organization there may be at work different mechanisms of evolution. There is, however, at least one case that gives us a clear example of this alternative. It is the origin of mitochondria in primitive eukaryotic cells.

For a long time it has been assumed that mitochondria were the result of a gradual evolution from within the cell, but then it was found that they originated by the incorporation of whole cells into other cells by endosymbiosis. Those two types of cell had been in existence for millions of years before the symbiosis event, and all their components had been copied at each generation, and had been subject to evolution by natural selection. Their coming together in symbiosis, however, was a process that took place *at the cellular level*. It was the cells acting as whole systems that gave origin to endosymbiosis. Their components had to be "compatible" with endosymbiosis, but in no way had been selected for that purpose. Endosymbiosis, in short, is a mechanism that exists only at the cellular level, not at the molecular level, and represents therefore a distinct mechanism of evolution.

In the case of the organic codes, the situation is somewhat intermediate between the molecular and the cellular level. The genetic code, for example, is at the same time a supramolecular system and a subcellular one. All its molecular components must be inherited and copied individually, and yet a code is necessarily a collective entity. The important point is that coding, like endosymbiosis, simply does not exist at the molecular level. It is a mechanism that belongs to the supramolecular level just as endosymbiosis belongs to the cellular level. There is no doubt that copying is absolutely necessary for coding, but the important point is that it is not *sufficient* for it. Coding cannot be reduced to copying because they are mechanisms that belong to different levels of organization.

4 Natural Selection and Natural Conventions

Natural selection is based on molecular copying, more precisely on the indefinite repetition of a process of molecular copying in a world of limited resources, and this means that *natural selection would be the sole mechanism of evolution if molecular copying were the sole basic mechanism of life.*

As a matter of fact, this *could* have happened. If living systems could have been made entirely of RNA enzymes and RNA genes, only the copying of RNA molecules would have been necessary, and natural selection could indeed have been the sole mechanism of evolution. But that is not what happened. Long before the origin of the first cells, proteins were being made on the primitive Earth, and proteins, unlike genes, could not be made by copying. The manufacture of proteins required codemakers, not copymakers. It required two independent recognition processes at each step, not one, and above all it required the rules of a code. In an RNA world, in short, molecular copying – and therefore natural selection – *could* have been enough, but in a world where proteins exist there must necessarily be natural conventions, and these cannot be reduced to natural selection because coding cannot be reduced to copying.

There is, however, another scenario where we could say that natural selection has *virtually* been the sole mechanism of evolution. If no other organic code had appeared on Earth after the genetic code, we would have to conclude that copying has been the sole mechanism of change *for almost 4 billion years*, and natural selection could legitimately be regarded as the sole mechanism of evolution for almost the entire history of life. In this case, the origin of the genetic code, at the beginning of that history, and the origin of the cultural codes, at the end of it, could be regarded as two extraordinary exceptions, and natural selection would remain *in practice* the sole mechanism of evolutionary change.

But the genetic code is not the only code of life. There are many other organic codes in Nature, and this means that they came into being in the course of evolution. This in turn means that copying and coding operated throughout the whole history of life, and gave different contributions to it, which makes us realize that there have been two distinct mechanisms of evolution. It also makes us realize that natural selection and natural conventions had complementary roles: natural conventions account for the discontinuities of the history of life whereas natural selection explains the gradual transformations that took place in between.

Evolution, in short, was not produced only by natural selection but *by natural selection and by natural conventions* (Barbieri, 1985, 2003), which in no way is a belittlement of natural selection. It is only an extension of it.

5 Codes and Macroevolution

The role of the organic codes in the history of life can be appreciated by underlining that their origins are closely associated with the great events of macroevolution. *Any time that a new organic code came into being, something totally new appeared in Nature, something that had never existed before.*

The origin of the genetic code, for example, made it possible to produce proteins with specific sequences and to pass them on indefinitely to other systems. That gave origin to *biological specificity* and to *heredity*, the most fundamental of life's properties. The origin of the genetic code, in short, was also the origin of protein-based life, i.e. of life-as-we-know-it.

Similar considerations apply to the other organic codes. The signal transduction codes, for example, allowed primitive systems to produce their own signals and therefore to separate their *internal* space from the *outside* environment. That was a precondition for the origin of *individuality*, and in particular for the origin of the cell.

Another great innovation was brought about by the codes of splicing, because the appearance of a complete set of splicing rules brought something unprecedented into being. Splicing requires a separation *in time* between transcription and translation and that was a precondition for their separation *in space*, i.e. for *the origin of the nucleus*. The defining feature of the eukaryotes, in other words, was made possible by the origin of the splicing codes.

Many other eukaryotic innovations were brought into existence by a whole stream of organic codes. The cytoskeleton codes, for example, allowed the cells to build their own scaffoldings, to change their own shapes and to perform their own movements. The origin of embryos was also associated with organic codes because typical embryonic processes like *cell determination, cell adhesion, cell migration,* and *cell death* have all the qualifying characteristics of codified phenomena (Barbieri, 1998, 2003).

In the case of embryonic development, furthermore, we have entirely new codes before us. The correspondence is no longer between two types of molecules, like genes and proteins or first and second messengers, but between molecules and *cell states*. The determination of the body axes, for example, is obtained by a link between molecules and *cell memory*. The body axes are the same in all triploblasts, but their molecular determinants are of countless different types, which show that there is no necessary correspondence between molecules and cell states. This means that the link between molecular determinants and cell states can only be realized by codes that we can refer to as *body pattern codes*.

The major events in the history of life, in short, went hand in hand with the appearance of new organic codes, from the first cells all the way up to multicellular life, and this suggests a very deep link between codes and evolution. It suggests that *the great events of macroevolution were made possible by the appearance of new organic codes* (Barbieri, 1998, 2003).

6   The Contribution of the Codes

The history of life has been "punctuated" by the appearance of new organic codes and it has been deeply shaped by their characteristics. Five of them are particularly important.

1. *Discontinuities* The evolution of the individual rules of a code can take an extremely long time, but the "origin" of a new code corresponds to the appearance of a "complete" set of rules and that is a sudden event. The great evolutionary novelties produced by a new code, therefore, appeared suddenly in the history of life. This is a new explanation of the discontinuities that paleontology has documented, and shows that natural selection and natural conventions had complementary roles. Natural conventions account for the discontinuities of the history of life whereas natural selection explains the gradual transformations that took place in between.

2. *Invariance* The genetic code appeared at the beginning of the history of life and has remained substantially the same ever since. The same apply to the deep codes that define prokaryotes and eukaryotes. Once in existence they have not been changed, despite the fact that all the molecular components of a code must be inherited and are subject therefore to the chance variations of the copying mechanism and to the long term results of that mechanism, i.e. to natural selection and to neutral drift. The fact that the deep organic codes have been conserved for billions of years suggests that their conservation has been the top priority of every living systems. Everything else could be changed except the rules of the basic codes of life. While morphological structures did rise and fall countless times, the "deep" organic codes have never been removed. This tells us that they truly are the *fundamentals* of life, the invariants that persist while everything else is changing.

3. *Additivity* A new organic code has never abolished previous codes. The genetic code has not been removed by the signal transduction codes and neither of them has been supplanted by the splicing codes. A new code has always been *added* to the previous ones, which shows that new codes do not originate by the transformation of previous codes. Once in existence, organic codes do not tend to change, and the origin of a new code is always the origin of an entirely new set of rules.

4. *Stability* The genetic code is present in all living creatures, but the other organic codes appeared in increasingly smaller groups. The greater the number of codes the smaller the number of species that possess them. This shows that living systems coexist whatever is the number of their codes. Eukaryotes did not remove prokaryotes, and metazoa did not remove unicellular eukaryotes. Every organic code, in short, represents a stable form of life.

5. *Complexity* The addition of new organic codes to a living system can rightly be regarded as an *increase of complexity* of that system. The structural complexity of some organisms did diminish in time, as many cases of simplification clearly show, but the complexity of the codes has never been lowered. Even the animals which lost or reduced the greatest number of parts, in order to lead a parasitic life, have conserved all the fundamental codes of animal life. The number of organic codes is therefore a new measure of biological complexity, and probably it is more fundamental than all other parameters which have been proposed so far.

7   The Contribution of Natural Selection

Life is essentially a *manufacturing* activity, and molecular biology tells us that this activity is based on the mechanisms of copying and coding. Life, in short, is *artifact-making by copying and coding*. This conclusion may appear to give importance only to *internal* factors, as if the environment had almost no role to play, but that is not the case. The concept that life is artifact-making gives at least three major roles to the environment.

To start with, it is the environment that provides the building blocks for the manufacturing activity of the living systems. All components of life come from the environment and eventually go back to it, which means that any living system is totally dependent on its surrounding world.

The second point is that it is the environment that decides whether the structures manufactured by copying and coding are viable or not. Copying and coding have the potential to create an unlimited number of artifacts, but not all of them actually work in the real world. Copying and coding propose, but in the end it is the environment that disposes of their products.

The third point is that the environment is not only the place where living systems exist. It is also the place that living systems tend to become adapted to. We have learned from Darwin that in a world of limited resources not all organisms can survive and a process of selection is bound to take place between them. The survival can be a matter of luck, but in general it is the degree of adaptation to the environment that gives the best chances of success. The result is that organisms tend to become more and more adapted to their environment.

The process of adaptation allows organisms to become increasingly capable to cope with their surrounding world, and therefore to reduce the distance that separates them from *reality*. Natural selection can be regarded, therefore, as a process that allows organisms to incorporate at least some amount of reality into their constitution, even if the gap between internal and external reality can never be abolished.

Francois Jacob has expressed this concept with admirable clarity: "If the image that a bird gets of the insects it needs to feed its progeny does not reflect at least some aspects of reality, there are no more progeny. If the representation that a monkey builds of the branch it wants to leap to has nothing to do with reality, then there is no more monkey. And if this did not apply to ourselves, we would not be here to discuss this point" (Jacob, 1982).

8   Common Descent

Darwin's greatest contribution to Biology was probably the theory of Common Descent, the idea that "all the organic beings which have ever lived on this Earth may be descended from some one primordial form" (Darwin, 1859). In fact, when Theodosius Dobzhansky (1973) wrote that "Nothing in biology makes sense except in the light of evolution," it was Common Descent that he had in mind. The idea that all creatures of the present are linked to all creatures of the past, is indeed the greatest unifying theme in Biology, the concept that we use as an Ariadne's thread to reconstruct the history of life.

Common Descent, however, is not a single theory because it is compatible with different mechanisms of evolution. In order to find out the truth about Common Descent, therefore, we need to know the actual mechanisms that gave origin to biological objects in the course of time. How did novelties appear on Earth? Did new objects arise *by natural selection alone*, or *by natural selection and by natural conventions*?

If evolution took place only by natural selection, we would have to conclude that nothing similar to the origin of the genetic code happened again in the 4 billion years of life's history. But we know that many other organic codes exist in life, and this means that there have been many other *origins*, because any new organic code gives origin to unprecedented structures. We have therefore two very different versions of Common Descent before us. Evolution by natural selection implies *Common Descent with a Single Origin*, whereas evolution by natural selection and by natural conventions leads to *Common Descent with Multiple Origins*. (This is not the old theory that *cells* originated many times, because the multiple origins are referred to *codes* not to cells).

The idea that natural conventions bring absolute novelties into existence is equivalent to saying that life has not lost its creative power in the course of time. The origin of embryos, the origin of the mind, or the origin of language, for example, do not seem to be less of a novelty than the origin of the cell. The theory of Common Descent with Multiple Origins makes us realize that absolute novelties appeared not only at the beginning, but throughout the entire history of life, and this is not a belittlement of the traditional theory of Common Descent. It is only an extension of it.

9   Conclusion

There are two great facts about life that shed a new light on the mechanisms of evolution. The first is that genes and proteins are produced by copying and coding, by transcription and translation, i.e. by two fundamentally different mechanisms. The second is that natural selection is the long-term result of copying, i.e. of only one of those two basic mechanisms. This implies that coding is a distinct mechanism of molecular change and, in the long run, a distinct mechanism of evolutionary change. The difference between the two mechanisms, furthermore, is not limited to genes and proteins, and we find it at all the other levels of organization. It is the difference that exists between heredity and metabolism, between genetics and development, between individual change and collective change, between information and meaning.

Embryologists have traditionally maintained that development is a novelty-generating mechanism in its own rights, but now the time has come to qualify this long standing claim and turn it into a testable proposition. The solution proposed here is that the novelty-generating property of development comes from specific *codes of development* and is distinct from natural selection because coding is distinct from copying.

It may be pointed out that embryonic development has been explained by the *genes* of development without invoking any *codes* of development, so what is the point of such codes? The point can perhaps be illustrated by a comparison with protein synthesis. It is certainly true that there are genes for all the molecules of protein synthesis, but could we really explain that process without the genetic code? The genes of translation do not make the code of translation redundant, and the same is true for the genes of all the other codes.

We realize in this way that there is more to development than the genes of development. The great challenge is precisely the codes of development because they are the whole of which the genes are a part. We simply cannot study life without studying its codes, and we cannot understand evolution if we do not realize that it has been shaped by natural selection and by natural conventions with largely complementary roles: natural conventions account for the great novelties of the history of life whereas natural selection explains the gradual transformations that adapted those novelties to the real world and allowed them to survive.

The long-standing claim of embryology has been that development cannot be explained by natural selection alone, but now the time has come to generalize it. It is the whole of life that cannot be explained by natural selection alone because there have been natural conventions in all steps and stages of evolution, from the first cells all the way up to embryos and then to brains, minds, and finally consciousness.

References

Alberts B, Bray D, Lewis J, Raff M, Roberts K, Watson JD (1994) Molecular Biology of the Cell. 3rd edn. Garland, New York

Arthur W (2004) Biased Embryos and Evolution. Cambridge University Press, New York

Barbieri M (1985) The Semantic Theory of Evolution. Harwood Academic Publishers, London/ New York

Barbieri M (1998) The Organic Codes. The basic mechanism of macroevolution. Rivista di Biologia-Biol Forum 91:481–514

Barbieri M (2003) The Organic Codes. An Introduction to Semantic Biology. Cambridge University Press, Cambridge

Berridge M (1985) The molecular basis of communication within the cell. Sci Am 253:142–152

Callebaut W, Rasskin-Gutman D (eds) (2005) Modularity: Understanding the Development and Evolution of Natural Complex Systems. MIT Press, Cambridge, Massachusetts

Carroll SB (2005) Endless Forms Most Beautiful: The New Science of Evo-Devo. W W Norton, New York

Darwin C (1859) On the Origin of Species by Means of Natural Selection. Murray, London

Davidson EH (2001) Genomic Regulatory Systems: Development and Evolution. Academic Press, San Diego, California

Dobzhansky T (1973). Nothing in biology makes sense except in the light of evolution. Am Biol Teach 35:125–129

Gabius H-J (2000) Biological information transfer beyond the genetic code: the sugar code. Naturwissenschaften 87:108–121

Gabius H-J, André S, Kaltner H, Siebert H-C (2002) The sugar code: functional lectinomics. Biochimica et Biophysica Acta 1572:165–177

Gamble MJ, Freedman LP (2002) A coactivator code for transcription. Trends Biochem Sci 27(4):165–167

Gilbert SF (2006a) The generation of novelty: the province of developmental biology. Biol Theory 1(2):209–212

Gilbert SF (2006b) Developmental Biology, 8th edn. Sinauer, Sunderland, Massachusetts

Gould SJ (2002) The Structure of Evolutionary Theory. Harvard University Press, Cambridge, Massachusetts

Holland PWH (1999) The future of evolutionary developmental biology. Nature 402(Suppl.): C41–C44

Jacob F (1982) The Possible and the Actual. Pantheon Books, New York

Jenuwein T, Allis D (2001) Translating the histone code. Science 293:1074–1080

Maynard Smith J, Szathmáry E (1995) The Major Transitions in Evolution. Oxford University Press, Oxford

Readies C, Takeichi M (1996) Cadherine in the developing central nervous system: an adhesive code for segmental and functional subdivisions. Develop Biol 180:413–423

Richards EJ, Elgin SCR (2002) Epigenetic codes for heterochromatin formation and silencing: rounding up the usual suspects. Cell 108:489–500

Ruse M (2006) Forty years a philosopher of biology: Why EvoDevo makes me still excited about my subject. Biol Theory 1(1):35–37

Shapiro L, Colman DR (1999) The diversity of cadherins and implications for a synaptic adhesive code in the CNS. Neuron 23:427–430

Strahl BD, Allis D (2000) The language of covalent histone modifications. Nature 403:41–45

Sutherland EW (1972) Studies on the mechanism of hormone action. Science 177:401–408

Trifonov EN (1987) Translation framing code and frame-monitoring mechanism as suggested by the analysis of mRNA and 16s rRNA nucleotide sequence. J Mol Biol 194:643–652

Trifonov EN (1989) The multiple codes of nucleotide sequences. Bull Math Biol 51:417–432

Trifonov EN (1996) Interfering contexts of regulatory sequence elements. Cabios 12:423–429

Trifonov EN (1999) Elucidating sequence codes: three codes for evolution. Annal NY Acad Sci 870:330–338

Turner BM (2000) Histone acetylation and an epigenetic code. BioEssay 22:836–845

Turner BM (2002) Cellular memory and the histone code. Cell 111:285–291

Wilkins AS (2002) The Evolution of Developmental Pathways. Sinauer, Sunderland, Massachusetts

Part 2
The Genetic Code

Chapter 3

Catalytic Propensity of Amino Acids and the Origins of the Genetic Code and Proteins

Ádám Kun[1,2], Sándor Pongor[3,4], Ferenc Jordán[2,5], and Eörs Szathmáry[1,2,6]

Abstract The origin of the genetic code is still not fully understood, despite considerable progress in the last decade. Far from being a frozen complete accident, the canonical genetic code is full of patterns that seem to open a window on its evolutionary history. In this chapter we rethink the hypothesis that the primary selective force in favour of the emergence of genetic coding was the added value by amino acids to the RNA world in the form of increased catalytic potential. We identify a novel pattern in the genetic code suggesting that the catalytic propensity of amino acids has considerably shaped its structure. This suggestion complements older ideas arguing in favour of a driving force to build the smallest stable oligopeptide structures, such as hairpins (β-turns stabilized by small β-sheets). We outline experiments to test some of the proposals.

1 Introduction

As Crick et al. (1976) noted, 'the origin of protein synthesis is a notoriously difficult problem'. This remark refers to protein synthesis in translation using the genetic code. When thinking about difficult evolutionary transitions (cf. Maynard Smith and Szathmáry, 1995) it is rewarding to break down the problem into steps that are more readily soluble by evolution and easier to understand for us. The idea

[1] Biological Institute, Eötvös University, Budapest

[2] Collegium Budapest, Institute for Advanced Study, 2 Szentháromság utca, H-1014 Budapest, Hungary

[3] International Centre for Genetic Engineering and Biotechnology, Padriciano 99, 34012 Trieste, Italy

[4] Bioinformatics Group, Biological Research Center, 6726 Szeged, Hungary

[5] Animal Ecology Research Group of HAS, Hungarian Natural History Museum, Budapest

[6] Parmenides Center for the Study of Thinking, 14a Kardinal-Faulhaber-Str, D-80333 Munich, Germany

of an RNA world (e.g. Gilbert, 1986) is important because it separates the problem of life's origin from the origin of translation. The origin of the genetic code by itself seems to be burdened by a dual difficulty: no meaningful proteins without the genetic code, and no genetic code without the appropriate proteins (especially synthetases). Fortunately, there is a way out: we know that selected RNA molecules (aptamers) can specifically bind amino acids, and charge them to RNA either in *cis* or in *trans* (discussed below). The fact that peptidyl transfer in the ribosome is catalysed by RNA rather than proteins (Moore and Steitz, 2002, Steitz and Moore, 2003) supports the view that there was a way out from the RNA world into ours, aided by RNA itself.

Yet these exciting developments leave the nature of positive selection for the genetic code obscure. Polypeptides must attain a critical size and complexity before they can serve structural and catalytic functions in a rudimentary way. Söding and Lupas (2003, p. 837) called attention to the fact that, consonant with the RNA world scenario, the first polypeptides (supersecondary structures) would have been unable to attain stable conformation by themselves, as witnessed even by contemporary ribosomal proteins: 'The peptides forming these building blocks would not in themselves have had the ability to fold, but would have emerged as cofactors supporting RNA-based replication and catalysis (the 'RNA world'). Their association into larger structures and eventual fusion into polypeptide chains would have allowed them to become independent of their RNA scaffold, leading to the evolution of a novel type of macromolecule: the folded protein.' The path from the RNA world presumably went through a marked RNA-polypeptide phase. Corollary to this is the notion that modern metabolism is a 'palimpsest' of the RNA world (Benner et al., 1989) and that evolution of modern protein aminoacyl-tRNA synthetases may shed little direct light on the very origin of the genetic code if the ancient form of the code was implemented by ribozymes rather than proteins; at most the evolution of protein synthetases could have been partly analogous to that of RNA synthetases (Wetzel, 1995).

What could have been the force that drove life out of the RNA world? The end result is clear: proteins in general are much more versatile catalysts than RNA, partly by the virtue of the greater catalytic potential of 20 amino acids as opposed to 4 nucleotides (e.g. Szathmáry, 1999). In general, replicability and catalytic potential are in conflict: they prefer smaller and larger alphabets, respectively (Szathmáry, 1991, 1992). But evolution has no foresight: one cannot rationalize a transition by noting that the end result is fitter than the starting point: a more or less smooth path on the adaptive landscape must be found.

Some time ago, one of us proposed an idea (Szathmáry, 1990) how this could have been possible by still keeping catalysis by amino acids in focus. In short, some amino acids could have been utilized as cofactors of ribozymes in a metabolically complex RNA world. According to this scenario, amino acids were linked to specific short oligonucleotides (called handles) by ribozymes, in a manner that followed the logic of the genetic code: one type of amino acids was allowed to be charged to different handles, but each particular handle with a specified sequence was charged with one type of amino acids only. Szathmáry (1990) proposed that

2 Catalytic Propensity of Amino Acids and Organization of the Genetic Code

Catalytic propensity of amino acids (Fig. 2), collected from catalytic sites of known enzymes, are taken from Bartlett et al. (2002), who argued that the sample is representative.

Amino acids with the highest values gather in column A and (with smaller values) column G (Fig. 3). Is this pattern due to chance, or is it significant?

List of catalytic residues were obtained from the Catalytic Site Atlas of EMBL (Porter et al., 2004). Only literature-based entries pertaining to amino acids were used (residues inferred from sequence homologies and non-amino acid residues, such as metal ions and cofactors are left out of our analysis). In total, there were 5845 catalytic residues. The distribution of amino acids among the catalytic residues is markedly different from the frequencies of amino acids found in peptides (Bartlett et al., 2002). We performed a randomization test as follows. We took the biosynthetically restricted random set (Fig. 4) as defined by Freeland et al. (2000), which rests on the potential importance of the co-evolution theory (Wong, 1975) of the genetic code (code assignment was influenced by biosynthetic kinship of amino acids) and the observation that amino acids belonging to the same biosynthetic family tend to share the same first codon letter (i.e. they are in the same row of the table; Taylor and Coates, 1989).

Alternative tables of the genetic code were generated according to Freeland et al. (2000), limiting the number of possible alternatives to 6.48×10^6, compared to the 20, $\approx 2.43 \times 10^{18}$ totally random codes. Each of the 6.48×10^6 code tables was analysed according to the following procedure. First, the list of amino acids is ordered according to catalytic frequency in active sites. The place they occupy in

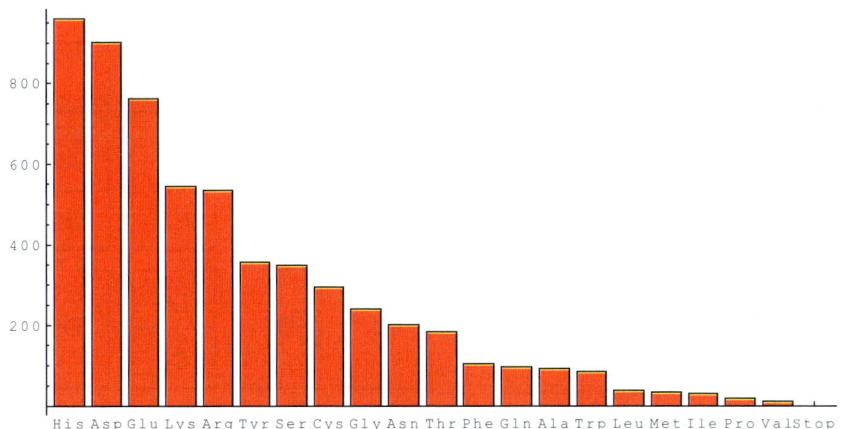

Fig. 2 Catalytic propensity of amino acids in catalytic sites of known enzymes. (From Bartlett et al., 2002.)

			Middle letter				
			U	C	A	G	
First letter	U	Phe		Ser	Tyr	Cys	U
							C
					Stop	Stop	A
						Trp	G
	C	Leu		Pro	His	Arg	U
							C
					Gln		A
							G
	A	Ile		Thr	Asn	Ser	U
							C
					Lys	Arg	A
		Met					G
	G	Val		Ala	Asp	Gly	U
							C
					Glu		A
							G

(Third letter)

Fig. 3 Catalytic propensities and β-turn propensities superimposed on the genetic code. Only the highest values are shown (very high catalytic propensity: red, moderately high catalytic propensity: pink, highest turn propensities: green frame)

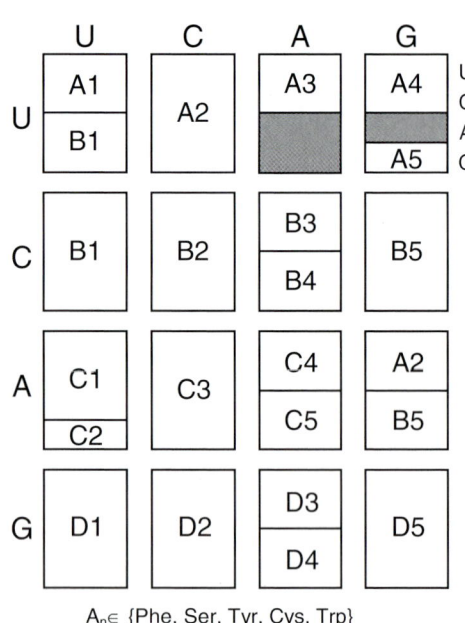

Fig. 4 The set of possible codes constrained by biosynthetic kinship (Freeland et al., 2000). In a randomized code any amino acid from set A_n can occupy any single position in the table, but only from A1 to A5. There are 6.48×10^6 possible alternative codes

$A_n \in$ {Phe, Ser, Tyr, Cys, Trp}
$B_n \in$ {Leu, Pro, His, Gln, Arg}
$C_n \in$ {Ile, Met, Thr, Asn, Lys}
$D_n \in$ {Val, Ala, Asp, Glu, Gly}

this list is assigned to them as a rank. In case of a tie the average of the ranks are assigned to each amino acid having the same value. With regard to catalytic frequencies, only serine (or Phe, Tyr, Cys, and Trp in alternative codes) appears twice in the list for being in two columns, and thus has the same catalytic frequency. Sum of the ranks belonging to amino acids present in the same column of the genetic code were squared, and then summed. This procedure is identical to the calculation employed in the Kruskal-Wallis test (Zar, 1998), which is a non-parametric test employed in testing differences between multiple groups. It is similar to one-factor ANOVA, except normality of the data is not required. The pattern that amino acids segregate according to columns of the genetic code (in this order) is statistically significant (p = 0.0107) (Fig. 5), in agreement with the cluster analysis of catalytic propensities (Fig. 6).

We have performed a similar test of propensities for the β-turns (Prevelige and Fasman, 1989) having been taken from the EMBOSS programme (Rice et al., 2000), β-sheets (Muñoz and Serrano, 1994) and α-helices (Muñoz and Serrano, 1994): The values for the β-turns (p = 0.0059) and β-sheets (p = 0.028) have significant columnar organization *at the level of single columns*.

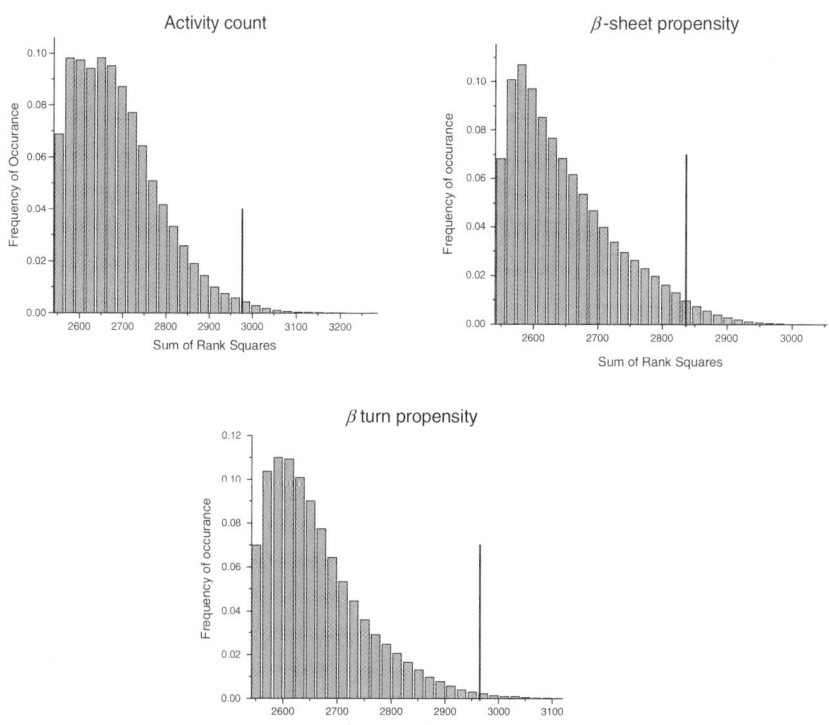

Fig. 5 Randomization test for the columnar organization of three amino acid properties in the genetic code. Thin pole indicates the position of the canonical genetic code. See text for further explanation

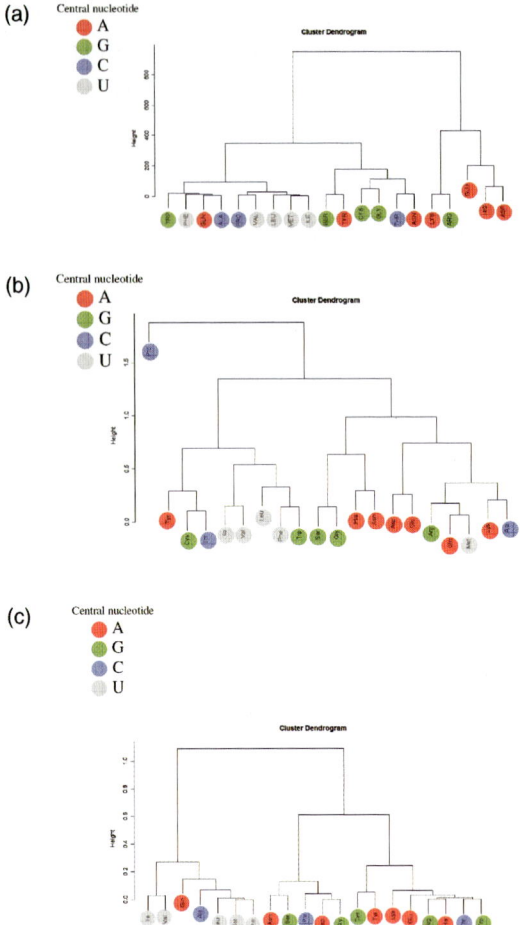

Fig. 6 Cluster analysis of amino acid propensities. (a) Catalytic propensity, (b) α-helix and β-sheet building propensities (taken together), (c) β-turn forming propensity. The hierarchical clustering was based on an Euclidean distance calculated between amino acid properties, and was carried out with the complete linkage clustering algorithm (Hartigan, 1975) as implemented in the R package, version 2.2.0 (From R Development Core Team, 2004)

It is clear that the chemically most 'exciting' amino acids are the catalytically most important ones (Fig. 7), and that the central purine bases play an exclusive role in their coding.

Our suggestion is that *the introduction of the first amino acids into the genetic code coincided with the order of decreasing catalytic importance*. This presumes that in the ribo-organisms the highly catalytic amino acids were available either in

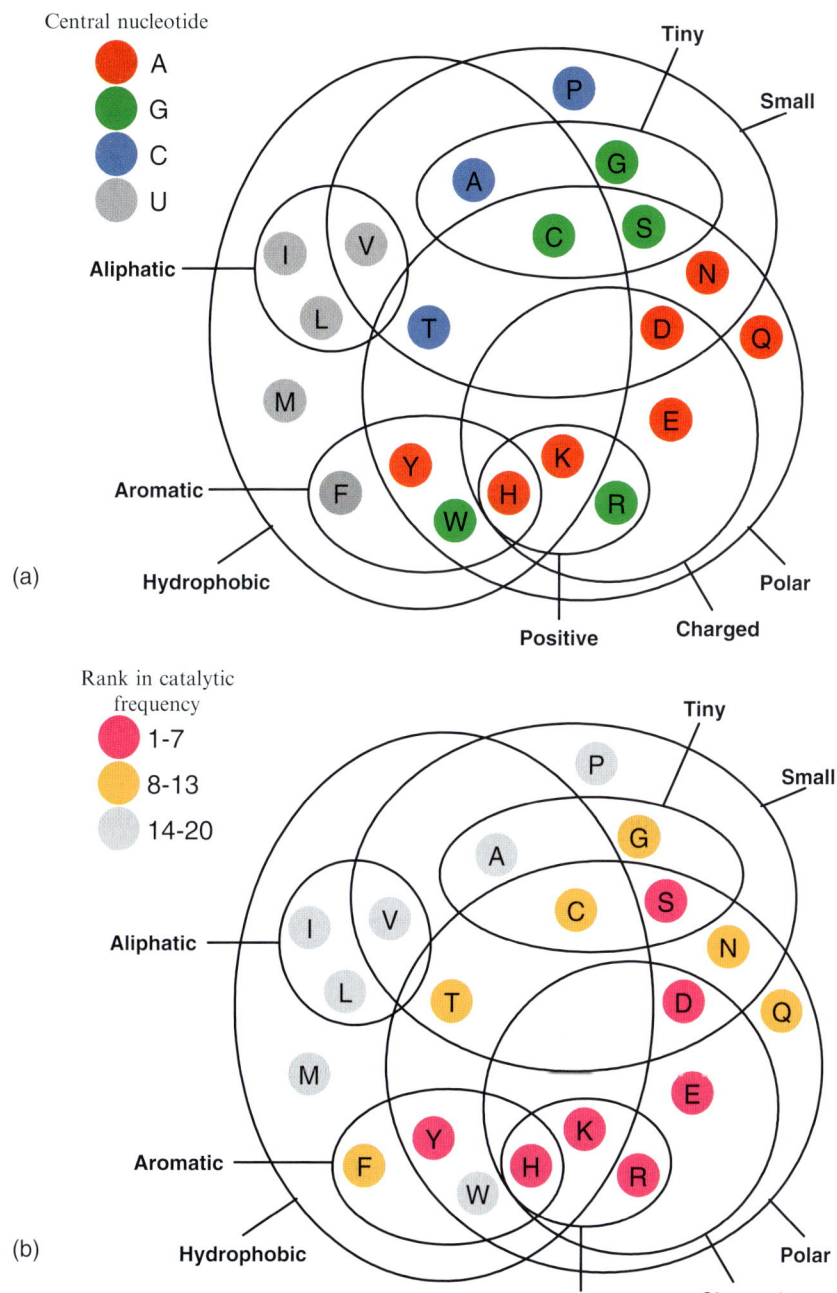

Fig. 7 Venn diagrams of amino acids (chemical sets from Taylor, 1986). (a) Distribution of amino acids based on the middle letter of the genetic code, (b) distribution of amino acids according catalytic frequency ranks and chemical properties

the medium or as a result or internal synthesis. As noted by Wong and Bronskill (1979), ideas about amino acid availability in the 'primordial soup' are inadequate when one considers origin of the genetic code. Indeed, if the RNA world was metabolically complex (which seems likely: Benner et al., 1989) then a protracted period of co-evolution of ribozymes, membranes, and metabolism is likely to have taken place (Szathmáry, 2007). Nevertheless it is important that, with the exception of lysine and arginine, all catalytically important amino acids seem to have at least some prebiotic plausibility (Miller, 1986), including histidine (Shen et al., 1990). Lysine has two different, complicated biosynthetic routes in modern organisms (Berg et al., 2003), so for the time being it is safer to assume that it is a very late invention. We propose that its role and position in an interim genetic code could have been taken by arginine (see section on protein appearance). We believe that arginine goes back to the RNA world, supported by its recognition by RNA aptamers with codonic binding sites (Knight and Landweber, 2000).

As discussed above, the ancient charging enzymes are assumed to have been ribozymes. Specific aptamers for the amino acids Arg, Ile, Tyr, Gln, Phe, His, Trp, and Leu have been selected by now. According to the 'escaped triplet theory', triplets overrepresented in aptamer binding sites for amino acids became part of the modern genetic code (Yarus et al., 2005). Noteworthy in this regard is that *in vitro* generated RNA aptamers contain – in a statistically important way – anticodonic and, to a lesser degree, codonic binding sites for these amino acids (Caporaso et al., 2005). Although aptamers for Asp and Glu have not yet been selected, it is likely that divalent metal ions could neutralize the repulsion between RNA and these negatively charged amino acids and, ultimately, aptamers will be selected with success (R. Knight, personal communication, email, 2007).

Finally, it is important to mention that RNA molecules can charge amino acids either in *cis* (Illangasekare et al., 1995) or *trans* (Lee et al., 2000). Even the phosphate anhydride activation reaction of amino acids is feasible by RNA (Kumar and Yarus, 2001).

3 The Anticodon Hairpin as the Ancient Adaptor

The simplest form of evolutionary continuity is provided by conservation of the anticodon hairpin (stem and loop) of tRNAs as the most ancient adaptor, which was charged at position 37 of the modern tRNA molecule, adjacent to the 3'-end of the anticodon (Woese, 1972). The CCH hypothesis has adopted this view (Szathmáry, 1996, 1999). As discussed in those papers, such a hairpin offers an ideal transient binding by ribozymes through complementary structures (e.g. in the form of 'kissing' hairpins). Here we deal with two questions: the nature and synthesis of the chemical bond between the adaptor and the amino acid, and the growth of the tRNA molecule to its present form.

As suggested by Woese (1972) and Wong (1991), the link must have been established by a stable *N*-bond. Inspection of relevant contemporary modified bases (Fig. 8) suggests that the nature of this primordial form was through a carbamoyl group.

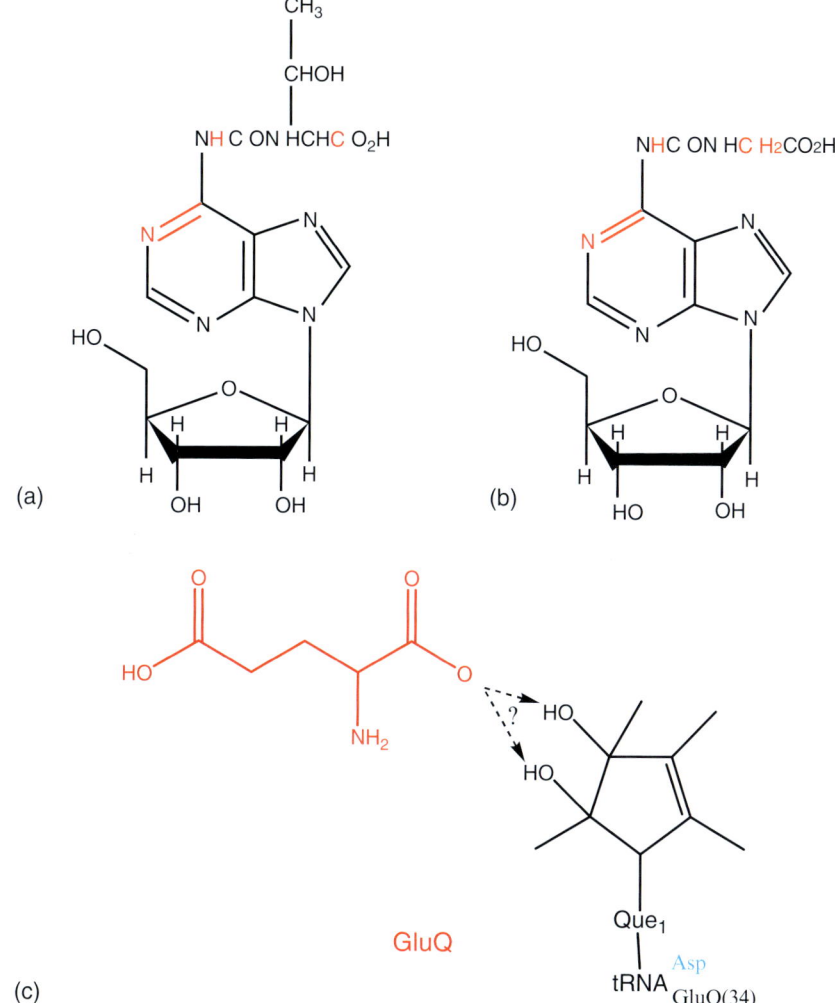

Fig. 8 Nucleosides modified by amino acids in tRNA molecules (from Grosjean et al., 2004). (a) N6-threonylcarbamoyladenosine (lu6 A. N6-hydroxynorvalylcarbamoyladenosine); (b) N6-glycylcarbamoyladenosine; (c) glutamylqueuosine

Regarding the early plausibility of this link, we call attention to the experimental investigations of Taillades et al. (1998) who suggested that *N*-protected *N*-carbamoyl-α-amino acids rather than free α-amino acids formed in the primitive hydrosphere, which serve as eminent starting points also for peptide synthesis. This type of metabolism may have been present even in the RNA world. Remarkably, a formal peptide bond arises by the coupling between the amino acid and adenine through the carbamoyl link!

If the original charging site was in the anticodon loop, then there must have been a stage in evolution when amino acids were still charged to the old position in this loop and, *at the same time*, already to the new 3′-end of the full tRNA molecule (Szathmáry, 1999). Some of the initial charging persisted as modern tRNA modifications, provided translation evolved using tRNA molecules still charged in the loop (Woese, 1972). Put differently, translation has adapted to these molecules being charged, which partly explains why removal of these modifications disturbs translation today. It is remarkable that the prediction of dual charging of tRNA by synthetases turned out to be correct for tRNA^{Asp} in *Escherichia coli* (Dubois et al., 2004). A paralog of a glutamyl-tRNA synthetase charges Glu to tRNA^{Asp} at position 34 (wobble) to queuosine (Fig. 8c). This example is quite suggestive, even if the chemistry and exact location is different from what we consider important for the CCH scenario. We predict that other synthetase paralogs with tRNA modifying activity will be found in the future. In agreement with the view advocated here, Grosjean et al. (2004, p. 519) comment that 'this modified nucleoside might be a relic of an ancient code'.

We wish to comment further on position 37. It is adjacent to the third base of the anticodon, which is complementary to the first base of the codon. As mentioned above, there is good correlation between first codon base identity and amino acid biosynthetic family membership (Taylor and Coates, 1989). If it is true that position 37 was an important ancient charging site, then the third anticodon base was closest to it. If some of the amino acid transformations (analogous to the modern tRNA: synthetase system) took place in the context of the ancient adaptor–ribozyme synthetase relation, then in agreement with the co-evolution theory (Wong, 1975) we suggest that some amino acids could have been biosynthetically transformed while bound to the ancient adaptor (the anticodon hairpin). The immediate and amino acid-specific neighbourhood on the adaptor would have been position 36, to which the transforming ribozymes would have been sensitive. The correlation of the third anticodon base with biosynthesis could be a relic of the most ancient genetic code and adaptor charging.

We should also explain how the ancient adaptor could have grown to its current size (tRNA). Again, position 37 seems to convey a message. It is between position 37 and 38 that tRNA genes for Glu, His, Gln, and initiator Met has been found in the archeon *Nanoarchaeum equitans* (Randau et al., 2005a,b). The splicing endonuclease splices other intron-coding tRNA genes as well as the transcripts of the two halves of the split tRNA genes (Randau et al., 2005c). The latter is made possible by overhangs at the 3′ and at the 5′-ends of the two transcripts, respectively. This archeon remains the only known organism that functions with split tRNA genes, so this is likely to be a derived (rather than ancestral) phenomenon, in contrast to the interpretation of Di Giulio (2006).

Nevertheless, an adjacent position (between 36 and 37) seems to be important. In Eubacteria there is a Group I self-splicing intron in the same position in several tRNA genes (Reinhold-Hurek and Shub, 1992; Haugen et al., 2005). It is this type of intron that can bind arginine with codonic binding sites (Yarus, 1989). Szathmáry (1993) accepted the idea of Ho (1988) that Group I introns had once been primordial synthetases. This idea still seems promising to us, and experimental attempts at demonstrating such a function in some (mutant) version would be welcome. Preservation of these introns could be due to fortuitous self-insertion (reverse self-splicing) of these molecules into some anticodon loops (Szathmáry, 1993).

Regarding the evolutionary growth of the anticodon hairpin to a full-blown tRNA, we think that ideas resting simply on single, major hairpin duplication (e.g. Widmann et al., 2005) are remote from what we want because half a tRNA is much bigger than a single anticodon loop. This question warrants careful investigation; here we merely call attention to the fact that Bloch et al. (1985) and Nazarea et al. (1985) found statistical evidence for tandem repeats of units of length 8–10 (centred around 9) in both tRNAs and rRNAs. Note that the Group I intron splits the anticodon hairpin into one piece of 10 nucleotides and another one of seven nucleotides. We do not yet know what this could mean. Yet, the most direct evidence for the primitive ancestry of the anticodon arm is that the anticodon arms of tRNAs with complementary anticodons are also complementary, which is not true for the acceptor stem (Rodin et al., 1993).

Last but not least, we point out that the proposed ancestral mechanism for adaptor charging can explain a strange feature found in the acceptor stem. There seems to be a vestigial anticodon–codon pair in the 1-2-3 position, and opposite to in the tRNA acceptor stem (Rodin et al., 1996). The same investigation revealed that there are several tandem repeats in tRNAs, e.g. those of the –DCCA motif (D is the overhanging 'discriminator base' at the 3′end) and its complementary sequence. Szathmáry (1999) presented a somewhat artificial scenario for the evolutionary growth of the anticodon arm to a tRNA with Rodin's anticodon–codon pair in the acceptor stem. Finally, we mention a resolution of the apparent conflict between the primitive ancestry of the anticodon arm and the idea that the anticodon-binding part of present day synthetases is regarded younger than the part binding the acceptor stem (Woese et al., 2000). Whereas ribozymes charged tRNAs both at the old (anticodon) and at the new (acceptor stem) positions by the cognate amino acids, most of the emerging protein synthetases charged them only at the new site.

4   Towards the Appearance of Proteins

One can imagine two ways to build up proteins: (i) to start with a more or less structural role of maybe otherwise 'boring' oligo/polypeptides, which later became complemented by slowly emerging catalytic potential; or (ii) to introduce catalytically highly promising amino acids which later become complemented by structural supports that would ultimately fold without the help of RNA. We prefer the second alternative, since (as explained above) it offers a straightforward way of the appearance of coding before translation, and it provides a substantial and immediate selective advantage in the RNA world.

But it is a good question to ask how single amino acids could have grown to polypeptides in this scenario. Presumably, first dipeptides would appear that would be kept in place after formation of the peptides bond between two adjacent CCH molecules (cf. Szathmáry, 1999), which would then result in a dipeptide bound to one of the adaptors, an intermediate that is identical to Wong's (1991) peptidyl-tRNA; the other adaptor would be liberated. Further growth can be envisaged under selection for improved enzymatic activity, but then the burning question arises: what would keep the growing polypeptide in a stable/useful conformation? One possibility is that it is binding to the RNA 'scaffold' of the ribozyme, as mentioned previously (Söding and Lupas, 2003). Another would be the *build-up of the smallest possible foldable structures, which are the β-turns stabilized by short β-sheets* (Lesk, 2001). Knowing the opportunistic nature of evolution, we would not exclude either possibility. Following a suggestion by Orgel (1977), Jurka and Smith (1987) argued that *the first β-turns were encoded by RRN codons, which include, with exception of His, all the catalytically most important amino acids!* But the second most important group of amino acids for the β-turns is the YRN group, which includes His, and the two add up to NRN, the last two columns of the code (Fig. 3). Dendrogram Fig. 6c confirms this idea strongly. Amino acids with the NRN pattern are also *mediocre α-helix and β-sheet builders* (Fig. 6b), so experimentation in that direction was not totally excluded either. Data (Prevelige and Fasman, 1989) show that proline is the third strongest loop builder, which makes it the only real exception to the NRN rule (Ser has also an AGY codon).

It is known that one can make enzymes with fewer than 20 amino acids. Walter et al. (2005) managed to evolve a chorismate mutase built of nine amino acids only: Arg, Asp, Glu, Asn, Lys, Phe, Ile, Leu, and Met. Noting the redundancies Asp/Glu and Ile/Leu, the enzyme could be probably even more simplified in the future. Remarkably, its active site (Fig. 9) is built of RRN codon-type

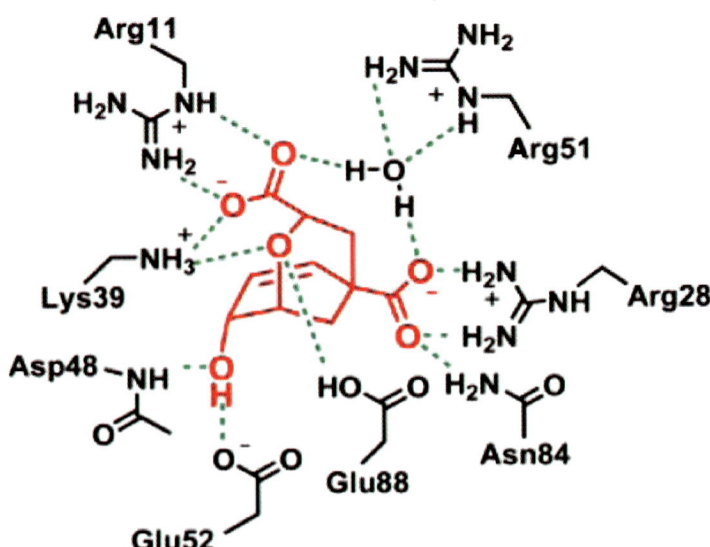

Fig. 9 Proposed active site of a chorismate mutase built of just nine amino acids. (Walter et al., 2005)

amino acids only! Adding amino acids of the NUN codon-type allows the set to build proper α-helices; amino acids with NCN codons are not required at all. Evolution of this enzyme is in rather good agreement with the theoretical estimate that the minimum number of amino acids to fold a protein is around ten (Fan and Wang, 2003).

The next amino acids, to further stabilize the β-turns with β-sheets were amino acids with NYN codons (Jurka and Smith, 1987). The proposal by Di Giulio (1996) that the genetic code was driven by the need to form β-sheets is in our view secondary to the catalytic propensity/β-turn-driven primary evolution, but independently important to build the scaffolds for the catalytic structures. But selection for structures in this order gives α-helices for free since the code is virtually complete. We suggest that *the multiplicity of catalytically boring amino acids in the genetic code is explained by selection for fine-tuning of the 3D structure of the scaffolds to optimize the geometric arrangements of the active sites.*

This scenario is supported by the correlation coefficients between the amino acids properties (Table 1).

There is a weak negative correlation between catalytic and α-helix propensity, so the latter structures cannot arise based on amino acids selected for catalysis; it is easiest to go for the β-turns. From these, one can go by evolution (amino acid vocabulary extension) in the direction of either the β-sheets (slightly favoured) or the α-helices, but presumably not in both.

We performed a network analysis of the BLOSUM amino acid substitution matrix (Henikoff and Henikoff, 1992) in order to see along which lines amino acid vocabulary extension/replacement could have been most likely (Fig. 10).

The most common substitutions occur within the (Lys, Arg), (Ile, Val), and (Phe, Tyr, Trp) sets. The firs dyad is clearly a catalytically very important one, and this is another reason why we suggest that Arg replaced Lys before LUCA. Note the remarkable role of His in these plots also. The catalytically most important amino acid His builds the bridge via the Tyr-His-(Asn, Gln) link between the catalytically unimportant and important (internally well connected) clusters (Fig. 10c,d)! We have specific RNA aptamers (Caporaso et al., 2005) for the whole bridge (none for Asn, but the bridge is still functional via Gln), so we suggest that it has been built in the RNA world.

Table 1 Correlation between pairs of amino acid properties related to the construction of active sites and scaffolds

Trait pair	Correlation coefficient
Activity, α-helix	−0.15049
Activity, β-sheet	0.31893
Activity, β-turn	0.38407
α-helix, β-sheet	0.42399
α-helix, β-turn	0.64948
β-sheet, β-turn	0.67883

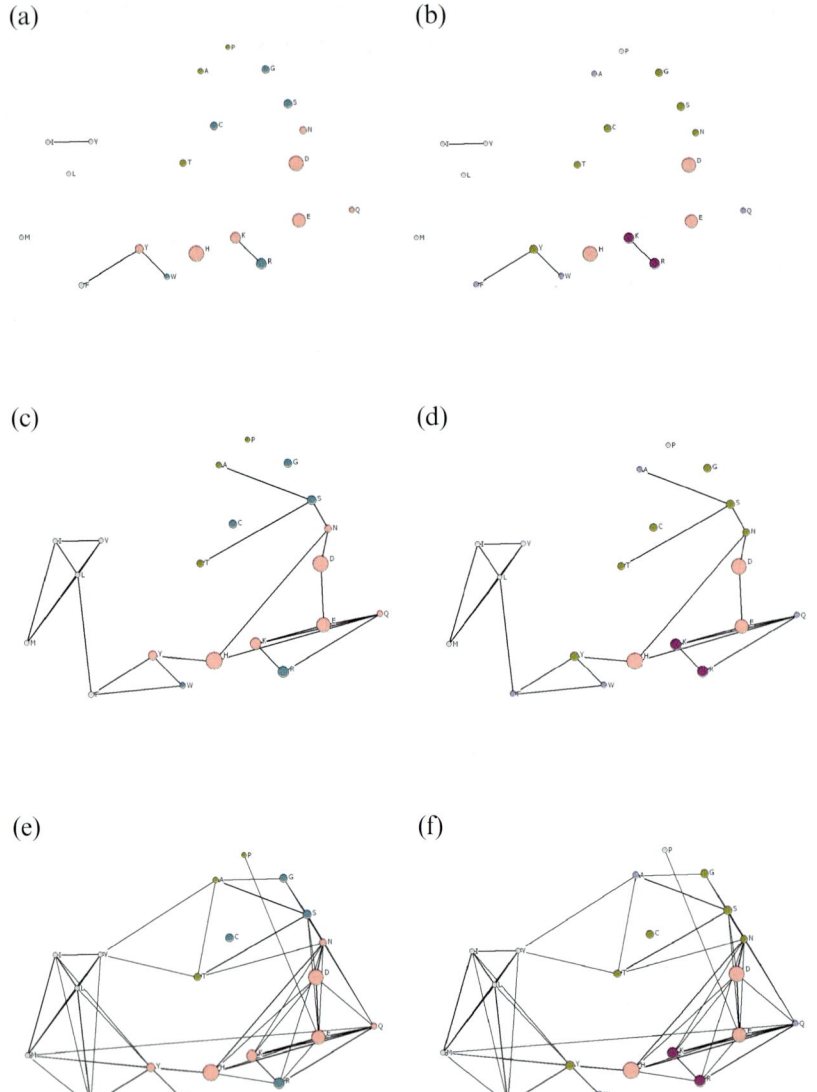

Fig. 10 Connectivity of the amino acid substitution network based on the BLOSUM 62 matrix (Henikoff and Henikoff, 1992). Colour codes of amino acids refer to those of Fig. 7a (a, c, e) and Fig. 7b (b, d, f). Substitution data have been transformed: we added 6 to the originals given by Henikoff and Henikoff (1992). For clarity, loops given in the main diagonal are not shown. We illustrate the network with different lower thresholds (minimal frequency values) for the transformed substitution data: 9 in a and b, 7 in c and d, and 6 in e and f (9 is the strongest value, i.e. it was equal to 3 in the original data matrix). Note that the network is undirected. Drawn by UCINET. (From Borgatti et al., 2002)

5   Towards an Experimental Test of the CCH Hypothesis with Catalytically Important Amino Acids

As noted before, the strongest experimental support in favour of the 'amino acids as cofactors' idea is the successful histidine-dependent RNA-cleaving enzyme made of DNA (Roth and Breaker, 1998). One would like to demonstrate the usefulness of the CCH hypothesis by the evolved usage by ribozymes of amino acids linked to handles (i.e. the free anticodon arm).

To this end one could repeat the above experiment with histidine linked to position 37 of tRNA[His] by the carbamoyl bond. For this and the forthcoming experiments we suggest usage of the *in vitro* compartment selection technique of Griffiths (Agresti et al., 2005).

A perhaps more appealing test would be to go for a metabolic reaction. The successful in vitro selection of an NADH-dependent alcohol dehydrogenase ribozyme by Tsukiji et al. (2004) is an excellent starting point (Fig. 11); it is also highly relevant for the idea of a metabolically complex RNA world (Benner et al., 1989).

First, a variant of the ribozyme that would work in *trans* should be selected in compartments for a different substrate, such as malate. Malate dehydrogenase has Asp and His in its active centre. In the second step ribozymes that could utilize

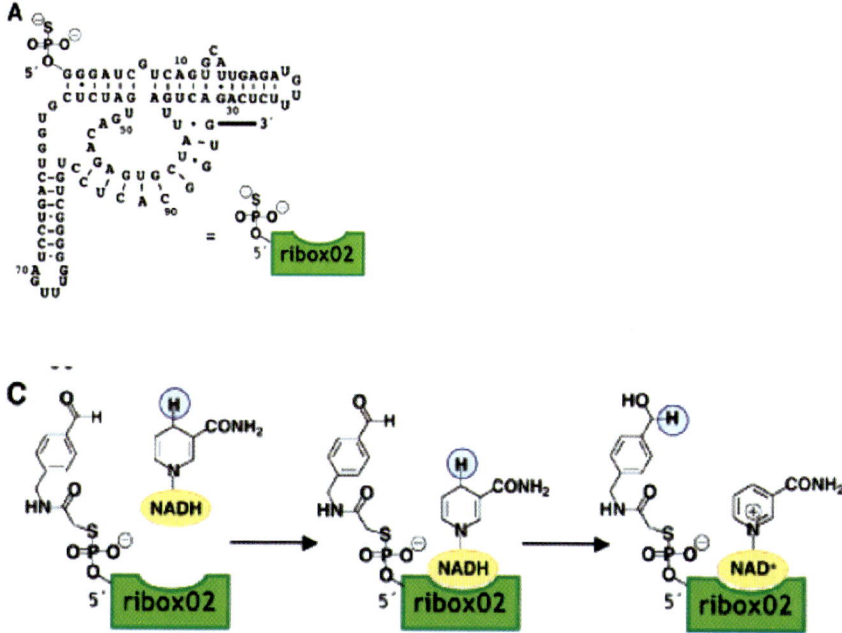

Fig. 11 Structure and action (on benzaldehyde) of an alcohol dehydrogenase ribozyme. (From Tsukiji et al., 2004)

either of these amino acids in free form should be selected. Finally, ribozymes that could use the CCH-amino acid should be selected. If successful, all the ribozymes should be compared for structure and activity. Ultimately a trial using both amino acids should be attempted. This would be an exciting experiment for the origin of the genetic code and the increase in biological complexity.

Acknowledgements This work was supported by the Hungarian Scientific Research Fund (D048406) and by the National Office for Research and Technology (NAP 2005/KCKHA005). We thank L. Patthy for discussion. The help of Paolo Sonego with hierarchical clustering is gratefully acknowledged. FJ is fully supported by Society in Science: The Branco Weiss Fellowship, ETH Zürich, Switzerland.

References

Agresti JJ, Kelly BT, Jaschke A, Griffiths AD (2005) Selection of ribozymes that catalyse multiple turnover Diels-Alder cycloadditions by using *in vitro* compartmentalization. Proc Natl Acad Sci USA 102:16170–16175

Bartlett GJ, Porter CT, Borkakoti N, Thornton JM (2002) Analysis of catalytic residues in enzyme active sites. J Mol Biol 324:105–121

Benner SA, Ellington AD Tauer A (1989) Modern metabolism as a palimpsest of the RNA world. Proc Natl Acad Sci USA 86:7054–7058

Berg JM, Tymoczko JL, Stryer L (2003) Biochemistry, Fifth Edition. W. H. Freeman, San Francisco

Bloch DP, McArthur B, Mirrop S (1985) tRNA-rRNA sequence homologies: evidence for an ancient modular format shared by tRNAs and rRNAs. BioSystems 17:209–225

Borgatti SP, Everett MG, Freeman LC (2002) Ucinet for Windows: Software for Social Network Analysis. Analytic Technologies, Harvard

Caporaso JG, Yarus M, Knight R (2005) Error minimization of coding triplet/binding site associations are independent features of the canonical genetic code. J Mol Evol 61:597–607

Crick FHC, Brenner S, Klug A, Pieczenik G (1976) A speculation on the origin of protein synthesis. Orig Life 7:389–397

Di Giulio M (1996) The β-sheets of proteins, the biosynthetic relationships between amino acids, and the origin of the genetic code. Orig Life Evol Biosph 26(6):589–609

Di Giulio M (2006) *Nanoarchaeum equitans* is a living fossil. J Theor Biol 242:257–260

Dubois DY, Blaise M, Becker HD, Campanacci V, Keith G, Giege R, Cambillau C, Lapointe J, Kern D (2004) An aminoacyl-tRNA synthetase-like protein encoded by the *Escherichia coli yadB* gene glutamylates specifically tRNA^{Asp}. Proc Natl Acad Sci USA 101:7030–7035

Fan K, Wang W (2003) What is the minimum number of letters required to fold a protein? J Mol Biol 328:921–926

Freeland SJ, Knight RD, Landweber LF, Hurst LD (2000) Early fixation of an optimal genetic code. Mol Biol Evol 17:511–518

Gilbert W (1986) The RNA world. Nature 319:618

Grosjean H, De Crécy-Lagard V, Björk GR (2004) Aminoacylation of the anticodon stem by a tRNA-synthetase paralog: relic of an ancient code? Trends Biochem Sci 29:519–522

Hartigan JA (1975) Clustering Algorithms, Wiley, New York

Haugen P, Simon DM, Bhattacharya D (2005) The natural history of group I introns. Trends Genet 21:111–119

Henikoff S, Henikoff JG (1992) Amino acid substitution matrices from protein blocks. Proc Natl Acad Sci USA 89:10915–10919

Ho CK (1988) Primitive ancestry of transfer RNA. Nature 333:24

Illangasekare M, Sanchez G, Nickles T, Yarus M (1995) Aminoacyl-RNA synthesis catalyzed by an RNA. Science 267:643–647

Jurka J, Smith TF (1987) β-turn driven early evolution: the genetic code and biosynthetic pathways. J Mol Evol 25:151–159

Porter CT, Bartlett GJ, Thornton JM (2004) The catalytic site atlas: a resource of catalytic sites and residues identified in enzymes using structural data. Nucleic Acid Res. 32:D129–D133

Knight RD, Landweber LF (2000) Guilt by association: the arginine case revisited. RNA 6:499–510

Kumar RK, Yarus M (2001) RNA-catalyzed amino acid activation. Biochemistry 40:6998–7004

Lee N, Bessho Y, Wei K, Szostak JW, Suga H (2000) Ribozyme-catalyzed tRNA aminoacylation. Nat Struct Biol 7:28–33

Lesk AM (2001) Introduction to Protein Architecture. Oxford University Press, Oxford

Maynard Smith J, Szathmáry E (1995) The Major Transitions in Evolution. Freeman, Oxford

Miller SL (1986) Current status of the prebiotic synthesis of small molecules. Chem Scr 26B:5–11

Moore PB, Steitz TA (2002) The involvement of RNA in ribosome function. Nature 418:229–235

Muñoz V, Serrano L (1994) Intrinsic secondary structure propensities of the amino acids, using statistical phi-psi matrices. Comparison with experimental scales. Proteins 20:301–311

Nazarea AD, Bloch DP, Semrau AC (1985) Detection of a fundamental modular format common to transfer and ribosomal RNAs: second-order spectral analysis. Proc Natl Acad Sci USA 82:5337–5341

Orgel LE (1977) β-Turns and the evolution of protein synthesis. In: Bradbury EM, Javaherian K (eds) The Organization and Expression of the Eukaryotic Genome. Academic Press, London, pp. 499–504

Prevelige JP, Fasman G (1989) Chou-Fasman prediction of the secondary structure of proteins: the Chou-Fasman-Prevelige algorithm. In: Fasman G (ed.) Prediction of Protein Structure and the Principles of Protein Confoimation. Plenum, New York, pp. 391–416

Randau L, Munch R, Hohn MJ, Jahn D, Söll D (2005a) *Nanoarchaeum equitans* creates functional tRNAs from separate genes for their 5′- and 3′-halves. Nature 433:537–541

Randau L, Pearson M, Söll D (2005b) The complete set of tRNA species in *Nanoarchaeum equitans*. FEBS Lett 579:2945–2947

Randau L, Calvin K, Hall M, Yuan J, Podar M, Li H, Söll D (2005c) The heteromeric *Nanoarchaeum equitans* splicing endonuclease cleaves noncanonical bulge-helix-bulge motifs of joined tRNA halves. Proc Natl Acad Sci USA 102:17934–17939

R Development Core Team (2004) A Language And Environment For Statistical Computing. R Foundation for Statistical Computing, Vienna, Austria

Reinhold-Hurek B, Shub DA (1992) Self-splicing introns in tRNA genes of widely divergent bacteria. Nature 357:173–176

Rice P, Longden I, Bleasby A (2000) EMBOSS: the European molecular biology open software suite. Trends Genet 16:276–277

Rodin S, Ohno S, Rodin A (1993) Transfer RNAs with complementary anticodons: could they reflect early evolution of discriminative genetic code adaptors? Proc Natl Acad Sci USA 90:4723–4727

Rodin S, Rodin A, Ohno S (1996) The presence of codon-anticodon pairs in the acceptor stem of tRNAs. Proc Natl Acad Sci USA 93:4537–4542

Roth A, Breaker RR (1998) An amino acid as a cofactor for a catalytic polynucleotide. Proc Natl Acad Sci USA 95:6027–6031

Shen C, Yang L, Miller SL, Oró J (1990) Prebiotic synthesis of histidine. J Mol Evol 31:167–174

Söding J, Lupas AN (2003) More than the sum of their parts: on the evolution of proteins from peptides. BioEssays 25:837–846

Steitz TA, Moore PB (2003) RNA, the first macromolecular catalyst: the ribosome is a ribozyme. Trends Biochem Sci 28:411–418

Szathmáry E (1990) Useful coding before translation: the coding coenzymes handle hypothesis for the origin of the genetic code. In: Lukács B. et al. (eds) Evolution: from Cosmogenesis to Biogenesis. KFKI-1990-50/C, Budapest, pp. 77–83

Szathmáry E (1991) Four letters in the genetic alphabet: a frozen evolutionary optimum? Proc R Soc Lond B 245:91–99

Szathmáry E (1992) What determines the size of the genetic alphabet? Proc Natl Acad Sci USA 89:2614–2618

Szathmáry E (1993) Coding coenzyme handles: A hypothesis for the origin of the genetic code. Proc Natl Acad Sci USA 90:9916–9920

Szathmáry E (1996) Coding coenzyme handles and the origin of the genetic code. In: Müller A, Dress A, Vögtle F (eds) From Simplicity to Complexity in Chemistry – and Beyond. Part I. Vieweg, Braunschweig, pp. 33–41

Szathmáry E (1999) The origin of the genetic code: amino acids as cofactors in an RNA world. Trends Genet 15:223–229

Szathmáry E (2007) Coevolution of metabolic networks and membranes: the scenario of progressive sequestration. Phil Trans R Soc B DOI: 10.1098/rstb.2007.2070

Taillades J, Beuzelin I, Garrel L, Tabacik V, Bied C, Commeyras A (1998) N-carbamoyl-alpha-amino acids rather than free alpha-amino acids formation in the primitive hydrosphere: a novel proposal for the emergence of prebiotic peptides. Orig Life Evol Biosph 28:61–77

Taylor FJ, Coates D (1989) The Code within the codons. Biosystems 22:177–187

Taylor WR (1986) The classification of amino acid conservation. J Theor Biol 119:205–218

Tsukiji S, Pattnaik SB, Suga H (2004) Reduction of aldehyde by a NADH/Zn^{2+}-dependent redox active ribozyme. J Am Chem Soc 126:5044–5045

Walter KU, Vamvaca K, Hilvert D (2005) An active enzyme constructed from a 9-amino acid alphabet. J Biol Chem 45:37742–37746

Wetzel R (1995) Evolution of the aminoacyl-tRNA synthetases and the origin of the genetic code. J Mol Evol 40:545–550

Widmann J, Di Giulio M, Yarus M, Knight R (2005) tRNA creation by hairpin duplication. J Mol Evol 61:524–530

Woese CR (1972) The emergence of genetic organization. In: Ponnamperuma, C. (ed.) Exobiology. North-Holland Publishing, Amsterdam, The Netherlands, pp. 301–341

Woese CR, Olsen GJ, Ibba M, Soll D (2000) Aminoacyl-tRNA synthetases, the genetic code, and the evolutionary process. Microbiol Mol Biol Rev 64:202–236

Wong JT (1975) A co-evolution theory of the genetic code. Proc Natl Acad Sci USA 72:1909–1912

Wong JT (1991) Origin of genetically encoded protein synthesis: a model based on selection for RNA peptidation. Orig Life Evol Biosph 21:165–176

Wong JT Bronskill PM (1979) Inadequacy of prebiotic synthesis as origin of proteinous amino acids. J Mol Evol 13:115–125

Yarus M (1989) Specificity of arginine binding by the Tetrahymena intron. Biochemistry 28:980–988

Yarus M, Caporaso JG, Knight R (2005) Origins of the genetic code: the escaped triplet theory. Annu Rev Biochem 74:179–198

Zar JH (1998) Biostatistical Analysis (4th Edition). Prentice Hall

Chapter 4

Why the Genetic Code Originated: Implications for the Origin of Protein Synthesis

Massimo Di Giulio

Abstract A theory is developed on the origin of the first codified protein, that is to say the appearance of the first messenger RNA (mRNA). This theory hypothesises a central role in enzyme catalysis played by peptidyl-RNAs, i.e. complexes of peptides and RNAs, in a very early evolutionary stage of genetic code origin. The evolution of peptidyl-RNA would then lead to the origin of peptidyl-tRNA, a key intermediary in the origin of protein synthesis. The appearance of the first mRNA might be justified by the need of the evolving system for unique and directional interactions between peptidyl-RNAs, which had now become peptidyl-tRNA-like molecules. The appearance of the first mRNA, would therefore be the result of the optimisation and channelling of interactions between peptidyl-tRNA-like molecules.

1 Introduction

There are a number of ways of considering the origin of the genetic code, which can basically be grouped into two broad viewpoints: (i) an early origin, perhaps even related to the origin of life (Mackinlay, 1982); and (ii) a late origin (Di Giulio, 1998) characterising only the later phases of the origin of life or, rather, the evolution of the last universal common ancestor (LUCA) (Di Giulio, 2001).

It seems more sensible to me to consider genetic code origin as representing the terminal phases of the evolution of the origin of protein synthesis and having nothing to do with the origin of life. Therefore, its origin should not be early on. This is justified by the fact that I am unable to find a plausible reason for the origin of the genetic code unless in an already complex system, and this evolutionary stage seems to be very late on in the origin of life (Di Giulio, 1998, 2001). Furthermore, in order to trigger a selective pressure that may eventually result in the origin of the genetic code, I think that certain circumstances need to exist.

Laboratory of Molecular Evolution, Institute of Genetics and Biophysics 'Adriano Buzzati Traverso', CNR, Via P. Castellino, 111, 80131 Naples, Italy, e-mail: digiulio@igb.cnr.it

M. Barbieri (ed.), *The Codes of Life: The Rules of Macroevolution.*   59
© Springer 2008

Firstly, a permanent necessity to evolve, and hence codify, something truly important, such as the evolution of enzymatic catalysis because without catalysis there is no life. Secondly, catalysis itself must have been performed, at least in some of its evolutionary stages, by molecules that have something to do with genetic material because, otherwise, the link to the genetic code would be difficult to achieve. In this case, the evolution of enzymatic catalysis may even have come to a halt at a rudimental stage, perhaps leading inevitably to the extinction of life itself or at least severely limiting its complexity. In other words, the replicating molecules must, at some evolutionary stage, have had something to do with catalysis-performing molecules because otherwise the link between these two worlds would have been difficult to achieve. All the above introduces the topic discussed in the following section.

2 Peptidyl-tRNA-like Molecules were the Centre of Protocell Catalysis and the Fulcrum for the Origin of the Genetic Code

The peptidyl-tRNA molecule, the key intermediary in the origin of protein synthesis, has no function per se nowadays but is merely an intermediary for achieving protein synthesis. How can peptidyl-tRNA have evolved if it performs no real direct function and clearly cannot, therefore, have been selected? It is more than reasonable to think that, at the origin of its evolution, this intermediary performed all the catalysis needed by the protocell (Wong, 1991; Wong and Xue, 2002; Di Giulio, 1997, 2003). In other words, given that peptidyl-tRNA performs no function per se, in order to justify its origin we can hypothesise that it originally performed the function now performed by proteins because, otherwise, its selection evolution could not be justified. Its origin can be justified by assigning it an essential role in protocell catalysis so that peptidyl-tRNA performs, in an evolutionary sense, the main function of its 'products' – the proteins. In this way if peptidyl-tRNA-like molecules performed all enzymatic catalysis in the protocell (Wong, 1991; Wong and Xue, 2002; Di Giulio, 1997, 2003) then we can understand their evolution in that by acquiring a specific function they become molecules on which natural selection might have acted.

Another argument, independent of the former and in agreement with the idea that peptidyl-tRNA-like molecules were crucial to and characteristic of a long period of the evolutionary history of protocells, results from the consideration that these complexes between a peptide and RNA must have been central to the evolution of the genetic code. More simply, the origin of the first mRNA might have been the result of the optimisation and channelling of the interactions between peptidyl-tRNA-like molecules because these interactions intuitively seem the only ones which could, if guided by an RNA acting in *trans*, result in the origin of the first mRNA and ultimately lead to the origin of the genetic code. It therefore seems to me that it is only the interactions between peptidyl-tRNA-like molecules which ensures the simplicity, and at the same time the complexity that could originate the genetic code. This is why I have decided to investigate a model based on the evolution of these interactions between peptidyl-tRNA-like molecules and see whether or not a plausible evolutionary scheme can be found for the appearance of the first mRNA.

3 The First 'Messengers RNAs' Codified Successions of Interactions Between Different Peptide-RNAs

If, as seems to be the case, a stage was reached in which all protocell catalysis was performed by covalent complexes of peptides and RNAs (peptide-RNA) (Wong, 1991; Di Giulio, 1997; Wong and Xue, 2002; Di Giulio, 2003), then any improvement in peptide-RNA synthesis must have had a highly positive effect on protocell activity. More specifically, the improvement of peptide-RNA synthesis must have passed through certain stages. The most elementary must have been a direct interaction between two different peptide-RNAs (Orgel, 1989; Wong, 1991; Di Giulio, 1994, 1997, 2003), obviously mediated by hydrogen bonds between bases which could lead, for instance, to a lengthening of the peptide attached to one of the RNA or to a simple and different peptidation of an RNA (Fig. 1, stage 1) and therefore to an improvement in catalysis.

However, by direct interaction between peptide-RNAs (two or more than two), these syntheses would undergo a veritable improvement only when some peptide-RNAs started to be used as coordinators (templates) of the successions (sequences) of interactions between different peptide-RNAs (Fig. 1, stage 2). The RNA component of some peptide-RNAs probably began to favour (codify) the successions of interactions between peptide-RNAs (Fig. 1, stage 2). It seems to me, that this is the oldest form of messenger RNA (mRNA) that can be identified. In other words, the ancestor of mRNA could be an RNA codifying the successions of interactions between different peptide-RNAs. Clearly, in the early evolutionary stages, these interactions took place using regions that were not necessarily short and contiguous on the template RNA (Fig. 1, stage 2). In order to optimise these interactions between peptide-RNAs on the template of an RNA, evolution must have favoured the template RNAs that clearly used short and contiguous regions of the template RNA (Fig. 1, stages 3 and 4). Therefore, the first form of 'genetic code' can be seen in these early 'mRNAs' which codified these sequences of interactions between peptide-RNAs in the form of a message which, at its beginning, might have been highly heterogeneous in the sense that it could use a number of different bases (2, 3, 4, 5, or more) to codify one of these interactions (Fig. 1, stage 4). Therefore, at the dawn of the genetic code, codification was not performed only by three nucleotides.

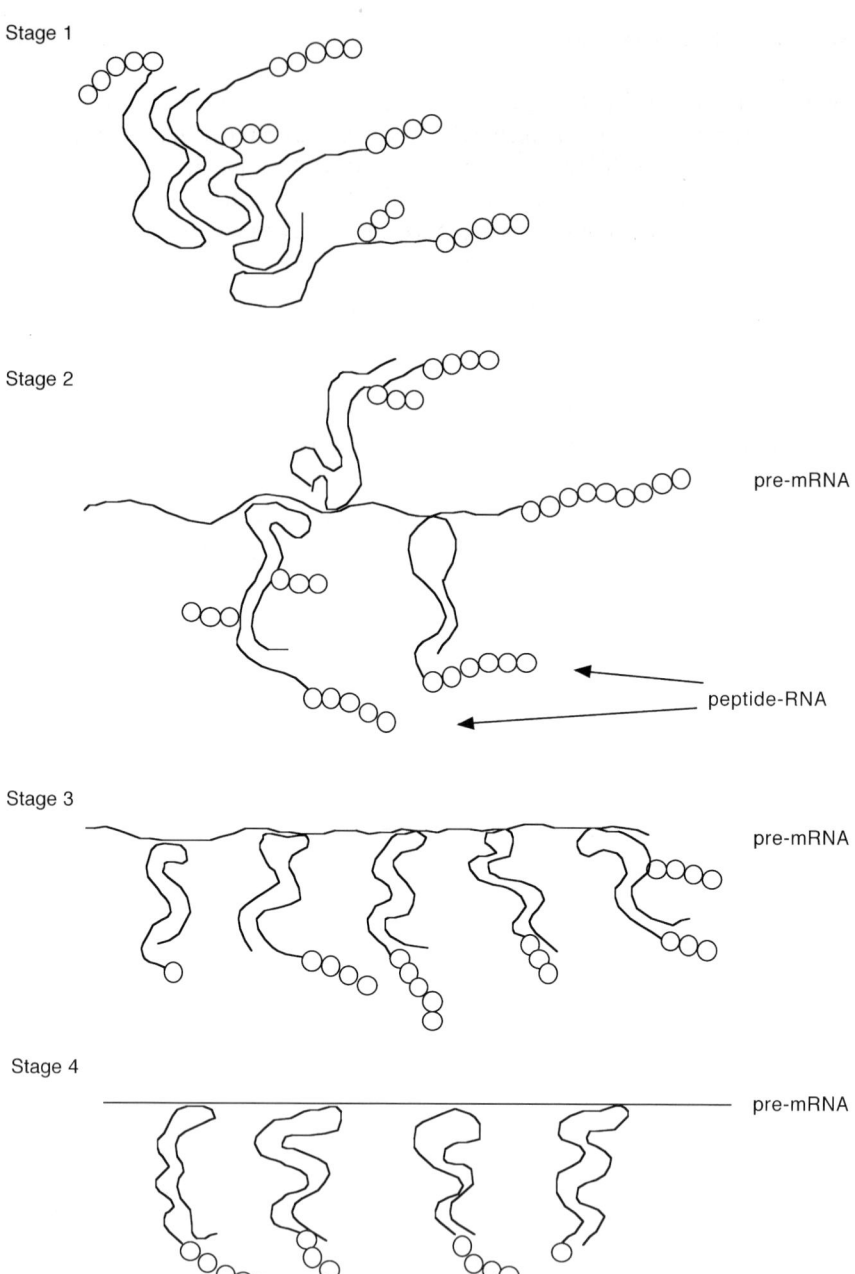

Fig. 1 This shows various evolutionary stages of the origin of protein synthesis and the first mRNA. Circles indicate single amino acids, and hence peptides, while the RNAs are symbolised by curved or straight lines. In the last two stages each short segment indicates a single pairing of nucleotides or single nucleotides, while the tRNA-like molecules are indicated using L-shaped structures. The protoribosome is not represented. Every single stage shows the most characteristic molecules for that stage, sometimes in a number higher than that which were simultaneously interacting, and this is particularly true for stages 7 and 8

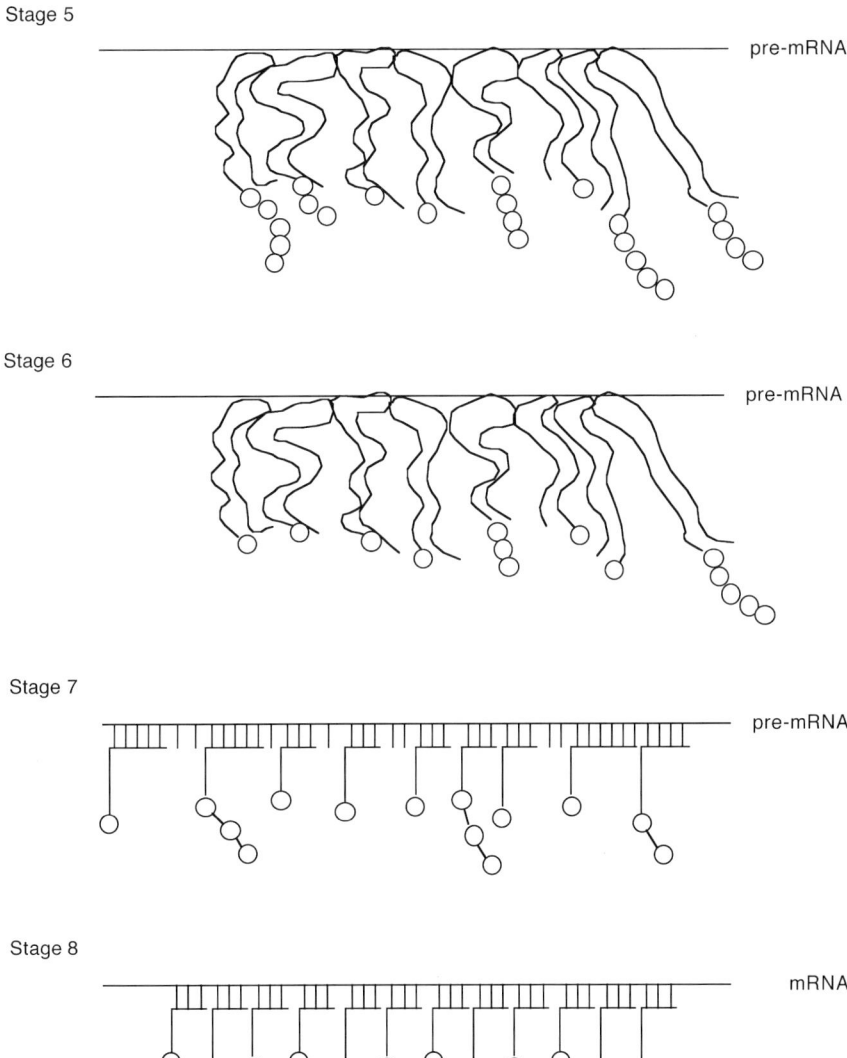

Fig. 1 (continued)

4   The Birth of the First mRNA

We must now show how an evolutionary stage in which the RNA templates codified for the interactions between peptide-RNAs (Fig. 1, stages 3 and 4) lead on to mRNAs proper (Fig. 1, stage 8).

It is clear that for RNAs (pre-mRNAs) which codified the successions of interactions between peptide-RNAs it would have been more convenient to have these codifications concentrated on rigorously contiguous regions as this would have made the interactions between peptide-RNAs more efficient. Therefore, we expect an evolutionary stage in which the pre-mRNA molecules evolved (probably through intensive use of genetic recombination between pre-mRNA molecules) the codification of the successions of interactions between peptide-RNAs so that each of these codifications was strictly contiguous to the next one on the pre-mRNAs (Fig. 1, stage 5). In other words, at this evolutionary stage the pre-mRNAs had evolved a message primarily made up of 'words' of code that were contiguous to one another (Fig. 1, stage 5) but of different lengths.

It is equally clear that the evolutionary stages envisaging interaction between pre-mRNAs and peptide-RNAs (Fig. 1, stage 2 on, but perhaps also in stage 1) triggered a selective pressure to favour the origin of protoribosome, that is to say a structure, which was somehow able to favour all these interactions between pre-mRNAs and peptide-RNAs. Here I will not address the evolution of ribosome except in using the word protoribosome when necessary, given that the model I analyse here justifies its evolution.

We have thus reached an evolutionary stage in which pre-mRNAs used contiguous 'words' of different-length nucleotides to codify the individual interactions between peptide-RNAs (Fig. 1, stage 5). We must now show the crucial phase in the appearance of the first mRNA, that is to say, from pre-mRNA (Fig. 1, stage 5) to mRNA proper (Fig. 1, stage 8).

It seems clear to me that if the synthesis of peptide-RNAs had to be improved and, in particular, make the sequences of peptides in these complexes more reproducible, then at this evolutionary stage (Fig. 1, stage 5) it would have been necessary to trigger a selective pressure that would evolve tRNA-like molecules carrying a single amino acid because it is only the addition of a single amino acid that can faithfully reproduce a precise sequence of a peptide or a polypeptide. Indeed, the addition of more than one amino acid at a time to a peptide in order to reproduce a given sequence may create certain difficulties that are not encountered by adding a single amino acid residue at a time to the growing peptide: in this way the reproduction of the peptide, residue by residue, started to evolve (Fig. 1, stage 6). Moreover, it can be hypothesised that tRNA-like molecules carrying a single amino acid were no longer directly involved in catalysis as this role was performed by their 'products'. Finally, and paradoxically, it is also possible that the selection to reproduce a specific sequence of a peptide might never have actually been operative; nevertheless, the evolution of tRNA-like molecules carrying a single amino acid might have evolved by means of specific peptidation of the catalytic RNA. In other words, the RNAs might have behaved as a scaffold on which to attach the amino acids and, therefore, transform them into more active catalysts.

This might have had repercussions on the sequences of the pre-mRNAs and on protoribosome in the sense that particular classes of pre-mRNAs as well as a particular type of protoribosome, more suitable for acting with these new pre-mRNAs and with the tRNA-like molecules carrying a single amino acid, might have evolved. These particular classes of pre-mRNAs might, for instance, have been made up of words of code that were similar in length but not necessarily equal, for example with a prevalence of three-letter words of code (codons) but also many long words comprising four, five, six, or more nucleotides. In any case, all these words of code were mostly recognised by specific tRNA-like molecules carrying a single amino acid residue but not only by these molecules in the sense that there still existed at this stage tRNA-like molecules carrying more than one amino acid (dipeptides, tripeptides, etc.) (Fig. 1, stage 7).

Finally, it seems to me that the evolution from stage 7 in Fig. 1 to the translation of mRNA proper (Fig. 1, stage 8) can be conjectured by simply postulating the evolution of a particular subclass of pre-mRNA, that is to say mRNAs (with triplets, i.e. codons) codifying only for precursor amino acids (Ala, Gly, Ser, Asp, and Glu) as predicted by the co-evolution theory of the origin of the genetic code (Wong, 1975; Di Giulio and Medugno, 1999). The only specification to be made is that, initially, these mRNAs had to represent only a tiny fraction of all the translatable RNAs and that these took an incredibly long time to replace all the pre-mRNAs. Even if it is a question of transforming a pre-mRNA codifying for a message with a heterogeneous reading, i.e. mostly made up of triplets but with interspersed words of four, five, six, or more bases, into an mRNA with words solely of triplets, I think that this is not the cause of strong evolutionary discontinuity if these mRNAs were, as already said, very few in number at the beginning of their evolution and codified only for the five precursor amino acids because: (i) in the pre-mRNA population there might have existed some with marked mRNA characters because protoribosome might also favoured the evolution of just one particular type of pre-mRNA; (ii) many pre-mRNAs, rather than evolving, i.e. transforming into new mRNAs, might have been replaced by new mRNAs which were only partially in relation with the original pre-mRNAs; (iii) a pre-mRNA editing mechanism, such as the addition or removal of bases (Landweber and Gilbert, 1994), might have favoured the evolution of the first mRNAs. Furthermore, the codification limited only to the five precursor amino acids must have taken place very slowly and might initially have been favoured by their use only for very few and not necessarily fundamental functions. However, it seems to me that the evolution of enzymatic catalysis, that is to say, the reproducibility of some polypeptide-RNAs can still be considered the main adaptive theme characterising this evolutionary stage and leading to the codification of only the precursor amino acids. The replacement of the entire population of pre-mRNAs by mRNAs (Fig. 1, stages 7 and 8) must have been favoured by the hypothesis that catalysis, at this evolutionary stage, was still fairly rudimental and this clearly favoured the predominance of mRNAs which were able to better reproduce polypeptides because they used only the addition of a single amino acid residue at a time to the growing polypeptide (Fig. 1, stage 8).

In conclusion, the replacement of pre-mRNAs by the emerging mRNAs could not have been the cause of an insurmountable evolutionary discontinuity if the time available was enormous (perhaps billions of years) and if the substitution mechanism was above all based on an evolution ex novo of mRNAs which replaced the pre-mRNAs and only partially evolved from the latter: the duplication of messages and their recombination must have played a determining role in the origin of the first mRNA.

5 A Prediction of the Model

The model here discussed for the origin of protein synthesis and the first mRNA (Fig. 1) does not necessarily seem to be worthless because it makes predictions that can be falsified. The main prediction the model makes is that some molecules characterising above all its final stages (Fig. 1) could still be present in the form of molecular fossils in contemporary cells. In other words, we cannot exclude the possibility that some of the molecules predicted by the model still exist, such as a tRNA-like molecule carrying a peptide and still taking part in protein synthesis in this form. As already discussed (Di Giulio, 2003), the tmRNA molecule (Muto et al., 1998) might be a similar fossil and a witness to the evolutionary stages discussed in the present model (Fig. 1). This strengthens the hypothesis that there might exist other and perhaps more interesting and intriguing fossils.

6 Conclusions

As argued here, we can say that the reason why the genetic code originated must be sought only in the evolution of the codification of enzymatic catalysis mediated by peptidyl-tRNA-like molecules, whose interactions, channelled and optimised by pre-mRNAs, resulted firstly in the origin of protein synthesis and finally in the origin of the genetic code through the birth of the first mRNA codifying only for precursor amino acids, as predicted by the co-evolution theory of genetic code origin. There would thus have been very little that was truly deterministic in this evolution.

Acknowledgements I wish to thank Pier Luigi Luisi and Pasquale Stano for an interesting discussion during a trip from Erice to Palermo airport. I also thank Simone Cossu for drawing the figure.

References

Di Giulio M (1994) On the origin of protein synthesis a speculative model based on hairpin RNA structures. J Theor Biol 171:303–308
Di Giulio M (1997) On the RNA world evidence in favor of an early ribonucleopeptide world. J Mol Evol 45:571–578
Di Giulio M (1998) Reflections on the origin of the genetic code: a hypothesis. J Theor Biol 191:191–196

Di Giulio M (2001) The non-universality of the genetic code the universal ancestor was a progenote. J Theor Biol 209:345–349

Di Giulio M (2003) The early phases of the genetic code origin: conjecture on the evolution of coded catalysis. Orig Life Evol Biosph 33:479–489

Di Giulio M, Medugno M (1999) Physicochemical optimization in the genetic code origin as the number of codified amino acids increases. J Mol Evol 49:1–10

Landweber LF, Gilbert W (1994) Phylogenetic analysis of RNA editing: a primitive genetic phenomenon. Proc Natl Acad Sci USA 91:918–921

Mackinlay AG (1982) Polynucleotide replication coupled to protein synthesis: a possible mechanism for the origin of life. Origins Life Evol Biosph 12:55–66

Muto A, Ushida C, Himeno H (1998) A bacterial RNA that functions as both a tRNA and an mRNA. Trends Biochem Sci 23:25–29

Orgel LE (1989) The origin of polynucleotide-directed protein synthesis. J Mol Evol 29:465–474

Wong JT (1975) A co-evolution theory of the genetic code. Proc Natl Acad Sci USA 72:1909–1912

Wong JT (1991) Origin of genetically encoded protein synthesis a model based on selection for RNA peptidation. Orig Life Evol Biosph 21:165–176

Wong JT, Xue H (2002) Self-perfecting evolution of heteropolymer building blocks and sequences as the basis of life. In: Fundamentals of Life. Editions scientifiques et medicales Elsevier SAS, Paris

Chapter 5

Self-Referential Formation of the Genetic System

Romeu Cardoso Guimarães[1], Carlos Henrique Costa Moreira[2], and Sávio Torres de Farias[1]

Abstract Formation of the genetic code is considered a part of the process of establishing precise nucleoprotein associations. The process is initiated by transfer tRNA (tRNA) dimers paired through the perfect palindromic anticodons, which are at the same time codons for each other; the amino acid acceptor ends produce the transferase function, in a manner similar to the reaction occurring in ribosomes. The connections between nucleic acids and proteins are bidirectional, forming a self-feeding system. In one direction, proteins that are resistant to degradation and efficient RNA binders stabilize the tRNAs that are specifically involved with their production; in the other direction, these tRNAs become fixed with the correspondences which are the amino acid codes. Replication of the stabilized tRNAs becomes elongational, forming poly-tRNAs, the precursors of the mRNA strings (genes), and of ribosomes. The linear order in the gene sequences follows the temporal succession of the encoding of tRNA pairs. The whole encoding process is oriented by the tRNA pairs. The core sequence of proteins shows the predominant aperiodic conformation and the anticodonic principal dinucleotides (pDiN) are composed of two purines or two pyrimidines: (1a) Gly/Pro; (1b) Ser/Ser; (2a) Asp, Glu/Leu; (2b) Asn, Lys/Phe. Members of the following pairs, with pDiN composed of a purine and a pyrimidine [(3a) Ala/Arg; (3b) Val/His, Gln; (3c) Thr/Cys, Trp; (4) Ile, Met/Tyr, and iMet/Stop], are added, respectively, to the mRNA heads/tails. It is indicated that: (a) The last universal common ancestor populations could, at some early stages, be composed of lineages bearing similar genetic codes, due to the simple and highly deterministic character of the process; (b) genetic information was created during the process of formation of the coding/decoding subsystem, inside a proto-metabolic system already producing some amino acids and tRNA-like precursors; (c) genes were defined by the proteins that stabilized the system, as memories for their production.

---

[1] Dept. Biologia Geral, Inst. Ciências Biológicas, Univ. Federal de Minas Gerais, 31270.901 Belo Horizonte MG Brasil, e-mail: romeucg@icb.ufmg.br

[2] Dept. Matemática, Inst. Ciências Exatas

M. Barbieri (ed.), *The Codes of Life: The Rules of Macroevolution.*
© Springer 2008

1   Introduction

Cellular physiology is didactically described in the top-down tradition, from genes to the phenotype (Alberts et al., 2002). This description is similar to that of man-made factories, with designers or programmers (the genes), and workers (the RNAs and proteins that build the phenotype). The Darwinian thought placed a directing agency in the interactions between organisms and the environmental contexts but the main focus remained in the genetic variants that allowed some phenotypes to survive the interactions while others were selected out. We remain with the task of explaining how such a genetic system first came into being. An apparent paradox arises, of how could the planners start their enterprise just gathering together into the factory some workers-to-be who never experienced their jobs, which would be similar to slave-hunting practices. A bottom-up perspective is taken in this work, devising a mechanism for the spontaneous origin of the biotic system, inside which the genetic processes arose. The spontaneity of processes may be categorized as self-organizational and genes are considered the memory part of the system. Sections 1 and 2 present some basic concepts about the living system components, relevant to understanding the genetic code. The model is presented in Sections 3–6, with some technical details of the process of formation of the coding/decoding system. The derived conceptual implications comprise Section 7.

2   The Biotic World

2.1   Strings and Folding

The basic constituents of the living system are nucleic acids and proteins so that cells are called nucleoprotein systems. These "noblest" components are strings, polymers where the monomers are, respectively, nucleotides and amino acids. The nobility in these strings comes from the specificity in the order of the monomers in the sequences, which bring about the main functions of the system. Amino acids are fundamentally 20, at the lower range of the number of letters in the symbolic alphabets of written languages.

The chains may be very long and are intricately folded to acquire specific spatial configurations. The plasticity of the tri-dimensional arrangements is a consequence of the string constitution and is a main factor in the regulatory and adaptive behavior of the entities. The main influences in the acquisition of the spatial architectures are the sequences in the polymers – the primary structure – and their interactions with water (Alberts et al., 2002). The folding is obtained through weak associative bonds, guaranteeing the dynamic character of the structures. In nucleic acids, the main structuring rule is the base pairing that builds the double helices; in RNA, single-stranded loops are frequent. The secondary structures are more varied in proteins, due to the small size and the diversity of amino acids. The complexity of the sequence organization required to guarantee the protein conformations is ordered from the simplest (aperiodic) coils, where most of the interactions of the amino acids are directed towards the environment, to the turns, with short-range interactions among the neighbor amino acids, then to the helices, with more regular interactions along a segment of the chain, and to the strands that, in order to form sheets, have to produce compatible segments in distant portions of the chains.

2.2 Hydropathy and Cohesiveness

The reactivity of molecules towards water, the most abundant constituent of cells, is designated hydropathy. Hydrophilic parts of molecules mix well with water and the hydrophobic parts repel or are repelled by water (Kyte and Doolittle, 1982). Nucleotides are amphiphilic, containing the hydrophilic sugars and phosphates and the more hydrophobic nucleobases. Amino acids span the whole hydropathy range so that proteins depict complex organization. Protein aggregates tend to form globules with membrane-like surfaces, for which the contribution of lipids may be accessory. The same principles governing the internal organization and folding of the macromolecules apply to the formation of associations between proteins or between proteins and nucleic acids. The nucleoprotein system is tightly associated: proteins and nucleic acids are both very sticky and form aggregates, the components communicating with each other in almost contiguity.

2.3 Networks and Stability

The cellular components form a large communication network (Barabási and Oltvai, 2004). The transformations involving both the macro- and micromolecules are of the same kind, justifying the extension of the term metabolism to the whole network and the description of the cell as a metabolic system (Guimarães et al., 2007 and references therein). In spite of so many interesting attributes that support the proposition of metabolic networks as original in forming living systems (Kauffman, 1993) there is an enormous difficulty in modeling or obtaining them experimentally. Polymers with replication abilities are considered a necessary prerequisite for the origin of biosystems (Orgel, 2002).

Chemistry and especially biochemistry study mainly stable objects. Stability of a molecule means also its abundance and this is a driving force in chemical reactions, together with the energetic gradients. The constituents of cells, however, are not particularly stable. The synthesis of a peptide bond or of an ester bond is accompanied by the formation of water molecules but, in the aqueous environment, the polymers are continually being forced into the reverse reaction, of de-polymerization through hydrolysis. So, a characteristic of the biologic system is the requirement for continuous replacement of the damaged or decaying polymers. The consumption of components creates another driving force on metabolism, accentuating the mass gradient in the anabolic direction. The driving force of consumption of products is considered more important for the organization of metabolism than the availability of precursors or substrates, in the sense of being pathway-specific, and the latter are categorized mainly as prerequisites for the anabolic reactions.

Only the DNA molecule can be efficiently repaired (see Zenkin et al., 2006, for the transcriptional repair) and this is the reason for its widespread double-strandedness. RNA is less stable than DNA due to its partial base pairing and the presence of two hydroxyls in the sugar. When RNAs are more stable, this character is partially obtained through nucleotide modifications but mainly through the protection provided by associated proteins that may be intrinsically stable. The need for re-synthesis of sequences makes memory structures necessary; these are provided by the nucleic acids. A template RNA may serve the translation of various copies of proteins, and a segment of DNA or RNA the replication, transcription, or reverse transcription of various copies of their complements. Cells could survive and evolve only when they acquired the capacity to stabilize the templates and to re-synthesize polymers in excess of the degradation rates.

2.4   The Ribonucleoprotein (RNP) World and Prebiotic Chemistry

Studies on cellular origins became more complicated when it was realized that DNA probably was not an early molecule. Deoxyribonucleotides are derived from ribonucleotides and DNA is vastly less reactive than RNA, not able to participate efficiently in metabolic reactions (Orgel, 2002). To the contrary, cellular RNAs are known to catalyze metabolic transformations – ribozymes, the most remarkable being the ribosomal peptidyl transferase activity (Yusupov et al., 2001). There is in vitro evidence even for self-replicating RNAs (Hughes et al., 2004). The context outlined is of an early RNP living world, that later became enriched with DNA, therefore acquiring a stable memory and separating this function from the strictly metabolic ones, but the hypothesis of the early existence of a pure RNA world does not seem plausible. At times when prebiotic ribozymes may have been abundant so should have been the prebiotic peptides and protein-type catalysts.

The prebiotic chemical scenario points to the abundance of some amino acids and scarcity of nucleotides. The latter are complex molecules while amino acids are small. The most abundant amino acids recovered from syntheses under plausible early Earth conditions are among the simplest, Gly and Ala (Miller and Lazcano, 2002). The easiest way to obtain peptidic and nucleotidic polymers so far advocated, albeit with short sizes and low productivity is based on mineral surfaces, especially from clays (Cairns-Smith, 1982). The plausible scenario is that a mineral order can be transferred to the polymers and these could be more easily adjusted to one another in the formation of aggregates since they were based on a common origin (Baziuk and Sainz-Rojas, 2001). A problem remains with the prebiotic metabolism, wherefrom a sufficient supply of monomers could be guaranteed. A variant of the "genetic takeover" hypothesis is acceptable, stating that nucleic acids became later replicated through means independent from the minerals that first guided their polymerization (Ferris, 2002). The necessary character of the prebiotic metabolic system that survived in the form of cells is that it was apt to produce internally the genetic subsystem.

3   The Coded Biotic World

In cells, proteins are polymerized in amino acid sequences following the order of the nucleotides in a template RNA. Such strict correspondence could be easily understood if some kind of fit existed between nucleic acid surfaces or pockets and the amino acids. This stereo-chemical hypothesis is not presently in favor for the generation of the whole code but it could have been responsible for some of the correspondences (Yang, 2005; Yarus et al., 2005). In fact, the correspondences are mediated by adaptor RNA molecules, the transfer RNAs (tRNAs), each one carrying a specific amino acid and bearing a code, a triplet of nucleotides (an anticodon) that matches by base pairing a triplet in the template mRNA (the codon) (Ibba et al., 2000; Barbieri, 2003).

The matrices of codes and anticodes are shown in Table 1. The largest matrix of anticodes is the eukaryotic, containing 46 tRNAs. However, such complexity is not

Table 1   The matrices of the genetic coding system

A

Central	U		C		G		A	
	Quadrants				Quadrants			
	Homogeneous pDiN YY				Mixed pDiN YR			
5' U	UUU	Phe, F	UCU	Ser, S	UGU	Cys, C	UAU	Tyr, Y
	UUC	Phe, F	UCC	Ser, S	UGC	Cys, C	UAC	Tyr, Y
	UUG	[Leu]	UCG	Ser, S	UGG	Trp, W	UAG	Stop, X
	UUA	[Leu]	UCA	Ser, S	UGA	Stop, X	UAA	Stop, X
5' C	CUU	Leu, L	CCU	Pro, P	CGU	Arg, R	CAU	His, H
	CUC	Leu, L	CCC	Pro, P	CGC	Arg, R	CAC	His, H
	CUG	Leu, L	CCG	Pro, P	CGG	Arg, R	CAG	Gln, Q
	CUA	Leu, L	CCA	Pro, P	CGA	Arg, R	CAA	Gln, Q
5' G	GUU	Val, V	GCU	Ala, A	GGU	Gly, G	GAU	Asp, D
	GUC	Val, V	GCC	Ala, A	GGC	Gly, G	GAC	Asp, D
	GUG	Val, V	GCG	Ala, A	GGG	Gly, G	GAG	Glu, E
	GUA	Val, V	GCA	Ala, A	GGA	Gly, G	GAA	Glu, E
5' A	AUU	Ile, I	ACU	Thr, T	AGU	Ser, S	AAU	Asn, N
	AUC	Ile, I	ACC	Thr, T	AGC	Ser, S	AAC	Asn, N
	AUG	iMet						
	AUG	Met, M	ACG	Thr, T	AGG	[Arg]	AAG	Lys, K
	AUA	Ile, I	ACA	Thr, T	AGA	[Arg]	AAA	Lys, K
	Mixed pDiN RY				Homogeneous pDiN RR			

(continued)

Table 1 (continued)

B

Central	A		G		C		U	
	Homogeneous pDiN RR				Mixed pDiN YR			
3′ A	AAG	Phe	AGG	Ser	ACG	Cys	AUG	Tyr
	AAC	[Leu]	AGC	Ser	ACC	Trp	(AUC)	X
	AAU	[Leu]	AGU	Ser	(ACU)	X	(AUU)	X
3′ G	GAG	Leu	GGG	Pro	GCG	Arg	GUG	His
	GAC	Leu	GGC	Pro	GCC	Arg	GUC	Gln
	GAU	Leu	GGU	Pro	GCU	Arg	GUU	Gln
3′ C	CAG	Val	CGG	Ala	CCG	Gly	CUG	Asp
	CAC	Val	CGC	Ala	CCC	Gly	CUC	Glu
	CAU	Val	CGU	Ala	CCU	Gly	CUU	Glu
3′ U	UAG	Ile	UGG	Thr	UCG	Ser	UUG	Asn
	UAC	iMet						
	UAC	Met	UGC	Thr	UCC	[Arg]	UUC	Lys
	UAU	Ile	UGU	Thr	UCU	[Arg]	UUU	Lys
	Mixed pDiN RY				Homogeneous pDiN YY			

C

Central	A	G	C	U
	Homogeneous pDiN RR		Mixed pDiN YR	
3′ A	AA	AG	AC	AU
3′ G	GA	**GG**	GC	GU
3′ C	CA	**CG**	**CC**	CU
3′ U	UA	UG	UC	UU
	Mixed pDiN RY		Homogeneous pDiN YY	

(A) The standard genetic code matrix. The order of bases follows the increasing hydrophobicity: U<C<G<A. The 3′ base of the triplets may be generically designated by W (wobble) or N. In the main initiation codon (iMet) and the other less frequent initiators (not shown), the wobble position is the 5′ base. In square brackets, the dicodonic components of the hexacodonics Arg and Leu. The principal dinucleotides (pDiN) of the triplets exclude the W position and identify the boxes. These are grouped in quadrants and sectors: pDiN with two R or two Y are called homogeneous; those with an R and a Y are called mixed.

(B) The standard genetic anticode matrix. The number of triplets is taken from eukaryotes but these have 5′ inosine instead of guanosine in a half of the boxes. The order of bases follows the increasing hydrophilicity: A<G<C<U. The 5′ base A of the triplets is rarely present. In the initiation anticodon, the wobble position is the 3′ base. In parenthesis, the absent Stop anticodons.

(C) Symmetries in the genetic anticode principal dinucleotide matrix. Anticodons are simplified to the 16 pDiN, which form 8 complementary pairs. The axes are formed by the core (with G- and C-only pDiN; bold) and the tips (A- and U-only pDiN; underlined) boxes. Simple boxes, corresponding to one attribution only, are read as the core and the non-axial with central R. Complex boxes, corresponding to more than one attribution, are the tips and the non-axial with central Y.

necessary for decoding the full set of 64 codons. There are simpler matrices that rely more on the wobbling abilities of 5′ G and U than on the retention of all kinds of bases at the 5′ position. The bacterial anticodes are smaller, the archaeal intermediate and the vertebrate mitochondrial is the most intensely reduced, to the total of 23 tRNAs, including the iMet (initiator; Osawa, 1995). The maximum size of the anticode is due to the absence of base A in the 5' position and of the three tRNAs corresponding to the Stop codons.

3.1   Hypotheses of Early Translation

Attempts at developing a plausible explanation for the formation of the translation system have not been satisfactory so far (Trevors and Abel, 2004). A problem with the direct early translation models stems from their assumption that preexisting RNAs served as templates (early mRNAs) for the alignment of aminoacyl-tRNAs (ac-tRNA; or with some early form of the tRNAs) and that this system gradually acquired efficiency and functionality. Following this premise, based on the traditional evolutionary genetics, the learning process would have been very slow, through trials, errors, and some successes that became fixed (Poole et al., 1998; Tamura and Schimmel, 2003).

The template could be efficiently translated by an early specific ac-tRNA only if it had a homogeneous or repetitive sequence. This is not an interesting start, but it could be supposed that useful variations would be superimposed on this, if they were coherent in the template and in the new tRNAs; the necessary coherence in the evolution is the main problem. It is more plausible that the early template was heterogeneous but it could be efficiently translated only by a nonspecific set of tRNAs, producing the correspondent highly variable proteins. This system would evolve by selection of the interesting and functional proteins and mRNAs, again in coherence. Specific tRNAs would not be adequate for translation of heterogeneous mRNAs due to the frequent occurrence of nonsense segments in the template.

Various attempts have been made to envisage the constitution of an early small set of amino acids that would also fit a coherent region of the genetic code matrix. This set should also ideally correspond to some interesting protein property so that a functional system would be constructed, to become a seed for further evolution. The proposals tend generally to concentrate on and vary somewhat upon the set localized on the 3' C row of the anticodon matrix (Val, Ala, Gly, Asp, and Glu; Lehmann, 2002; Klipcan and Safro, 2004).

We will not detail all the argumentation involved in this choice, but most of the important parameters that have been proposed in support of it are also interesting for understanding the code organization and can be summarized as follows:

- The amino acids in the 3' C row of the anticodon matrix belong generally in the category of the simpler and smaller. This produces consistency with some of the most abundant and stable molecules obtained in prebiotic syntheses.
- None of them are, in the more widely accepted biosynthesis routes of amino acids, derived from others belonging in the code. They are all originated directly,

or with a low number of transformations, from the most basic of the metabolic processes, the routes of glycolysis, of the pentose shunt and of the Citrate Cycle. This line of thought states that the early attributions were those of amino acids originating from the basic routes of biosynthesis and that, with complexification of the routes, producing new amino acids in families of derivation, the codes attributed to them also formed families of similarity. So, there was a coevolution in the derivation of both amino acids and their codes (Wong, 2005).

- The amino acid set is heterogeneous enough to produce peptides with a variety of properties, thus opening the way for the ample functionality of proteins (Oba et al., 2005).
- The anticodon set is rich in G + C bases, guaranteeing higher thermal stability in the pairings with the codons, and homogeneous at the 3′ C, making easier the mutual transformations inside the set, so that an earlier single starting code would have had less difficulty in producing the others.
- The set of present day catalysts of the formation of the aminoacyl-tRNAs, the ac-tRNA synthetases (aRS), is rich in class II enzymes, and these are suggested to have generally preceded class I. In our model, only the three last amino acids in this set (Gly, Asp, Glu) are fixed early.

4   The Self-Referential Model

Our study of the formation of the genetic code (Guimarães et al., 2007, and earlier references therein) allowed the proposition of a procedure for obtaining synthesis of proteins that is considered plausible and consistent with the present day knowledge on cellular processes, and can dispose of the requirement for an early mRNA to be translated. It is called self-referential to the tRNAs and based on a simple type of symmetry produced by the dimerization of tRNAs through the complementary anti-codons (Grosjean et al., 1986; Grosjean and Houssier 1990). We will start with a presentation of the model and then give an account of its consistency with the main attributes that have been found adequate and necessary to fit the evolutionary para-digm, e. g. of going from simple to complex arrangements and of providing a driving force or a phenotype with fitness value that can be selected for.

The process is based on a small machine-like system (Fig. 1). Its simplicity offers a fast and high probability mode of evolution, to fill the matrix in a few steps. Such expediency is necessary in the light of the estimated short time span

---→

Fig. 1 (continued) The poly-tRNAs may acquire different configurations: some become the ribosomal RNAs and others form aligned anticodons, whose copies are codon strings (mRNA). The meaningful (to the system being formed) association is selective and specific (cognitive): proteins should be intrinsically stable and efficient binders, and should bind the same carriers that were involved in their production. (**3**) *Ribosomal translation*. Protein synthesis is directed by strings and the carriers become the decoding system. Different types of cytosolic dimers of tRNAs (both uncharged, one of them charged, both charged), and their relative concentrations, may have regulatory functions

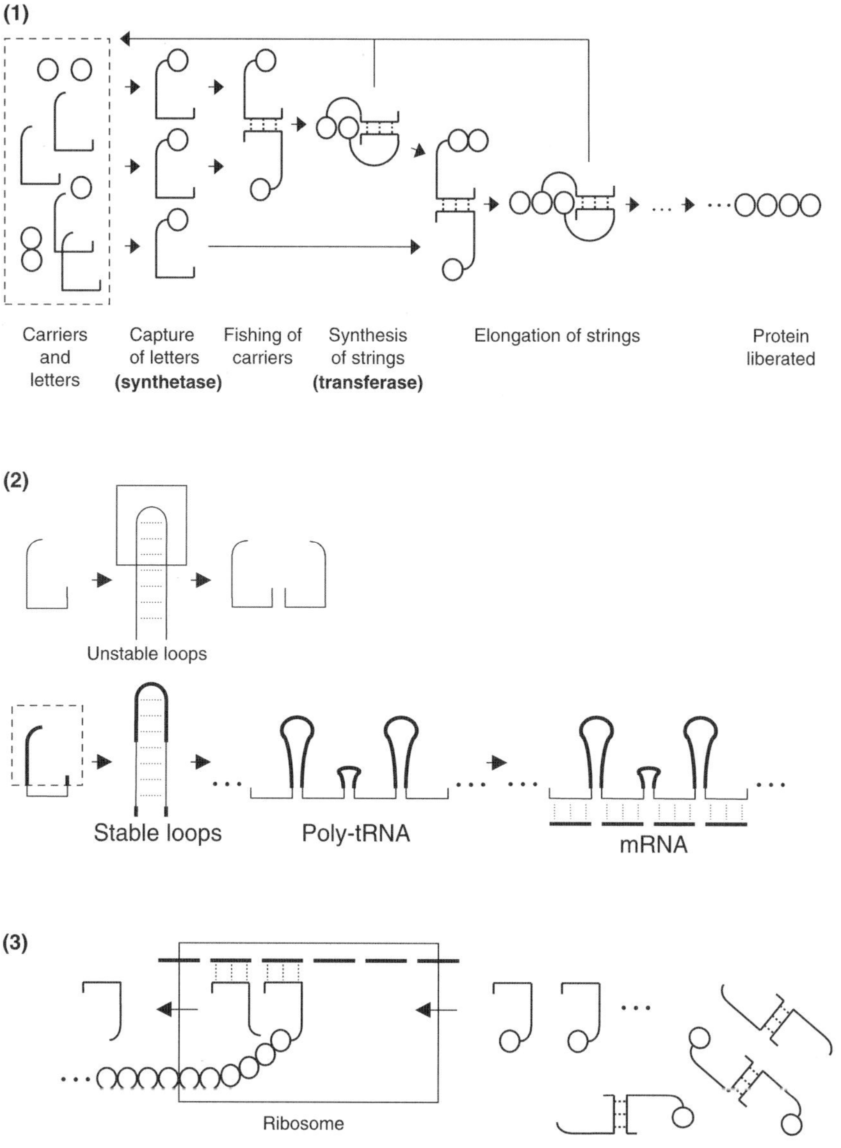

(1)

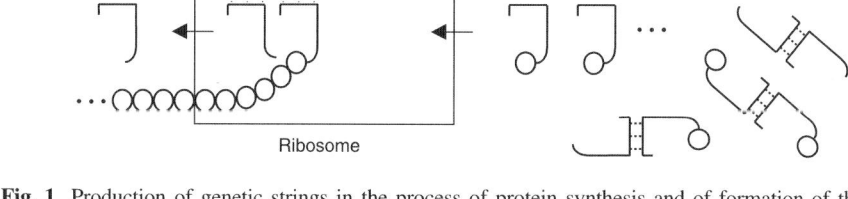

| Carriers and letters | Capture of letters **(synthetase)** | Fishing of carriers | Synthesis of strings **(transferase)** | Elongation of strings | Protein liberated |

(2)

Unstable loops

Stable loops   Poly-tRNA   mRNA

(3)

Ribosome

Fig. 1 Production of genetic strings in the process of protein synthesis and of formation of the code. (**1**) *Protein synthesis directed by tRNA dimers*. The system is self-referent, to the tRNAs (carriers). The amino acids (letters; circles) are fished in couples through tRNA dimerization (fishing). Catalytic activities are in boldface. After each transferase reaction with synthesis of a peptide bond, an uncharged tRNA goes back to the pool and the peptidyl-tRNA is elongated. (**2**) *Binding of proteins to the carriers and formation of elongated tRNA strings*. For replication the tRNA molecule is extended and duplicated. Since loops are labile (thin, boxed), without stabilization by proteins they break, originating copies of the original molecule. When the loops are stabilized (thick, dashed box) replication becomes elongational and chains of carriers (poly-tRNAs) are produced.

available for the origin of life on Earth (Poole et al., 1998). The slow part of the process is sent back to the origin of replication and of the tRNAs. The simplicity and the highly deterministic character of the process suggest that the Last Universal Common Ancestor populations could, at some early stages, be composed of lineages bearing similar genetic codes. Simulations of the probabilities of the different codes inside these populations are being conducted (Farias et al., in preparation).

4.1   The Pools of Reactants: tRNAs and Amino Acids

The main ingredients of the coding system are of three kinds – the tRNAs, the amino acids and the catalysts of their union, accomplishing the aRS function. This function was initially obtained with the participation of, e.g. the tRNAs and metals, the RNA component being possibly also ribozymic, or prebiotic peptides. The tRNA pool contains the full set of 64 anticodons; with replication it is possible to generate them easily. The early amino acid pool is probably not full, according to the list of amino acids obtained by chemical synthesis under conditions imitating the supposed early Earth conditions (see Trifonov, 2004; in the decreasing rank order: GALVDEIPS, plus marginally T) or considered not derived from other amino acids belonging in the code (see Wong, 2005; only the rank order is different: GASDEVLI, plus marginally P and T).

4.2   Stages in the Formation of the Coding System

The process is divided into three stages (Fig. 1): (1) dimer-directed protein synthesis; (2) primitive mRNA translation; and (3) maturation of the mRNA structure. A precise stepwise succession of pairs of tRNAs being recruited and fixed with correspondences, to compose and to be integrated in the system, is delineated.

4.2.1   Protein Synthesis Directed by tRNA Dimers

Various kinds of dimers may be formed, including the ones not relying upon the pairing through the anticodons (e. g. Martinez-Giménes and Tabarés-Seisdedos, 2002), and many of them could be producing oligo- or polypeptides concomitantly. Proteins produced by each type of dimer would be largely repetitive, composed by either one amino acid, such as the Ser, which is coded by complementary pDiN, or up to many, in cases where each tRNA in the dimer would demonstrate greater affinity for a different amino acid and lower affinities for others.

Our model says that the dimers formed by pairing of tRNAs through the anticodons of the perfect palindromic kind were those demonstrating higher stability, therefore more apt for the task of performing the transferase activity (see Guimarães et al., 2007). If all of them were at work concomitantly upon the early set of amino acids, the model needs to invoke a self-stimulatory molecular selection process mediated by the proteins, choosing which ones would be fixed at each moment of the formation of the code.

The properties of the proteins produced by the pool of dimers would be able to explore the whole space of possibilities opened by the types of amino acids available and the semi-repetitiveness of each of them. The earliest self-stimulatory process relevant for the start of the RNP-based protein synthesis would depend on a few of these properties, the main one being the ability of proteins to bind to the RNAs. When the RNA-binding protein is also stable against degradation, it would protect the RNA and form a stable RNP. The RNA component should continue being adequate for replication and for participating in the synthetase and transferase reactions.

The set of correspondences in this stage is the GPS group of amino acids, attributed to the NCC:NGG and the NGA:NCU sets of triplets. It is noteworthy that there should be no need for external compartmentalization to facilitate the nucleoprotein binding when the nascent proteins are composed of amino acids with strong RNA-binding properties; they are not released and stay bound to the tRNAs, with immediate cohesiveness.

4.2.2 Translation of Primitive mRNA and the Triplet Coding

This is the most complex stage, at the transition from the dimer-directed to the translational mode of protein synthesis, with separation of the anticodon and codon functions. The set of nine amino acids corresponding to the pDiN of the homogeneous sector form the core structure of proteins with predominantly aperiodic conformation: five of the amino acids are characteristic of coils and turns (GPSDN), three of α-helices (DLK). Ribosomes and mRNAs are derived from poly-tRNA strings, which may be elongated after the RNA stabilization is guaranteed. Evidence indicating the derivation of rRNA from tRNA has been presented by Bloch et al. (1984, 1989). The synthetase function of proteins is developed, accompanied by the establishment of the hydropathy correlation (Farias et al., 2007).

The change from dimer-directed to templated protein synthesis introduced modifications in the way the stability of the tRNA couples active in the transferase function is obtained and helped in the restriction of the coding to the anticodon triplets. In the paired anticodon configuration, the base pairs in the loops holding the tRNAs together should point to one surface with the anticodons in the middle and other bases flanking the triplets might be involved in the pairing (Grosjean and Houssier, 1990). In the ribosomal side by side configuration, the anticodon/codon pairing becomes restricted to the three bases at the center of the loop, and the flanking bases could pair laterally with the neighbor tRNA (see Smith and Yarus, 1989).

4.2.3 Maturation of the Generalized mRNA Structure

The tRNAs belonging to the mixed pDiN sector are integrated into the system next. Dimers do not penetrate ribosomes but dictate that new attributions should follow the pairing of tRNAs carrying them. The NRY attributions are ligated to the 5' ends of mRNAs and the NYR to the 3' ends, forming the generalized mRNA structure 5' (NRY) (NRR, NYY) (NYR) 3' (Tables 2 and 3). It is indicated that the codes corresponding to the tRNAs of the homogeneous pDiN sector are ligated *in tandem*, following the succession of the stages, and that the staggered mode of addition of the codes corresponding to the tRNAs of the mixed pDiN sector is due to their belonging already in the mRNA or DNA mechanisms.

The chronology proposed for the fixation of the genetic code correspondences obeys the rules that the tRNAs were recruited in palindromic pairs and that the tRNAs with homogeneous pDiN were fixed earlier than the tRNAs with mixed pDiN. It is possible that some kind of precoding dictated by direct interactions between amino acids and oligonucleotides, belonging in the RNA world, might be relevant to the process, but we cannot find any link between such proposals (e.g. Seligmann and Amzallag, 2002) and the standard code. They are based mostly in thermodynamic considerations, possible stereochemical affinities, specific amino acid binding to some RNAs, not tRNA-like, but containing segments similar to

Table 2 The necktie model for the structure of primeval protein domains. The sequence of the primeval mRNA module combines the chronology of incorporation of attributions to the genetic code, according to the stages (1a–4) of the self-referential model, and the proposition of the universal loop-and-lock domains of proteins (Sobolevsky and Trifonov, 2005, 2006). It is indicated that the codes corresponding to the tRNA pairs of the homogeneous pDiN sector are ligated *in tandem* and compose a loop, while those of the mixed pDiN sector are added in a staggered mode, the NRY to the *N*-ends and the NYR to the *C*-ends, originating a mini-loop-and-lock structure. The sequence is presented in the linear form in Table 3

			2a	2a			
Homogeneous							
pDiN sector			Asp	Glu			
	1b		Ser		Leu	2a	
1b	Ser					Asn	2b
1a	Pro					Lys	2b
	1a		Gly		Phe	2b	
			||		||		
Mixed	3a		Ala		Arg	3a	
pDiN	3b		Val		His	3b	
sector					Gln	3b	
	3c		Thr		Cys	3c	
					Trp	3c	
					X	—4	
	4		Ile		Tyr	4	
	4		Met				
N-end	—4		iMet		X	—4	C-end

Table 3 A model for the generalized structure of proteins based on the properties of amino acids in the quadrants of the genetic anticode matrix. The stages of the stepwise model (second row) are arranged according to a linear protein chain. The core sequence contains the attributions in the successive pairs of the homogeneous pDiN sector (stages 1a–2b). The attributions of the RY pDiN are added to the N-ends (stages 4 to 3a), and the attributions of the YR pDiN are added to the C-ends (stages 3a–4)

Quadrant	The NRY boxes of the mixed pCiN sector; the N-end set					The homogeneous pDiN sector; the middle set									The NYR boxes of the mixed pDiN sector; the C-end set					
Stage	4	4	3c	3b	3a	1a	1a	1b	2a	2a	2a	2b	2b	2b	3a	3b	3b	3c	3c	4
Amino acid	Met	Ile	Thr	Val	Ala	Gly	Pro	Ser	Asp	Glu	Leu	Asn	Lys	Phe	Arg	His	Gln	Cys	Trp	Tyr
Hydropathy / value	Phob / 0.059	Phob / 0.029	Apath / 0.438	Phcb / 0.069	Mod. phob / 0.258	Apath / 0.423	Apath / 0.677	Apath / 0.508	Phil / 0.962	Phil / 0.935	Phob / 0.066	Phil / 0.809	Phil / 1.000	Phob / 0.000	Phil / 0.982	Apath / 0.573	Phil / 0.841	Phob / 0.111	Mod. phob / 0.325	Mod. phob / 0.361
pDiN / hydropathy	CAU UAU / 0.199	RAU UAU / 0.199	NGU / 0.433	NAC / 0.189	NGC / 0.403	NCC / 0.928 (Outlier)	NGG / 0.196 (Outlier)	NGA RCU / 0.098:1 (Outlier)	RUC / 0.931	YUC / 0.931	NAG / 0.059	RUU / 0.931	YUU / 0.931	RAA / 0	NCG / 0.564	NUG / 0.596	NUG / 0.596	RCA / 0.312	CCA / 0.312	RUA / 0.303
Amino acid size	Large / 105	Large / 111	Med. / 61	Large / 84	Small / 31	Small / 3	Small / 32.5	Small / 32	Med. / 54	Large / 83	Large / 111	Med. / 56	Large / 119	Large / 132	Large / 124	Large / 96	Large / 85	Med. / 55	Large / 170	Large / 136
aRS class	I	I	II	I	II	II	II	II	II	I	I	II	I,II*	II*	I	II	I	I	I	I
Stabilization	1	6	2	1	2	1	1	2	4	4	9	5	8	9	8	7	5	3	9	9
Protein end	N	Both	N	N	Both	Both	N	Both	–	Both	Both	C	C	C	C	–	–	C	–	C
Protein conformation	α-helix	β-strand	β-strand	β-strand	α-helix	Aperiodic	Aperiodic	Aperiodic	Aperiodic	α-helix	α-helix	Aperiodic	α-helix	β-strand	α-helix	α-helix	α-helix	β-strand	β-strand	β-strand
Nucleic acid binding	R1	D2	D1	R4	D2	R7	R4	R1	D2	D2	R5	–	R3	R3	D2	D2	D1	D1	D1	D2
	D0	R2	R0	D3	R1	D2	D2	D0	R1	R1	D3	D2	D2	D2	R2	R1	R1	R0	R1	R2

The amino acid characters are summarized below.

(A) Characters related to aRS class specificity.

(1) Amino acid hydropathy (from Farias et al., 2007). Hydrophobics (<0.111, FIMLVC) are acylated in the class I mode, including the atypical PheRS. Hydroapathetics (0.423 – 0.677, GPSTH) are acylated by aRS class II. The moderately hydrophobic (0.258 – 0.361, AWY) and the hydrophilic (>0.809, NQEDRK) are heterogeneous with respect to the aRS classes. The outliers belong to aRS class II and to the homogeneous pDiN sector. The other attributions in this sector conform to a regression with $r^2 = 0.99$. Attributions of the mixed pDiN sector build a regression line ($r^2 = 0.85$) with steeper inclination.

(continued)

Table 3 (continued)

(2) Amino acid size (in Å, from Grantham, 1974). Small (GASP, <32.5) and medium (DCNT, 54–61) are characteristic of aRS class II, including the Cys, charged by the bifunctional ProCysRS class II of some organisms. The 12 large amino acids (>83) are characteristic of aRS class I, except for His and for the two atypical (∗) acylation systems: for Lys (class II in some organisms and atypically accepting the anticodonic 5′ Y triplets of complex boxes) and Phe (acylating the ribose 2′ OH, which is typical of class I enzymes). These two amino acids are also the largest of the homogeneous pDiN sector and of class II, and the extremes of hydropathy.

(3) pDiN hydropathy contrast. The lowest contrast between the pDiN of the homogeneous type is CC − GG = 0.732. The highest contrast between the pDiN of the mixed type is UG − AC = 0.407. It could be reasoned that it would be easier to start a process of hydropathy fitting utilizing the pDiN of the homogenous type, due to their high contrasts, and that the start with the pDiN of the mixed type would have been more difficult, requiring high discriminatory power. Nonetheless, the outliers reside in the homogeneous sector.

(B) Characters related to protein properties.

(1) Metabolic stabilization (half-life) of proteins dictated by the amino acid residing in their N-ends (adapted from Varshavsky, 1996). Grade 1 is the strongest stabilizer, grade 9 the strongest destabilizer. Mixed pDiN sector: four of the five amino acids in the N-end set are strong stabilizers (grades 1 and 2); the C-end set has none of these and is rich in the destabilizers. In the middle set, of the homogeneous pDiN sector and the first stages of the construction of the code, the chronological sequence corresponds to starting with the strong stabilizers and ending with the strong destabilizers.

(2) Statistical preference of amino acids for being located in the N-ends (two first positions) or the C-ends (two last positions) of proteins (adapted from Berezovsky et al., 1997, 1999). The statistical significance shows clear preferential locations at the protein ends only for ten amino acids but is entirely consistent with the metabolic stabilization grades. These data (B1, B2) are also in full agreement with the amino acids residing in the boxes where the specific punctuation triplets are. The overall consensus indicates that the polar distribution of the amino acids in proteins constitutes a nonspecific punctuation system, based on the stabilization properties of the amino acids. This system is already shown by the middle set. With the addition of the mixed pDiN sector, the specific punctuation system was established, located on top of and in accordance with the nonspecific system.

(3) Preferential participation in protein conformations (from Creighton, 1993). All amino acids characteristic of aperiodic strings (GPSDN) belong in the homogenous pDiN sector. This has only three of the eight characteristic of α-helices (ELKARHQM) and one of the seven (FVTCWIY) characteristic of β-strands.

(4) Preferential participation in conserved sites of nine RNA-binding and eight DNA-binding motifs or proteins. The monotonic basic motifs, rich in Lys and Arg, were not considered. Six of the eight amino acids, characteristic of RNA-binding motifs (GPSLKFVM) belong in the homogeneous pDiN sector. Nine of the ten amino acids, characteristic of DNA-binding motifs (EAHTC) or occurring equally in both types of motifs (RQWIY), belong in the mixed pDiN sector.

their correspondent codons or anticodons (Yarus et al., 2005), or models for possible ribozyme activities that could have generated the amino acid transformations (Szathmáry, 1999; Copley et al., 2005).

We also find it difficult to envisage an RNA-based mechanism that would have distinguished and directed the early fixations to the tRNAs with homogeneous pDiN and the late ones to the tRNAs with mixed pDiN. It is more plausible to admit, based mainly on the hydropathy correlation being established in dependence of protein properties (see below), that the correspondences were fixed as codes by the self-stimulatory effects of the protein binding and stabilization of the RNAs that were producing them. It is suggested that the early tRNAs of the homogeneous pDiN sector had more repetitive and simpler sequences than those of the mixed pDiN sector, and were then more adequate for being bound to the early repetitive and simpler proteins produced in the dimer-directed stage. The rationale based on the simplicity of the interacting partners is also pointed at by (a) the simple sequence character of the palindromic triplets (lateral bases repeated), as compared to the other types, and (b) the nondirectionality that they allow for the interactions (the lateral bases may be equally the start or the endpoints in an interaction).

On the amino acid side, simplicity has always been a main theme in guiding the proposals for the early attributions of the code. These are usually chosen as subsets from the lists of amino acids obtained by chemical synthesis under conditions imitating the supposed early Earth conditions. Considering the first nine amino acids in the ranks shown above, we see that two thirds of them belong in the sector of homogeneous pDiN, corresponding to the first three pairs of our model (GPSDEL). On the protein side, our prediction is that a core sequence should be composed by the nine amino acids of the homogeneous pDiN sector and should acquire predominantly the aperiodic conformation. This may be considered certified by the data of Sobolevsky and Trifonov (2005, 2006). The amino acid composition of the 21 most abundant types of conserved octamers belonging in the universal loop-and-lock structures of proteins presents the rank order G(DLT)SAP(KR), from which two thirds belong in the homogeneous pDiN sector (GPSDLK).

4.3 The tRNA Dimers Orient the Entire Process

In the next sections, data are presented to demonstrate that the description of the entire genetic code through the dimer-oriented rationale is physiologic. The consistency of the results with biochemical data and evolutionary prescriptions is surprising in the light of the constraints imposed by the necessity of attributions having to be inserted into the system obeying the tRNA pairs, which is considered an argument in favor of the model. Our main novelty concerning the role of tRNA dimers is the proposition of the palindromic dimer-directed protein synthesis in the first stage and that, in later stages, dimers do not enter the ribosomes but orient the entrance of new attributions. This rationale incorporates the earlier studies of Miller et al. (1981) and Yamane et al. (1981), showing that dimers could be regulatory to protein synthesis, modulating the availability of tRNAs for translation, and of Smith and Yarus (1989), showing that the neighbor tRNAs in the P and A sites of ribosomes interact side by side via the bases lateral to the anticodons in the loop, which is considered a different kind of dimer. The non-ribosomal dimer-directed transferase activity could be experimentally tested, either utilizing present day tRNAs (possibly too large and rigid for facilitating the reaction) or the various kinds of mini-tRNAs that have been used as acceptors for the aRS function or for spontaneous aminoacylation (see Beuning and Musier-Forsyth, 1999), but containing anticodon-like loops, able to dimerize.

4.4   Processes Forming the Code

Three mechanisms are discerned in the formation of the codes. Two are primary: one for the attribution of correspondences between triplets and amino acids, the other for the generation of the specific punctuation signs. The mechanisms forming the primary attributions are similar in the sense that they involved the fishing of a second attribution by the first, dictated by the triplet pairing rules. (a) For amino acid coding, anticodonic pairings directed the attribution to the second tRNA that dimerized with the first; (b) for the localization of the termination signs, their tRNAs formed pairs with the initiation codon and competed with the initiation anticodon for this codon, therefore being deleted (see Table 4). The specific punctuation was developed after all amino acid attributions were completed, which is consistent with the specific punctuation being the most complex of the translation mechanisms. The third mechanism (c) is called secondary, for the generation of the hexacodonic attributions, derived from expansion of the specificities of the class I ArgRS and LeuRS. The expansions of Arg into the Y<u>CU</u> and of Leu into the Y<u>AA</u> triplets were established after the formation of their respective primary tetracodonic attributions (N<u>CG</u> and N<u>AG</u>), so that Ser was originally octacodonic (N<u>GA</u> and N<u>CU</u>), and Phe was originally tetracodonic (N<u>AA</u>).

4.5   Amino Acid Coding

It is known that tRNAs can form dimers through the pairing of complementary anticodons. The thermal stability of the dimers is high, equivalent to the formation of about seven base-pairs in the common RNA helices (Grosjean et al., 1986; Grosjean and Houssier 1990). It is indicated that the type of mini-helix formed either has a peculiar stability by itself or receives additional support, either from special base modifications or from other bases in the anticodon loop, besides the anticodon triplets. Thermal stability of dimers of present day tRNAs is also not much influenced by the G + C contents of the triplets, but there are indications that early coding might have been influenced by this character (Ferreira and

Cavalcanti, 1997), its main feature being that all boxes at the core of the matrix are simple. Our model requires a perfect palindromic topology in the triplet pairs and obeys only partially the thermodynamic stability principle: in each of the sectors, the initial boxes are at the core and the last ones to be filled with attributions are at the tips.

The tRNA dimers are considered proto-ribosomes and proto-mRNAs. In ribosomes, the two tRNAs are front to back in the A and P sites, guided by the codons in the mRNA (Fig. 1). In the dimers, the anticodon loops are associated head to head through a mini-helix and the anticodons are simultaneously codons for each other, making unnecessary the presence of an external template, and the tRNA acceptor ends hang to different sides. In both cases the two tRNAs are held together in a stable structure so that the acceptor ends are placed in contact and the transferase reaction facilitated. This reaction is driven towards peptide synthesis due to the peptide bond being covalent and the dimer association dynamic, dependent only on hydrogen bonds. Polymerization of peptides works as a sink, providing a suction force to the system. This force works as long as there are dimers of ac-tRNAs. Dimers of an ac-tRNA and a non-acylated tRNA may be inhibitory when the concentration of the latter is higher. When only one synthetase is available, acylating one tRNA type, the system may be de-repressed when a second synthetase arises with specificity for the other tRNA in the pair. In this way, dimerization is also a driving force for the fixation of catalysts that can acylate the second member of a pair. In this double-catalyst situation, there are difficulties in deciding which came first since the concentrations of the members of the pair will fluctuate and tend to get equilibrated; any decision on which came first relies upon external factors. A limited complementary behavior of the amino acids being recruited into the code is seen, when the amino acids follow the hydropathies of the tRNAs in the pairs.

4.6 The Palindromic Triplets and Pairs

The type of pairs found meaningful for initiating the coding process is configured as perfect palindromes, containing the same bases in the extremities, which we shall call the fishing triplets (see Guimarães et al., 2007). Among all types of triplet pairs, the palindromic configuration is the one guaranteeing full and long-lasting single-strandedness which is a requisite for their ability to produce stable dimers through the formation of the mini-helices. Considering the restriction of accepting only the perfect standard base pairs, it is indicated that the palindromic coding was developed before the exclusion of A from the 5′ position of the triplets. The exclusion of 5′ A became necessary in the complex boxes due to its wide wobbling possibilities that would produce ambiguity in their decoding. In our scheme, the mechanism of 5′ A exclusion may have been initiated as early as in the stage of incorporation of the two acidic amino acids, sharing the UC box. The remaining 5′ G solved the ambiguity problem and was enough for decoding. The later extension of the 5′ A exclusion to all boxes may have been due to its benefit to the regulatory mechanisms, when some perfect palindromes (A<u>NA</u>:U<u>NU</u>) yielded to imperfect ones (G<u>NA</u>:U<u>NU</u>), with the consequent acceptance of G:U pairings.

4.7 Steps in the Coding at Each Box

After (a) the initial coding of the fishing triplets, the perfect palindromic triplets in each of the paired boxes, the (b) 5′ degeneracy was developed and accepted by the synthetases and the ribosomal decoding mechanisms, integrating all triplets in a box (see Table 5). This function is accessory to the main one provided by the pDiN. The full wobbling provided by 5′ U would be sufficient for decoding all kinds of codons with the same pDiN, as in the mitochondrial anticodes, but the extended anticode of eukaryotes utilized more extensive 5′ details. (c) When complex boxes were developed, the first occupier of the box receded to 5′ R and conceded 5′ Y to the new occupier(s). These concessions follow a consistent rationale and indicate that all 5′ Y attributions are late in relation to the 5′ R in the respective boxes (see below, Variant codes).

4.8 Proteins Organized the Code

Peptides formed through the mechanism of tRNA dimerization are partially organized. The set of amino acids coming from a tRNA pair have properties correlated with those of the anticodons. The next set in the succession may have characteristics independent from the previous one and a collection of independent pairs will compose a large pool of repetitive peptides each containing one or more types of amino acids. Details of further organizational steps require examination of real proteins and of the genetic code structure, to model a succession of dimers that is relevant to physiology and to the formation of the code. The model also provides a mechanism for the entrance of templates and ribosomes into the system. Various imprints of protein structure and properties, and of the mechanism of protein synthesis, were detected in the structure of the genetic code, indicating that a part of the general structure of the matrix was configured in dependence of protein properties.

4.9 Stages Indicated by the Hydropathy Correlation

The long known correlated distribution of hydropathies of amino acids and of the types of triplets was the first evidence showing that the correspondence between amino acids and triplets is not entirely arbitrary. It was reexamined considering the hydropathies that the amino acids present as residues in proteins (Farias et al., 2007).

Previous studies utilized the hydropathies of amino acid molecules in solution and could not offer a rule to understand how the correlation was established and how to explain the deviations (Lacey and Mullins, 1983).

Our study showed that 19 of the attributions conformed to a wide correlation area and 4, belonging in the homogeneous pDiN sector, were identified as outliers from the correlation (Gly-CC:Pro-GG and Ser-GA:Ser-CU; the GPS group; Table 3). The steepness of the regression line angle of inclination grows from the other attributions in this sector (DELNKF; 43°) to the mixed pDiN sector (64°). Taking the GPS set as the first amino acids to become encoded (stage 1), it is indicated that peptides with this simple constitution were not able to produce the correlation. This was established when the peptides had a richer amino acid complement, with the entrance of the other amino acids of the homogeneous sector (stage 2). When amino acids of the mixed sector entered (stage 3), completing the full set of encoded amino acids, the correlation became stronger, indicating greater sensitivity of the fitting mechanisms. Demarcation of stage 4 (Ile, Met/Tyr) was not dependent on the hydropathy correlation data but highlights the installation of the specific punctuation system.

The palindromic pairs are the only types possible inside the set of outliers that can join the two Ser boxes and accommodate the Gly and Pro boxes without ambiguities; this was the first indication of the meaningfulness of the palindromic pairs. The outliers are presently charged by class II aRS (the class of enzymes typically acylating the 3′ OH of the terminal adenosine of tRNAs). The other amino acids in the homogeneous sector (stage 2) are charged by one couple of aRS class I (the enzymes typically acylating the 2′ OH of the terminal adenosine of tRNAs; the pair of boxes with Glu and Leu, together with AspRS class II), and the couple of atypical synthetases (the pair of boxes with Lys and Phe, together with AsnRS class II). The end result of the aRS class distribution in the homogeneous sector is five class II, the two atypical, and two class I, in contrast with the mixed sector, where there are three class II and eight class I. The more hydropathy-sensitive character of the mixed sector is related to the enrichment in class I enzymes and may derive from their mode of docking on the tRNA acceptor stem (see below, aRS class characterization).

The correlation was established by the protein catalysts but each sector of tRNAs produced a characteristic high affinity and specificity, whose superposition formed the wider correlation area. The best fit produced was with class I and the triplets of intermediate hydropathies (mixed pDiN); the correlation established by class II enzymes does not discriminate pDiN types. Both sectors exploited the full hydropathy range but the homogeneous, more the hydrophilics than the hydrophobics; the mixed sector contains all the moderately hydrophobic and most of the hydrophobics. Hydroapathetics are all (GPSTH) typical class II. Hydrophobics are all acylated at the 2′ position of the ribose: Cys plus all at the central A column (class I and the atypical PheRS class II).

It is indicated that when the catalysts unite two substrates in a product, the substrates should be hydropathically compatible and coherent so that they can be adequately placed in contact thereby obtaining the facilitated reaction. This interpretation suggests that at the time of fixing the attributions, either the anticodons or some correlates of them in the acceptor arm (possibly akin to the operational codes; see Schimmel, 1995) participated in the substrate contacts, which is reminiscent of the physicochemical hypothesis. Direct tests of the first possibility are problematic, since the anticodons are presently far from the aminoacyl-adenosine synthesis site and some of them do not even interact directly with the synthetases.

4.10   Selection in the Regionalization of Attributes

An alternative hypothesis would suggest evolutionary minimization of errors or optimization of the distances between the properties of amino acids and of triplets. The mechanism suggested by this hypothesis is that various distributions of attributions once existed, produced by point mutations that changed slightly the triplet character. When this change would code for an amino acid with properties very different from the original attribution, such coding would be strongly selected against. The end result of this process would be the observed regionalization of the attributions, so that similar amino acids would be attributed to similar codes and the changes produced by point mutations usually would not change drastically the amino acid character. The only tests possible of this hypothesis are simulations of the evolutionary process, and they do show that the present distribution of attributions is among the best for minimization of errors (Knight et al., 1999). Otherwise, the hypothesis does not refer to mechanisms of origins of the attributions.

   We propose that both catalyst-driven and selective optimization hypotheses are complementary and refer to different aspects or moments of formation of the code. Our attempt is to maximize biochemical mechanistic explanations and to reduce the more vague propositions of natural selection, based only in the considerations that the present state of the system is optimized, and that other variants once existed but tended to produce worse products (phenotypes) and were selected out. The outlier attributions are considered the earliest to be fixed, based on the premise that the catalysts responsible for them had properties different from the ones that produced the correlated attributions.

4.11   Protein Structure and Nucleic Acid-Binding

The dynamics in the succession of stages of formation of the code follows a symmetry that builds the picture of a levorotatory windmill (Fig. 2). The extreme stages correspond to starting with the single class II-only pair of boxes (Gly with Pro) and ending with the single class I-only pair (Ile and Met with Tyr). Stage 1 contains three small and hydroapathetic amino acids in the hydropathy correlation outlier attributions (GPS). Five of the six amino acids in the final stages (3c and 4) are hydrophobic (CWIMY), only Thr being hydroapathetic. In the intermediate stages,

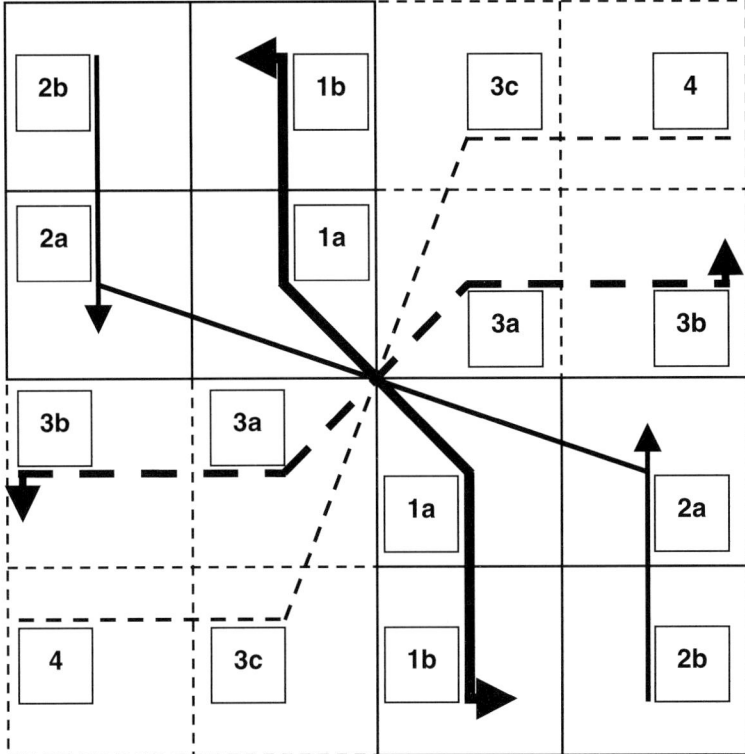

Fig. 2 The symmetry in the succession of integration of pairs into the system builds the dynamic picture of a levorotatory windmill. The drawing follows the matrix in Table 1B. The hydropathy outliers (flaps [1a] Pro:Gly; [1b] Ser:Ser [Arg]) conserve the central G and C, and the remaining pairs of the homogeneous pDiN sector are central A and U (flaps [2a] Asp, Glu:Leu; [2b] Asn, Lys:Phe [Leu]). Pairs 3a and 3b conserve the 3' G and C (flaps [3a] Ala:Arg; [3b] His, Gln:Val), and the remaining pairs of the mixed pDiN sector are the 3' A and U (flaps [3c] Thr:Cys, Trp; [4] Ile, Met:Tyr)

hydropathies are correlated with the anticodon complementarity. The main characters found relevant for formation of the code were the acquisition of metabolic stability by the proteins and their ability to bind to RNA, wherefrom an RNP system could be developed.

4.12   Protein Stability and Nonspecific Punctuation

The fundamental property of protein metabolic stability (half-life; Varshavsky, 1996), which depends strongly on the amino acid residing at the *N*-ends (the head), was found compatible with the frequency of amino acids residing on the *N*-ends of most proteins (Berezovsky et al., 1997, 1999). Accordingly, the amino acids that destabilize the proteins, when residing in the N-ends, were found to be concentrated in the C-ends (the tail). The net result of these studies is that present day proteins demonstrating higher stability show this polar distribution of the amino acids. These properties correlate with and support the staging proposed for the genetic code (Table 3). Such relationships indicate that the property of metabolic stabilization of proteins is primordial and intrinsic to the amino acids. Therefore, their polar distribution in proteins and locations in the code matrix were dictated by this property, in the same way as the protein degradation mechanisms were subsequently adjusted to the amino acid properties.

The model is valid both for the chronological order of amino acid encoding and for the generation of the polar organization of protein sequences. When such order is inscribed in the code, it may be considered another punctuation system: start the sequences with stabilizing amino acids and direct the destabilizers to the tails. This is called a nonspecific punctuation system with respect to the variety of amino acids satisfying the rules and in contrast with the traditional system, which is specific to one tRNA at initiation and to three codons at termination. It is indicated that the nonspecific preceded the specific punctuation system and that this was superposed on the former: Met is a strong stabilizer and the Stop codons belong in boxes containing amino acids that are strong destabilizers or preferred in the C-ends.

The nonspecific system can be identified in the homogeneous sector alone. The chronologic succession of entry of the nine amino acids in this sector corresponds to the construction of peptides demonstrating the polar organization of the sequences, which is the character of fully structured small proteins. It can be read from Table 3 that the three amino acids in stage 1 contribute to form stable protein heads and that the amino acids of stage 2b should not be incorporated earlier, to the cost of destabilizing the peptides due to the properties of Lys and Phe; the amino acids of stage 2a show heterogeneous properties. In the mixed sector, the other protein-stabilizing amino acids are concentrated in the NRY quadrant (MVTA) and the other destabilizers in the NYR quadrant (HRWY). The chronology of entry of the attributions belonging in the homogeneous sector corresponds to their order in a string but those of the mixed sector were added to the primordial string of the homogeneous sector in a different way, the NRY being placed in the N-ends and the NYR in the C-ends. A modification of one of the NRY (iMet) was responsible for starting the formation of the specific punctuation system.

4.13   Specific Punctuation

The puzzle set forth by observing that the two pDiN, the CAU of Met, utilized for elongation, and the CAU of iMet, utilized for initiation, coincide complementarily or directly with the pDiN of the boxes where the Stop signs reside (UA and CA, respectively) inspired a more detailed search for the links between the tRNAs involved in the

Table 4 Conflicts between tRNAs directed the localization of the Stop signs

The similarity between the triplet formed by the slipped pDiN of the initiation anticodon plus the first base of the second anticodon and the anticodons of the tRNAs corresponding to the Stop codons indicates that the latter competed with the former for the initiation codons. The conflict was solved with the exclusion of the tRNAs corresponding to the Stop codons. The second codons are listed according to the criteria of having 5′ R and of corresponding to amino acids that are strong stabilizers of the N-ends of proteins against degradation

		Initiation			Elongation	
	First		Second		Third	
	iMet		Val, Ala, Gly, Met, Thr, Ser^{CU}		Any	
Codons 5′	W U G		R Y, G W		N N W	
Anticodons 3′	U A C		Y R, C W		N N W	
Anticodons at the Stop hemiboxes 3′ A Y Y	A U U		X from the Tyr box		[Deleted and substituted	
	A U C		X from the Tyr box		[by the protein	
	A C U		X from the Trp box		[Release factors	
	A C C		Trp		[Retained	

whole punctuation system (Table 4). The recognition of the tRNA^{iMet} by the initiation system, with the wobble position in the 3′ extremity, is indicated by the observation that various Start codons may be accepted, with variation in the 5′ position (NUG). The second codons shown were selected according to the criteria of having 5′ R and of corresponding to amino acids that are strong stabilizers of the N-ends of proteins against degradation. It is shown that the initiation system is built upon a configuration of the two first codon/anticodon pairs where the pDiN are contiguous, forming a tetra-nucleotide without the possibility of interruption by a wobble pair (codons NUGRNN).

In the anticodon triplets corresponding to the Stop codons there is a constant 3′ A, identical to the central A of the initiation anticodon, both forming a standard base pair with the central U of the initiation codon. This is indicated to be the main source of the competition between the initiation anticodon and the anticodons corresponding to the Stop codons. These conflicts led to the exclusion of the latter tRNAs and their substitution by the protein release factors. The 3′ A is preceded by two Y. The central Y forms a pair with the 3′ G of the initiation codon and the 5′ Y forms a pair with the 5′ R of the second codon. This mode of coding of termination is indicated to be a necessary consequence of the installation of the initiation mechanism based on the slipped pDiN; when the code matrix is full, conflicts between anticodons competing in the initiation process will automatically arise. Other characters of the Trp tRNA or of its recognition by the termination system were developed so that it could be retained with avoidance of the conflicts, which is consistent with data in Rodin et al. (1993).

4.14   Nucleic Acid-Binding

The compilation of the amino acids which are preferred in the conserved sites of
nucleic acid-binding motifs (see Guimarães et al., 2007) relative to the ones
showing up in the nonconserved sites in the same motifs (Table 3), was able to
identify the homogeneous sector of the code with the RNA-binding ability
(GPSLKF, plus V and M of the mixed sector) and the mixed sector with the DNA-
binding (AHTC, plus E of the homogeneous sector) or with the ability for binding
both kinds of nucleic acids (RQWIY). Most of the RNA-binding motifs are
highly enriched in Gly, some of them being rich in Pro. Among the nine amino
acids in the homogeneous sector, six are preferred in RNA-binding motifs; only
Glu is preferred in DNA-binding motifs, and Asp and Asn do not show up in the
conserved sites of nucleic acid-binding sequences. Among the eleven amino acids
of the mixed sector, only Val and Met are added to the list of the preferred in
RNA-binding sequences. It can be said that the homogeneous sector belongs in
the RNP-world functions and the mixed sector in the fully developed nucleoprotein world.

4.15   Protein Conformations

An order of increasing complexity of protein conformations can be also correlated
to the succession of stages of amino acid entry into the code (Table 3). The full set
of amino acids preferred in coils and turns (GPSDN) is completed in the homogeneous sector, which is poor in the amino acids preferred in the strands that form the
β-sheets (FVTCWIY). Amino acids preferred in α-helices (DLKARHQM) are also
more frequent in the mixed sector than in the homogeneous sector.

4.16   Amino Acid Biosynthesis and Possible Precodes
at the Core of the Matrix

The criterion of obedience to the routes of biosynthesis of amino acids has been
advocated by various authors as a guide for defining the succession of their entry
into the code (see Davis, 1999; Wong, 2005). Nonetheless, the precise precursor-
derived relationships are not entirely consensual. We simplified this rationale, saying that the restrictions should refer only to the most basic of the rules of derivation:
amino acids which are consensually recognized as derived from others belonging
in the code should not precede the precursors. Some of the nonderived amino acids
are considered precursors to biosynthesis families: the S family (GCW), the D family (NKTIMR), and the E family (QPRK); F is precursor to Y or derived earlier in

the same route forming Y; H may be considered nonderived, coming directly from modification of ribulose-5-P, or derived from E or Q. Accordingly, S cannot be preceded by G, C, or W; D by N, T or M; T by I; E by Q, P or H; F by Y; R and K cannot be preceded by either one of D or E. Other amino acids not belonging in these families are derived directly from the glycolysis pathway (VLA) or are the most abundant prebiotically (GA).

Our staging is entirely compatible with the biosynthesis derivation rules, with the single exception of having Pro in stage 1 while its precursor Glu is placed in stage 2a. Fortunately, obedience to this requirement contributed positively to the staging, placing only Ser and Gly in stage 1, and both coherently octacodonic, Pro (the final code) having substituted Gly (a precode) in the GG box at stage 2a. This restricted set of stage 1 amino acids is also backed by their biosynthetic relatedness, Ser being the biosynthetic precursor to Gly (Davis, 1999). Ser is placed among the most interesting prebiotic amino acids by Nanita and Cooks (2006) due to its peculiar ability to form clusters with chiral selection. The last of the nonderived amino acids to enter the code are His and Val, in stage 3b.

Another consideration derived from the biosynthesis rationale also contributed positively to the staging, namely the proposal (see Osawa, 1995) that Arg can be considered an "intruder," substituting for a previous amino acid in its codes. The first proposal for the predecessor to Arg was of Ornithine, which is in the biosynthesis routes for Arg, but various others have been added to a list of putative predecessors (see Jiménez-Montaño, 1999). In the attempt to gather the most of biochemical precedents, adding the least of theoretical novelties, the list should be simplified to contain only amino acids already belonging in the code. We consider the calculations (see Osawa, 1995) on codons that are more frequently used in proteins than are predicted from the codons available in the code (Lys > Asp > Glu > Ala; Lys is the most overused) and on the codons that are less used in proteins than predicted (Ser > Leu > His > Pro > Arg; Arg is the most underused). From these calculations, Lys would be a good candidate due to its basicity, but we propose it to have been Ala, in spite of the different biochemical character but based on the palindromic pairing mechanism. The early coding of Ala would have been octacodonic (NGC:NCG), in the same manner as the other pair of boxes in the core (NCC:NGG) are considered to have coded for Gly. Ala has also been considered a "filler" in protein sequences (see Osawa, 1995), relatively neutral due to its small side chain. Our proposal helps to explain the data on the overuse of Ala codons in present proteins: it had a greater number of codons earlier but had them reduced after the Arg intrusion. The substitution of Ala-CG (in the precode) by Arg (the final code) occurred in stage 3a.

These suggested precodes could be tested in studies of variant codes (see below). There are reports of losses of Arg codons with still unknown destinations. They could also satisfy other authors proposing both Gly and Ala to have been early in the code, due to their being the most abundant amino acids in abiotic syntheses; their abundance might have forced their aminoacylation by tRNA dimers, resulting in the formation of the octacodonic attributions. Accordingly, the whole core of the matrix and the whole set of the hydropathy outliers would have passed though a fully octacodonic precode stage.

4.17   Biosynthesis of Gly and Ser Driven by Stage 1 Protein Synthesis

The model concentrates on the proposition of a succession of steps that fill the matrix with the correspondences as they are in the standard code. Nonetheless, we may also consider other possible functions of the proteins being made. Among the whole population of dimers engaged in protein synthesis, some were being recruited and fixed into the code but others could be contributing in parallel to the enrichment of the metabolic system. The model cannot evaluate all these accessory components but it can suggest a new perspective for interpreting the proposition of the coevolution between the formation of the codes and the development of amino acid biosynthesis routes, which is of protein synthesis working as a pulling force for the development of amino acid biosynthesis. When the transferase function works as a sink of amino acids, any developments able to regenerate the amino acids being consumed will be favored.

When the GPS group is being utilized for the first codings, it is indicated that the first biosynthetic routes to be more strongly pulled will be those generating them. Our specific proposition on this (Farias and Guimarães, in preparation) is for the biosynthesis of Gly directly from one-carbon units, the Gly synthase utilizing CO_2 for the carboxyl and another one-carbon unit, carried by the tetrahydrofolate pathway, for the α-carbon. The synthesis of Ser would come through the utilization of two Gly molecules and the Serine hydroxymethyltransferase. These routes are known to be active under conditions of scarcity of fermentable substrates (Sinclair and Dawes, 1995). The pathways usually considered in the studies of the coevolution hypothesis (Davis, 1999; Klipcan and Safro, 1994) are based on the free availability of fermentable substrates, where the precursors to these amino acids are of three-carbon units as in the glycolysis pathway, Ser being derived from 3-phosphoglycerate and Gly being derived from Ser. Our model also indicates that Asp was recruited into the code before Glu, suggesting that the next biosynthesis routes to be fixed utilized two-carbon units (e.g. acetate derived from the catabolism of Gly and Ser) to generate Oxaloacetate directly, such as in the Glyoxylate Cycle or in the reverse Citrate Cycle. Only after the development of the full Citrate Cycle there would be free availability of α-Keto-Glutarate, from which Glu and then Pro are derived.

5   The Proteic Synthetases

The tRNAs are considered better guides for developing a model for the structure of the code on the basis of their forming an evolutionarily more conserved class of molecules when compared to proteins in general, that demonstrate enormous plasticity and high openness to regulatory modulation. On another side, the overall plasticity inherent to all kinds of molecular interactions, including those involving nucleic acids, only rarely approaching 100% specificity, leads to spreads that inevitably conduct to the formation of networks. The study of the genetic code will still need to consider other components of RNA plasticity, such as the role of the nonpalindromic tRNA pairings, of the high anticodon degeneracy of eukaryotes, the richness of base modifications in tRNAs, and the differential usage of codons and of anticodons.

The synthetases are nowadays one per amino acid. They are grouped in two classes, each composing a homology family bearing a conserved domain for interaction with the acceptor arm of tRNAs and for their charging with the amino acid. The acceptor arms of tRNAs have some identity sites (operational codes; Schimmel, 1995) but other identity sites may be spread along the tRNA sequences, at places other than the major code, the anticodon triplet. Besides the conserved domain characteristic of the aRS class, other domains are responsible for binding and recognition of the specific tRNAs. Class II docks on the acceptor arm of RNAs through the major (more external) groove of the double helix; class I docks through the minor groove, reaching more directly the bases. Through this double approaching mode, the identity sites on the acceptor arms may be fully explored (Pouplana and Schimmel, 2001). The docking mechanism has a biochemical correspondence with the site of acylation on the hydroxyls of the ribose in the terminal adenosine of tRNAs: class I acylates typically the 2′ OH and class II typically the 3′ OH. A part of the network is formed when the two aRS classes share the tRNA set defined by one pDiN (Table 5). The topology of this network will form the core of a larger one, incorporating the variety of other functions and interactions of the aRS in the cellular network (Quevillon et al., 1997; Simos et al., 1998; Ibba et al., 2005).

Amino acids form groups of chemical or structural relatedness, partially correlated with the aRS classes and groups (see also Pouplana and Schimmel, 2001). Class II enzyme pockets accept seven of the eight small or medium amino acids, only Cys being medium and class I. The twelve large amino acids are typical of class I (nine of them) and of the mixed sector (eight). The four large amino acids in the homogeneous sector are the couple of Glu and Leu (class I in the UC:AG pair) and the largest, Lys and Phe, correspond to the atypical acylation systems (in the UU:AA pair).

The anticodonic 5′ Y are typical of class I, irrespective of the aRS class at 5′ R, and of the specific punctuation. The 5′ Y location is the result of concessions from the first occupiers of the boxes: (a) From class I to class I or punctuation: Ile conceded to Met and this to iMet; Cys to Trp and this to X; Tyr to X; (b) From class II to class I: His to Gln; Asp to Glu; Asn to the Lys class I of some organisms (many Archaea and a few Bacteria); (c) From class II to the class I expansions: Phe to [Leu]; Ser-CU to [Arg].

Table 5 The sharing of the tRNAs in a box by the different aRS classes generates a network system. The sectors are similar in being filled from the core, with both boxes simple, to the tips, with both boxes complex, passing through the non-axial pairs, with one box simple and one complex, these sharing the two aRS classes. The sectors differ in: the homogeneous pDiN core is class II-only and the tips share the two classes; the mixed pDiN core has both aRS classes and the tips are class I-only. The full tetracodonic degeneracy of the first occupier of the Phe, His, Cys and Tyr boxes is not strictly necessary due to both the fishing and the final triplets being 5′ R; it is necessary in the Ser^{CU}, Asp, Asn, and Ile boxes due to the fishing triplets being 5′ Y and the final ones 5′ R

Fishing triplet	Degeneracy		Concession aRS class			Box type
Homogeneous pDiN sector						
1a						
Gly CCC	NCC			II		Simple
Pro GGG	NGG			II		Simple
1b						
Ser AGA	NGA			II		Simple
Ser UCU	NCU	RCU		II		Complex
		YCU	[Arg]		I	
2a						
Asp CUC	NUC	RUC		II		Complex
		YUC	Glu		I	
Leu GAG	NAG				I	Simple
2b						
Asn UUU	NUU	RUU		II		Complex
		YUU	Lys	II*	I	
Phe AAA		RAA		II²′		Complex
		YAA	[Leu]		I	
Mixed pDiN sector						
3a						
Ala CGC	NGC			II		Simple
Arg GCG	NCG				I	Simple
3b						
His GUG		RUG		II		Complex
		YUG	Gln		I	
Val CAC	NAC				I	Simple
3c						
Thr UGU	NGU			II		Simple
Cys ACA		RCA		II (ProCys)	I	Complex
		YCA	Trp CCA		I	
			Stop UCA			
4						
Ile UAU	NAU	RAU	UAU		I	Complex
		YAU	Met CAU		I	
			iMet CAU			
Tyr AUA	RUA				I	Complex
	YUA		Stop			

5.1   The Atypical Acylation Systems

Two acylation systems, for Phe and Lys, are atypical, each in a different way. It is indicated that the atypical character is consequent to the large size of Phe and Lys relative to the class II enzyme pocket and to the amino acids being at the extremes of hydropathies (Table 3), and that the development of the atypical behavior was consequent to a historical event of class I duplications being not available at the times of fixation of these two attributions while class II enzymes were available and adopted the large amino acids. The atypical couple was fished by the last tRNA pair of the homogeneous sector. The PheRS is class II but acylates the 2′ OH of the terminal adenosine of the tRNA, which is the class I mode of activation. The high hydrophobicity of Phe required a conformational change in the enzyme, to achieve its peculiar mode of acylation. The LysRS is class II in some organisms (Eucarya and most Bacteria) and class I in others (Ibba et al., 1997), where a class I duplication was available at the time of the incorporation of Lys to the system. The class I LysRS fulfills the class I or punctuation homogeneity of all attributions to 5′ Y of complex boxes; the 2′ acylation, typical of class I enzymes, fulfills the homogeneity of all attributions to the central A column. LysRS class II is the only enzyme of the class to adopt the 5′ Y triplets in complex boxes. It is concluded that both the class II PheRS and LysRS should be considered atypical.

The Selenocysteine (Sec) and Pyrrolysine (Pyl) attributions are punctual additions to the amino acid repertoire, called recoding (Baranov et al., 2003). Stop codons occurring internally in some mRNAs are decoded via suppressor tRNAs and utilize specific charging systems. They are also cases of atypical location of aRS class II on 5′ Y anticodons. The AGU Stop codon is decoded by a tRNA^{Sec} and charged by the normal SerRS class II, and the Ser-tRNA^{Sec} is transformed into the Sec-tRNA^{Sec}. The GAU Stop codon is decoded by a tRNA^{Pyl} and charged by either a PylRS or a ternary complex formed with LysRS class I and class II (Polycarpo et al., 2004).

5.2   Regionalization and Plasticity of the Synthetases

The regularities detected in the distribution of the aRS classes in the matrix are shown in Table 3. The combinations with the least deviations are: the central A plus the YR quadrant, typical of class I, deviants being only the HisRS class II and the atypical PheRS; the central G plus the YY quadrant, typical of class II, deviants being only the GluRS class I, the LysRS class I of some organisms and the ArgRS expansion. The contribution of synthetase classes to the building of an architecturally integrated network derives also from their specificities for the central purines, which do not distinguish the sectors: class II unites all central G boxes and class I the central A boxes. Further contributions derive from their spreads, which were mostly due to the central Y ambiguity. The spreads become the norm rather than errors or deviations.

A large gap is highlighted between what can be proposed for the constitution of the primeval protein modules, peptides with the predominant aperiodic conformation, composed by the amino acids of the homogeneous sector, and the complex organization of the aRS, that will be difficult to fill. Various changes in the composition and genomic organization of the synthetase sets are being discovered, the majority occurring in the Archaea, which may help in tracing earlier states of the code. An intriguing feature of the collected examples is the high number of occurrences involving members of the families of amino acids derived biosynthetically from the acidics of the NUC box: Glu (Gln and Pro) and Asp (Asn and Lys). Arg also enters the list due to being derived from either one of the acidic amino acids and supposed to have had a predecessor.

Some of the intermediate steps may be called expansions of the aRS specificities. In some organisms, synthetases may accept tRNAs which, in the standard code, are charged by a different enzyme. The paradigmatic cases are of the aRS for the two acidic amino acids: AspRS may accept the tRNAs for Asn to form Asp-tRNA^{Asn} and this will later be transformed into Asn-tRNA^{Asn} by an amidation enzyme; a similar pathway is followed for the formation of Gln-tRNA^{Gln} from Glu-tRNA^{Gln}. There are many bacterial lineages that still keep the Glu-tRNA^{Gln} pathway for obtaining Gln, and it has been proposed that a separate GlnRS arose first in the eukaryotic lineage, later being transfected to some of the bacterial groups (Skouloubris et al., 2003). Another instance of a synthetase with expanded specificity but which remained fixed as such in the standard code is the MetRS that also charges the tRNA^{iMet}. Some archaea lack a separate CysRS and the charging of the tRNA^{Cys} is achieved by a class II ProRS, which is ambiguous or bi-functional (ProCysRS; Stathopoulos et al., 2000; Yarus 2000). Another form of bi-functionality is the fusion of the ProRS and GluRS into a single polypeptide, in most eukaryotes (Berthonneau and Mirande, 2000).

5.3   Specificity and Timing the Entrance of Synthetases

The scenario displayed by our staging model indicates a faster encoding by class II duplications, which predominate strongly in the homogeneous sector. Class I enzymes predominate in the mixed sector but their numbers only equilibrate with the class II numbers in the last stage. The wide-range asynchrony of the two aRS classes has a counterpoint in the short-range concerted duplications inside each of the classes, indicating the occurrence of coupled historical inductions possibly related to the entrance of the tRNAs in pairs.

The specificity and spread of the synthetase classes is indicated to run strictly through the pairs of columns. Class II occupies fully the central G and central C columns, including the proposed early occupier of the CG box (Ala) and the first occupiers of the CA and CU boxes (presently receded to the 5′ G triplets), respectively Cys and Ser, the first through the dual-specificity ProCysRS. Class I occupies fully the central A and central U columns. The homogeneity in the central A column includes the atypical PheRS, refering to its 2′ mode of acylation. In the complex central U boxes, class I corresponds to the second occupiers, in the 5′ Y triplets.

The spread due to class II central Y ambiguity was to the first occupiers of the central U column: UG (His), UC (Asp), and UU (Asn), plus the atypical pair UU (Lys)/AA (Phe). The dispersion of class I replaced various early class II attributions, such as the dicodonic expansions of Arg and Leu, and the homogeneous sector ended up with a greater mixture of characters than the mixed sector. Nonetheless, the homogeneous sector maintained a neat regularity in the attribution of enzymes of the same class to all tRNA pairs: the hydropathy outliers occupy two pairs with class II; the other two have either the couple of class I (EL) or the couple of atypical acylation systems (KF). We take these regularities, clearer in the homogenous than in the mixed sector, to indicate that characters of the paired tRNAs may have guided the fishing of aRS of the same class. Such regularity was partially eroded in the mixed sector. Couples of aRS class I can be seen only in the UG (Gln)/AC (Val) and in the AU (Ile, Met)/UA (Tyr) pairs, while the CA (Cys, Trp)/GU (Thr) and the CG (Arg)/GC (Ala) pairs became class-discordant after the class I displacement of the previous class II occupiers of the CA (ProCysRS) and of the CG (AlaRS) boxes. The latter two substitutions were the main symmetry-breaking events to the configuration of the code.

The mechanism of formation of the code by the fishing of complementary anticodons is entirely consistent with the aRS class specificity for the complementary central bases, only adding a further specificity, namely that the complementary triplets are of the perfect palindromic kind, united diagonally in the matrix. This regularity is not immediately apparent from the plain observation of the overall distribution of synthetases in the matrix. Only two diagonally paired boxes are unambiguous, with synthetases of the same class: class II in the NGG/NCC pair (stage I), class I in the NAU/NUA pair (stage 4).

6   Evolutionary Code Variants and the Hierarchy of Codes

Variant codes are alterations affecting all proteins of the organisms or organelles, with redistribution of the preexisting degeneracy. No case is known of full loss of any of the attributions, so that the standard configuration of the code is preserved, and our model survives also this test. The general mechanism is of a decoding system developing expanded capability (most frequently due to posttranscriptional modification of tRNAs, that become able to decode new codons, plus their acceptance by the synthetases, or ribosomal changes), substituting the previous meaning of one or more codons (the donor codons). The variants are considered to have developed after the standard code was formed, due to being dependent on changes in various components of the decoding system, which require complex genomes, containing sets of tRNA-modifying enzymes, besides ribosomes and the aRS sets. In each type or occurrence of a change, the expansion of the decoding system may be preceded or followed by the loss of the former meaning of the donor codon. When it is preceded by the loss of a codon (e.g. in genomes with strongly biased base compositions, such as in the mitochondria and in the firmicute bacteria) or of its meaning (e.g. loss of tRNA-modifying enzymes), it may be said that the expansion is an event of compensation for the loss. Genomic minimization or simplification is also indicated to be a causal mechanism, evidenced in mitochondria and in the firmicute-derived mycoplasmas.

It is indicated that the changes observed (see Guimarães et al., 2007) were those that could be tolerated, occurring upon attributions that are more expendable and less crucial to physiology. Attributions not tolerating losses, that became essential to all organisms and organelles, are in three of the complex boxes (Phe, [Leu]; His, Gln; Asp, Glu) and in the five simple boxes of the nonhexacodonic attributions (Pro, Gly, Val, Ala, Thr). The high prevalence of changes in the punctuation boxes is possibly due to the complexity and plasticity of the punctuation mechanisms, involving the protein factors. The high frequency of changes in the box containing the dicodonic components of the hexacodonics Ser and [Arg] indicates that these attributions are less crucial to physiology and more expendable than other codes.

We indicate that the functional hierarchy corresponds to a temporal hierarchy: attributions fixed earlier became more tightly integrated to other components of the physiological network (as hubs), and therefore more difficult to change. The later attributions would be more loosely coupled to the network, therefore more expendable. The more widely connected earlier attributions would be also more apt for adopting extra codes. Evidences for the physiological and temporal hierarchy are: (a) The vast majority of the donors are the 5′ R codonic attributions; there are only two losses of 5′ Y codons (AGY Ser). Such instability of the 5′ codons (or 5′ Y anticodons) may be among the forces resulting in the maintenance of the two types of 5′ bases in the cellular anticodes. (b) Stop codons changed most frequently to amino acid attributions, there being only two occurrences of the reverse path (AGR [Arg] or UCA Ser changing to Stop). (c) Codons with 5′ R changed to the 5′ Y meanings (UGA Stop to Cys, UAA Stop to Tyr, AGR [Arg] to Ser, AAA Lys to Asn), there being no example of the reverse path. (d) There are also changes between the 5′ Y (UGA Stop to Trp, the AUA exchanges between Ile and Met). (e) Other examples of changes between different boxes are also from late fixations conceding to earlier ones (AGR [Arg] to Gly, AGY Ser to Gly, CUG Leu to Ser), there being one example of the reverse path (CUN Leu to Thr). The general panorama of the changes can be interpreted as loss of complexity of the matrix or reversal to simpler or earlier states.

7 Discussion

7.1 The Systemic Concept of the Gene

Nucleic acid molecules may be taken as mere physical entities, very interesting in themselves, especially due to their ability for the templating of replicas, but this property is not exclusive to them. Being a gene is a very different attribute, derived from the embedment of nucleic acid molecules in a biotic system where, besides replication, they can template for the synthesis of proteins through a code.

If this were the only process involved in their participation in the bio-world, the outcome would be dispersive: functions of the proteins would conduct mostly to interactions with the outer environment. Our investigation of the process of formation of the code is rooted on the perspective that protein functions should also feed back positively upon the nucleic acids, thereby providing for a double link between genes and proteins, and both links being of equal weight. When RNAs were demonstrated to be better candidates for the primeval nucleic acids, instead of the intrinsically more stable DNA molecules, the definition of the principal function of early proteins could be pinpointed to stabilization of the RNA.

These processes involving molecular affinities, replication and differential physical stability should be considered in the realm of the self-organization of systems (see the categorization of systemic approaches by Di Giulio (2005)). The extension of the Darwinian concept of natural selection to the molecular world in the origins of life only adds confusion; it should be reserved for the definitely biologic process of differential reproduction of cells and organisms. In the present context, it is sufficient to say that self-stabilizing and self-constructive systems grow faster and remain longer, or the term "molecular selection" should be adequately quoted.

In fact, we succeeded in showing that the genetic code organization presents clear signs of the relevance of the mechanism of RNA stabilization by proteins for fixation of the early attributions. This process established a stable nucleus (fixation of the GPS group of amino acids) around which the whole system could have developed. The temporal order of the stages in the model indicates the order of fixation of the paired boxes considering their final configuration in the code. The perspective of starting with a core, which grows outwards in all directions, differs from many of the technological models, attempting to sequentially lengthen and branch the chains of reactions in the clean test-tube chemistry. The process obviously includes the minimization of deviations from chemical norms or specificities, possibly starting with less specific and advancing to more specific catalysts, and may be modeled through the engineering optimization principles.

We would like to consider that the process of formation of the RNP genetic system as modeled here may be called self-cognitive, the term describing the reflexive and stimulatory association of the protein products to exactly the same RNAs involved with their production. The term cognition is derived from human communication affairs, among themselves or between humans and the environment, and spread into the realm of zoology. However, a correlate of it is frequently used by biochemists, relative to the molecular specificities of interaction, when saying that a substrate with high affinity to an enzyme, and this, are cognate to each other. When the binding of a stable protein occurred to a noncognate RNA – which did not participate in its synthesis – a self-maintaining system would not be formed. So, cognition is indicated to be at the basis of the productive and positive consequence or result of an interaction.

When we said (Pardini and Guimarães, 1992) that "the system defines the gene," still based on observation of the large amount of evidence of the ambiguity of genetic sequences, we were duly asked (J. D. Watson, 1995) to clarify the workings of the system. We can now offer some molecular details. The genes are defined by the two-way processes that construct the genetic system: they produce proteins that are meaningful to them, that is, when the proteins help them to become stable and to be integrated in a system. The nucleoprotein system is constructed by the circular or reflexive association, through the coding (digital, letter by letter) and through the stabilizing connections (analogic, based on sequence patterns), and both are important to the same degree.

It happens as if the nucleic acids were, surprised, telling to the proteins – "you are my life!" – at the same time when the proteins were, equally surprised, telling to the nucleic acids – "you are my genes!" The definition of the producer by the product, in the present case, of the genotype (memory with code) by the phenotype (meaningful product), is just one specific case of a general process. For instance, in the building of some sentences in human languages, words are added sequentially but the precise meaning of the sentence is only reached when the last words are known; the last words are needed so that the first may be adequately chosen and correctly understood. The connections between the initial and final segments are multiple, simultaneous, and entangled in the circular configuration.

The former protein-first "or" nucleic acid-first hypotheses are now changed into proteins "and" genes together. When asking about the formation of systems, the question of which component came first is not relevant. The intelligent question is how the components became associated and integrated in a system. A piece of nucleic acid may be a gene for the systems that are able to accept it productively, but not for those where that piece is not productive, due to deficiencies in any of the processes of the circular associations. So, the definition of the gene is relative to its belonging in a system, not absolute to a piece of nucleic acid.

7.2   Stability, Abundance and Strings as Driving Forces

We appreciate the description of the forces of nature by the vacuum or the topologic metaphor, as if matter is drained or sucked down towards the eye of a sink or falls down along the slope of a hill, thereby reaching a more stable state. There is no need to discriminate some specific more important pushing or pulling forces, the best being to indicate the existence of a difference or a gradient from a less to a more stable state, and both types of forces are important. The argument allows a rationalization of the polymerization of proteins or of nucleic acids being effective for the maintenance of metabolic cycles or networks.

In the origins of the genetic code, the consideration is clear that formation of peptide bonds works as a sink of aminoacyl-tRNAs, against their being maintained as tRNA dimers. Another instance of the fundamental participation of the force of stability or abundance in driving the evolution of the system is at the origin of the template strings for translation. The coding system passed from the stage where, in the tRNA dimers, codons and anticodons resided in the same structures while, after the stabilized poly-tRNAs could work as or produce the template strings (mRNAs), the codon function became separated from that of anticodons.

In the maintenance of metabolic networks, the forces propelling the system may be considered to converge on the environmental sink of diluted products. The metabolic fluxes are kept on going only when the products are being continuously transformed so that they do not accumulate and would force the reactions in the reverse direction; they are either stored in a different form or structure, or are secreted/excreted from the biotic compartment.

The formation of long RNA strings is not only a possibility opened by the stabilization of RNAs by proteins. Their production is enforced by the greater efficiency of elongational replication and of templated protein synthesis. Stable strings work as sinks or suction forces that consume tRNAs and amino acids. Their elongation drives the fixation of the dispersed components of the network into mRNAs, ribosomes, and proteins, with dissipation of the whole space of codes to fill the matrix. The formation of strings also introduced other forces contributing to the structuring and coordination of the system. When tRNAs are replicated separately and independently from each other, the formation of the dimers in solution is dependent on mutual affinities and on the facilitation by external compartmentalization but also subject to fluctuations in concentrations, that may be inhibitory if not equilibrated. At elongational replication, the complementary tRNAs are linked *in tandem*, lowering the requirements on external compartmentalization, and are synthesized coordinately, with lower probability of generating conflictive concentrations.

7.3 Origins of the Genetic System and of Cells

In considering other possible functions of the proteins being made during the process of formation of the code, which should contribute to the enrichment of the metabolic system, it is justified here to concentrate on the role of protein synthesis as a force driving the fixation of the pathways of amino acid synthesis. This leads to a reinterpretation of the proposition of the coevolution between the formation of the codes and the development of biosynthesis routes (Wong, 2005). Instead of saying that the availability of amino acids drove the chronology of their incorporation into the code, it is indicated that: (a) amino acids were recruited into the code according to their participation in protein composition, structure, and functions in building the nucleoprotein system; (b) development of amino acid biosynthesis routes was driven by the consumption of the amino acids at protein synthesis.

The self-referential model for the origin of the genetic code is at the same time a model for the formation of the integral genetic system and of cells, and could also be called "the nucleoprotein aggregate" or the "ribosomal" model. It starts with the tRNA-directed protein synthesis. Proteins associate with the tRNAs and lead to the formation poly-tRNAs from which ribosomes and mRNAs (protein-coding genes) are derived. Metabolic routes are developed, driven by the consumption of amino acids at their incorporation into proteins. The protein aggregates build a growing globule, whose surface properties are improved by the aggregation of lipids.

7.4   Memories for Self-Production

The relative autonomy of bio-systems with respect to environmental fluctuations, allowing them to maintain stable phenotypes and identity along the passage of time and of the generations, requires the possession of memory structures – structures providing for the regenerative abilities. We would like to consider that both types of memories discussed in this study of the genetic code formation are of the same basic kind of cyclic structures – *memory in cycles*. Cyclical routes inside the metabolic networks, besides being important due to the recycling and self-activation properties are also similar to attractors, due to being activated at high frequency by the network.

In the origin of the coding system, the stabilizing role of the proteins upon the nucleic acids composes a cycle of self-stimulation and the best term for the role of the nucleic acids is that of a memory structure – *memory in strings*. Mechanisms of this kind could be interesting for robotic engineering: the outcomes would be programmed with binding elements that could help them to attach to the producers and cognitively reinforce their connections, so that a string memory is created and maintained. This is different from the memories in Artificial Neural Networks. It will be interesting to look for counter-examples, of systems demonstrating self-referring properties that lack some kind of memory.

When the role of genes is established as a type of memory, produced inside and by the system, it becomes clear that the distinction between genotype and phenotype is only didactic and descriptive of the top-down approach to physiology. It could be considered that genes are an integral part of the metabolic system and that this is the basic unit of the living; the phenotype is the whole system, including its memory structures, be they in the "classic" part of metabolism or in the "molecular biology" part.

7.5   What is Life

The definition of life proposed earlier (see Guimarães et al., 2007) can now be made clearer: Life is the process of stabilization and self-construction depicted by individualized metabolic systems. The material and architectural aspect resulting from the self-constructive process is highlighted. Cells or organisms with suspended metabolism, frozen or dried, or dead and fossilized, can be recognized on morphological bases alone. It is proposed that any "self" process depends on the formation of memory systems and become irrevocably bound to them, and to their evolution. Memories are one step further from simple stability properties, allowing for repetitive processes.

The string memory, when built in the form of nucleic acids, and more stably in the form of DNA, acquired a striking expansive property. Stabilized elongation of chains, mainly through the process of incorporation of duplicates, with variations, is among the most important evolutionary processes, leading to increased genome sizes and to the large diversity. Functionality comes about through the coding machinery and the products feeding back upon the producers. Further expansion is obtained through the processes of multiple coding by the genetic sequences and of the multiple and plastic functions of the products in the metabolic networks.

The two expansive components, the genetic and the proteic, each have some attributes exclusive to them and with some autonomy of their mechanisms and dynamics, in spite of being kept irrevocably interdependent and integrated. The differences between the two components are clear when their community-forming processes are compared. Parts of the genetic memories may be interchanged and recombined between organisms, from the segmental horizontal transfers, now plainly utilized for production of modified genomes, to the sexual transfer of whole genomes. The workings of the DNA memory compartment are widely blind to the quality or source of pieces coming in. The networks and phenotypes communicate more plastically and variably, mostly through chemical means – the prionic-type transfers being possibly limited – and behaviors, but also up to the linguistic symbols. Networks and phenotypes are the places where expressions of the memory are put to tests, at their integration with the other components of the networks and at the interactions of the phenotypes with the environment. The expansive character of the living systems may be compared to the Big Bang of cosmology. It is indicated that the biotic DNP world, mainly composed of unicellular forms, occupied the whole Earth surface in a very fast and almost explosive Big-Bio-Bang.

7.6 Information

It seems impossible to refrain from using the term information in the biologic realm, and advisable to clarify and propose how it could be best understood and applied. It is utilized in two complementary contexts, referring to the internal workings of the cell and to the interactions between the cells or organisms and the environment. The term is taken from human affairs (as with the term cognition), but also in different contexts, when we communicate with others or interact with the environment, and is extendable to all interactions of living beings with others or with the environment.

A general definition of information can be taken as the property of a pattern that can be distinguished from noise. This is equivalent to saying that information is a difference (between a clear pattern and a noisy one) that makes a difference, and making a difference implies that the observer is able to discern what matters and what does not matter. The distinguisher is an observing or interpreting system, so that information is the relational quality of an object, defined by the system and relative to it. The pattern has some meaning to the system because it has receptors and processors that invoke reactions in the system.

The adequacy in the connection between the workings of the system and the inputs comes from the possibilities opened by the system's receptors. The induction of a reaction (meaning) by an interaction (information) is only possible when there is some kind of preparedness of the receptor to the input – the receptor discriminates what is and what is not informational. This is equivalent to saying that the input has some previously defined – evolutionarily adjusted – meaning to the receptor. In biochemical terms, it is said that the receptor has some kind of affinity to the input, or that there is some kind of cognitive interaction. Objects display a variety of properties, distinguishing them from other objects or from noise, but only a subset of these is utilized in each kind of interaction, which is the informational subset of inputs for that particular interaction. Some interactions are not productive because there is no fitting between the interagents, the inputs having no meaning to the receptors they reached. The same inputs may fit other receptors or the same ones under different contexts and then induce a reaction, indicating that the information is not an absolute property of the inputs or of the receptors, but a property of their mutual relations. So, information is a concept a posteriori to an interaction and is linked to the possible outcomes of it. A contact is informational when some reaction results.

The term "transfer of information" may be misleading when implying that something energetic or material is being transferred, which is not always the case. After some meaningful contacts, the effector, agonist, or messenger molecule may remain integrally preserved, indicating that the informational interactions involved only some minimal and transient exchanges of influences (Ricard, 2004). The technological treatment of the process of transferring signals from sources to receptors, through channels that may introduce noise (Shannon, 1948), is compatible with the molecular concept. When effectors and receptors interact, their contacts are mediated by weak bonds, very sensitive to modulation by a variety of physicochemical interferences (noise), from the electronic and thermal motions to the intermingling of small molecules (water, ions, etc.), inside the nanometric channel that separates the interacting molecules. However, the Shannonian concept of information explicitly leaves aside the question of meaning.

The application of the term is quite clear in the human communication context, through languages, but there are some gradations and quantitative possibilities in the outcome of the events, that need to be qualified for clearness. Some prescriptive or normative instructions are almost dictatorial with respect to the responses invoked, as in a master–slave relationship or in digital computing. Others may be more democratic and respectful of the receptors, which may have some freedom to interpret the instructions and react accordingly. Some prefer to reserve the term informational to the interactions of the latter kind, the former being too mechanicist for their taste.

In the cellular context, there are also both kinds of reactivity. The almost dictatorial type is seen in the flow of string information, from genes to the primary sequence of proteins, through the genetic decoding machine. The order of nucleotides or codons in a template becomes translated almost automatically into that of amino acids in the protein. The more democratic type is seen in all other components of the plastic metabolic network, from gene regulation to the interactions between the cell and the environment (Markos, 2002). This is a job of mainly the proteins, when they mediate the interactions of the system with the environment, gene activation or repression, the processing of the transcripts, the timing and quantity of translation, so configuring the bulk of the phenotype.

In our model for the process of formation of the genetic code, string information arises during the process, when protein synthesis passes from the tRNA dimer-directed to the template-directed or translational mode, and this occurs in dependence of the protein properties that helped molding and structuring the code. In this context, protein properties define the character of the working system and the genes are just the memories, adequate for the maintenance of the system. The genetic memory is a store of string information for repetitive production of the primary structure of proteins.

Acknowledgements Financial support from FAPEMIG and CNPq to RCG, and from PET/MEC to CHCM; doctoral fellowship from CAPES to STF.

References

Alberts B, Jonhson A, Lewis J, et al. (2002) Molecular Biology of the Cell. Garland Publishing, New York

Barabási AL, Oltvai ZN (2004) Network biology: understanding the cell's functional organization. Nat Rev Genetics 5:101–113

Baranov PV, Gurvich OL, Hammer AW, et al. (2003) Recode 2003. Nucleic Acids Res 31:87–89

Barbieri M (2003) The Organic Codes – An Introduction to Semantic Biology. Cambridge University Press, Cambridge

Baziuk VA, Sainz-Rojas J (2001) Catalysis of peptide formation by inorganic oxides: high efficiency of alumina under mild condition on the earth-like planets. Adv Space Res 2:225–230

Berezovsky IN, Kilosanidze G, Tumanyan VG, et al. (1997) COOH-terminal decamers in proteins are non-random, FEBS Lett 404:140–142

Berezovsky IN, Kilosanidze G, Tumanyan VG, et al. (1999) Amino acid composition of protein termini are biased in different manners. Protein Eng 12:23–30

Berthonneau E, Mirande M (2000) A gene fusion event in the evolution of aminoacyl-tRNA synthetases. FEBS Lett 470:300–304

Beuning PJ, Musier-Forsyth K (1999) Transfer RNA recognition by aminoacyl-tRNA synthetases. Biopolymers 52:1–28

Bloch DP, McArthur B, Widdowson R, et al. (1984) tRNA-rRNA sequence homologies: a model for the generation of a common ancestral molecule and prospects for its reconstruction. Origins Life Evol Biosph 14:571–578

Bloch DP, McArthur B, Guimarães RC, et al. (1989) tRNA-rRNA sequence matches from inter- and intraspecies comparisons suggest common origins for the two RNAs. Brazilian J Med Biol Res 22:931–944

Cairns-Smith AG (1982) Genetic Takeover and the Mineral Origins of Life. Cambridge University Press, Cambridge

Copley SD, Smith E, Morowitz HJ (2005) A mechanism for association of amino acids with their codons and the origin of the genetic code. Proc Natl Acad Sci USA 102:4442–4447

Creighton TE (1993) Proteins: Structures and Molecular Properties. Freeman, New York

Davis BK (1999) Evolution of the genetic code. Prog Biophys Mol Biol 72:157–243

Di Giulio M (2005) The origin of the genetic code: theories and their relationships, a review. BioSystems 80:175–184

Farias ST, Guimarães RCG (in preparation) An autotrophic scenario based on glycine synthesis from CO_2 and NH_4 is compatible with the self-referential model for the genetic code origin

Farias ST, Manzoli J, Guimarães RC (in preparation) Formal structure of the genetic code formation according to the self-referential model

Farias ST, Moreira CHC, Guimarães RC (2007) Structure of the genetic code suggested by the hydropathy correlation between anticodons and amino acid residues. Origins Life Evol Biosph 37:83–103

Ferreira R, Cavalcanti ARO (1997) Vestiges of early molecular processes leading to the genetic code. Origins Life Evol Biosph 27:397–403

Ferris JP (2002) From building blocks to the polymers of life. In: Willian Schopf (ed.) Life's Origin: The Beginnings of Biological Evolution. University of California Press, Los Angeles, pp. 113–139

Grantham R (1974) Amino acid difference formula to help explain protein evolution. Science 185:862–864

Grosjean H, Houssier C (1990) Codon recognition: evaluation of the effects of modified bases in the anticodon loop of tRNA using the temperature-jump relaxation method. In: Gehrke CW, Kuo K C. T. (eds) Chromatography and Modification of Nucleotides. Elsevier, Amsterdam, The Netherlands, pp. A255–A295

Grosjean H, Houssier C, Cedergren R (1986) Anticodon-anticodon interactions and tRNA sequence comparison: approaches to codon recognition. In: Knippenberg PH, Hilbers CW (eds) Structure and Dynamics of RNA. Plenum, New York, pp. 161–174

Guimarães RC, Moreira CHC, Farias ST (2007) A self-referential model for the formation of the genetic code. Th Biosemiotics (in press)

Hughes RA, Robertson MP, Ellington AD, et al. (2004) The importance of prebiotic chemistry in the RNA world. Curr Opin Chem Biol 8:629–633

Ibba M, Morgan S, Curnow AW, et al. (1997) A euryarchaeal lysyl-tRNA synthetase: resemblance to class I synthetases. Science 278:1119–1122

Ibba M, Becker HD, Stathopoulos C, et al. (2000) The adaptator hypothesis revisited. Trends Biochem Sci 25:311–316

Ibba MP, Rogers TE, Samson R, et al. (2005) Association between archaeal Prolyl- and Leucyl-tRNA synthetases enhances tRNA[Pro] aminoacylation. J Biol Chem 280:26099–26104

Jiménez-Montaño MA (1999) Protein evolution drives the evolution of the genetic code and vice-versa. Biosystems 54:47–64

Kauffman SA (1993) The Origins of Order – Self-Organization and Selection in Evolution. Oxford University Press, New York

Klipcan L, Safro M (2004) Amino acids biogenesis, evolution of the genetic code and aminoacyl-tRNA synthetase. J Theor Biol 228:389–396

Knight RD, Freeland SJ, Landweber LF (1999) Selection, history and chemistry: the three faces of the genetic code. Trends Biochem Sci 24:241–247

Kyte J, Doolittle RF (1982) A simple method for displaying the hydrophatic character of protein. J Mol Biol 157:105–132

Lacey JC Jr., Mullins DW Jr (1983) Experimental studies related to the origin of the genetic code and the process of protein synthesis - A review. Origins Life Evolution Biosph 13:3–42

Lehmann J (2002) Amplification of the sequences displaying the pattern RNY in the RNA world: The translation → translation/replication hypothesis. J Theor Biol 219:521–537

Markos A (2002) Readers of the Book of Life – Contextualizing Developmental Evolutionary Biology. Oxford University Press, Oxford

Martinez-Giménez JA, Tabarés-Seisdedos R (2002) On the dimerization of the primitive tRNAs: implications in the origin of the genetic code. J Theor Biol 217:493–498

Miller SL, Lazcano A (2002) Formation of the building blocks of life. In: Willian Schopf (ed.) Life's Origin: The Beginnings of Biological Evolution. University of California Press, Los Angeles, pp. 78–109

Miller DL, Yamane T, Hopfield JJ (1981) Effect of tRNA dimer formation on polyphenylalanine biosynthesis. Biochemistry 20:5457–5461

Nanita SC, Cooks RG (2006) Serine octamers: cluster formation, reactions, and implications for biomolecule homochirality. Angewandte Chemie Intl Edn 45:554–569

Oba T, Fukushima J, Maruyama M., et al. (2005) Catalytic activities of [GADV]-protein world for the emergence of life. Origins Life Evol Biosph 35:447–460

Orgel LE (2002) The origin of biological information. In: William Schopf (ed.) Life's Origin: The Beginnings of Biological Evolution. University of California Press, Los Angeles, pp. 140–155

Osawa S (1995) Evolution of the Genetic Code. Oxford University Press, New York

Pardini MIMC, Guimarães RC (1992) A systemic concept of the gene. Genetics Mol Biol 15:713–721

Polycarpo C, Ambrogelly A, Bérubé A, et al. (2004) An aminoacyl-tRNA synthetase that specifically activates pyrrolysine. Proc Natl Acad Sci USA 101:12450–12454

Poole AM, Jeffares DC, Penny D (1998) The path from the RNA world. J Mol Evol 46:1–17

Pouplana LR, Schimmel P (2001) Two classes of tRNA synthetases suggested by sterically compatible dockings on tRNA acceptor stem. Cell 104:191–193

Quevillon S, Agou F, Robinson JC, et al. (1997) The p43 component of the mammalian multisynthetase complex is likely to be the precursor of the endothelial monocyte-activating polypeptide II cytokine. J Biol Chem 272:32573–32579

Ricard J (2004) Reduction, integration and emergence in biochemical networks. Biol Cell 96:719–725

Rodin SN, Ohno S, Rodin A (1993) Transfer RNAs with complementary anticodons: could they reflect early evolution of discriminative genetic code adaptors? Proc Natl Acad Sci USA 90:4723–4727

Schimmel P (1995) An operational RNA code for amino acids and variations in critical nucleotide sequences in evolution. J Mol Evol 40:531–536

Seligmann H, Amzallag GN (2002) Chemical interactions between amino acid and RNA: multiplicity of the levels of specificity explains origin of the genetic code. Naturwissenschaften 89:542–551

Shannon CE (1948) A mathematical theory of communication. Bell Syst Tech J 27:379–423, 623–656

Simos G, Sauer A, Fasiolo F, et al. (1998) A conserved domain within Arc1p delivers tRNA to aminoacyl-tRNA synthetases. Mol Cell 1:235–242

Sinclair DA, Dawes IW (1995) Genetics of the synthesis of serine from glycine and the utilization of glycine as sole nitrogen source by *Saccaromyces cerevisiae*. Genetics 140:1213–1222

Skouloubris S, Ribas de Pouplana L, Reuse H., et al. (2003) A noncognate aminoacyl-tRNA synthetase that may resolve a missing link in protein evolution. Proc Natl Acad Sci USA 100:11296–11302

Smith D, Yarus M (1989) tRNA-tRNA interactions within cellular ribosomes. Proc Natl Acad Sci USA 86:4397–4401

Sobolevsky Y, Trifonov EN (2005) Conserved sequences of prokaryotic proteomes and their compositional age. J Mol Evol 61:1–7

Sobolevsky Y, Trifonov EN (2006) Protein modules conserved since LUCA. J Mol Evol 63:622–634

Stathopoulos C, Li T, Longman R, et al. (2000) One polypeptide with two aminoacyl-tRNA synthetase activities. Science 287:479–482

Szathmáry E (1999) The origin of the genetic code: amino acids as cofactors in an RNA world. Trends Genetics 15:223–229

Tamura K, Schimmel P (2003) Peptide synthesis with a template-like RNA guide and aminoacyl phosphate adaptor. Proc Natl Acad Sci USA 100:8666–8669

Trevors JT, Abel DL (2004) Chance and necessity do not explain the origin of life. Cell Biol Intl 28:729–739

Trifonov EN (2004) The triplet code from first principles. J Biomol Structure Dynamics 22:1–11

Varshavsky A (1996) The N-end rule: functions, mysteries, uses. Biochemistry 93:12142–12149

Watson JD (1995) Personal communication

Wong JTF (2005) Coevolution theory of the genetic code at age thirty. BioEssays 27:416–425

Yamane T, Miller DL, Hopfield JJ (1981) Interaction of elongation factor Tu with the aminoacyl-tRNA dimer Phe-tRNA:Glu-tRNA. Biochemistry 20:449–452

Yang CM (2005) On the structural regularity in nucleobases and amino acids and relationship to the origin and evolution of the genetic code. Origins Life Evol Biosph 35:275–295

Yarus M (2000) Perspectives: protein synthesis - unraveling the riddle of ProCys tRNA synthetase. Science 287:440–441

Yarus M, Caporaso JG, Knight R (2005) Origins of the genetic code: the escaped triplet theory. Ann Rev Biochem 74:179–198

Yusupov MM, Yusupova GZ, Baucom A, et al. (2001) Crystal structure of the ribossome at 5.5Å resolution. Science 292:883–896

Zenkin N, Yuzenkova Y, Severinov K (2006) Transcript-assisted transcriptional proofreading. Science 313:518–520

Chapter 6

The Mathematical Structure
of the Genetic Code

Diego L. Gonzalez

Abstract In this chapter a new mathematical theory of the genetic code is presented, based on a particular kind of number representation: a non-power binary representation. A mathematical model is constructed that allows the description of many known properties of the genetic code, such as the degeneracy distribution and the specific codon-amino acid assignation, and also of some new properties such as, for example, palindromic symmetry (a degeneracy preserving transformation), which is shown to be the highest level of a series of hierarchical symmetries. The role of chemical dichotomy classes, which varies between purine–pyrimidine, amino–keto, and strong–weak following the position of the bases in the codon frame, is shown. A new characterization of codons, obtained through the parity of the corresponding binary strings in the mathematical model, together with the associated symbolic structure acting on the codon space, is also illustrated. Furthermore, it is shown that Rumer's classes (or degeneracy classes) can be obtained symbolically from the two first letters of a codon by means of an operation, which is identical to that of parity determination from a structural point of view. On this basis, the existence of a third dichotomy class sharing the former properties can be hypothesized.

Two main facts related to this new theory are also discussed: first, the intrinsic parity of codons in the number representation strongly suggests the existence of error detection/correction mechanisms based on such coding. These suggested mechanisms should work on the basis of the same general principles used in man-made systems for the transmission of digital data, like those associated with CDs, DVDs, wireless, and cellular telephone technologies. The hypothesis that, at an implementation level, such error correction processes may be based on principles borrowed from the theory of non-linear dynamics is discussed, placing them within the more general context of genetic information processing; second, the existence of a strong mathematical ordering inside the genetic code complicates to some extent the framework for the explanation of the origin of the code (and consequently

St. George School Foundation and National Research Council of Italy,
e-mail: diego.gonzalez@cini.ve.cnr.it

M. Barbieri (ed.), *The Codes of Life: The Rules of Macroevolution.*
© Springer 2008

of the origin of life): if it is difficult to understand the origin of the code within a necessarily short evolutionary time, it is still more puzzling to comprehend how a high degree of ordering has been attained in such a short time if a clear biological advantage cannot be associated with this mathematical structure. It must be remarked that the few other mathematical approaches aimed at describing the organization of the genetic code point to similar paradoxical conclusions (see also Chapter 7, this volume).

1 Introduction

In *The Major Transitions in Evolution*, John Maynard Smith and Eörs Szathmáry (1995) wrote: 'the origin of the code represents the most perplexing problem in evolutionary biology; the existing translational machinery is at the same time so complex, so universal, and so essential that it is hard to see how it could have come into existence, or how life could have existed without it.' This sentence has been cited many times because it is difficult to describe better the essential paradox about the translational apparatus. The standard genetic code is one of the most universal features of life. It is shared by almost every kind of organism, from bacteria to humans, with only a few minor exceptions. In an evolutionary framework, this universality implies that the standard code was used by a common ancestor of all forms of life, and thus the code assumed its actual form in a very short period of time at or very near the supposed beginning of life on Earth, but in any case, prior to the oldest discovered microfossils, i.e. bacterial and cyano-bacterial structures found in Archean apex cherts of the Warrawoona Group in Western Australia (Mojzsis et al., 1996; Schopf, 1993; Weberndorfer, 2002), dated to be at least 3,465 million years old.

Moreover, the standard genetic code seems to be at present an almost frozen version (Crick, 1968). This is because a minimal change, for example, in the amino acid meaning of a single codon will necessarily produce deleterious consequences for the organism involved. Even a single point mutation in a protein can cause severe diseases. This is the case, for example, for the point mutation in the synthesis of haemoglobin leading to the severe human disease microcytemy. An organism of medium complexity synthesizes at least several thousands of proteins (of the order of 50,000—60,000 in man following the latest results of human genome analysis); thus a change in the meaning of a single codon in the genetic code will necessarily produce changes in many different proteins: all those using this codon for the coding of a specific amino acid in the polypeptide chain, which are typically at least of the order of several hundreds in number. To ensure that not one of these modifications in many different key proteins should be fatal for the organism involved seems a very difficult if not an impossible task. Of course, if the original code were very much simpler than the present one, the task should be more plausible. Alternative explanations for different possible ways of facilitating code changes have been proposed; for example, the case in which a particular codon becomes totally absent on all proteins of an organism. If such an eventuality happens, it is possible to change the meaning of this particular codon without producing harmful consequences. This is a neutral mutation, meaning that it does not carry an associated biological advantage. This last fact, together with the low probability that a particular codon be completely exhausted in the entire protein library of an organism, makes it highly improbable that significant changes of the genetic code should be produced in this way. Regarding the case of primitive codes, due to the universality of the present one, any proposed simpler code must necessarily be a predecessor of the code of the common ancestor; thus the evolutionary period required to shape it is included in the short period allowed for the universal standard genetic code to acquire its present form.

Returning to the main characteristics of the code as a fundamental feature of the translational machinery of life, as well as universal, the genetic code is also complex and essential. It is essential because without the code the synthesis of proteins on the basis of genetically stored information is not possible. It is also complex because, despite its apparent simplicity (neglecting by the moment the associated highly complex synthesis machinery), it can lead to an astronomical combinatorial complexity; this complexity is analogous to the possible number of games that can be played on a 64-square chessboard, or, following the mythological origin of the game of chess, to the number of grains of wheat requested from the king as a reward by its inventor: starting with one grain on the first square, and duplicating the quantity for each succeeding chessboard square up to 64; curiously, this problem was first solved by Fibonacci using the fact that the sum of the first n terms of the series is equal to $2^{n+1} - 1$ (Geronimi, 2006) (this result is the basis of the univocal property of the power binary number representation system, as is shown below). The Fibonacci formula for $n = 64$ gives 18,446,744, 073,709,551,616, i.e. approximately 2×10^{19}. This great number – to the dismay of the king, there was no sufficient wheat in the kingdom to provide the due reward for the ingenuity of the chess inventor – is in any case insignificant if compared with the number of possible genetic codes. Assuming as an independent statement that the 64 codons code all the 20 different amino acids plus the termination signal indicating the end of protein synthesis with at least one codon, that is, ignoring for the moment the biochemical level of complexity, some 10^{83} arbitrary codes can be generated; this is a purely combinatory problem (see, e.g. Chechetkin, 2003). Thus, the probability of existence of the actual code as a casual event produced in a single evolutionary step is of the order of $1/(10^{83}) = (10^{-83})$, i.e. almost zero. As a comparative example it may be recalled that the total number of elemental particles in the universe is estimated at about 10^{80}; about 10^{60} times the grains of wheat requested by the chess inventor. This number of possible codes refers only to static complexity, that is, does not involve the combinatorial complexity of the genetic information encoded by the genetic code. In fact, the degeneracy of the code allows the existence of different sequences of bases encoding the same protein. As an example, a not very large protein, consisting say of 200 amino acids, can be coded in approximately 3^{200}, that is roughly 10^{95}, different ways.

Considering these enormous combinatorial numbers it is licit to ask a very fundamental question: why is the code as it is? Why is a particular code preferred over 10^{83} other possibilities? Why is a particular codon sequence of 200 amino acids in length preferred over its 10^{95} alternatives? In order to attempt to answer these questions it is necessary to characterize in some quantitative way the complexity of the code. As said before, code complexity can be understood at least at two different levels of description: the first refers to the biochemical complexity of the code, i.e. the code is implemented by a complex synthesis machinery interpreting codons of exactly three non-overlapping mRNA bases as different amino acids and/or start and stop synthesis signals. Moreover, the bases and amino acids utilized in the code are a subset of those available in Nature exhibiting specific chemico-physical properties; and so on[1] (Anderson et al., 2004; Mac Donaill, 2002). A second level of complexity refers, instead, to the internal structure of the code, that is, to the description of the degeneracy distribution: the number and the identity of codons that are assigned to every amino acid and to the initiation and termination signals (Jimenez-Montaño, 1999). Due to the almost frozen character of the present standard code we cannot expect a high degree of internal ordering at the degeneracy level; for the reasons described above the evolutionary forces that have tailored the code have had very little evolutionary time to produce a strong internal organization. Furthermore, it has been demonstrated that the code is in many senses arbitrary, that is, a given tRNA can be modified to bind with different codons, thus changing the meaning of the code (Anderson et al., 2004); this is also shown by the existence of variants of the standard genetic code (Watanabe and Suzuki, 2001; Knight et al., 2001). If the code is to a great extent arbitrary at a chemical level, what can be the biological advantage of a strong internal structure not depending on biochemical constraints? Which can be the hidden evolutionary forces leading to such a tailoring of the code?

This article hopes to contribute to this understanding by showing a deep internal mathematical structure of the genetic code, which is based on the redundant representation of integer numbers by means of binary strings. This structure reveals the existence of symmetry properties of the genetic code whose uncovering may contribute to the understanding of the organization of the genetic information regarding the use of synonymous codons for the coding of specific amino acids in a protein chain. The applications of number theory in the modelling of complex

---

[1] In fact, it has been demonstrated that amino acids and bases on one hand, and the number of bases used for amino acid coding on the other, can be arbitrary. For example, in Anderson et al., 2004, an unnatural amino acid is assigned to a four bases length codon (AGGA) without altering the synchronization of the frame reading and thus maintaining more or less invariant the rest of the composition of a given protein in *Escherichia coli*. Assignation of different amino acids to, for example, nonsense codons, allows for the incorporation of two unnatural amino acids, demonstrating that the code can be expanded arbitrarily in the number of amino acids used for protein synthesis (before this experiment restricted to assignment changes in nonsense codons), but also in the coding rules affecting, for example, the number of bases used to define a codon.

systems have seen a significant growth in the scientific literature of the last decades. This flourishing is largely associated with the role played by number theory in the mathematical description of many important phenomena in the theory of dynamical systems, and in several technological applications (Schroeder, 1986). In fact, the principal motivation for the present work has been the search for hidden deterministic mechanisms of error detection-correction within the genetic machinery. It is hypothesized that these mechanisms are directly related with the described mathematical ordering of the genetic code, which, in turn, is grounded in the dynamic properties of the translational apparatus. The results are very promising because, as is shown herein, strong analogies between the organization of the genetic code and of man-made codes for digital data transmission have been found. The present work, thus, aims to be a contribution towards the understanding of both the hidden rules of organization of degeneracy in the genetic information through a mathematical model of the code, on one hand, and of the associated functionality of this ordering, probably connected with a deterministic error detection/correction mechanism, on the other. This insight may contribute, in turn, to shedding light on the causes leading to the actual structure of the universal genetic code.

2   A Biochemical Communication Code Called the 'Standard Genetic Code'

In what sense is the genetic code a code in the usual meaning of the word? If we search in a dictionary for the word 'code', we find at least two different meanings that can be applied to the case of the genetic code:

1. System for communication (binary coding of telegraph, etc.)
2. A system used for secrecy or brevity of communication.

Moreover, we find two related verbs: to encode, that is, to translate a message into a code, and to decode, that is, to extract meaning from symbols with a procedure which is in some sense the inverse of the encoding one. It is obvious that a code is a product of human intelligence and that encoding is performed with the above-defined aims, that is, for communication and, at a more sophisticated level, for communication secrecy and/or efficiency of information transmission (compression). Surely, the most prevalent man-made code, which has allowed for the cultural development of mankind, is represented by oral and written languages. Through language, the world of thoughts and the physical world are connected by means of graphical signs and/or sound phonemes. To an arbitrary sequence of letters, i.e. an arbitrary word, is assigned one or more definite meanings. A dictionary describes such a connection in a given language: it specifies the meaning of a word using words of the same language. If we trust in the external reality of the physical world, it is trivially evident that the convention giving a meaning to a word could be changed for many or all of the words in a dictionary; this is the case of a different language. We say *home* in English or *casa* in Spanish but we remain confident that both words have the same meaning. A bilingual dictionary establishes the correspondence between words in one language and the words in a different language having the same meaning. With an entry in one language we can know the meaning in the other language: a translation is performed. This is the case of the genetic code: the genetic code is a bilingual translation dictionary with a reduced set of words and meanings. One difference from the bilingual translation dictionary, which operates in the world of meanings, is that the genetic code connects two otherwise weakly interacting biochemical worlds: the world of nucleic acids, where relevant biological information is stored, and the world of proteins, the essential chemical bricks for cellular metabolism. Thus the genetic code is truly a code in the sense of communication theory: it is a set of arbitrary symbols used for the scope of communication, i.e. for information transmission. In a typical communication process (Fig. 1a) there are five main components: an information source, a transmitter which mainly for practical reasons also encodes the information, a communication channel through which the information is sent, a receiver which usually decodes the information following definite rules depending on the encoding, and an information sink. Regarding the synthesis of proteins, we can identify some of these principal components of a communication process: the information is encoded as a sequence of codons in mRNA, which are sent through the nuclear membrane for protein synthesis. This information is read by the decoder/receiver, consisting of the tRNAs and the ribosome synthesis machinery, following the rules of the genetic code. Finally, the proteins, which represent the biological meaning of the encoded message, are used for the cell metabolism: the information sink. However some pieces are missing: the information source is unknown, as it is the encoding part of the transmitter; the information is available in an already encoded format (see Fig. 1b). As a consequence, the decoding step is slightly different from usual communication systems where the information sources

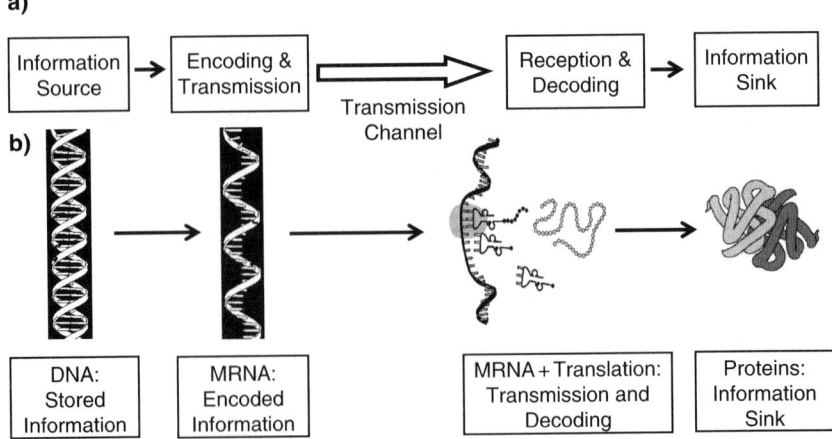

Fig. 1 (a) Generic communication process; (b) simplified information flux for protein synthesis

and sinks pertain to the same world. Going into the details of the coding strategy, using an analogy with language we can say that words in the mRNA world are composed of three letters picked out from an alphabet of four letters corresponding to the four possible bases along the mRNA chain, that is, Uracil, Cytosine, Adenine, and Guanine (U, C, A, G). Such a sequence of three letters – an elementary word – is called a codon and there are 64 possible different codons, which are the possible words of length 3′ using four different letters in any possible order including repetitions: $4 \times 4 \times 4 = 64$. On the amino acid side, instead, there are 20 different amino acids (see also the interesting paper by Hayes (1998), about the history of the discovery of the standard genetic code and the different original hypothesis about its organization). Because 64 different codons encode 20 different amino acids plus the termination signal, some amino acids are necessarily coded by more than one codon. This fact leads us to the redundancy and degeneracy properties of the code, which are described in detail below. In language too we find the same kind of redundancy, because there exist words that are synonyms. Some elemental means of error correction are connected to redundancy and to the concept of distance between words in human languages (see Chapter 18, this volume). In the case of the genetic code we also have synonymous codons, but these correspond to different words in a language (mRNA codons) that have the same meaning in the other language (amino acids), and which are listed in the bilingual dictionary (the genetic code).

In this contribution we shall study the genetic code from the point of view of communication and coding theory. Now that we have established that the genetic code is truly a code in the sense assigned to the word in communication theory, we attempt to discover the internal mathematical organization of the code, if any. The finding of different levels of internal structure in the genetic code needs to be related to the functionality of the mechanisms that are mediated by the code itself, that is mainly protein coding. However, if highly structured patterns are found, which is the case here, such an organization of the genetic information may carry a deeper meaning: perhaps it can contribute to revealing the basic mechanism for storage and utilization of the genetic information in other stages of the genetic machinery, such as replication and transcription. Some aspects related to the organizational structure of the genetic code were clear soon after its discovery. However, complete descriptions of such order are still missing and their meaning remains unclear.

From a fundamental point of view, we need to search for the structural organization of the code at a mathematical level. Thus, we search for mathematical structures describing in a closed and complete way the known properties of the genetic code, mainly the two levels of degeneracy distribution; this kind of study can contribute to clarifying the origin of the code and the functionality of the biological processes mediated by it. The existence of a strong mathematical organization of the code is demonstrated here, and it is guessed that this mathematical organization implies a coding strategy exploited by suitable non-random error detection/correction mechanisms. It is thought also that such mechanisms may operate on the basis of principles borrowed from the theory of dynamical systems and, in particular, based on the description of their associated symbolic dynamics.

3   Specifying the Two Levels of Degeneracy of the Standard Genetic Code

The genetic code is a *discrete* code, that is, a *correspondence* connecting two different *discrete* and *finite* sets of objects or elements: the domain (codons) and the codomain (amino acids). A *discrete* code is equivalent to some digital or numerical code in the sense that the elements of both sets can be numbered, that is, represented or labelled with numbers; a discrete mapping, thus, can be considered as a mapping between the numbers assigned to the elements in the two sets. Such a labelling may represent a more suitable way to treat the coding problem from a computational point of view. In this sense many attempts have been made to represent the genetic code numerically. The main idea is to take advantage of the small number of elements that characterize the code: the 64 codons, on one hand, and the 20 amino acids plus the stop signal, on the other. Digital computers work with bits, which are considered the smallest quantity of information. A yes or a no (a dichotomic question), an off or an on (the states of an electrical or electronic switch), a dot or a line (the Morse code), are all different equivalent expressions of a two symbol *binary system*. Particularly relevant in this context is the *binary representation system*, which allows the representation of an arbitrary number with sequences of only two digits, 1 and 0. Defining ordered strings of bits, we can construct words, or binary strings of arbitrary length, say n. With a binary string of length n we can represent 2^n different numbers (or objects), thus, the alphabet of codons, which consists of only four bases: U, C, A, and G, can be exactly represented with binary strings of length 2, since $2^2 = 4$; for example, U = 01, C = 00, A = 10, G = 11. Consequently, the codons or elemental words of the genetic alphabet, are formed by all the possible combinations of three of these bases, and thus can be represented by binary strings of length 6, as $2^6 = 64$. Of course, this binary numbering says nothing about the particular assignation of a binary string to a specific codon; in principle there is complete freedom of choice. If such a binary representation of the genetic code may be useful for computational purposes, it does not provide additional information about the code organization: it is neither degenerate nor redundant. For this reason, some authors have tried to answer a more general question: does the digital (or almost binary) nature of the genetic code reflect the existence of a deeper structural order, which can be expressed in mathematical terms? This idea was first expressed by George Gamow in a letter to the discoverers of the double helix structure of DNA, James D. Watson and Francis Crick: 'If your point of view is correct each organism will be characterized by a long number written in quad system with figures 1, 2, 3, 4 standing for different bases ... This would open a very exciting possibility of theoretical research based on combinatory and the theory of numbers! ... I have a feeling that this can be done. What do you think?' This question is not solely of theoretical interest; the standard genetic code is shared among almost every form of life on Earth and this fact is necessarily related to its functional meaning. In turn, the genetic code needs to reflect at some structural level these functional capabilities.

In this sense, among the interesting unsolved questions regarding the genetic code, the one related to amino acid code degeneracy is particularly puzzling. Why are amino acids not univocally represented? And why does the degeneracy of the representation differ from amino acid to amino acid? It should be noted that in terms of modern information theory, redundancy in the data-coding side of a communication system implies that errors, usually produced during the transmission process, can be detected and eventually corrected at the data-decoding side of such a system (Sweeney, 2002; Yockey, 1992, *see also* chapter 17 this volume).

In the following, a new strategy for explaining degeneracy of the genetic code is presented: the use of number representation systems that are inherently redundant. This approach allows one to investigate the mathematical structure of the genetic code at two complementary levels related to degeneracy and redundancy: that of the *degeneracy distribution*, i.e. the number of different codons that represent a given amino acid, and that of the *codon distribution*, i.e. the specific codons that are assigned to a given amino acid once its degeneracy is specified.

The two sets defining the genetic code are of different cardinalities: 64 codons on one side, and 20 amino acids plus the termination signal, on the other. This implies that the code is redundant and degenerate. The term degeneracy originates in particle physics and refers to the existence of different quantum states having the same energy. Analogously, we can say that a given amino acid is degenerate when it can be assembled in a protein starting with different codons at the mRNA level. In fact, we have in this case a similar situation to that in quantum physics because we can see that there exists an energy degeneracy corresponding to the decoding states leading to a specific amino acid (the codon–anticodon energy interaction plus the energetic contribution of the related translational mechanism) (Lehmann, 2002; Jimenez-Montaño, 1999). The term redundancy, instead, has a linguistic origin and refers to the excessive repetition of words in a sentence. The repeated words have similar or identical meaning and thus the idea can be applied to codons that code for the same amino acid; in fact, following this language analogy, these redundant codons are also said to be *synonymous*. Thus degeneracy corresponds to a viewpoint from the perspective of the codomain of the mapping defining the genetic code, i.e. the set of amino acids. It means that, given a particular amino acid, once it is decoded at the synthesis site, it becomes impossible to know which specific codon encoded it: *a given amino acid is encoded by more than one codon*. From a mathematical point of view, a mapping with these characteristics is said to be *non-invertible*. Redundancy, instead, corresponds to a viewpoint from the domain, i.e. the set of the 64 codons. A group of codons specifying the same amino acid is said to be *redundant*. Specifically, degeneracy and redundancy are quantified by the numerals of the corresponding subsets defining them. For example, Threonine is a degeneracy-4 amino acid because there exists a subset of four redundant codons coding for it (ACU, ACC, ACA, ACG). At this point we can try to identify the existence of a mathematical structure inside the code by observing that this intrinsic redundancy and degeneracy can be characterized at two independent levels, i.e. the *degeneracy distribution* and the *specific codon assignation*.

3.1   Degeneracy Distribution

The genetic code maps the 64 codons into 20 different amino acids plus the punctuation signal of stop that indicates the end of amino acid translation in the synthesis of a given protein. From this point of view the standard genetic code, and also all the other known versions of the code, are a *surjective* (all elements in the codomain come from at least one element in the domain) *and non-injective* (some elements in the codomain are degenerate) *mapping between sets of different cardinality*[2] (the number of elements in the domain and codomain are different). For a given redundant code, for example, the genetic code, we have a particular degeneracy distribution, that is, we have some possible degeneracy numbers (redundant codons) together with the associated number of elements (amino acids) sharing this specific degeneracy. The distribution of degeneracy in the standard genetic code is displayed in Table 1. This table represents the first property of the genetic code that we want to describe in mathematical terms: the *degeneracy distribution*. Before presenting a new approach to this problem we shall discuss the second property, which is independent of the first, and that we shall also attempt to describe later in mathematical terms.

For various reasons – the contiguity of codons coding a given amino acid, the decoding properties of the tRNAs, the degeneracy observed in the last base of the codon but not in the first or second, etc – the degeneracy distribution is usually considered within quartets, that is, inside groups of four codons sharing their first two letters. A quartet is in fact the maximal group that can be decoded by a given tRNA having a U in the first anti-codon base. This is the point of view adopted by the majority of authors (Rumer, 1966; Lehmann, 2002; *sh*Cherbak, 1993; Karasev and Stefanov, 2001; Jimenez-Montaño, 1999).

Table 1   Degeneracy distribution for the standard genetic code (coinciding with the degeneracy distribution of the euplotid nuclear version shown in Table 4)

No of amino acids sharing the same degeneracy	Degeneracy, i.e., no. of codons coding the same amino acid
2	1
9	2
2	3
5	4
3	6

[2] In rigorous terms, in some cases the code may be one-to-many. For example, the AUG codon that specifies Methionine represents also the signal for translation initiation, or start. Moreover, the UGA codon, which specifies a termination signal in the standard genetic code, may also code a selenium-containing 21st amino acid, Selenocystein (Hatfield and Gladyshev, 2002), depending on sequence context and the presence of concurrent biochemical conditions. In any case, the former ambiguities concern amino acids and punctuation signals, avoiding, thus, a possible amino acid–amino acid conflict in translation.

Table 2 Degeneracy distribution inside quartets of the euplotid nuclear version of the genetic code

No. of amino acids sharing the same degeneracy	Degeneracy, i.e. no. of codons coding the same amino acid
2	1
12 = 9 + 3	2
2	3
8 = 5 + 3	4

Thus a second table specifying the degeneracy distribution inside quartets for the standard genetic code is shown in Table 2. The main difference is that the three amino acids of degeneracy 6, Arginine, Leucine, and Serine, are considered as divided in two subgroups of degeneracy 4 and 2. Thus, each one of these amino acids contributes with a subgroup of degeneracy 2 and a subgroup of degeneracy 4, augmenting by 3 the degeneracies 4 and 2 shown in Table 1, and degeneracy 6 disappears. We show in Table 2 the degeneracy distribution inside quartets for the *euplotid nuclear* version of the genetic code.

3.2 Codon Distribution

Suppose that a particular amino acid has a given degeneracy, for example 2. This means that two different redundant codons code for the same amino acid, but which codons? It is clear that the degeneracy does not change if we change the two specific codons assigned to this amino acid choosing them from the other 62 existing ones: once we pick out specific codons for all the degeneracies we have specified a particular *codon assignation* for the specific *degeneracy distribution* of Table 2. This codon assignation represents a second level of complexity: once a degeneracy distribution is assigned we must fill out the code with specific codons, that is, we must specify the *codon assignation* in the degenerate code. In Table 3 such a codon assignation for the case of the standard genetic code is given as a translation table with three different entries corresponding to four different letters (U, C, A, G).

In Table 4 we give also the *euplotid nuclear* version of the genetic code. This version differs from the standard genetic code only in the assignation of the variable codon UGA, which in the latter case is assigned to Cysteine instead of to the stop signal.

4 A Mathematical Description of the Standard Genetic Code

Table 2 summarizes the degeneracy distribution inside quartets of the standard genetic code, which we aim to describe. For doing so we will use a particular kind of whole number positional representation called a non-power representation.

Table 3 The standard genetic code

	U	C	A	G	
	UUU Phe	UCU Ser	UAU Tyr	UGU Cys	U
U	UUC Phe	UCC Ser	UAC Tyr	UGC Cys	C
	UUA Leu	UCA Ser	UAA Stop	UGA Stop	A
	UUG Leu	UCG Ser	UAG Stop	UGG Trp	G
	CUU Leu	CCU Pro	CAU His	CGU Arg	U
C	CUC Leu	CCC Pro	CAC His	CGC Arg	C
	CUA Leu	CCA Pro	CAA Gln	CGA Arg	A
	CUG Leu	CCG Pro	CAG Gln	CGG Arg	G
	AUU Ile	ACU Thr	AAU Asn	AGU Ser	U
A	AUC Ile	ACC Thr	AAC Asn	AGC Ser	C
	AUA Ile	ACA Thr	AAA Lys	AGA Arg	A
	AUG Met	ACG Thr	AAG Lys	AGG Arg	G
	GUU Val	GCU Ala	GAU Asp	GGU Gly	U
G	GUC Val	GCC Ala	GAC Asp	GGC Gly	C
	GUA Val	GCA Ala	GAA Glu	GGA Gly	A
	GUG Val	GCG Ala	GAG Glu	GGG Gly	G

Table 4 Euplotid nuclear version of the genetic code; it differs of the standard genetic code only in the assignation of the codon UGA, which in this case is assigned to Cysteine instead of to the Stop signal

	U	C	A	G	
	UUU Phe	UCU Ser	UAU Tyr	UGU Cys	U
U	UUC Phe	UCC Ser	UAC Tyr	UGC Cys	C
	UUA Leu	UCA Ser	UAA Stop	UGA Cys	A
	UUG Leu	UCG Ser	UAG Stop	UGG Trp	G
	CUU Leu	CCU Pro	CAU His	CGU Arg	U
C	CUC Leu	CCC Pro	CAC His	CGC Arg	C
	CUA Leu	CCA Pro	CAA Gln	CGA Arg	A
	CUG Leu	CCG Pro	CAG Gln	CGG Arg	G
	AUU Ile	ACU Thr	AAU Asn	AGU Ser	U
A	AUC Ile	ACC Thr	AAC Asn	AGC Ser	C
	AUA Ile	ACA Thr	AAA Lys	AGA Arg	A
	AUG Met	ACG Thr	AAG Lys	AGG Arg	G
	GUU Val	GCU Ala	GAU Asp	GGU Gly	U
G	GUC Val	GCC Ala	GAC Asp	GGC Gly	C
	GUA Val	GCA Ala	GAA Glu	GGA Gly	A
	GUG Val	GCG Ala	GAG Glu	GGG Gly	G

In order to understand positional non-power representation systems we describe first the usual positional power representation systems; in such systems an ordered sequence of numbers (called digits) are weighted with values that depend on the position of these digits along the sequence (Ore, 1988); the represented number is obtained by adding together these weighted terms. In usual number representation systems, that is, positional power representation systems, the positional weights grow as the powers of some integer number called the base of the system. For example, bases 10 and 2 characterize, respectively, the *decimal* and the *binary* power representation systems. For power systems, the digits are allowed to vary in the range from 0 to $n - 1$, where n represents the system base. For the decimal system this range goes from 0 to 9. For instance, the number **735** can be obtained as,

n		$10^2 = 100$	$10^1 = 10$	$10^0 = 1$
735		7	3	5

$$735 = 7 \times 10^2 + 3 \times 10^1 + 5 \times 10^0$$

The fact that the digits are limited to the value of the base minus one ensures the one-to-one character of the representation: a number is represented by only one combination of digits and vice versa. However, we are interested in redundant representation systems. Redundancy can be obtained in two ways: (a) allowing the digits to go over their allotted range, or (b) decreasing to some extent the values of the positional weights (the ordered powers of the base for power representation systems). This latter case corresponds to the so-called non power representation systems in which the positional weights grow more slowly than the power of some base, for example, than the powers of 2 in a non-power *binary* system.

We develop in the following this second possibility (to our knowledge the degeneracy of the genetic code cannot be described by the other possibility of allowing the digits to go over their range). The usual binary representation corresponds to $n = 2$ (Ore, 1988), so the positional weights correspond to the powers of 2 and the relative coefficients can take only the values 0 and 1 (their allowed range). For example, the number 15 is represented in the binary system as **1111**,

n		$2^3 = 8$	$2^2 = 4$	$2^1 = 2$	$2^0 = 1$
15		1	1	1	1

that is, **15 = 1**.8 + **1**.4 + **1**.2 + **1**.1 If the positional values grow more slowly than the powers of two the representation is in general complete (all numbers from zero to the sum of all positional weights are representable) but redundant (a given number may be represented by more than one binary string). An interesting example of this is the Fibonacci representation (Zeckendorf, 1972) for which the positional weights are successive Fibonacci numbers, which take their name from Leonardo Pisano, known also as Fibonacci, the Italian mathematician who discovered them. Fibonacci numbers form a series in which the nth number of the series is obtained as the sum of its two predecessors, i.e. $F_n = F_{n-2} + F_{n-1}$, with the initial condition $F_1 = 1$ and $F_2 = 1$. The Fibonacci representation of order 6 (a 6-bit word size) uses the first six Fibonacci numbers, 1, 1, 2, 3, 5, and 8. The number 15, for example, is represented by the following three binary strings:

n		8	5	3	2	1	1		8	5	3	2	1	1		8	5	3	2	1	1
15		1	1	0	1	0	0		1	1	0	0	1	1		1	0	1	1	1	1

$$15 = 1.8 + 1.5 + 1.2 = 1.8 + 1.5 + 1.1 + 1.1 = 1.8 + 1.3 + 1.2 + 1.1 + 1.1$$

We have highlighted in grey the concept of parity of a binary string: even strings are in grey, odd strings are in white. Parity is defined taking into consideration the number of 1's in a given string: an even number of 1's gives an even string and an odd number of 1's an odd string. The order-6 Fibonacci representation shares some properties with the genetic code, for example, it is redundant (the number 15 is represented by three binary strings) and there are exactly 21 numbers represented by 64 binary strings (this last number depends uniquely on the binary word's length). However, the degeneracy of the order-6 Fibonacci system does not describe the actual degeneracy of the genetic code. For example, there are no amino acids with degeneracy 5 in the standard genetic code, but there are two numbers with degeneracy 5 in the Fibonacci representation system, as can be seen in Table 5.

At this point, a question naturally arises: does there exist a non-power binary representation – with length 6 words – with the same degeneracy distribution as the genetic code? A first answer to this question can be given taking into account very general properties of non-power representations: for the degeneracy distribution at large (presented in Table 1) there is no solution; but for the degeneracy inside quartets (presented in Table 2) there exists a unique solution. The demonstration of the first assertion can be performed by observing that any non-power representation system exhibits a very general property of the degeneracy distribution, i.e., palindromy. In fact, degeneracy is a palindromic function of the represented number: the represented numbers r and $R - r$, where R is the maximum integer that can be represented, share the same degeneracy. Palindromically related pairs are numerically represented by a complement-to-one operation: the palindrome of a given string is obtained by replacing all the digits 0 by 1 and 1 by 0. A consequence of the palindromic property is that the subset of represented numbers with a given degeneracy has a cardinal number that is even, except for the case in which $(R + 1)$ is odd. In such a case, the degeneracy of the central number, $R/2$, is shared by an odd number of represented numbers. In Table 5 it can be observed that this is so for the Fibonacci representation; there is only an odd degeneracy value (i.e. a 5 in the left column) corresponding to the redundancy 4 (right column). Thus, there exist five different numbers which are represented each by four different binary strings; between them we find the central number of the representation, i.e. $(21 + 1)/2 = 11$.

Table 5 Redundancy table for the Fibonacci non-power representation

Numbers sharing the same degeneracy	Degeneracy, i.e. no. of binary strings coding the same number
2	1
4	2
8	3
5	4
2	5

As a consequence, not all possible degeneracy distributions can be obtained using non-power representations; the degeneracy distribution must satisfy the palindromic property, i.e. it should be possible to order the redundancy numbers in a list consisting of a first part of non-decreasing values and a second part of non-increasing ones, in such a way that the list is identical to itself when read in the reverse direction.

As the full degeneracy distribution of the genetic code (Table 1) cannot be ordered in this way, there does not exist a non-power representation system sharing this degeneracy distribution. The degeneracy inside quartets, however, can be ordered in a palindromic table; a possible (not unique) palindromic ordering is shown in Table 6.

Table 6 Palindromic representation of the genetic code (euplotid nuclear version). In the first column are listed the degeneracies while in the second are the corresponding amino acids. Trp is represented by only one codon, Phe by two, and so on. Of course, any permutation of the amino-acids sharing a given degeneracy, for example, Phe with Tyr, maintains the palindromic structure of the list, i.e. there is also degeneracy in the degeneracy numbers

Degeneracy # (codons)	Coded amino acid
1	T Trp
2	F Phe
2	Ter (Stop)
2	Y Tyr
2	L Leu(2)
2	H His
2	Q Gln
3	C Cys
4	S Ser(4)
4	P Pro
4	V Val
4	L Leu(4)
4	R Arg(4)
4	G Gly
4	A Ala
4	T Thr
3	I Ile
2	E Glu
2	D Asp
2	R Arg(2)
2	N Asn
2	K Lys
2	S Ser(2)
1	M Met

4.1 A Particular Non-Power Number Representation System as a Structural Isomorphism with the Genetic Code Mapping

The palindromic ordering given in Table 6 shows that the degeneracy distribution of the genetic code is compatible with the degeneracy distribution of a non-power representation system. There are not many theoretical results about non-power representation systems (see, e.g. Wolfram, 2002) and, to the author's knowledge, it has not been demonstrated that any palindromic degeneracy list is amenable to such a kind of system. In more mathematical terms we can say that palindromy of the degeneracy distribution is a necessary condition, but it is not known if it is also sufficient.

For the case of the genetic code we have proceeded by trial and error to find a unique solution for the set of positional weighting values of a length 6 binary non-power representation system satisfying the distribution of Table 2, which implies also necessarily matching the degeneracies shown in Table 6. The values of the non-power positional weights are: [1, 1, 2, 4, 7, 8], and this solution, shown in Table 7, is unique up to trivial equivalence classes (Gonzalez and Zanna, 2003; Gonzalez, 2004). It may be remarked that there exists a multiplicative equivalence: all the positional weights can be multiplied by an integer (a scale factor) without affecting the degeneracy distribution (but breaking the contiguity of represented numbers). Allowing the representation of negative numbers, it is also possible to change the sign of one or more positional weights while conserving the degeneracy distribution, and maintaining contiguity of the represented numbers. Finally, it is interesting to note that there does not exist additive invariance, that is, the addition of the same integer to all the weights changes the degeneracy distribution.

The degeneracy distribution corresponding to this unique solution is shown in Tables 7, 8, and 9. It is easy to see on inspection that Tables 8 and 9 are identical, respectively, to Tables 2 and 6. This result is remarkable because the exact degeneracy distribution of the genetic code has been obtained with a one-step mathematical procedure: the non-power representation of the whole numbers from 0 to 23 by means of binary strings of length 6 (a total of 64) and using a unique set of six positional weights [1, 1, 2, 4, 7, 8].

However, this result does not represent per se a model of the genetic code.

The genetic code maps the 64 codons into 24 elements (amino acids inside quartets plus the stop signal) while the non-power representation maps the 64 length-6 binary strings into the first 24 whole numbers: the two mappings have the same degeneracy distribution.

Thus, from a mathematical point of view, the genetic code and the non-power representation are connected by a *structural isomorphism* (they are two mappings sharing the same logical structure). But it is known from modelling theory that a structural isomorphism does not necessarily represents a model; it can be a coincidence, albeit a highly fortunate one (the probability of a random origin of a much less restricting property of the code, known as the Rumer's transformation, which is treated in Section 6, has been estimated as being very low indeed, $P = 3.09^{-32}$ (Zhaxybayeva, 1996)).

Table 7 Non-power representation of the first 23 whole numbers by length 6 binary strings and positional weights 1, 1, 2, 4, 7, 8 The number 7, for example, can be represented by three different binary strings, i.e. (010000; 001101; 001110); also the parity of strings has been shown: even strings are given in grey

Represented number	Length 6 binary strings																							
	8	7	4	2	1	1	8	7	4	2	1	1	8	7	4	2	1	1	8	7	4	2	1	1
0	0	0	0	0	0	0																		
1	0	0	0	0	0	1	0	0	0	0	1	0												
2	0	0	0	0	1	1	0	0	0	1	0	0												
3	0	0	0	1	0	1	0	0	0	1	1	0												
4	0	0	1	0	0	0	0	0	0	1	1	1												
5	0	0	1	0	0	1	0	0	1	0	1	0												
6	0	0	1	1	0	0	0	0	1	0	1	1												
7	0	1	0	0	0	0	0	0	1	1	0	1	0	0	1	1	1	0						
8	0	0	1	1	1	1	0	1	0	0	1	0	0	1	0	0	0	1	1	0	0	0	0	0
9	1	0	0	0	0	1	1	0	0	0	1	0	0	1	0	1	0	0	0	1	0	0	1	1
10	1	0	0	1	0	0	1	0	0	0	1	1	0	1	0	1	0	1	0	1	0	1	1	0
11	0	1	1	0	0	0	0	1	0	1	1	1	1	0	0	1	0	1	1	0	0	1	1	0
12	1	0	1	0	0	0	1	0	0	1	1	1	0	1	1	0	0	1	0	1	1	0	1	0
13	0	1	1	0	1	1	1	0	1	0	0	1	1	0	1	0	1	0	0	1	1	1	0	0
14	0	1	1	1	1	0	0	1	1	1	0	1	1	0	1	0	1	1	1	0	1	1	0	0
15	1	1	0	0	0	0	1	0	1	1	0	1	1	0	1	1	1	0	0	1	1	1	1	1
16	1	1	0	0	0	1	1	1	0	0	1	0	1	0	1	1	1	1						
17	1	1	0	0	1	1	1	1	0	1	0	0												
18	1	1	0	1	0	1	1	1	0	1	1	0												
19	1	1	1	0	1	1	1	1	0	1	1	1												
20	1	1	1	0	0	1	1	1	1	0	1	0												
21	1	1	1	1	0	0	1	1	1	0	1	1												
22	1	1	1	1	0	1	1	1	1	1	1	0												
23	1	1	1	1	1	1																		

Table 8 Degeneracy distribution for the binary non-power number representation with positional weights [1, 1, 2, 4, 7, 8] (recall that the corresponding positional weights for the usual univocal binary system of length 6 are the first six ordered powers of two, i.e.,[1, 2, 4, 8, 16, 32]

Numbers sharing the same degeneracy	Degeneracy,i.e. no. of binary strings coding the same number
2	1
12	2
2	3
8	4

Table 9 Explicit correspondence between the 64 length-6 binary strings and the first 24 whole numbers for the non-power number representation with positional bases [1, 1, 2, 4, 7, 8] (compare with Table 6)

Degeneracy no. (length-6 binary strings)	Coded whole numbers
1	0
2	1
2	2
2	3
2	4
2	5
2	6
3	7
4	8
4	9
4	10
4	11
4	12
4	13
4	14
4	15
3	16
2	17
2	18
2	19
2	20
2	21
2	22
1	23

In order to construct a model it is necessary to establish links between the two pairs of sets defining the applications: inside the domains (codons and length-6 binary strings), and inside the codomains (the amino acids and the whole numbers of the representation). Section 5 is devoted to demonstrating that such links can be established, and thus, that a true mathematical model of the genetic code can be constructed.

5   A Mathematical Model of the Genetic Code

As we have mentioned, the euplotid nuclear variant of the genetic code differs from the standard one only in the assignation of the UGA codon (see Tables 3 and 4). But before discussing variants of the genetic code we investigate first some properties that are universally shared. These properties lead us to consider the *euplotid nuclear* version as the more symmetrical one, and for this reason it is used to develop the mathematical model of the genetic code.

5.1 Symmetry Properties

A remarkable property of the genetic code is the degeneracy of the last letter of a codon regarding pyrimidine exchange, that is, exchange between U and C. In other words, if a given amino acid is encoded by a codon NNU, where N stands for any of the four bases U, C, A, G, the codon NNT encodes the same amino acid (see Table 10). Still more remarkable is the fact that all known variants of the genetic code, including 10 nuclear and 16 mitochondrial ones (Watanabe and Suzuki, 2001; Knight et al., 2001), respect this symmetry. This implies that 32 codons are degenerated by the exchange of U and C in their last letter (two groups of 16 codons each are so defined). An inspection of Table 6 shows that this same symmetry is displayed by the non-power representation system (see Table 7). The binary strings xxxx01 and xxxx10 – where x stands for either binary number, 0 or 1 – always encode the same whole number: in such a way 32 degenerate binary strings are defined forming two groups of 16 strings each. It must be remarked that, in an arbitrary coding, there is no need for any connection between the degeneracy distribution (the global property of Table 2) and symmetry (a local property defining 16 degenerate pairs of codons). Furthermore, there is no biochemical reason for this degeneracy, there being identified at least one way to recognize U but not C by a specific tRNA as a third codon base; in fact, xo⁵U (a hydroxymethyluridine derivative) in the first anticodon position may decode U, A, or G, but not C, in the third position of a codon (Watanabe and Suzuki, 2001).

Moreover, the specific degeneracy of the representation system is a consequence of the non-power positional weights chosen, and these are univocally determined by the global degeneracy distribution of the genetic code. That is: (i) we start with a degeneracy table for the standard genetic code; (ii) we find a unique solution for a mathematical isomorphism based on a non-power positional number representation describing this degeneracy; (iii) this choice imposes an internal symmetry in the model; and finally (iv) this internal symmetry is also exhibited by all the known versions of the genetic code, closing in this way a modelling circle: the structural isomorphism is indeed a model of the genetic code as we demonstrate in the following. As a biological consequence we find that the genetic code is blind to the codon's pyrimidine last letter because of its redundancy distribution (or vice versa), a fact not noted previously (the global degeneracy of the code is uniquely linked to its internal organization).

This property thus shows the way for the mathematical modelling of the genetic code because we have established links between one half of the codons (codons ending in pyrimidine) and one half of the binary strings of the model (those of the form xxxx01 or xxxx10) sharing the same symmetry properties (see Table 11). The former result produces an immediate consequence for the attempt to complete the links between codons and strings: the 32 remaining binary strings are of the form xxxx00 or xxxx11 and this means that codons of the form NNR, where R stands for a purine (A or G), are necessarily coded by these kind of strings.

This assertion can be further refined by observing that the strings 000000 and 111111 (the degeneracy-1 strings) necessarily code the degeneracy-1 amino acids,

Table 10 Essential degeneracy in the third letter of pyrimidine-ending codons (3rd letter U or C) showed for the standard genetic code but shared by all known versions of the genetic code (10 nuclear and 16 mitochondrial)

		U	*C*	*A*	*G*	
		UUU **Phe**	UCU **Ser**	UAU **Tyr**	UGU **Cys**	*U*
U		UUC **Phe**	UCC **Ser**	UAC **Tyr**	UGC **Cys**	*C*
						A
						G
		CUU **Leu**	CCU **Pro**	CAU **His**	CGU **Arg**	*U*
C		CUC **Leu**	CCC **Pro**	CAU **His**	CGC **Arg**	*C*
						A
						G
		AUU **Ile**	ACU **Thr**	AAU **Asn**	AGU **Ser**	*U*
A		AUC **Ile**	ACC **Thr**	AAC **Asn**	AGC **Ser**	*C*
						A
						G
		GUU **Val**	GCU **Ala**	GAU **Asp**	GGU **Gly**	*U*
G		GUC **Val**	GCC **Ala**	GAC **Asp**	GGC **Gly**	*C*
						A
						G

Table 11 Equivalence between binary strings and pyrimidine-ending codons

Length-6 strings (either)						Type of codon (either)		
x	x	x	x	0	1	N	N	U
x	x	x	x	1	0	N	N	C

i.e. Methionine and Triptophan (see Table 12). These two amino acids are coded by codons ending in G, i.e. AUG and UGG. Thus a final G is coded in one case by a string ending in 00 and in the other by a string ending in 11. Consequently 11 or 00 endings alone do not suffice for determining the G or A ending character of a codon. But if we take into account the parity of the strings this indeterminacy can be resolved. Recall that parity can be defined as the parity of the number of ones in a string, for example, 000000, has 0 ones being thus even; 111111 has 6 ones being also even.

Thus we can assert that strings ending in 00 or 11 with even parity code a final G in the corresponding codon, and that, by exclusion, the A ending codons are coded by strings ending in 00 or 11 but with odd parity (see Table 13).

Following the degeneracy table of the genetic code, the pair of degeneracy-3 elements also needs to be univocally assigned. This pair corresponds in the standard code to Ile and Stop. But we can observe that a more symmetric case corresponds to the *euplotid nuclear* version of the genetic code in which the Stop codon UGA is assigned to Cysteine.

Thus, the degeneracy-3 pair of amino acids is composed of Cysteine and Isoleucine as shown in Table 14. For the *euplotid nuclear* version degeneracy-1 and

Table 12 G-ending codons corresponding to the degeneracy-1 amino acids Trp and Met (this last representing also the start signal)

		U	C	A	G	
		TTT Phe	TCT Ser	TAT Tyr	TGT Cys	U
U		TTC Phe	TCC Ser	TAC Tyr	TGC Cys	C
		TTA Leu	TCA Ser	TAA Stop	TGA Cys	A
		TTG Leu	TCG Ser	TAG Stop	TGG Trp	G
C		CTT Leu	CCT Pro	CAT His	CGT Arg	U
		CTC Leu	CCC Pro	CAC His	CGC Arg	C
		CTA Leu	CCA Pro	CAA Gln	CGA Arg	A
		CTG Leu	CCG Pro	CAG Gln	CGG Arg	G
A		ATT Ile	ACT Thr	AAT Asn	AGT Ser	U
		ATC Ile	ACC Thr	AAC Asn	AGC Ser	C
		ATA Ile	ACA Thr	AAA Lys	AGA Arg	A
		ATG Met	ACG Thr	AAG Lys	AGG Arg	G
G		GTT Val	GCT Ala	GAT Asp	GGT Gly	U
		GTC Val	GCC Ala	GAC Asp	GGC Gly	C
		GTA Val	GCA Ala	GAA Glu	GGA Gly	A
		GTG Val	GCG Ala	GAG Glu	GGG Gly	G

Table 13 Equivalence between binary strings and purine-ending codons, which also takes into account the string's parity (see text)

Length-6 strings (either)						Parity	Type of codon		
x	x	x	x	1	1	Odd	N	N	A
x	x	x	x	0	0	Even	N	N	G

degeneracy-3 amino acids are associated in pairs defining entirely symmetric quartets (Table 15).

In turn, these quartets are related by a degeneracy-conserving transformation that consists of the exchange of U↔A in the first letter and U↔G in the second, as shown also by the arrows in Table 15. As there are only two degeneracy-3 amino acids, the corresponding codons can be assigned to the strings defining the degeneracy-3 numbers in the non-power model following the rules developed up to now, as shown in Table 13. In so doing, an important symmetry of the genetic code having a precise mathematical counterpart is evidenced: *palindromic* symmetry. In the string space this symmetry is shown by a complement to one operation, that is a 0↔1 and 1↔0 exchange in the binary digits. The fact that this operation changes strings of one of the quartets into strings of the palindromic associated quartet can be easily demonstrated by remembering, first, that the two amino acids of degeneracy 1 are assigned to the strings 111111 and 000000, which are evidently complementary as shown in Table 16 and, secondly, by inspection of the strings shown in Table 17.

The palindromic symmetry involves not only the two quartets analysed before but also, as is shown below, every quartet of the genetic code. This powerful

Table 14 Degeneracy-3 amino acids in the euplotid nuclear version of the genetic code

		U	C	A	G	
U		TTT Phe	TCT Ser	TAT Tyr	TGT Cys	U
		TTC Phe	TCC Ser	TAC Tyr	TGC Cys	C
		TTA Leu	TCA Ser	TAA Stop	TGA Cys	A
		TTG Leu	TCG Ser	TAG Stop	TGG Trp	G
C		CTT Leu	CCT Pro	CAT His	CGT Arg	U
		CTC Leu	CCC Pro	CAC His	CGC Arg	C
		CTA Leu	CCA Pro	CAA Gln	CGA Arg	A
		CTG Leu	CCG Pro	CAG Gln	CGG Arg	G
A		ATT Ile	ACT Thr	AAT Asn	AGT Ser	U
		ATC Ile	ACC Thr	AAC Asn	AGC Ser	C
		ATA Ile	ACA Thr	AAA Lys	AGA Arg	A
		ATG Met	ACG Thr	AAG Lys	AGG Arg	G
G		GTT Val	GCT Ala	GAT Asp	GGT Gly	U
		GTC Val	GCC Ala	GAC Asp	GGC Gly	C
		GTA Val	GCA Ala	GAA Glu	GGA Gly	A
		GTG Val	GCG Ala	GAG Glu	GGG Gly	G

Table 15 Degeneracy conserving transformation, i.e. exchange U↔A in the first letter and U↔G in the second letter of the codon. The transformation connects two quartets involving the two degeneracy-1 (Met and Trp) and the two degeneracy-3 (Ile and Cys) amino acids (euplotid nuclear version)

		U	C	A	G	
U		TTT Phe	TCT Ser	TAT Tyr	TGT Cys	U
		TTC Phe	TCC Ser	TAC Tyr	TGC Cys	C
		TTA Leu	TCA Ser	TAA Stop	TGA Cys	A
		TTG Leu	TCG Ser	TAG Stop	TGG Trp	G
C		CTT Leu	CCT Pro	CAT His	CGT Arg	U
		CTC Leu	CCC Pro	CAC His	CGC Arg	C
		CTA Leu	CCA Pro	CAA Gln	CGA Arg	A
		CTG Leu	CCG Pro	CAG Gln	CGG Arg	G
A		ATT Ile	ACT Thr	AAT Asn	AGT Ser	U
		ATC Ile	ACC Thr	AAC Asn	AGC Ser	C
		ATA Ile	ACA Thr	AAA Lys	AGA Arg	A
		ATG Met	ACG Thr	AAG Lys	AGG Arg	G
G		GTT Val	GCT Ala	GAT Asp	GGT Gly	U
		GTC Val	GCC Ala	GAC Asp	GGC Gly	C
		GTA Val	GCA Ala	GAA Glu	GGA Gly	A
		GTG Val	GCG Ala	GAG Glu	GGG Gly	G

symmetry connects quartets with the same degeneracy distribution and strings related by the complement-to-one operation. We postpone the complete analysis of this symmetry until the discussion of the placement of the remaining amino acids into the number representation map shown in Table 7.

Table 16 Binary strings corresponding to G-ending codons of the quartets AU and UG

#	Length-6 strings						Parity	Type of codon			Amino acid
0	0	0	0	0	0	0	Even	A	U	G	Met
23	1	1	1	1	1	1	Even	U	G	G	Trp

Table 17 Binary strings assigned to the degeneracy-3 amino acids following the rules previously established: xxxx01 or xxxx10 are U- or C-ending codons; codons ending in A correspond instead to xxxx00 or xxxx11 strings with odd parity (an odd number of ones). Observe that the sum of the numbers represented equals 23 and that this corresponds to the complement-to-one operation at the binary strings level (palindromic symmetry)

#	Length-6 strings						Parity	Type of codon			Amino acid
	0	0	1	1	0	1	Odd	A	U	U	
7	0	0	1	1	1	0	Odd	A	U	C	Ile
	0	1	0	0	0	0	Odd	A	U	A	
	1	1	0	0	1	0	Odd	U	G	U	
16	1	1	0	0	0	1	Odd	U	G	C	Cys
	1	0	1	1	1	1	Odd	U	G	A	

5.2 Degeneracy-6 Amino Acids

5.2.1 Coding of the Degeneracy-6 Amino Acids Arg (R) and Leu (L)

Observing Table 7 we can see that, following the rules previously enounced, there exist two degeneracy-2 numbers (4 and 19) that represent 2 A-ending codons. Because there do not exist degeneracy-2 amino acids coded by two codons ending in A, these two numbers must represent the degeneracy-2 part of some degeneracy-6 amino acids. This observation implies the existence of two strings pertaining to degeneracy-4 numbers representing G-ending codons, because the only amino acids with two A-ending codons are Leu and Arg, and both also possess two G-ending codons. These numbers exist, i.e. 11, and 12, (Table 7) and thus necessarily pertain to these degeneracy-6 amino acids (Arg and Leu); the third degeneracy-6 amino acid, Serine, has the degeneracy-2 group of codons ending in U and C. As a consequence, the degeneracy-6 palindromic amino acids Arg and Leu are univocally placed in the mathematical representation, coded by numbers 4, 11, 19, and 12; purine-ending codons of these two amino acids are coded by two disjoint mathematical objects, but also the mathematical objects code for disjoint - not contiguous in the usual table of the genetic code – codons.

Observe also that the involved numbers are palindromic (their sum equals 23), defining a new piece of the total palindromic transformation in the genetic code.

Such palindromic symmetry is expressed for the degeneracy-2 groups as a U↔A exchange in the first letter of the codons, and a U↔G exchange in the second. For the degeneracy-4 group, the symmetry is represented instead by only the second part of the former transformation, that is, a U↔G exchange. Observe that the first transformation coincides with the palindromic symmetry described for the quartets of degeneracy-3 and −1 amino acids (see Table 15).

5.2.2   Coding of the Degeneracy-6 Amino Acid Serine (S)

The degeneracy-6 amino acid Serine is instead represented by a normal degeneracy-4 group (with strings representing A- and G-ending codons) and a degeneracy-2 group with strings representing U- and C-ending codons. In order to place these groups in the representation it is convenient to begin the analysis with the parity of pyrimidine-ending codons and the full palindromic symmetry.

Firstly, we can observe that the two previous placements of codons associated with the palindromic symmetry correspond to a second letter of the codon U or G; secondly, part of the transformation consists of the exchange of these two letters in the second position; thirdly, the four groups of two strings representing U- or C-ending codons are of odd parity. Is thus natural to suppose that the same assertion might be true for the remaining strings representing the rest of the amino acids with U/C ending codons and with a U or a G in the second position. If this is true, it must be that the number representation should possess two and only two numbers of degeneracy 4 and two and only two numbers with degeneracy 2 satisfying this property. It can be seen by inspection of Table 7 that this is indeed the case and thus we can assign binary strings to the amino acids Phe, Ser[2], Ala, and Gly.

5.3   *The Mathematical Model*

Now the rest of the placements are easy because we have amino acids with second letter C that are all of degeneracy 4, and amino acids with second letter A that are all of degeneracy 2 (including the A- and G-ending group of two stop codons). Again it can be confirmed by inspection of Table 7 that we have the appropriate binary strings according to the former rules.

Some degree of ambiguity remains in these associations because we can impose only the parity and degeneracy rules developed up to now. Some amino acids, however, can be placed following numerical rules observed in the U and G middle base codon part of the code. For example, we observe that the numbers connecting degeneracy-2 and -4 groups of the degeneracy-6 amino acids Arginine and Leucine, are equivalent $\mod_7$. Extrapolating this result to the degeneracy-6 amino acid Serine, we need to assign to the group of degeneracy 2, represented by numbers 1 or 22, the degeneracy-4 groups described by the numbers 8 or 15, which satisfy

Table 18 Non-power model of the euplotid nuclear genetic code. In a typical column (headed by the second letter of the listed codons) we find: (i) at the left-most subcolumn, the integer number represented by the non-power system; (ii) at the central subcolumn, the corresponding length-6 binary strings; (iii) at the right-most subcolumn, the corresponding amino acid. Parity coding is also evidenced: in white are shown even codons while in grey are shown odd ones

		U			C			A			G			
U	1	000001	Phe	15	101101	Ser	18	110110	Tyr	16	110010	Cys	U
	1	000010	Phe	15	101110	Ser	18	110101	Tyr	16	110001	Cys	C
	4	001000	Leu	15	011111	Ser	2	000100	Ter	16	101111	Cys	A
	11	011000	Leu	15	110000	Ser	2	000011	Ter	23	111111	Trp	G
C	11	100101	Leu	14	011110	Pro	3	000101	His	12	011010	Arg	U
	11	100110	Leu	14	011101	Pro	3	000110	His	12	011001	Arg	C
	4	000111	Leu	14	101100	Pro	17	110100	Gln	19	111000	Arg	A
	11	010111	Leu	14	101011	Pro	17	110011	Gln	12	101000	Arg	G
A	7	001101	Ile	8	010010	Thr	5	001001	Asn	22	111110	Ser	U
	7	001110	Ile	8	010001	Thr	5	001010	Asn	22	111101	Ser	C
	7	010000	Ile	8	100000	Thr	21	111011	Lys	19	110111	Arg	A
	0	000000	Met	8	001111	Thr	21	111100	Lys	12	100111	Arg	G
G	13	101001	Val	9	100001	Ala	20	111010	Asp	10	010110	Gly	U
	13	101010	Val	9	100010	Ala	20	111001	Asp	10	010101	Gly	C
	13	011100	Val	9	010011	Ala	6	001011	Glu	10	100011	Gly	A
	13	011011	Val	9	010100	Ala	6	001100	Glu	10	100100	Gly	G

the $n_1 - n_2 = 0 \bmod_7$ requirement (observe that also both pairs satisfy the same property). Considering all these properties together we can make an attempt to place all the amino acids into the mathematical model. The result is shown in Table 18. We can also complete the general rules for determining the parity of codons; the set of palindromic transformations is compatible with the following parity rule for triplets ending in U or C: triplets ending in U or C and with total even parity have a C or an A (Amino type) as a second letter; those with total odd parity have a U or G (Keto type) as a second letter. This rule complements the former one for triplets ending on A or G and shows that all triplets of the code are parity marked: if the triplets end in A or G, the mark is coincident with that letter, the string being even for a final G and odd for a final A; for triplets ending in C or U, (Y), the parity is instead coded on the second letter of the triplet (see also Table 18).

We have implemented in some cases other symmetry requirements regarding the properties of the numbers involved; in order not to complicate excessively the discussion we omit these details here.

6   Palindromic Symmetry and the Genetic Code Model

As shown before, we have up to now uncovered a part of the palindromic degeneracy preserving symmetry. Observing Tables 15 and 18, is easily to understand its meaning as a transformation inside the genetic code connecting quartets with the same degeneracy distribution. On the mathematical-model side

Table 19 Palindromic symmetry; a degeneracy preserving transformation connecting different quartets of the genetic code

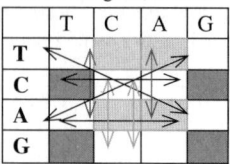

this symmetry is concisely represented by a simple operation: the complement to 1 of all the binary digits of a given string.

For completing the palindromic symmetry we need to place the remaining amino acids (four of degeneracy 4 and seven of degeneracy 2, plus the pair of codons representing the stop signal). This task is relatively easy because: (i) all the remaining degeneracy-4 aminoacids have a middle letter C while all the degeneracy-2 ones have a middle letter A; (ii) because on the model side there do not exist palindromic pairs of strings pertaining to a same number, there do not exist palindromic transformations inside a given quartet and thus a change in one or more letters of the codon, implying a quartet change, is needed. The most natural choice is to preserve the transformations uncovered until now but maintaining fixed the middle base.

By this choice we are able to place the amino acids Ser4, Thr, Tyr, Stop, Asn, and Lys; the associated transformation is represented by a exchange of U and A in the first letter of the codons. For the remaining amino acids, because of property (i) the unique possible choice is exchange of C and G letters in the first position of the codon. The complete symmetry is shown in Table 19.

Of course, at the end of this process, some remaining degeneracy still exists. The fundamental one is an indeterminacy with regard to the assignation of binary strings to quartets: the full set of strings assigned to a quartet can be exchanged with the full set of palindromic strings pertaining to a palindromic quartet. We shall not discuss here some numerical arguments for the final placements that we show in Table 18 representing our mathematical model of the genetic code. This is the most probable assignation taking into account our knowledge about the symmetry properties of the genetic code on one side, and the number representation system, on the other. As mentioned above, some room for ambiguity in this assignation still remains, but the main properties of the code are described and some additional uncovered symmetries are evidenced[3] (see also the following Section 7). The palindromic symmetry can be considered as the counterpart of the first discovered

[3] Some differences with the placement shown by Gonzalez (2004) are due to optimization regarding number theoretical properties that are roughly discussed here. All the differences (mainly inversion of palindromic amino acid pairs and variable ending assignation of Y-ending codons) are compatible with the degrees of freedom still remaining after codon assignation following the description in Sections 5 and 6, that is, observing the symmetry of the palindromic transformation and the other internal symmetries of the genetic code.

regularity in the degeneracy distribution of the genetic code: Rumer's transformation. The Russian theoretical physicist Y. Rumer (1966) observed that the genetic code could be divided into two halves, as represented in Table 20. One half corresponds to quartets of degeneracy 4 and the other to quartets containing degeneracies 3, 2, or 1. Rumer found that a global anti-symmetric transformation connects these two halves of the code. This transformation is represented by the following exchange between all the letters of a codon: U,C,A,G ↔G,A,C,U. We examine below the mathematical meaning of this transformation and its connection with the new concept of codon parity.

6.1 Parity of Codons

In the mathematical model, to every codon is assigned a parity bit that corresponds to the parity of the length-6 binary string shown in Table 18. Note that, as remarked above, the parity of the binary string can be computed by summing its symbols: an even number of ones leads to an even string and an odd number of ones to an odd string. We can observe also that the parity bit can be derived easily on the biochemical side by means of two complementary rules applied to codons: if the codon ends with A it is represented by an odd string; if it ends with G the string is even; if instead, the codon ends with U or C the parity bit is determined by the second letter of the codon, i.e. C or A in the second position defines an even codon; U or G defines an odd codon. As the complement-to-one operation does not change the parity, palindromic symmetry also preserves parity. Parity of codons and binary strings are shown in Table 18 where odd codons are denoted in grey.

At this point it is interesting to interpret parity as a non-linear algorithm operating on the chemical classes of the second and third position bases of the codon. This algorithm is illustrated graphically in Fig. 2. First we analyze the purine–pyrimidine class in the third letter. If the class is purine, the third letter alone determines the parity of the codon. If instead the third letter is a pyrimidine, we need to observe the middle letter of the codon: a Keto base determines an odd codon and an Amino base determines an even codon. We observe that parity induces a natural chemical dichotomy depending on the position of the base in the codon: for the third letter purine–pyrimidine matters, while for the middle letter Amino–Keto is the induced chemical dichotomy. It also needs to be noted that we are using a window of only two bases -the last two- in order to determine parity.

6.2 Rumer's Class

It is very surprising that Rumer's classification can also be presented in a completely analogous way. In Table 20 we show Rumer's classes, that is, the quartets of degeneracy 4 (given in grey) and the quartets of degeneracy-3, 2, and 1. It is easy

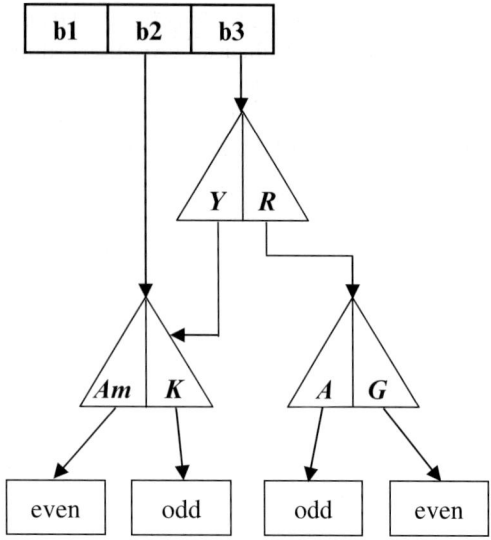

Fig. 2 Non-linear algorithmic representation of the parity class. Observe the induced chemical dichotomies Y-R in the last letter, and K-Am, in the middle one

Table 20 Rumer's classes in the genetic code; in grey are given degeneracy-4 amino acids while in white are represented degeneracy-3, -2, -1 ones

	U	C	A	G	
	TTT Phe	TCT Ser	TAT Tyr	TGT Cys	U
U	TTC Phe	TCC Ser	TAC Tyr	TGC Cys	C
	TTA Leu	TCA Ser	TAA Stop	TGA Cys	A
	TTG Leu	TCG Ser	TAG Stop	TGG Trp	G
	CTT Leu	CCT Pro	CAT His	CGT Arg	U
C	CTC Leu	CCC Pro	CAC His	CGC Arg	C
	CTA Leu	CCA Pro	CAA Gln	CGA Arg	A
	CTG Leu	CCG Pro	CAG Gln	CGG Arg	G
	ATT Ile	ACT Thr	AAT Asn	AGT Ser	U
A	ATC Ile	ACC Thr	AAC Asn	AGC Ser	C
	ATA Ile	ACA Thr	AAA Lys	AGA Arg	A
	ATG Met	ACG Thr	AAG Lys	AGG Arg	G
	GTT Val	GCT Ala	GAT Asp	GGT Gly	U
G	GTC Val	GCC Ala	GAC Asp	GGC Gly	C
	GTA Val	GCA Ala	GAA Glu	GGA Gly	A
	GTG Val	GCG Ala	GAG Glu	GGG Gly	G

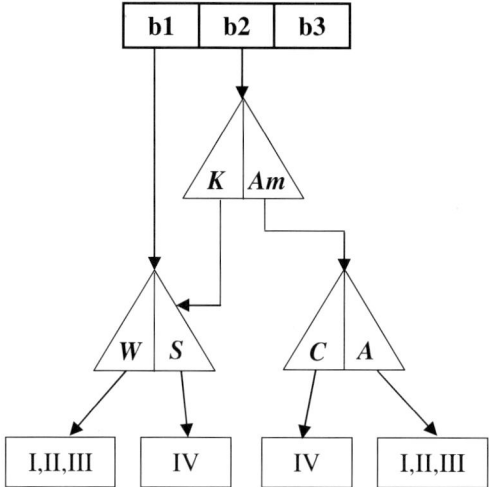

Fig. 3 Non-linear algorithmic representation of Rumer's class. Observe that the structure is completely analogous to that of parity class but acting on a window defined by the two first letters of the codon. The induced chemical dichotomies are K-Am in the middle letter (shared with the parity class algorithm) and W-S in the first one

to see that a non-linear algorithm with the same structure as the one for parity determination (Fig. 2) can be defined. Moreover, the analysis algorithm acts also on a window of two bases – the first two – and maintains the chemical dichotomy induced by parity determination in the middle base.

In fact, the middle base is shared by the two algorithms and thus, the right dichotomy class should be the Keto–Amino one. As can be seen in Fig. 3, this in indeed the case.

Rumer's classes are determined as follows: if the middle base is of the Amino type the class is immediately determined, i.e. a C defines class 4 codons while an A defines class-3, -2, and -1 codons. If instead the middle letter is of the Keto type, we need to observe the character of the first base. In this case, the chemical dichotomy is the Weak or Strong type. A Strong base in the first position defines class 4 codons while a Weak base defines class-3, -2, and -1 codons.

Thus, subsuming the former results, we have a natural means for the classification of the bases inside a codon following the three possible chemical dichotomies and depending on the position of the base in the codon: the third base following the purine–pyrimidine dichotomy, the second one the Amino–Keto, and the first one the Strong–Weak chemical dichotomy. This classification of bases allows for determination of Rumer's classes of degeneracy and the new parity class introduced by the numerical model of the genetic code. Rumer's transformation, i.e. U,C,A,G ↔G,A,C,U, maps any codon of the degeneracy-4 class to a codon of the degeneracy-3, -2, -1 class, thus, it is a degeneracy-breaking transformation (it reveals an anti-symmetric property of the degeneracy distribution).

Rumer's transformation is a global transformation acting on all the bases of the codon; however, the same effect is obtained if we restrict the transformation to the two first bases of the codon (only the two first bases (quartets) are determinant for the degeneracy class, the third one being irrelevant). Rumer's transformation can be viewed as a composition between the other two possible global transformations (excluding the identity and the transformations maintaining invariant two of the four bases).

These transformations are: U,C,A,G ↔A,G,U,C, and U,C,A,G ↔C,U,G,A. The first transformation exchanges bases inside the Strong and Weak dichotomy, and the second, inside the pyrimidine and purine one. Observe that Rumer's transformation corresponds to the exchange inside the third possible chemical dichotomy, the Keto and Amino one. It is interesting to note that the S-W or the Y-R transformation alone changes only one half of the codons of a given Rumer class. Moreover, following the parity rules presented above, it can be seen that the Y-R transformation changes the parity of a codon. This is completely analogous to Rumer's transformation: this transformation can be applied only to the two last letters of the codon with the same effect (the first letter is irrelevant for parity). This property has two interesting consequences; first, it can be hypothesized that a third dichotomy class, being determined by the last letter of a codon and the first of the next one, can be defined. Following the properties for the other two dichotomy classes, Rumer and parity, this class needs to be anti-symmetric under the action of the third chemical dichotomy transformation U,C,A,G↔A,G,U,C, (S-W). We have called this class the hidden class and results about it showing a possible connection with a graph theory of anti-codon classes will be published elsewhere. Moreover, the three anti-symmetric transformations together with the identity have a group structure, indeed, they form a 4-Klein group called Klein V group (Jimenez-Montaño, 1999; Négadi, 2004; Gusev and Schulze-Makuch, 2004). Second, denoted by p_p the probability of occurrence of an odd codon in a real coding sequence, the probability after application of the Y-R transformation will be $1 - p_p$. On the other hand, if p_R is the probability of Rumer degeneracy-4 class, because that the same transformation affects only one half of the codons, if parity and Rumer's classes are independent, the expected probability after application of the Y-R transformation will be one half. The departure from this value is a new indicator reflecting the independence of the two dichotomy classes. Moreover, the same applies for the other two transformations and the hidden class, opening the door to studying also correlations between dichotomy classes of different codons in a natural manner (Gonzalez et al., 2006).

7 A Complete Hierarchy of Symmetries Related to the Complement-to-One Binary Operation

We have shown that an important degeneracy-preserving symmetry exists in the genetic code: the palindromic symmetry. Moreover, we have shown that this symmetry is completely equivalent to a simple binary operation acting on the numerical model of the genetic code: the complement-to-one of length-6 binary strings. We have shown also that a perfect symmetry involving one half of the code exists, i.e. codons ending in U or C are indistinguishable. In general, whenever we have a symmetry, we have an associated conserved quantity. For the palindromic symmetry we have two conserved quantities: degeneracy and parity. For U↔C exchange, instead, the conserved quantity is the assigned amino acid. This symmetry can be viewed in the numerical model as a partial complement-to-one operation acting on strings of the form xxxx01 or xxxx10. However, there are other symmetries of the same kind, that is, a partial complement-to-one transformation with conserved quantity consisting in the assigned amino acid. The chemical meaning of these transformations is usually the exchange of base type in the last or second codon position (recall that palindromic transformation implies exchange in the first and second codon bases). In the following we show all these interesting hidden symmetries of the genetic code involving A- and G-ending codons.

7.1 A, G Exchanging Symmetry Involving 16 Codons (Non-Degeneracy-6, -3, and -1 Amino Acids)

From Table 21 is evident that the two kinds of strings represented are equivalents. Because they change parity they are associated with A↔G exchange. As many as 16 strings are involved by this symmetry and these are associated with A- or G-ending codons of amino acids having degeneracy 2, and 4.

7.2 A, G Non-Exchanging Symmetry of 8 Codons Pertaining to the Degeneracy-6 Amino Acids Leucine and Arginine

In Table 22, we observe that eight strings associated with the Leucine and Arginine amino acids are also equivalents. This symmetry does not change parity and for this reason is associated with A↔A and G↔G exchange between different degeneracy groups of the two amino acids, that is, is equivalent to exchange in the second letter of the codon: U↔C, for Leucine, and C↔A, for Arginine.

Table 21 Symmetry associated with amino acid invariance by A↔G exchange in the last letter of purine-ending codons. This symmetry involves a total of 16 codons pertaining to the non-degeneracy-6 class amino acids: i.e. degeneracy-4 Pro, Ala, Val, Gly, and degeneracy-2 Stop, Gln, Lys, and Glu. An x symbol means either a 0 or a 1 digit, which remains constant in the partial complement-to-one transformation

	8	7	4	2	1	1
Original string	x	x	x	0	1	1
Symmetric string	x	x	x	1	0	0

Table 22 Symmetry associated with U↔C or C↔A exchange in the second letter of codons involving different degeneracy groups of amino acids Leucine and Arginine

	8	7	4	2	1	1
Original string	x	x	0	1	1	1
Symmetric string	x	x	1	0	0	0

7.3 A↔G Exchanging Symmetry of 4 Codons Pertaining to the other Degeneracy-6 Amino Acid Serine and Its Palindromically Associated Amino Acid Threonine

From Table 23 it is also evident that the two kinds of represented strings are equivalent. Because they change parity they are associated with A↔G exchange in the last letter, as in Table 21.

Four strings are involved in this symmetry: A or G ending codons of degeneracy-6 amino acid Serine and its associated palindromic amino acid Threonine.

7.4 Four Remaining A, G, Ending Codons

The remaining codons are represented in Table 24. These are four isolated strings not displaying amino acid invariance under partial palindromic transformation.

Table 23 Symmetry associated with amino acid invariance by A↔G exchange in the last letter of purine-ending codons. This symmetry involves a total of four codons pertaining to the degeneracy-6 amino acid with pyrimidine ending second group and its palindromic degeneracy-4 amino acid: i.e. degeneracy-4 group of Ser and degeneracy-4 Thr

	8	7	4	2	1	1
Original string	0	x	1	1	1	1
Symmetric string	1	x	0	0	0	0

Table 24 Four individual codons pertaining to the purine-ending codons of degeneracy-3 and -1 amino acids, i.e. degeneracy-3 Cys (stop), Ile, and degeneracy-1 Trp, Met (start)

8	7	4	2	1	1
0	1	0	0	0	0
1	0	1	1	1	1
1	1	1	1	1	1
0	0	0	0	0	0

7.5 Other Symmetries

Table 25 shows the first part of the symmetry regarding pyrimidine↔purine exchange on degeneracy-4 and -3 classes: Y↔A, and Y↔G exchange inside degeneracy-4 quartets.

Four are U, C to A (Ser, Pro, Thr, Ala) and four are U, C to G (Leu, Arg, Val, Gly).

This is a parity changing transformation and for this reason pyrimidine even codons go to odd A-ending ones, and pyrimidine odd codons go to even G-ending ones.

Table 26 shows the second part of the symmetry regarding pyrimidine↔purine exchange on the degeneracy-4 and -3 classes. This involves other possible Y↔A or G exchange. Two are U, C to G (Ser, Thr) and two are U, C to A (Cys, Ile).

This is a non-parity-changing transformation and for this reason pyrimidine even codons go to even G-ending ones, and pyrimidine odd codons go to odd A-ending ones.

In Table 27 is shown the Y↔R non-parity-changing transformation involving the regular degeneracy-4 amino acids Val, Gly, Pro, Ala. As in Table 26, even codons go to G-ending ones, and odd codons go to A-ending ones. This second transformation indicates the doubly degenerative character of the symmetries involving purine↔pyrimidine exchange in the four regular amino acids.

Finally, Table 28 shows the symmetry connecting the two disjoint groups (degeneracy-4 and 2) of the degeneracy-6 amino acids Leu and Arg. This is a

Table 25 First part of the symmetry corresponding to Y↔R exchange in the last letter of codons pertaining only to the degeneracy-4 Rumer's class

	8	7	4	2	1	1
Original string	0	1	x	x	x	1
Symmetric string	1	0	x	x	x	1

	8	7	4	2	1	1
Original string	0	1	x	x	1	x
Symmetric string	1	0	x	x	0	x

Table 26 Second part of the symmetry corresponding to Y↔R exchange in the last letter of codons pertaining to the degeneracy-4 and -3 classes

	8	7	4	2	1	1
Original string	x	0	1	1	x	1
Symmetric string	x	1	0	0	x	0

	8	7	4	2	1	1
Original string	x	0	1	1	1	x
Symmetric string	x	1	0	0	0	x

Table 27 Doubly degenerative character of the Y↔R transformation involving the four regular degeneracy-4 amino acids Val, Gly, Pro, and Ala

	8	7	4	2	1	1
Original string	1	0	x	0	x	1
Symmetric string	0	1	x	1	x	0

	8	7	4	2	1	1
Original string	1	0	x	0	1	x
Symmetric string	0	1	x	1	0	x

Table 28 Y↔R transformation connecting the disjoint degeneracy-4 and -2 groups of the amino acids Arg, and Leu

	8	7	4	2	1	1
Original string	0	1	1	0	x	0
Symmetric string	1	0	0	1	x	1

	8	7	4	2	1	1
Original string	0	1	1	0	0	x
Symmetric string	1	0	0	1	1	x

parity-changing symmetry and thus connects pyrimidine odd codons of the degeneracy-4 group with the G-ending codon of the degeneracy-2 group.

7.6 Complement-to-one in the Seventh Position

8	7	4	2	1	1
x	1	x	x	x	x
x	0	x	x	x	x

This transformation is related to all the degeneracy-1,-3,-6, amino acids. Observing the numbers involved:

	Numbers ≤ 11				Palindromic numbers		
	#	Δ	#		#	Δ	#
1,3-Degeneracy Met-Ile-Trp-Cys	0	7	7		16	7	23
4+2-Degeneracy Phe-Thr-Ser^4-Ser^2	1	7	8		15	7	22
4+2 Degeneracy Leu-Arg	4	7	11		12	7	19

It is easy to recognize that this transformation identifies all the 'non-regular' amino acids, that is, all those having degeneracy 1, 3, or 6, and their palindromically related ones.

As a conclusion we can say that there is a clear influence of the global degeneracy properties on the local symmetries regarding palindromic codon exchange, last letter pyrimidine exchange, last letter purine exchange, and last letter purine↔pyrimidine exchange. Furthermore, purine exchange shows clearly that amino acids are divided in groups following their degeneracy properties and these groups are completely described by the numerical model; all amino acids of non-degeneracy-6, -3, and -1, define a first group including 16 codons. A second group includes the two degeneracy-6 amino acids with second subgroup with purine-ending codons (Leu, Arg) involving a total of eight codons. A third group includes the remaining degeneracy-6 amino acid Serine and its degeneracy-4 palindromic amino acid Threonine with a total of four codons. Finally, four separate codons correspond to the purine-ending codons of the degeneracy-1 and -3 amino acids. The fact that the stop codon of the standard genetic code, UGA, pertains to this class, i.e. is a free codon not involved in a symmetry exchange with any G-ending codon, may be related to the variability of this codon, which is assigned to different amino acids in different versions of the code. Such variability does not break the symmetry of the subjacent numerical coding: this codon is one of the four less involved in symmetry transformations (indeed subject only to the palindromic symmetry). Finally the complement-to-one operation in the seventh position identifies in one step all the 'non-regular' amino acids, the ones with degeneracy 1, 3, or 6 and the palindromically related ones (Met (Start), Ile, Trp, Cys (Stop), Arg, Leu, Ser, Thr, and Phe); in fact, one half of the complete genetic code composed of eight quartets. It can be remarked that the Start codon corresponds to the null string, i.e. (0,0,0,0,0,0).

8   Error Control and Dynamical Attractors: A High Level Strategy for the Management of Genetic Information?

We have shown, up to now, that the genetic code has a profound mathematical organization. The non-power representation system (Table 7) describes: (i) the degeneracy distribution (Table 2); (ii) most of the codon assignments (Table 18); (iii) a degeneracy-preserving transformation associated with palindromic symmetry (Table 19); (iv) a complete set of hierarchical symmetries including U and C exchange in the last codon letter (shared by all the known versions of the genetic code), and different partial symmetries regarding A, G, and Y, R, exchange (Section 7); (v) a structural rule which can be applied either to the determination of Rumer's classes or to the newly defined codon parity classes (Table 18 and Section 6); (vi) a suggestion of the existence of a third dichotomy class (the hidden class); (vii) a group of global transformations acting anti-symmetrically on the three dichotomic classes (in fact having the structure of a Klein V group if we include the identity transformation U,C,A,G ↔ U,C,A,G); (viii) evidence about the position-dependent role of the three possible chemical dichotomies, and many other properties mentioned parenthetically or in course of development (associated mainly with number-theoretical properties, the organization of redundancy, and the study of statistical properties of actual DNA sequences (Gonzalez et al., 2006). On the basis of these results two main observations can be made at a first glance: (i) a simple mathematical structure based on the most essential concept in number theory, that is, number representation, describes most of the structural properties of the genetic code without the use of any additional hypothesis; (ii) the uniqueness of the model's solution implies that the internal structure of the code is strongly related (indeed is a consequence) of the global degeneracy distribution (a fact not previously understood). Thus, assuming that the mathematical model indeed represents the code structure, it is licit to ask why the code is mathematically structured in this way. The 'minimal action' or economy principle of life is well known to biologists. Surviving competition is a truly expensive game and there is no room for luxurious accessories. If the code is mathematically organized this organization must necessarily reflect a functional meaning or utility. The proposal of the author is that the mathematical structure of the code is linked to the organization and management of the genetic information in terms very similar to the ones used in present man-made digital information transmission systems. One of the main issues related to this kind of system is the management of involuntary errors produced along the entire chain of information processing. No actual man-made system exists without appropriate subsystems dedicated to error detection and correction. Of course, such subsystems are based on methods that imply different degrees of mathematical organization of the data, and most of them utilize the concept of parity (of binary strings). Both elements, that is, mathematical organization and parity coding, arise naturally in the mathematical modelling of the genetic code.

In the following, we discuss briefly the possible links between structured methods of error detection/correction and the management of information at the different levels of the genetic machinery. Finally, we propose a possible way for implementing error control using general properties regarding the behaviour of non-linear dynamical systems.

The genetic code acts at the level of translation, that is, of protein synthesis controlled by genetic information. Once mRNA relative to a specific protein becomes available for translation, it is reasonable to suppose that the information carried is reliable. Indeed, the very existence of the genetic code points to this concept: a codon is translated as the corresponding amino acid and there is no reason for changing its meaning (there are only a few exceptions in which ambiguous codons can be translated with different meanings depending on the context[2]). Thus, the main issue regarding error protection in protein synthesis is related to translation accuracy. We need to ensure that a given codon is translated into the amino acid assigned by the genetic code. An additional problem is represented by the speed of protein synthesis; if we need many molecules of the target protein, for example, as a response to a vital cell signalling, the velocity of protein synthesis may be crucial for the cell's and hence the organism's survival. The ribosome is the biochemical machine charged with this task, and using the appropriate tRNAs can attain a sustained rhythm of several tens of amino acids per second in the synthesis of a given protein. This is a fantastic performance for a guided chemical reaction in which a correct chemical is selected sequentially from a soup where all the basic reactants are mixed and the differences in chemical free energies are very low. Our guess is that the mathematical organization of the code should be the theoretical basis for a system exhibiting such extraordinary capabilities in terms of synthesis speed and translation accuracy. The ribosome can be considered as a complex dynamical system. We know from general theory that dynamical systems can exhibit complex dynamics such as, for example, chaotic behaviour, which is necessary in order to ensure that high-complexity sequences may be produced by a dynamical system. We can think of the ribosome as a chaotic dynamical system synchronized by the sequence of symbols coming from the mRNA. Synchronization can ensure that the output of the system is identical to the input. In our case, inputs and outputs are of a different nature and are also of a discrete character, while the ribosome is a continuous system. But this is not an impediment, because it is very easy to convert a continuous dynamical system into a discrete one. The method is to induce a mapping (e.g. a Poincaré surface of section) and to define on this mapping a so-called symbolic dynamics (Hao and Zheng, 1998). In practical terms this signifies that the states of the iterated system are quantized following its position in phase space (the dynamical space on which the system evolves following a trajectory given a set of initial conditions). Thus, resuming, we can consider a synchronization of the translation system through a forcing of the ribosome with a sequence of symbols represented by the sequence of bases of the mRNA; the synchronized output is evident as an intermediate output consisting of the sequence of anticodons in tRNA, and the symbolic dynamics of the machine – the ribosome – is represented by the amino acids and the punctuation symbols. An iteration of the machine corresponds to adding a new amino acid to the polymeric chain of the protein. As the sequences can be very complex it is necessary that the behaviour of the machine indeed be chaotic. Moreover, it is known that chaotic systems can be synchronized, and for this two conditions usually need to be meet[4]: (i) the synchronized systems need to be identical, (ii) some control procedure needs to be implemented in the slave system in order to ensure synchronization. As we do not have a true master dynamical system in the translation process, the first requirement can be met only if the sequence of symbols is compatible with the dynamics of the slave system (the translational apparatus). In other words, the coding needs to 'know' the dynamic entanglement of the dynamical decoder (the mRNA coding sequence represents the output of a 'virtual' master system); this can be the ultimate purpose of redundancy (why a given sequence and not any other synonymous one is used). Moreover, redundancy symbols may act as true error signals used for maintaining the synchronization of the sequence of codons and the sequence of output symbols as required by the second point above; because redundancy is intrinsically related to the positional role of chemical dichotomies, this same mechanism may also be the basis of frame maintenance. In this way, the organization of the genetic information should reflect some fundamental properties related to the dynamics of the ribosome complex. Finally the importance of initial conditions for determining the actual behaviour of a given dynamical system must be noted. In order to maintain accuracy, at the beginning of the translation, the system needs to be initialized with an appropriate initial condition or 'seed'. The compatibility of the input sequence with the dynamics of the translation

---

[4]Two alternative methods have been described for transmitting information using chaotic systems, i.e. chaotic synchronization and control of symbolic dynamics (see, e.g. Bollt, 2003).

system ensures the short-range prediction of the apparatus, which is further controlled by the use of redundant information. This natural requirement is compatible with what is known regarding the existence of particular conserved regions associated with the start of protein translation, such as the Shine-Dalgarno domain in prokaryotes (May et al., 2006). This initialization of the translation system is also compatible with the usual scanning until founding of Kozak consensus in eukaryotes. The seed may be placed in the region immediately before that sequence including the Methionine (AUG) synthesis Start signal (Brown, 2002; Lewin, 2004). Given that the genetic code organization is directly involved only in the translation step, it cannot be excluded that error detection and correction will be performed in other instances of genetic information processing on the basis of the same organizational principles previously described. This role may be present as an informational guide for mechanisms of duplication and transcription (and conserved sequences are also involved, for example, in transcription initiation, a fact compatible with the initialization of a dynamical process, as mentioned above). However, at these two stages some more essential mechanisms may be present. In principle, any mechanism interacting with the double helix of DNA can resort to the information available in the two strands. Apparently, there is no new information in the second strand (ignoring of course the case of wrong complementary bases) but the complementary strand implements one of the three anti-symmetric Klein transformations reported in Section 6; thus, the complementary brand can be considered as a 'hardware' signal processing. Secondary information derived from this strand can be useful for error control. For coding regions, if there exists an organizational level related to the dynamics of protein translation, it may in principle be possible to correct 'meaning' errors, that is, sequences not allowable or 'forbidden' from a dynamical point of view should be rejected or restored (in fact, there also exist some forbidden sequences from the bio-chemical point of view, as shown, for example, by the absence in some cases of particular synonymous codons (Britten, 1993)). Observed mechanisms for DNA repair make evident their connection with information content. Two different mechanisms are utilized for repairing short DNA regions damaged in one strand depending on transcriptional activity: transcriptionally active regions damaged by UV light follow a repair pathway known as transcription-coupled repair, while non-coding DNA regions follow a different repair pathway known as global excision repair (Fuss and Cooper, 2006). If genetic information consists only of a linear sequence of bases, why are two different mechanisms for repairing coding and non-coding DNA regions used? An open problem related to these mechanisms is how to recognize that an error exists and, in such a case, which is the strand that needs to be cleaved. This is, indeed, a theoretical issue: there is not any apparent a priori reason for preferring one sequence over another. The only solution for this question is to have a mathematical template that produces some extremum (a maximum or a minimum) for the valid – correct – sequences. Of course, this extremum principle may have a physico-chemical counterpart, for example, as an energy minimization, and needs to be implimented by some 'extreme evaluator', for example, a protein complex that is part of some error-correction mechanism. The existence of, for example, the TFIIH complex which is involved in nucleotide excision repair, transcription-coupled repair, and normal transcription, points to the complexity of DNA repair mechanisms and to the possibility that absolute templates may be used for implementing such repair pathways depending on the sequence's information content. Dynamically allowed sequences observed with the mathematical organization induced by the genetic code may form the basis for such template-based recognition, at least for coding regions, with forbidden dynamical sequences indicating necessarily the presence of errors.

This mechanism can explain also the selection of particular sequences (the specific use of the sequence redundancy) because of the fact that such sequences may be compatible with the dynamics of the ribosome and consequently more robust to error translation (aiding in this way the sequence conservation). Another interesting point is that, if the synchronization mechanism is in some way sensitive to the chirality of a chemical, the former arguments apply to sequence selection favouring the privileged chiral form used by the dynamic mechanism of error correction.

The main problem faced by living beings is how to interact with an external world, or environment, which is essentially of infinite dimensionality. In order to interpret external environmental signals, and to react in consequence, life needs to implement a drastic dimensionality reduction process. Non-linear dynamical systems offer a natural way to cope with this problem: dynamical attractors, that is, the asymptotic trajectories of a dynamical system under specific external conditions (forcing), represent a low-dimensional response to that external stimulus which can be used for its characterization. Dynamical attractors are of low dimensionality and robust, that is, they persist under external (stimulus) and internal (system) perturbations. As life is also auto-reflexive, that is, it needs to understand itself, a dynamical-system approach for the management of genetic information may represent a high-level strategy able to cope with the problems related to its maintenance and to the survival of individual organisms in interaction with their living environments. Perhaps, this dynamical systems strategy may be extended paradigmatically to other ambits of life, as has been demonstrated, for example, for neural processing of sensorial information (Cartwright et al., 1999a, 2001, and references therein).

Following this line of reasoning, we end this work with an analysis from the highest possible conceptual level, which implies necessarily a conjectural character. The present work demonstrates that the genetic code is indeed strongly organized from a mathematical point of view. This fact raises an important general question: what is the functional meaning of such mathematical organization? In order to answer this fundamental question two characteristics of the organization give hints for further work: (i) number theory is present at its most fundamental level, i.e. integer number representation; (ii) as in technological systems for digital data transmission, parity coding also plays a relevant role. Both facts point to error detection-correction technologies and non-linear dynamics[5].

---

[5] The existence of parity coding determining a privileged set of bases for DNA has been determined in a different context by Donall A. Mac Donaill (2002). This mechanism was proposed by the same author as being responsible for evolutionary selection of the bases used at present for information storing in the complementary strands of DNA. Furthermore, the hypothesis of error detection/correction mechanisms based on organizational coding of the genetic information has been studied from a theoretical and experimental point of view by different authors (May, 2006 and references therein; Forsyke, 1981; Liebovitch et al., 1996; Rzeszowska-Wolny, 1983).

Dynamical systems theory and its consequences for the modelling of complex systems are, among the more recent approaches applied to this aim, one of the most promising ones. Dynamical attractors offer dynamical dimensionality reduction, which in turn is capable of describing (equivalent to high level understanding) complex behaviour in very simple terms (Gonzalez, 1987; Gonzalez and Rosso, 1997; Cartwright, 1999b). It seems that a step further in the comprehension of life needs a shift in existing paradigms. The enormous complexity of living beings needs the use of new high-level archetypes in order to construct conceptual and simulation models approaching the actual behaviour of complex biological systems. It is, in some sense, a situation very similar to the case of electronic technology. The analysis of present complex electronic circuits is not any longer possible solely on the basis of elemental circuit dynamics, or measuring voltages and currents in different points of the same. We need to understand first new archetypes for implementing definite functions which are the basis of complex circuits and which are accessible to simple modelling such as basic electrical measuring. Present day integrated circuits represent physically these archetypes carrying also a definite functional meaning, which is at the basis of their logical interconnection in forming more complex circuits and systems. The development of different nano-technologies, MEMS devices, and also organic and biologically hybrid experimental devices, is making this analogy together with the related intersection area ever more clear. Ultimately, dynamics is ubiquitous, and the theory of dynamical systems is now offering the best approach for understanding nature and life's functional organization in theoretical terms allowing for qualitative and quantitative modelling of real biological systems. As has been recently hypothesized, a complete theory of complexity should be constructed on the basis of a symbolic analysis description of coupled chaotic oscillators, which, through generalized synchronization phenomena gives rise to emergent key spatio-temporal structures (Corron and Pethel, 2003; Pethel and Corron, 2003). As has been suggested (Izhikevich, 1999; Hoppensteadt and Izhikevich, 1998), similar mechanisms may be used for processing and communication selectivity between spatially separated neurons of the nervous system. On the same grounds, this non-linear dynamic approach may open the door for theoretical modelling and practical implementation of methods for the interpretation and further control (as in the case of genetic therapy) of the flux of genetic information.

Acknowledgements I wish to thank Prof. Marcello Barbieri for his invitation to write this contribution and for continuous encouragement through the development of the work. I am also profoundly indebted to Dr. Julyan Cartwright for his useful suggestions and careful reading of the manuscript.

References

Anderson J.C., Wu N., Santoro S.W., Lakshman V., King D.S., and Schultz P.G., An expanded genetic code with a functional quadruplet codon, Proc. Natl. Acad. Sci. USA, 101(20), 7566–7571, 2004.

Bollt E., Review of chaos communication by feedback control of symbolic dynamics, Inter. J. Bifur. Chaos, 13(2), 269–283, 2003.

Britten R.J., Forbidden synonymous substitutions in coding regions, Mol. Biol. Evol., 10(1), 205–220, 1993.

Brown T.A., *Genomes*, 2nd edn, Wiley, New York, 2002.

Cartwright J., Gonzalez D.L., and Piro O., Nonlinear dynamics of the perceived pitch of complex sounds, Phy. Rev. Lett., 82(26), 5389–5392, 1999a.

Cartwright J., Gonzalez D.L., Piro O., and Zanna M., Teoria dei sistemi dinamici: una base matematica per i fenomeni non-lineari in biologia, Sistema Naturae., 2, 215–254, 1999b.

Cartwright J., Gonzalez D.L., and Piro O., Pitch perception: a dynamical-systems perspective, PNAS, 98(9), 4855–4859, 2001.

Chechetkin V.R., Block structure and stability of the genetic code, J. Theo. Biol., 222, 177–188, 2003.

Corron N.J. and Pethel S.D., Phys. Lett. A, 313, 192, 2003.

Crick F.H.C., The origin of the genetic code, J. Mol. Biol., 38, 367–379, 1968.

Di Giulio M., On the origin of the genetic code, J. Theor. Biol., 191, 573–581, 1997.

Fordsyke D., Are introns in-series error-detecting sequences? J. Theor. Biol., 93, 861–866, 1981.

Fuss J.O. and Cooper P.K., DNA repair: dynamic defenders against cancer and aging, PloS Biol., 4(5), 899–903, 2006.

Geronimi N., *Giochi Matematici del Medioevo*, Bruno Mondadori Editori, 2006.

Gonzalez D.L., Syncronization and Chaos in Nonlinear Oscillators, (in Spanish), PhD thesis, Universidad Nacional de La Plata, 1987, 393 pp.

Gonzalez D.L., Can the genetic code be mathematically described? Med. Sci. Monitor, 10(4), HY11–17, 2004.

Gonzalez D.L. and Rosso O.A., Qualitative modeling of complex biological systems, Proceedings of the Second Italian-Latinamerican Meeting of Applied Mathematics, Rome, 1997, pp. 132–135.

Gonzalez D.L. and Zanna M., Una Nuova Descrizione Matematica del Codice Genetico, Sistema Naturae, Annali di Biologia Teorica, 5, 219–236, 2003.

Gonzalez D.L., Giannerini S., and Rosa R., Detecting structure in parity binary sequences: error correction and detection in DNA, IEEE Eng. Med. Biol. Mag., 25(1), 69–81, 2006.

Gusev A.V. and Schulze-Makuch D., Genetic code: lucky chance or fundamental law of nature? Phy. Life Rev., 1, 202–229, 2004.

Hao B. and Zheng W., *Applied Symbolic Dynamics and Chaos*, World Scientific, Singapore, 1998.

Hayes B., The invention of the genetic code, Comp. Sci., 86(14), 8–14, 1998.

Hatfield D.L. and Gladyshev V.N., How selenium has altered our understanding of the genetic code, Mol. Cell. Biol., 22(11), 3565–3576, 2002.

Hoppensteadt F.C. and Izhikevich E.M., Thalamo-cortical interactions modelled by weakly connected oscillators: could the brain use FM radio principles? Bio Syst., 48, 85–94, 1998.

Izhikevich F.M., Weakly connected quasi-periodic oscillators, FM interactions, and multiplexing in the brain, SIAM, J. Appl. Math., 59(6), 2193–2223, 1999.

Jimenez-Montaño, M.A., Protein evolution drives the evolution of the genetic code and viceversa, BioSyst., 54, 47–64, 1999.

Karasev V.A. and Stefanov V.E., Topological structure of the genetic code, J. Theor. Biol, 209, 303–317, 2001.

Knight R.D., Freeland S.J., and Landweber L.F., Rewiring the keyboard: evolvability of the genetic code, Nat. Rev. Genet., 2, 49–58, 2001.

Lehmann J., Physico-chemical constraints connected with the coding properties of the genetic system, J. Theor. Biol., 202, 129–144, 2002.

Lewin, B., *Genes VIII*, Pearson/Prentice Hall, New Jersey, 2004.

Liebovitch L.S., Tao Y., Todorov A.T., and Levine L., Is there error-correcting code in the base sequence of DNA? Biophys. J., 71, 1539–1544, 1996.

May E.E., Vouk M.A., and Bitzer D.L., Classification of *Escherichia coli* K-12 ribosome binding sites, IEEE Eng. Med. Biol. Mag., 25(1), 90–97, 2006.

May E.E. (ed), Special issue on communication theory and molecular biology, IEEE Eng. Med. Biol. Mag., 25(1), 28–97, 2006.

Mac Donaill D.A., A parity code interpretation of nucleotide alphabet composition, Chem. Commun., 18, 2062–2063, 2002.

Mojzsis S.J., Arrhenius G., McKeegan K., Harrison T.M., Nutman A.P., and Friend C.R., Evidence for life on earth before 3,800 million years ago, Nature, 384, 55–59, 1996.

Négadi T., Symmetry groups for the Rumer- Konopel'chenko-*sh*Cherbak Bisections of the genetic code and applications, Internet Elect. J. Mol. Design, 3, 247–270, 2004.

Ore O., *Number Theory and its History*, Dover Publications, New York, 1988.

Pethel S.D., Corron N.J., Underwood Q.R., and Myneni, K., Phys. Rev. Lett. 90, 254101, 2003.

Rumer Y.B., About the codon's systematization in the genetic code, (in Russian), Proc. Acad. Sci. U.S.S.R. (Doklady), 167, 1393, 1966.

Rzeszowska-Wolny J., Is genetic code error correcting? J. Theoret. Biol., 104(4), 701–702, 1983.

Schopf J. W., Microfossils of the early Archean apex chert: new evidence of the antiquity of life, Science, 260, 640–646, 1993.

*sh*Cherbak V.I., The symmetrical architecture of the genetic code systematization principle, J. Theor. Biol., 162, 395–398, 1993.

Schroeder M., *Number Theory in Science and Communication*, Springer Verlag, Berlin, 1986.

Smith J.M. and Szathmáry E., *The Major Transitions in Evolution*, Oxford University Press, New York, 1995, p.81.

Sweeney P., *Error Control Coding: From Theory to Practice*, Wiley, New York, 2002.

Watanabe K. and Suzuki T., Genetic Code and its variants, in Encyclopedia of Life Sciences, Wiley, Chichester, 2001. Available at: http://www.els.net/doi:101038.npg.els.0000810

Weberndorfer G., Computational Models of the Genetic Code Evolution Based on Empirical Potentials, PhD thesis, Wien University, 2002.

Wolfram S.A., *A New Kind of Science*, Wolfran Media, Illinois, 2002.

Yockey H.P., *Information Theory and Molecular Biology*, Cambridge University Press, New York, 1992.

Zeckendorf E., Représentation des nombres naturels par un somme de nombres de Fibonacci ou de nombres de Lucas, Bulletin de la Societé Royale ds Sciences de Liege, 41, 179–182, 1972.

Zhaxybayeva O., Statistical estimation of Rumer's transformation of the universal genetic code, ISSOL'96 Conference on the Origins of Life, 1996.

Chapter 7

The Arithmetical Origin of the Genetic Code

Vladimir *sh*Cherbak

Abstract Physics and chemistry are indifferent to the internal syntax of numerical language of arithmetic and, in particular, to the number system that this language employs. All they require from arithmetic is quantitative data. Absence of a privileged numerical system inherent to an object must therefore be a necessary condition of its natural origin. Recent research, however, has found an exception. That object is the universal genetic code. The genetic code turns out to be a syntactic structure of arithmetic, the result of unique summations that have been carried out by some primordial abacus at least three and half billion years ago. The decimal place-value numerical system with a zero conception was used for that arithmetic. It turned out that the zero sign governed the genetic code not only as an integral part of the decimal system, but also directly as an acting arithmetical symbol. Being non-material abstractions, all the zero, decimal syntax and unique summations can display an artificial nature of the genetic code. They refute traditional ideas about the stochastic origin of the genetic code. A new order in the genetic code hardly ever went through chemical evolution and, seemingly, originally appeared as pure information like arithmetic itself.

1 Introduction

Our recent research has shown that the science of calculation utilizing the nine Hindu-Arabic numerals and zero is embodied in the genetic code. In addition to the well-known triplet-amino acid symbols, the genetic code stores internally the fundamental symbols of arithmetic (*sh*Cherbak, 2003). They are: the zero, the decimal place-value or positional number system, and numerous equilibrated summations of nucleons in amino acid. A numerical symbol of the "Egyptian triangle" and strings of the cooperative symmetry, semantically interpreted as a special message,

Department of Applied Mathematics, al-Farabi Kazakh National University,
71 al-Farabi Avenue, Almaty 050038, the Republic of Kazakhstan,
e-mail: genecodelab@kazsu.kz; genecodelab@hotmail.com

M. Barbieri (ed.), *The Codes of Life: The Rules of Macroevolution.*   153
© Springer 2008

complete a list. The long search for the place-value number system crowned by the zero belongs to the history of the human mind. That quest lasted many hundred years and demanded, according to Oswald Spengler's (1922) figurative expression, "A Brahmin's Sensitive Soul" to invent the supreme abstraction that is the symbol for nothing now known as zero. What kind of object is the genetic code that it is governed directly by this symbol and by a number system with a zero conception? There seems to be but one conclusion: the genetic code is itself a unique structure of arithmetical syntax. The arithmetical syntax is separated from natural events by the unbridgeable gap between the fundamental laws of nature and the abstract codes of the human mind (Barbieri, 2005). Chemical evolution, no matter how long it took, could not possibly have stumbled on the arithmetical language and initialized the decimalization of the genetic code. Physics and chemistry can neither make such abstractions nor fit the genetic code out with them. It seems that the genetic code appeared as pure information like arithmetic did.

2   A *Stony Script* and *Frozen Accident*

The deciphering of Egyptian hieroglyphics on the Rosetta Stone by Jean-François Champollion is the most relevant example of code-breaking (Dewachter, 1990). The stele had three scripts, two Egyptian languages – hieroglyphic and Demotic – and ancient Greek. Champollion's knowledge of Greek and Demotic syntaxes and semantics enabled him to decipher the Egyptian hieroglyphic alphabet. Long before the rediscovery of hieroglyphics, scholars were cognizant of the fact that these were not just natural notches on a stony surface; they had to be artificial linguistic signs. This fact determined Champollion's approach. Fundamentally, his approach is not to be seen as a banal metaphor but a direct prototype of the approach we utilize in this work. Here we are interested in the relations between the juxtapositioned known syntax of the decimal arithmetic and the still unappreciated genetic code.

Before we begin to look at these relations we should point out that there is a significant difference between the Rosetta Stone and the genetic code: the origin of the genetic code has no evident proof. Researchers who studied this universal alphabet of life did not have the advantages, which Champollion had. After George Gamow's (1954) seminal paper on the genetic code, scientists sought a chemical explanation for its structure. Yet, despite their best efforts, their research yielded a surprising result. There is no plausible chemical logic to couple directly the triplets and the amino acids. In other words, the principles of chemistry were not the sought essence of the genetic code. In such a situation, Francis Crick (1968) concluded that the genetic code should be viewed as a *frozen accident* and molecular mediators – transfer ribonucleic acids (tRNAs) – were responsible for the coupling. The mediators transformed the old schema a *certain amino acid has necessarily to go with certain triplet* into the new one – *any amino acid can go with any triplet.*

Alphabets of man-made written languages couple both meaning and some material carrier into linguistic symbols. This is an artificial convention instead of physics and chemistry that serves as a combining force to incorporate a particular meaning with a corresponding carrier. Therefore, conventionality in alphabets and mediation in the genetic code free both of them from any restriction imposed by the laws of nature so they each can produce every kind of symbol. However, the genetic code, with its same theoretical ability to make abstract symbols, including arithmetical ones, remains the *frozen accident*. It is certainly a correct conclusion if one is guided by chemical logic. Incidentally, in the view of chemical logic the scripts of the Rosetta Stone with all its alphabets, syntaxes, and semantics also seems to be nothing else but a *stony accident* similar to the *frozen accident* of the genetic code. In other words, in his chemical description of the genetic code, Crick unintentionally described properties of an artificial alphabet as they would appear if the alphabet could be described in such strange notions.

It is likely that from the very beginning some essential elements of information within the code have been overlooked. We contend that these elements are abstract symbols, syntax, and semantics, which are in accordance with the oldest man-made language – arithmetic. Champollion's fruitful juxtaposition of two languages allowed Egyptology to cast a glance on events that took place three and a half millennia ago. The same approach might provide us with a clue to the hidden message that the genetic code has carried within itself for at least three and a half billion years.

3   A "Language of Nature"

Arithmetic is at the core of mathematics. Together with mathematics, arithmetic is metaphorically referred to as a "language of nature," even if arithmetic, and in particular its decimal syntax, were conceived by, and always reside, in our mind. Only written versions of the arithmetic become apparent externally as static symbols of manuscripts or as dynamic symbols of abaci and computers. In this chapter we apply the known arithmetical syntax to the genetic code just as Champollion juxtaposed the ancient Greek to the Egyptian alphabet. We do not go beyond the limits of elementary arithmetic that everyone studied at primary school. Calculations done are simple; they can be carried out even manually. We only refer to a few well-known abstract entities of arithmetic listed in Fig. 1.

Arithmetic begins with zero and natural numbers. The zero is the supreme abstraction of arithmetic. Its use by any alphabet, including the genetic code, can be an indicator of artificiality. But how can one identify the abstract entity of nothing in practice? Well, there is a simple and elegant method to detect the zero. It is of common knowledge that the natural numbers are aligned into a natural series 1, 2, 3, ... and so on endlessly. The zero occupies its own abstract position on the only flank of the natural series 0, 1, 2, 3, ... One can look for zero in that position inside the newly systematized genetic code in Sections 11 and 13.

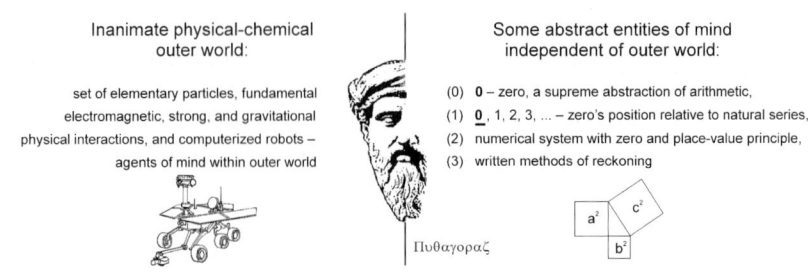

Inanimate physical-chemical outer world:	Some abstract entities of mind independent of outer world:
set of elementary particles, fundamental electromagnetic, strong, and gravitational physical interactions, and computerized robots – agents of mind within outer world	(0) **0** – zero, a supreme abstraction of arithmetic, (1) **0**, 1, 2, 3, ... – zero's position relative to natural series, (2) numerical system with zero and place-value principle, (3) written methods of reckoning

Πυθαγορας

Fig. 1 The inanimate outer world and living mind. The outer world is governed rigidly by immutable laws of nature. On the contrary, the human mind voluntarily makes its own conventional codes including the decimal syntax of arithmetic for the sake of information exchange and scientific activity. A vertical line demarcates lifeless substance from basic abstractions of arithmetic that are residents of mind. Computerized robots working in the outer world are equipped with instructions that use these abstractions in the form of written symbols.

The early Greek philosopher and mathematician Pythagoras looked for a numerical harmony in the Cosmos. His image here symbolizes abstract entities of mind and new arithmetical features of the genetic code. Pythagoras had proved a theorem that the sum of the squares of each of the two sides of any right triangle is equal to the square of the hypotenuse. The genetic code has within itself a numerical symbol of the "Egyptian triangle" that is by far the best-known example of a right triangle

The Hindus had devised a zero during the time when the text on the Rosetta Stone was written, i.e. almost 2000 years ago. They applied zero in the place-value number system and perfected it once and for all. The decimal system is a syntactic rule for symbolic writing of words whose meaning is the quantity of anything. These words are referred to as numbers. It is commonly accepted that people chose the radix or base ten for their number language voluntarily, being guided only by an anatomic feature – ten fingers. But this choice becomes more intriguing in the light of the decimal syntax discovered inside the code. Could it be that the decimal system through arranged decimally genetic code and through translated by such code genes predetermine the system's own choice recognized by people as a voluntary one?

It belongs to the world of culture, whereas any physical quantity described as a number is separated from this description. The counter assumption is that some "elements of numbers" are inherent to all things. This fancy is very ancient and belongs to the so-called Pythagorean doctrine. Modern science, however, shares Plato's views that numbers exist as abstract ideas apart from physical bodies. Such understanding establishes the distinction between mathematics and physics. Our subject matter skips knotty philosophical problems focusing instead on the syntax of the numerical language of arithmetic. Its symbolical notations done by the rule of some positional number system could only arise from information experimentations of the mind. Therefore, we can consider the positional system as indicator of artificiality. The fact that the zero is a component part of this system becomes an even clearer indicator.

Let us consider how indicators work. For instance, everyone uses digital signs and syntactic rules of arithmetic to do number notations and to perform certain procedures, say, summing up. Such calculations looking for quantitative properties of some physical object are routine acts. If some additions were originally performed in the decimal system, it would be an absurd idea to repeat the same calculations, say, in the binary system because important data would be lost. Indeed, physics and chemistry are indifferent to the internal syntax of arithmetical language. All they require from arithmetic is quantitative data. The absence of a privileged numerical system is therefore usually indicative of natural objects.

Computers are quite different in this matter. They are internally based on one privileged numerical system – binary – but communicate externally and can handle data in other systems. So, this time it is not an absurd idea at all to care about potential loss of data at investigation of a computer design when using systems other than the binary one. The presence of a privileged numerical system(s) in computers is unambiguously indicative of an information artifact. In a hypothetical case, if an issue of the origin of the robot-geologist in Fig. 1 was raised at the *Meridiani Planum* area on Mars, the zero-one system of the robot's microprocessors would instantly bring the issue to resolution. In a similar manner, the numerals written in the Egyptian scale of notation, say, in the phrase "10 gold crowns of the Pharaoh" would additionally confirm artificiality of the Rosetta Stone, if there were doubts to its origin. Of course, any given artificial object does not necessarily have inside itself a privileged numerical system. However, if this is the case, such a system indicates an indisputable artifact.

4   Prime Number 037

As regards the genetic code, one should juxtapose the abstract syntax of the arithmetical language with some well-defined sets of units in the genetic code. In other words, these units should be added and their sums within each set should be written temporarily as we are accustomed in decimals. Theoretically, some original number notations should replace these decimals as soon as we uncover the genetic code's primordial number system, if there is one. The art of number notation since the Babylonian sexagesimal system till our computer codes should help us in solving this problem. However, anyone who follows this approach would find unexpectedly that the choice among a variety of number systems is no longer necessary. Surprisingly, the genetic code really privileges a number system and, even more unusual, the system is the decimal one. This means that the decimal system is probably of the same age as the genetic code.

Arithmetic provided the genetic code with artifactural differences through a particular feature of one of its criterions for divisibility. As soon as one discovered the decimal system, all criterions of divisibility arose instantly and became attributes of the system. The decimals divisible by *prime number (PN)* 037 in Fig. 2 show the particular feature that attracts the attention of a reckoner looking for some numeration system in the genetic code. This feature exists in the decimal system and vanishes in other systems. For instance, the particular notations of decimals **111,**

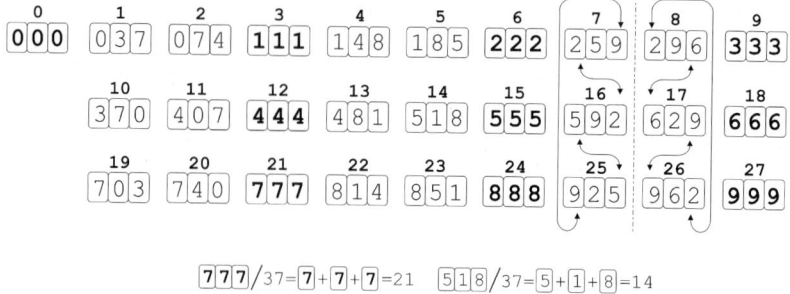

$$\boxed{7}\boxed{7}\boxed{7}/37=\boxed{7}+\boxed{7}+\boxed{7}=21 \qquad \boxed{5}\boxed{1}\boxed{8}/37=\boxed{5}+\boxed{1}+\boxed{8}=14$$

Fig. 2 The place-value decimal system represented through digital symmetry of the numbers divisible by *prime number (PN)* 037. This arithmetical syntactic feature is an innate attribute of the genetic code. The *PN* 037 notation with a leading zero emphasizes zero's equal participation in the digital symmetry. Numbers written by identical digits are devised by *PN* 037 × 3 = 111 and appear regularly. The sum of identical digits within each notation gives the quotients of these numbers divided by *PN* 037. Analogous summation for the numbers written by unique digits is equal to the central quotient in the column. This feature allows one to perform integer division by *PN* 037 through addition. Numbers written by unique digits are linked by cyclic permutations in the columns. The directions of permutations in the neighboring columns are mirror symmetrical.

Consider the decimal criterion for divisibility by *PN* 037 on decimal number 13901085 as an example. Use a triplet frame to apportion digits among digital triplets 139, 010, and 850. Note the triplet frame is indifferent about its three possible positions: insert zero(s) at the flank(s) to complete the digital triplets. The three-digit checksum of the digital triplets 139 + 010 + 850 is equal to **999**. The particular notation **999** symbolizes divisibility by *PN* 037. Make a carry from thousand's unit, if any, into unit's unit. This operation preserves both the three-digit notation of the checksum and the criterion symbolism. For example, decimal number 21902631 has the checksum 021 + 902 + 631 = 1554. The carry into unit's unit changes the checksum into 001 + 554 = **555**, which is the three-digit symbol of divisibility by *PN* 037. Therefore, only a three-digit adding machine is needed to reveal the divisibility by *PN* 037 of any number, irrespectively of how large this number would be. Add or take away one *PN* 037 additionally to those checksums that do not take the particular notation after the triplet summation. A subsequent appearance of the particular notation confirms the divisibility by *PN* 037, whereas its absence is evidence of indivisibility

222, and **333** look like ordinary numbers 157, 336, and 515 in the octal system. The feature not only excites our sensation of beauty but may simplify some computational procedures too. Moreover, decimalization of the genetic code may be a special case of the general computational power of genomes and their molecular machinery. In fact, the only reason for a number system to appear is for arithmetic calculations. We shall return to this issue in the final Section 17.

5   The Genetic Code Itself

Based on the above considerations, we present here a sort of a brief survey of the genetic code. The record in Fig. 3 is the universal genetic code. The genetic code is the only known alphabet that arose without participation of the human mind. What is more, the code's origin became a synonym of the origin of life. In fact,

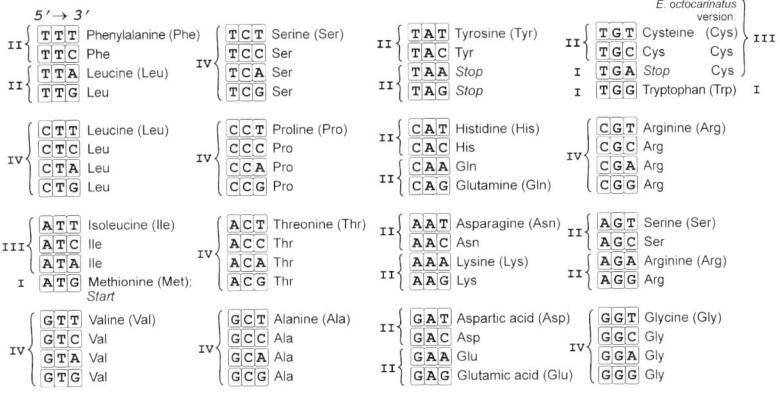

Fig. 3 The universal genetic code. The genetic code contains 64 triplets of four DNA nitrogenous bases Thymine, Cytosine, Adenine, and Guanine onto which 20 amino acids and 2 syntactic signs, *Start* and *Stop*, are mapped. There are two kinds of bases, i.e. pyrimidines (T and C) and purines (A and G). Each strand of DNA molecule and its axially oriented triplets of bases have the 5′ and 3′ ends. The unique direction of triplet reading is 5′ → 3′. The triplets that code for identical amino acids form a synonymic series denoted by initial brace. The names of the 20 amino acids and their trigram abbreviations are specified for each synonymic series. The amino acid molecules are shown in Fig. 4. The quantity of triplets in the series is referred to as degeneracy. The degeneracy is described by the Roman figures. There are the standard degeneracies IV, III, II, and I in the genetic code. Three amino acids serine (Ser), leucine (Leu), and arginine (Arg) have each two separate synonymic series with degeneracy II and IV, but not single series with degeneracy VI (Rumer, 1966). Beginning and termination of a protein synthesis is called *Start* and *Stop*. The sign *Start* is combined with amino acid methionine (Met). There is no amino acid for the sign *Stop* in the genetic code.

The universal genetic code has two triplets TGT and TGC that code for amino acid cysteine (Cys) and one triplet TGA that codes for *Stop*. There is another version of the genetic code conditionally called *Euplotes octocarinatus* code version (Marshal et al., 1967; Meyer et al., 1991; Grimm et al., 1998). Its triplets TGT, TGC, and TGA code for cysteine and are denoted by the closing brace. The *E. octocarinatus* code version is more symmetrical due to the same degeneracy within both 5′AT and 5′TG solid series. There are 24 synonymic series in the *E. octocarinatus* genetic code version. Both code versions are considered in this chapter

there is no way to write or read any gene when no code is available. Thanks to its immutability, the universal genetic code is the most accurate information messenger from the time of genesis to the present. It is generally accepted that the code only translates any gene from a triplet sequence of DNA into an amino acid sequence of a protein. This routine work, however, is not its unique capability.

A triplet as a universal codon is a primary attribute of the genetic code. A single hydrogen atom of the glycine side chain and the tryptophan double ring as well as an "emptiness" of syntactic sign *Stop* are coded universally by the triplet of bases. In fact, the constant length of the codon is a prerequisite to equip the code with regular degeneracy. In its turn, the degeneracy proves to be one of the crucial systematization parameters of the new order in the genetic code.

The degeneracy possesses numerical values. These are integer ordinals IV, III, II, and I. Two rules govern degeneracy. The first says that any degeneracy resides within a solid set of four triplets beginning with the same two bases. For example, glycine in the bottom right-hand corner is coded by the solid set of four synonymic triplets GGT, GGC, GGA, and GGG. It is said that degeneracy of glycine synonymic series equals IV. There are eight solid series with degeneracy IV in the genetic code. The second rule concerns the synonymic series of degeneracy II named the broken series. A pair of such series halves its four solid triplets so that these triplets form two pairs ending with either pyrimidines or purines. For example, the solid set in the top left-hand corner is halved between phenylalanine and leucine. Phenylalanine possesses the TTT and TTC pair ending with pyrimidines T and C. Another TTA and TTG pair ending with purines A and G is passed to leucine. There is also a kind of broken series with degeneracy III and I. Methionine and tryptophan each are coded by a single triplet ending with base G.

Note here that triplet ATG, which codes for methionine, is simultaneously a syntactic sign *Start*. It begins any protein gene. The semantically antisymmetrical syntactic sign *Stop* terminates genes. Its synonymic pair of triplets TAA and TAG carefully follows the second rule though this pair codes for "emptiness." It is not a coincidence that these two syntactic signs are special symbols among routine amino acid coding. Both these signs appear below as semantic cues inserted into the genetic code similar to the cartouches with Pharaohs' names inserted in the hieroglyphics of the Rosetta Stone (see Section 13.2).

6   Rumer's Transformation

Yuri Rumer (1966) was the first who saw an element of information hidden in the genetic code. Rumer gathered the triplets of the solid series into one set and the broken series into the other as shown in Fig. 4a. Both sets have the same number of triplets – 32 each. This halving provides an opportunity to map triplets in a *one-to-one* manner. A mapping was realized with the help of Rumer's unique transformation T ↔ G, C ↔ A or, in otherwise notation, TCAG → GACT. The transformation maps each triplet of degeneracy IV set onto a certain triplet within degeneracy III, II, and I set, and vice versa. As an example, consider the GGT triplet in the top left-hand corner. Rumer's transformation maps this triplet onto the TTG triplet that resides in the degeneracy III, II, and I set.

It is easy to see that there are two more transformations of the same type T ↔ C, G ↔ A and T ↔ A, C ↔ G or TCAG → CTGA and TCAG → AGTC. They each make half of the same that Rumer's transformation makes alone. Because of this, they are referred hereinafter to as 50% transformations. Though each of these three transformations can act independently or can even be absent, they constitute an ordered assembly (*sh*Cherbak, 1989a). One can arbitrarily substitute the actual degeneracy for another one to simulate hypothetical evolutionary changes in the genetic code. Substitutions in an overwhelming majority of cases result in the destruction of that assembly.

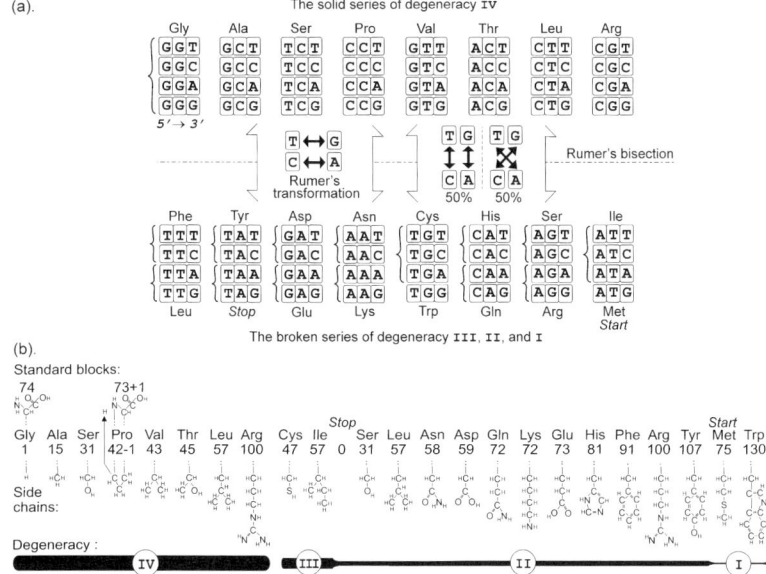

Fig. 4 Rumer's bisection and transformation of the genetic code and Hasegawa's and Miyata's antisymmetrical correlation of the degeneracy and amino acid nucleon numbers. (a) Rumer halved the 64 life-size triplets into two sets of degeneracy IV and III, II, and I. Each set maps its own 32 triplets onto the triplets of the opposite set either by Rumer's unique transformation or by two other 50% transformations. (b) The amino acid categorization by degeneracy. The amino acids are gathered correspondingly into two sets of degeneracy IV and III, II, and I. The degeneracy of these sets is denoted by the line of variable thickness. Generally, the smallest amino acids are concentrated in the set of highest degeneracy IV, whereas the mid-size and large amino acids occupy degeneracy III, II, and I set. The 20 amino acids are shown as free neutrally charged molecules. The amino acids have two specific component parts each. These parts are the standard block and individual side chain. They are covalently bound within a whole amino acid molecule. There is an imaginary cross-cut in the present research that virtually separates the standard blocks and the side chains of these 20 amino acids. Such virtual act reveals the new arithmetical order in the genetic code. In the figure we showed as an example for glycine (Gly) a standard block common for nineteen amino acids. The standard block of the amino acid proline (Pro) is the only exception. The arrow with a hydrogen atom represents an imaginary borrowing that standardizes the proline block. The bisected genetic code is represented by *E. octocarinatus* code version; the revealed regularities are also true for the universal genetic code

7 Hasegawa's and Miyata's Nucleons

The amino acids and syntactic signs followed the bisected triplets and gathered correspondingly into two own sets as shown in Fig. 4b. However, Rumer shelved the amino acid sets. Indeed, a mixture of diverse chemical structures together with syntactic signs *Start* and *Stop* is undesirable for a general arithmetical approach. One should substitute some common units for that mixture to apply arithmetic.

The relevance of amino acid mass and codon distribution had been recognized soon after the code decipherment and still remains in the sphere of active research interest (e.g. Schutzenberger et al., 1969; Di Giulio, 1989; Taylor and Coates, 1989; Chiusano et al., 2000; Downes and Richardson, 2002). Generally in the genetic code: the greater the degeneracy the smaller the amino acid size. This rather rough correlation leaves a gap for speculations that in the evolutionary history of the genetic code most of these small amino acids (due to their prevalence in number) had captured the biggest series with degeneracy IV. As shown in Section 12, this correlation is nothing but an external representation of the new arithmetical order in the genetic code. Nevertheless, Hasegawa and Miyata (1980) supplied the correlation with an integer-valued parameter – a nucleon number. They noted once again that the degeneracy and nucleon number of the 20 amino acids correlate antisymmetrically. After Hasegawa and Miyata, we have used a nucleon as an embodiment of arithmetical unit inside the genetic code.

"Nucleon" is the common name for two nuclear particles – a positive charged proton and an uncharged neutron. The most common and stable isotopes are taken to calculate the nucleon numbers of the 20 amino acids. For instance, the nucleon number of the glycine and tryptophan side chain equals 1 and 130, respectively. The only exception from the general structure of amino acids is provided by proline. It holds its own side chain with two bonds and has one less hydrogen inside the standard block. However, an imaginary borrowing of one nucleon from the proline side chain in favor of its block brings the block nucleon number to the standard $73 + 1 = 74$, whereas the side chain nucleon number becomes $42 - 1 = 41$. The syntactic sign *Start* is associated with the amino acid methionine whose nucleon number is 75. The nucleon number of the syntactic sign *Stop*, which has no associated amino acids, is designated as zero. Note that this assertion has introduced zero into the genetic code arithmetic. Both imaginary acts – the cross-cut and the borrowing – are an artificial obstacle insurmountable to natural events preventing them from establishing the new arithmetical order in the genetic code. Such virtualization of the genetic code acts regularly in what follows.

We have chosen the common arithmetical units and zero instead of the mixture of chemistry and linguistics. Finally, we should specify some particular sets for these units. Reasonably, we can assume that the genetic code as a whole is the best set for an initial research.

8   A Real-life Global Balance

An exact equilibration of the nucleon sums within the well-defined sets is one of the two standard representations of the artifactural differences of the genetic code. The unique global balance in Fig. 5 makes use of molecular residues of the amino acid standard blocks from an internal part of a protein and charged ions of side chains from the real-life environment of cytoplasm. The closed circle of the block

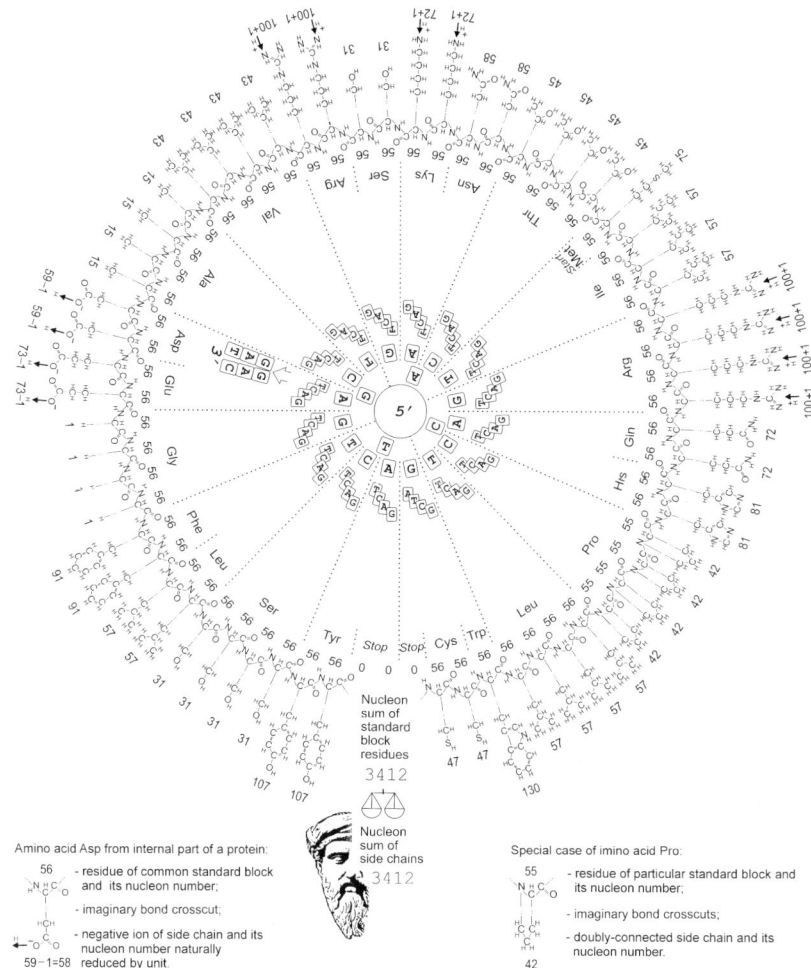

Fig. 5 The real-life global balance of the genetic code. The nucleon sum of the standard block residues and the nucleon sum of the ionized and protonated side chains draw up the balance. The amount of summands is determined by degeneracy. Pythagoras' image symbolizes the balanced summations. There is the well-known radial representation of the genetic code 64 triplets. The standard direction of reading 5′ → 3′ coincides with the direction outwards from the center. An example of reading is shown for two triplets coded for aspartic amino acid (Asp).

There are four amino acids whose side chains acquire charges in the real-life environment of cytoplasm. Aspartic acid and glutamic acid (Glu) are fully ionized, i.e. lose one proton each. Arginine (Arg) and lysine (Lys) are fully protonated, i.e. join one proton each. Both types of changes are denoted with arrows. To illustrate the use of graphical symbols there are the ionized aspartic acid and the particular proline (Pro) molecule at the bottom. The block residues from an internal part of a protein form a ring-shaped polymer using peptide bonds.

The real-life global balance is valid for the universal genetic code version. This balance has been found by the author and associates (Kashkarov et al., 2002) simultaneously with Downes and Richardson (2002) residues and their virtually cross-cut side chains draw up the balance. Such kind of equilibration is referred hereinafter to as the *block-to-chain* type.

Seems, this was the only way for the global balance to appear in the distant past. That was the same way we have revealed it now, i.e. by the 128 successive arithmetical summations. Assuredly, the nucleon number of an amino acid is an equivalent of its molecular mass. But it is hard to believe some natural events could balance the genetic code in another way than by the arithmetical summation. The detached amino acids never gather to be "weighed" at once. It is necessary to be "aware" of the actual degeneracy to repeat accordingly certain amino acid masses when weighing, e.g. the 57-nucleon isoleucine (Ile) side chain should be thrice repeated. Another result of those events should be in accurate cuts of bonds between amino acids blocks and chains with weighing of the ends that are easy to do in imagination, but not in reality. Intermediate weighing results should again be stored somehow in analog form. So when considering the origin of the nucleon equilibration in the genetic code, arithmetical summation is more realizable than physical weighing. Incidentally, it is shown in the sections below that the imaginary borrowing and some privileged numerical system would end the issue of the physical equilibration inside the code.

The closed ring of the block residues in Fig. 5 gives equal status to each of the 61 amino acids. The natural $-NH_2$ and $-COOH$ molecular groups on the opposite ends of the disclosed ring destroy the global balance. It is of common knowledge that there are circularly closed DNA molecules, but no circular proteins in a living cell. Because of such virtual closing the global balance can be placed among other virtual balances discovered throughout the genetic code in abundance. Nevertheless, the real-life chemical conditions are integrally embodied into the genetic code due to this global equilibration.

9   A Virtual Global Balance

In our considerations below we deal with the free neutrally charged amino acid molecules. The second standard representation of the artifactural differences is an accurate superimposition of the decimal syntax onto equilibrated nucleon sums. The virtual global balance is illustrated in Fig. 6 (we use the attribute "virtual" because of the imaginary borrowing). Both global equilibria have the same *block-to-chain* type, but the ways of balancing are different. The previous balance means that all the chains together cannot counterbalance all the blocks as soon as the nucleon number of a single block exceeds 56. This takes place in the virtual global balance. Its free neutrally charged amino acid molecules have the 74-nucleon standard blocks. This means that the chains all over the code can counterbalance only a part of their blocks. The genetic code elegantly coped with both exact balancing and uniform labeling of balanceable parts through decimalism.

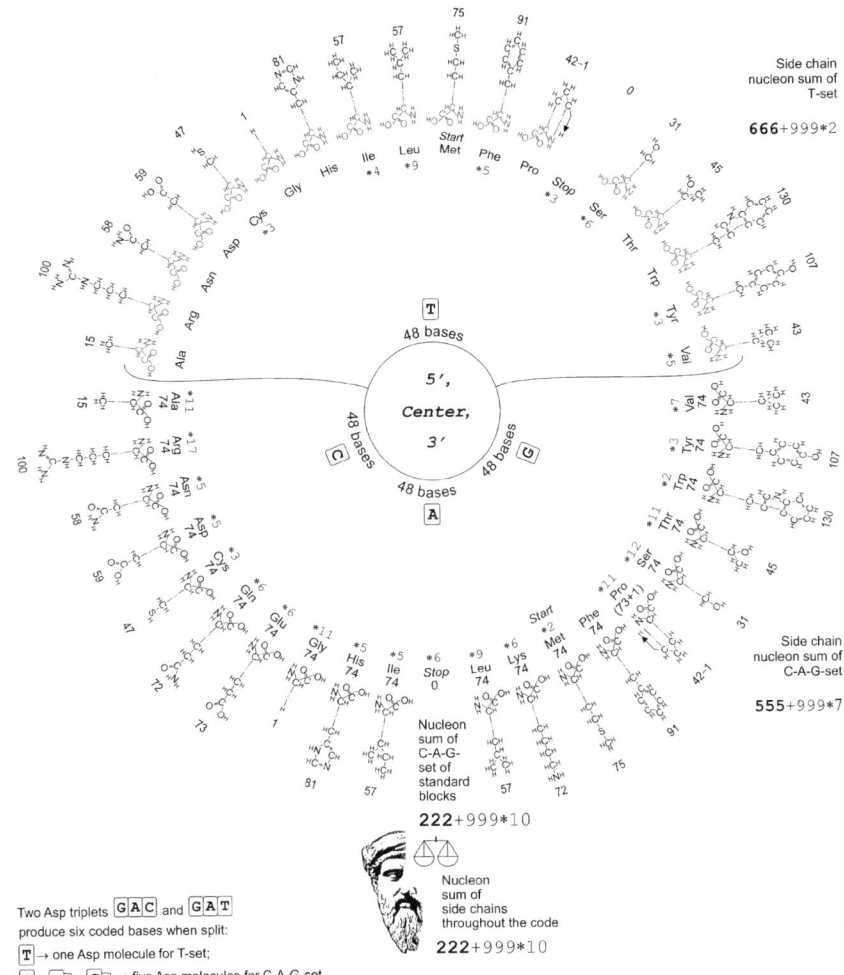

Fig. 6 The virtual global balance of the genetic code (*sh*Cherbak, 1999). There are the nucleon sum of an especially labeled part of the standard blocks and the nucleon sum of the side chains all over the code that draw up the balance. Pythagoras' image symbolizes balanced summations and decimalism. The alphabetized amino acids are gathered within the *T* set and *C-A-G* set. Besides the nucleon numbers of the standard block and side chains, the multiplier is assigned for each amino acid. Six separate bases composing two aspartic acid (Asp) triplets show the simple way to apportion the twenty amino acids between two sets.

In order to demonstrate the particular decimal notations, all numbers, divided by *PN* 037 and larger **999**, are written as decimal 2664 ≡ **666** + 999 × 2. According to the criterion of divisibility by *PN* 037 one can write down the number 2664 as three-digit sum 002 + 664 = **666** or represent shortened notation of the balance **222**-and-**222**; some register could keep multiplier 2 or 10, respectively, for 999.

The balance is valid for the universal genetic code version

The second global balance exists, so to speak, at a "quark" level. The "quarks" are the triplet bases. The 64 triplets – when they are split – make $64 \times 3 = 192$ separate bases of four types. The bases gather into four sets with 48 identical bases inside each set. The amino acids and syntactic signs are mapped onto these items by the simple rule that is demonstrated for the aspartic acid doubly degenerated series. Such rather abstract acts are justified by a unique fact. Only the T set alone and the joint C-A-G set show the particular number notations of decimals. These are their side chain nucleon sums **666** + 999 $\times$ 2 and **555** + 999 $\times$ 7, respectively. The genetic code labels in such manner the balanceable part of the standard blocks. The joint A-G-C set contains 138 standard blocks, whose nucleon sum is equal to **222** + 999 $\times$ 10. On the other hand, the nucleon sum of the amino acid side chains in both sets – the split code on the whole – is exactly the same number.

The decimalism refutes the traditional notion about a natural ordinariness of the genetic code. We should note here that the imaginary borrowing converts equilibrated nucleon sums into virtual quantities. A virtual quantity is out of any control of natural events. They would control only the natural 73-nucleon block and 42-nucleon chain of the proline molecule. For example, these blocks and chains change the global balance **222**-and-**222** in Fig. 6 into disordered decimals 211-and-234. The particular decimal notations would be also ruined as soon as the natural blocks and chains are in use. Therefore, any suggestion that natural events could balance and decimalize the virtual genetic code seems fairly improbable. Instead, arithmetic and its chief operation, addition, seem to remain the only tool capable of achieving this.

Champollion checked fidelity of interpretations of hieroglyphs when they had been repeated in another hieroglyphic context of the Rosetta Stone script. In our case, the juxtaposed arithmetic language and the genetic code have revealed the decimal syntax of primordial summations. This unprecedented phenomenon should be confirmed in other ordered representations or "contexts" throughout the genetic code.

10   Arithmetic in Gamow's "Context"

Gamow (1954) was the first who predicted the genetic code itself, its 64 base triplets, 20 canonical amino acids, and average degeneracy. Being a theoretician in the physical sciences, he searched for a universal formula of the genetic code. He stopped that search after biochemists had deciphered the code (Nirenberg et al., 1965). In the meantime, the universal formula exists and is described below.

The same numbers – 20 amino acids and 20 combinations of four bases, three at a time – encouraged Gamow to suggest the very first stereochemical model of the code origin. It turned out that his model is incorrect, but a remarkable coincidence remained. Now, Gamow's 20 combinations give rise to the numerous balances and their decimal syntax. The combinations – placed onto triangular substrates in Fig. 7 – become triplet makers now. Each combination makes its triplets – base after base – by circular motion or spin around the substrate in two opposite directions. A pair of circular arrows symbolizes these spins. New *Spin* → *Antispin* transformation as well as Rumer's and two 50% transformations apportion the 64 life-size triplets relative to the central axis (see for detail *sh*Cherbak, 2003).

According to Gamow's research approach, the combinations were grouped into three sets. This was done according to the composition of identical and unique bases, regardless of their type and position inside a triplet. The joint first and second sets of triplets having three identical and three unique bases show equilibrated nucleon sums **666 +** 999 of whole molecules in Fig. 7a. There is also a new *chain-to-chain* balancing mode in this balance with equilibrated sums 703, i.e. *PN* 19 × *PN* 037 (see Section 14).

The third set of triplets with two identical and a unique base are shown in Fig. 7b. There are two subsets bisected by the vertical central axis whose triplets have either two identical pyrimidine or two identical purine bases. These subsets form a balance with **999** nucleons in each equilibrated arms. The equilibrium appears again between the two sets of side chains. No standard blocks take part in the *chain-to-chain* type of balancing. However, the imaginary borrowing acts precisely even in absence of standard blocks.

Note that the vertical central axis has bisected the subsets in a way that disregards the type of unique base. Restoring symmetry the horizontal central axis bisects the same triplets using the type of unique base and, this time, disregarding the type of identical ones. Now, the side chain nucleon sum of the triplets whose unique base is pyrimidine equals to **888**. Therefore, decimalism is free to act without balance too.

Let us return to the **999**-and-**999** balance. Its right arm is thrice equilibrated. These three regularly organized arms consist of **333** nucleons each. One of the arms – complete line – has a cloned line written by synonymic triplets at the bottom of Fig. 10b. The cloned line of the synonymic triplets code for the same amino acids, hence, for the same **333** nucleons. The cloned line splits the corresponding subset into summands **333 + 777**. These summands can be represented as a balance **777 + 777** of the *block* + *chain-to-chain* type: add the **333** side chain nucleons to its own 6 × 74 = **444** nucleons in the standard blocks.

Besides Gamow's "context" there are a few standard representations among complementary life-size triplets categorized by total quantities of their hydrogen bonds. These quantities can be equal to integers 9, 8, 7, or 6. There are 24 triplets having seven hydrogen bonds in the genetic code. Their total sum of side chain nucleons can split regularly into balanced parts **333 + 333 + 333 + 333** or **444 + 444 + 444**.

Metaphorically, in our denary dialect of the arithmetical language we have written down the names of certain cardinal numerals of the genetic code. This is the answer the genetic code gave: These inscriptions are their inborn names. The names have been given them in the time of genesis. Now, in their native numerical language these names are reproduced in written form again because the current dialect turns out to be identical to that primordial numerical language.

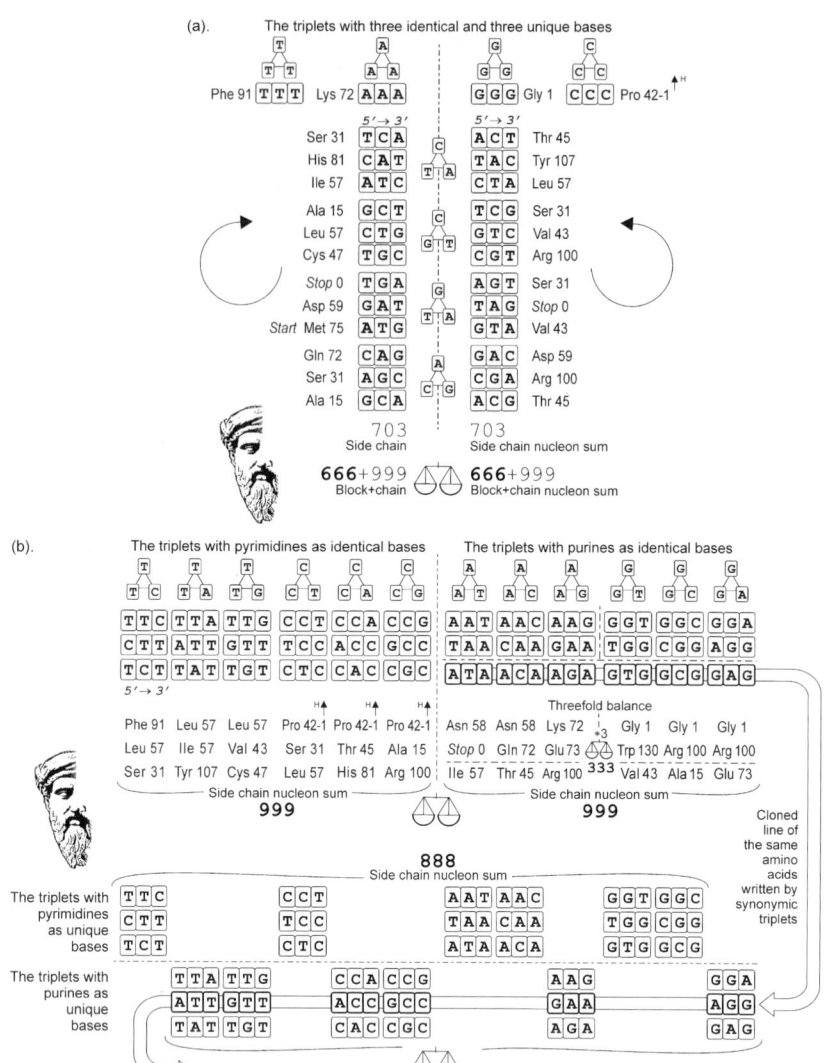

Fig. 7 Gamow's division of the genetic code. Pythagoras' images symbolize balanced summations and their decimal syntax. The proline imaginary borrowing turns balanced sums into virtual values.

Gamow's 20 combinations are placed on triangular substrates. Each of the combinations makes certain triplet(s). The combinations are divided into three sets depending on their base composition. Four combinations with identical bases make four triplets for the first set. The other four combinations with unique bases make two dozens of the triplets for the second set. Twelve remaining combinations make 36 triplets with two identical bases and a unique one and form the third set (*sh*Cherbak, 1996). (a) There is a balance that comprises the first and second sets. The *Spin* → *Antispin* transformation does not affect the triplets of the first set, but it apportions the triplets of the second set. The balance appears when Rumer's transformation is paired with 50% transformation T ↔ C, A ↔ G in the first set and with 50% transformation T ↔ A, C ↔ G in the second set. Both sets are coaxially placed in one of the alternative positions. The balance is true for the universal genetic code

11 The Systematization Principle

The systematization principle of the genetic code is a rule that arranges a code calligramme. A "calligramme" was invented in 1914 by the French poet Guillaume Apollinaire (1980). He arranged letters and words of his written texts in the two-dimensional space of a book sheet instead of a traditional line. Thereby he created the so-called calligrammes in which texts, semantics, visual images, and symmetries are joined together. A calligramme surpasses a common text in an information capacity and lucidity. Thanks to ancient calligraphers, Champollion got a ready calligramme of the Egyptian hieroglyphic text. Regarding the genetic code, at first glance there seems to be no recognizable calligramme. The best-known genetic code tabular and circular calligrammes in Figs 3 and 5 have, however, a parameter that people chose voluntarily – the conventional sequence order of four bases T, C, A, and G. Being guided by instructions outgoing from within the code, we should try to reproduce its primordial calligramme. Rumer's bisection in Fig. 4a appears to be a most useful half-finished product in that respect. We present its final upgrade in Fig. 8.

The centrosymmetrical flow chart shows equilibrium of the systematization principle formulation in Fig. 8a. The first object of systematization is the sets of the same degeneracy series. There are exactly four such sets in the genetic code according to four degeneracies IV, III, II, and I. The second object is the synonymic series of the same degeneracy inside each of these sets. Each of these objects is a component part of another one. The crossing arrows symbolize such interplay. There is only one general condition – the objects should be aligned with a monotonic and opposite directed changes of two parameters. These are the degeneracy number of the sets and the nucleon number of the amino acids or syntactic signs. To meet this condition the degeneracy numbers of four sets decrease from left to right in Fig. 8b. A line of variable thickness symbolizes this change. At the same time, the amino acid nucleon numbers – inside each set – are on the decrease in the opposite direction, i.e. from right to left. Diminution in size of amino acid's trigram abbreviations symbolizes this change.

The final calligramme is referred to as *the cooperative symmetry of the genetic code* (*sh*Cherbak, 1988). The calligramme is invariant under mirror symmetry. That means that the calligramme does not change if the opposite directions of the parameter changes were chosen. Note that the principle combines into a single whole Rumer's series bisection and Hasegawa's and Miyata's series alignment.

Fig. 7 (continued) (b) Regularly apportioned triplets of the third set demonstrate the decimal syntax and equilibration of different types. The right balanced arm of the **999**-and-**999** balance contains the triple balance. Its two of the three symmetrically subdivided arms are a pair of halved lines (a vertical dash-line indicates a halving), the third arm is a complete line (shown by wide bordering). The balances are true for the universal genetic code and its *E. octocarinatus* version

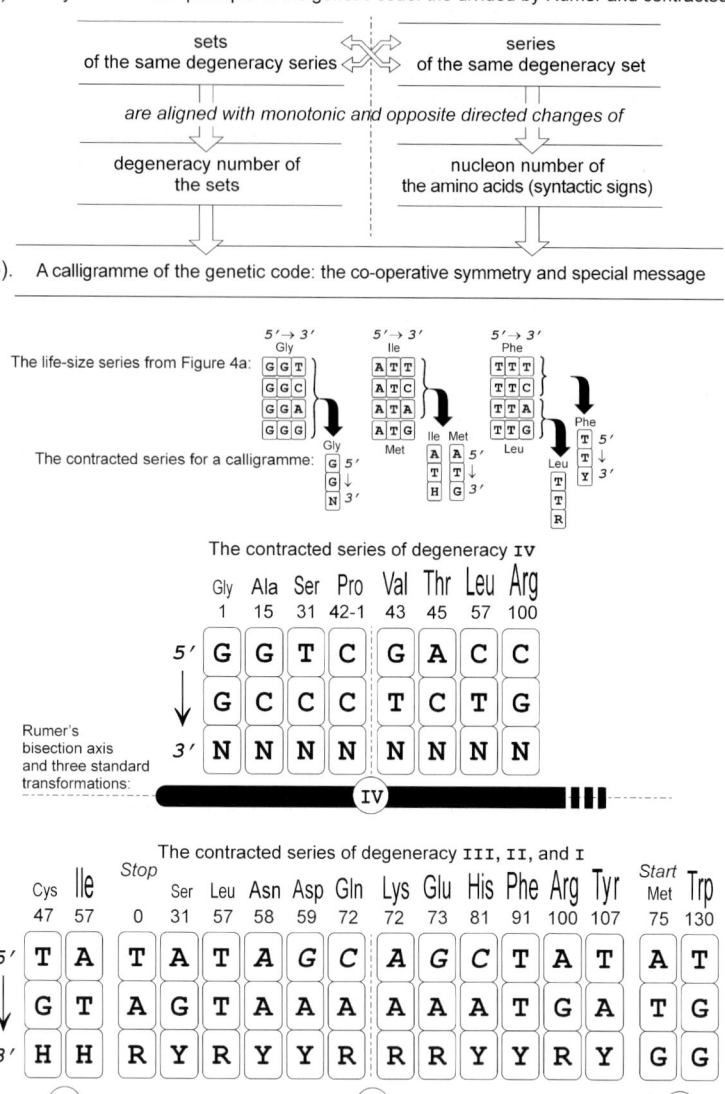

Fig. 8 The systematization principle and a calligramme of the genetic code (*sh*Cherbak, 1993a). (a) The flow chart of the systematization principle. (b) The principle requires contractions of the life-size series located in Fig. 4a. There are degeneracy-dependent 3′ triplet base contraction and 90° turn of the glycine (Gly), isoleucine (Ile), methionine (Met), phenylalanine (Phe), and leucine (Leu) synonymic series as an example. The synonymic series of glycine is contracted four times. The contraction symbol N = {T, C, A, G} substitutes for the four glycine 3′ triplet bases. There are also a thrice contraction symbol H = {T, C, A} and a pair of double contraction symbols for pyrimidines Y = {T, C} and purines R = {A, G} in the contracted genetic code. The methionine series retains its life-size. The 90° turn makes contracted series upright. The upright position of the contracted series describes graphically new symmetries of the calligramme. The genetic code is represented by its *E. octocarinatus* version

The zero of the *Stop* signs was previously trivial summands. Now, it forms the calligramme just as an ordinal number does. On that ground, the zero occupies its formally predetermined position at the beginning of the natural series and places its own triplet series at the flank of the degeneracy II set. We shall return to this important point when we discuss the cooperative symmetry.

12   The "Egyptian Triangle"

The 23 amino acids and one *Stop* sign form a basis of the calligramme. Lying under own contracted triplet series they arrange them in the cooperative symmetry. Let us consider this basis before we focus on the information content of the calligramme.

The amino acids and syntactic signs – i.e. those divided and set aside by Rumer – produce the numerical symbol of the "Egyptian triangle" and a new *block-to-chain* balance in Fig. 9. As before, the decimal syntax is exactly superimposed onto equilibrated nucleon sums.

We have interpreted so far the decimal syntax of separate numbers. However, in addition to this syntax, the genetic code displays also a combined symbol composed of three numbers with remarkable proportions. These numbers are $9 + 16 = 25$ $(3^2 + 4^2 = 5^2)$. It is worth noting that these numbers can exist as a symbol, say, in Pythagoras' mind, but never anywhere in the outer world. The reasons, which encourage us to interpret this symbol inside the code, are presented in Section 13. One possible interpretation is: "The 'Egyptian triangle'—the world-famous symbol of human cultural medium from ancient to our time." Its possible meaning could be: "Artifactural difference of the genetic code and symbolization of a two-dimensional space, i.e. a plane." Incidentally, the calligramme symbols in Fig. 8 are arranged flatly, but not in a one-dimensional string as we are accustomed.

The new *block-to-chain* balance at the bottom of Fig. 9 is similar to the previous global balance in Fig. 6, but is contrariwise organized. There, the fixed set of blocks balances all the code chains, whereas here, the fixed set of chains balances all their own blocks. As stated above, the nucleon sum of the standard blocks of the whole code is larger than that of the side chains. Chains are larger than blocks only in 7 out of 24 series. Only a limited set with the number of series less than 24 and containing mostly large amino acids is capable of balancing their own blocks. The new balance composed of large amino acids is located within the degeneracy III, II, and I set, but not within the degeneracy IV set. Such allocation is predetermined by simple algebra in Section 15. As a result, the concentration of small amino acids increases in the degeneracy IV set. Outwardly, this looks like the rough antisymmetrical correlation between the degeneracy and the nucleon numbers noted by Hasegawa and Miyata (1980). Therefore, the calligramme provides a simple explanation for that empirical observation which had been unexplainable till now. There are in addition to Rumer's bisection three more bisections among the contracted series in Fig. 10.

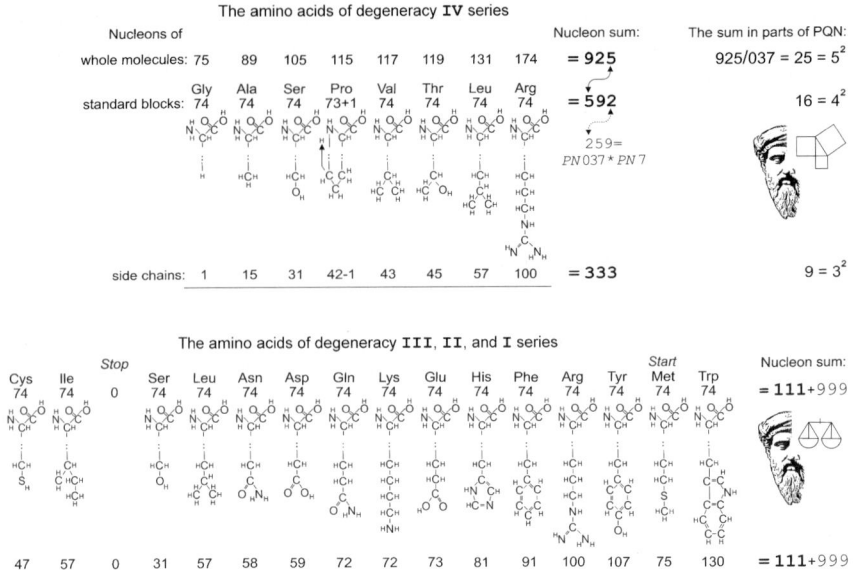

Fig. 9 The amino acid's basis of the cooperative symmetry (*sh*Cherbak, 1993b, 1994). Pythagoras' images symbolize the squares of the first three Pythagorean numbers, a new standard *block-to-chain* balance, and their decimal syntax. As before, the arrow near proline (Pro) indicates the virtualization act. The numerical symbol of the "Egyptian triangle" arises when corresponding nucleon sums appear in the parts of the *PN* 037. Let us consider the nucleon sum 925 as an example. The sought quantity of parts is equal to the quotient of 925 divided by the *PN* 037, i.e. to 25.

The amino acid's basis is represented by the *E. octocarinatus* code version; the revealed arithmetic regularities are also true for the universal genetic code

Note that transition from the life-size triplets in Figs. 5, 6, and 7 to the contracted series in Fig. 9, 10, 11, and 12, and vice versa, is a difficult computing task. There are joint contractions or expansions in four different scales in this transition. The arithmetic used should be quite refined to keep intact numerous equilibrations and their decimal syntax at both ends of the transition. This could be considered as one more proof for the great computing skills of the primordial abacus.

13   The Message

The artificial nature of decimalism has encouraged us to suggest a possible semantic interpretation of the four strings of symbols and view them as an intellectual message (*sh*Cherbak, 1989d). Evolving Crick's and Orgel's (1973) ideas, Marx discussed a theoretical possibility of such interpretation in 1979. Note that the genetic code is an information messenger with limited storage space. Because of

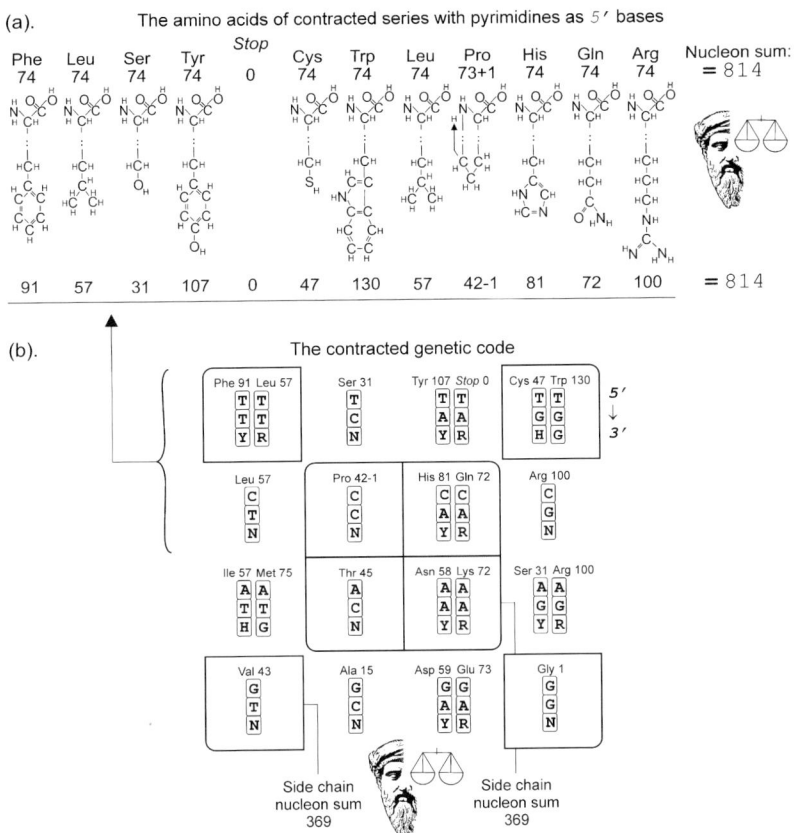

Fig. 10 Three additional bisections of the contracted genetic code. As before, the bisected series are interconnected by Rumer's and two 50% transformations. The virtualization acts as usual. The total nucleon number of side chains throughout the contracted genetic code is equal to **444** + 999. (a) The first 5′ bisection produces a new standard balance of the *block-to-chain* type. This balance is composed of contracted series with pyrimidines (Thymine and Cytosine) in the 5′ base positions. The initial brace denotes its position in the contracted code. Because of the previous *block-to-chain* balance in Fig. 9 only this equilibration could be algebraically independent orderliness. (b) The contracted genetic code (see its initial table in Fig. 3). The second 5′ and unique *center* TG-CA bisection divide the genetic code symmetrically. The framed code series are common parts of both bisections. These parts create balances of the *chain-to-chain* type. The balanced sums have noteworthy decimal notation. Mitochondria deviate from the universal code. It should be noted that none of the numerous deviations (Jukes and Osawa, 1990) disturbs this balance of the common parts. The bisected genetic code is represented by the *E. octocarinatus* code version; the revealed regularities are also true for the universal genetic code

this, the message within the code might be a forwarding address to permanent residences of a message body. Message symbols inside the code are recorded in a highly ordered written language. According to the biological standard of reading, we start interpretation from 5′ strings.

13.1 Two 5′ Strings

The short string in Fig. 11a demonstrates mirror, translation, and inversion symmetries. In more detail, the string bases are invariant under a combined operation of mirror symmetry and inversion of the *Base → Complementary base* type. A pair of the *RRYY* quartets forms a minimum pattern of the translation symmetry (see their black-and-white substrates on the right side). Such combined operation is inherent to the canonical base pairing in DNA molecules. In the one-dimensional space of the string, therefore, the four central bases and the central axis seem to represent the compact information symbol of a DNA molecule. A possible semantic interpretation for the short 5′ string: "A deoxyribonucleic acid—the universal keeper of genetic information". Its possible meaning: "Symbolization of a material carrier of the special message and an accentuation of Guanine-Cytosine complementary pair."

The same three symmetries arrange the long string in Fig. 11b. The pair of flanking *TATAT* sequences is mirror symmetrical. A pair of the central *AGC* triplets forms a minimum pattern of the translation symmetry. The inversion and the absence of it, i.e. identical transformation, show regular interdependence between the 5′ and 3′ bases of the degeneracy II set. This regularity being projected to the antitriplet of transfer RNA formalizes the well-known wobble pairing rule (Crick, 1966; *sh*Cherbak, 1989b).

The *TATAT* sequence could symbolize a so-called TATA box, which is the well-known DNA consensus sequence located in the promoter region of many eukaryotic genes. The TATA box is involved in binding an enzyme for making a messenger RNA from a DNA template to initiate a protein synthesis. A possible interpretation of the *TATAT* sequence: "A Goldstein-Hogness TATA box—a starting-point of transcription". Its possible meaning: "Symbolization of a biologically active genomic DNA." Still, there is no clear interpretation of two *AGC* triplets symmetrically arranged in the long string.

13.2 Two Center Strings

The short *center* string (see its black-and-white purine–pyrimidine substrates) and the long *center* string in Fig. 12a are mirror symmetrical relative to the central axis each. The base sequence of the short string and its projection into the coaxially positioned sequence of the long string are interconnected by the inversion symmetry. The coaxial symmetry shows itself in the unique geometric shape of the calligramme that had been made by the multiscale contraction. The shape and its coaxial symmetry decay in the life-size scale.

The purine–pyrimidine mirror symmetry of the short string is replaced by its semantic variant for a pair of the base triplets *CCC* and *TCT* flanked with the single Guanine bases *G*. Each of these triplets is a palindrome. The term "palindrome" is

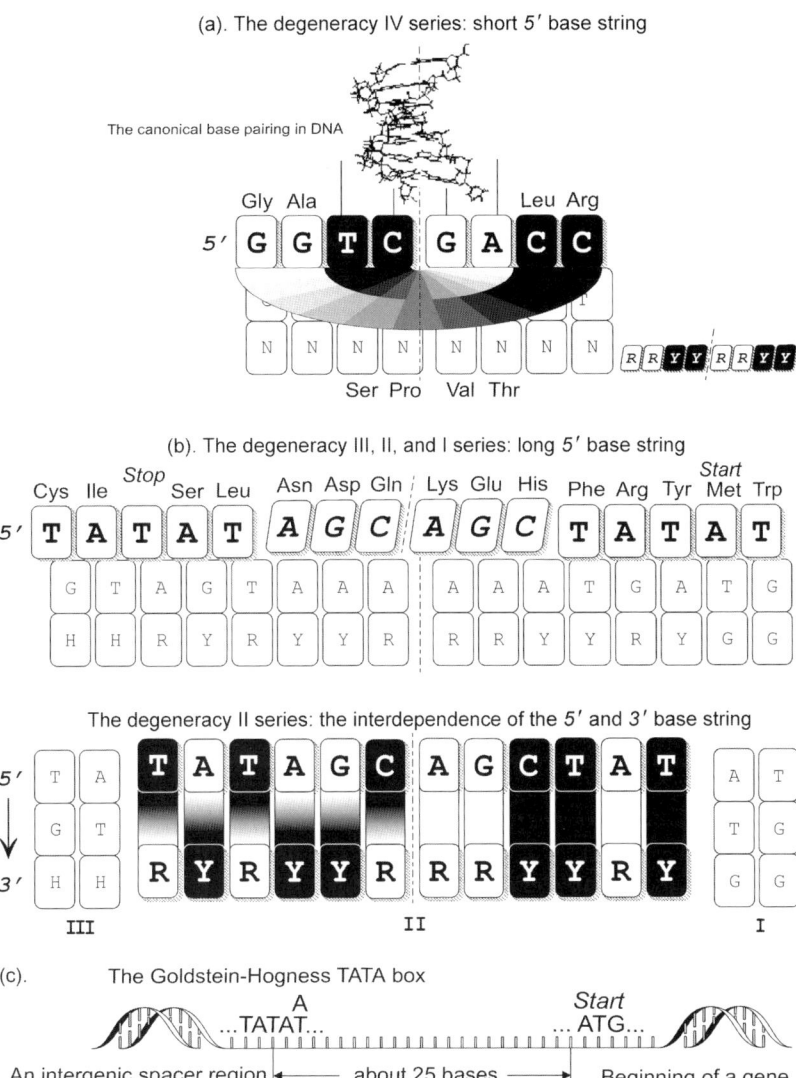

(a). The degeneracy IV series: short 5′ base string

The canonical base pairing in DNA

Gly Ala   Leu Arg

5′ G G T C G A C C   R R Y Y R R Y Y

N N N N   N N N N

Ser Pro   Val Thr

(b). The degeneracy III, II, and I series: long 5′ base string

Cys Ile *Stop* Ser Leu   Asn Asp Gln Lys Glu His   Phe Arg Tyr *Start* Met Trp

5′ T A T A T *A G C* *A G C* T A T A T

G T A G T A A A   A A A T G A T G

H H R Y R Y Y R   R R Y Y R Y G G

The degeneracy II series: the interdependence of the 5′ and 3′ base string

5′   T A   T A T A G C A G C T A T   A T

3′   G T
     H H   R Y R Y Y R R R Y Y R Y   T G
                                                  G G

III   II   I

(c).   The Goldstein-Hogness TATA box

A   *Start*
...TATAT...   ... ATG...

An intergenic spacer region ◄──── about 25 bases ────► Beginning of a gene

Fig. 11 The calligramme of the cooperative symmetry and its interpretation as a special message (*sh*Cherbak, 1988, 1989b, 1989d). There is a pair of the 5′ base strings in the calligramme. The strings are written by the bases that occupy the same positions in the contracted triplet series. The strings should be read horizontally, across the standard biological direction of reading 5′ → 3′. (a) The molecular structure of DNA symbolizes a possible semantics of the short string. (a, b) Both 5′ strings are ordered by the common set of symmetry operations. The common set incorporates three classical symmetries. These are the mirror symmetry (denoted by the central axes), translation symmetry (denoted by italicized letters and skew frames) and inversion of the *Base → Complementary base* or *Pyrimidine base* ↔ *Purine base* type (denoted by color gradient whose black and white colors symbolize the pyrimidine and purine bases, respectively). (c) A part of genomic DNA and the TATA box consensus sequence that might be a possible semantics of the long string

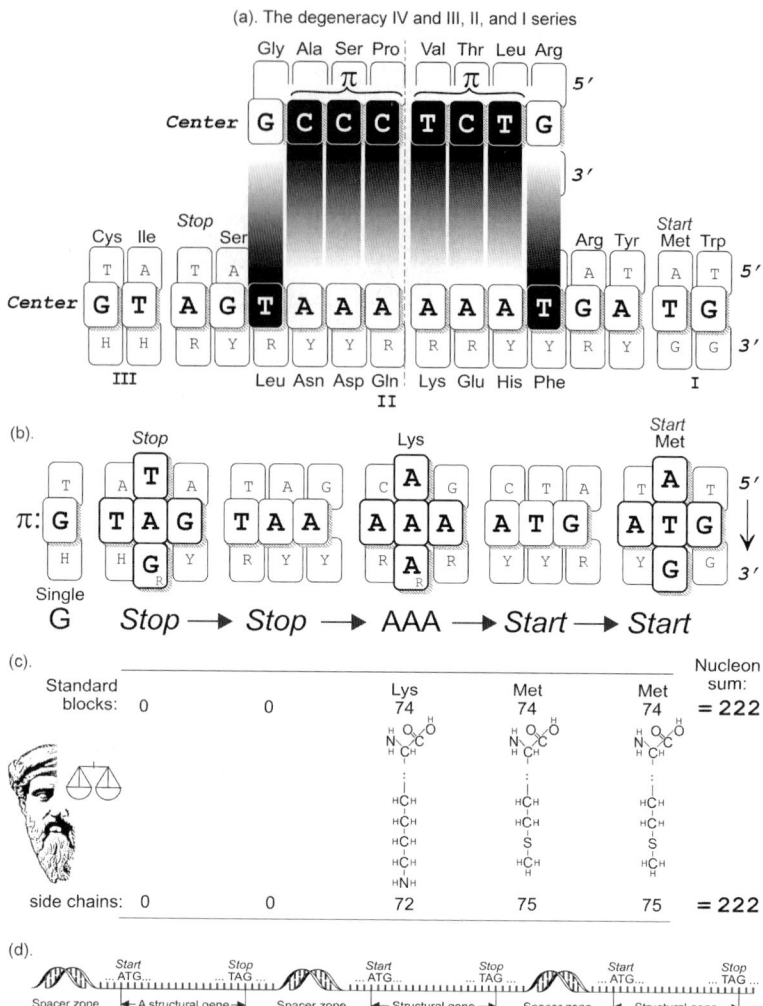

Fig. 12 The calligramme of the cooperative symmetry and its interpretation as a special message (*sh*Cherbak, 1988, 1989c, 1989d). There is a pair of *center* base strings. The graphical denotations are the same as in Fig. 11. (a) Both *center* strings are ordered by the common set of symmetry operations. The common set incorporates two classical symmetries and a new semantic symmetry. These are the mirror symmetry, inversion of the *Pyrimidine base* ↔ *Purine base* type, and the palindromic symmetry (designated by π). (b) The palindrome text of the long *center* string. Wide bordering shows crossing of synonymous triplets of the calligramme and palindrome text. (c) A new standard balance of the *block-to-chain* type revealed by the palindrome semantics. Another reading frame position "gt AGT AAA AAA TGA tg" duplicates the same side chain nucleon sum **222** as well as two triplet-palindromes CCC CTC together with one single G having 150 nucleons. (d) A part of genomic DNA and a possible meaning of the palindrome text. The genetic code in Figs 11 and 12 is represented by its *E. octocarinatus* version; the revealed regularities are true for this version

used in the literary sense as a line of characters, which are centrosymmetrical and whose semantics is invariant with respect to the directions of reading. One could interpret this symmetry-preserving palindromic feature as "An indicator of similar palindromic structure within the long string." The point is that both *center* strings have – similar to the 5′ strings above – a common set of symmetry operations, and now the palindromic symmetry is among these operations.

As predicted, the palindrome of the long string is represented in the triplet reading frame starting with the single Guanine base *G* at the flank (Fig. 12b). The palindrome has remarkable semantics expressed by the syntactic signs *Stop* and *Start*. These signs are in semantically antisymmetrical positions with respect to the central homogeneous triplet *AAA*. The palindrome uses all syntactic signs of the *E. octocarinatus* code version: two termination signs *Stop* and one twice-repeated initiation sign *Start*. Note there is a semantical symmetry of two nonmaterial linguistic antonyms. Such symmetry could be made and perceived only by the mind.

There are three possible positions of the reading frame in the palindrome. The unique semantical symmetry is not the only reason why the current position gets the better of its two other ones. The current position is fixed as the preferable one also by semantically dependent symmetrical crossing of the upright triplet series in the calligramme body and their abstract images in the palindrome text. One of two reading directions associated with this crossing becomes a preferable one. Note the crossing of the *Stop* signs is made directly by the zero symbol. The written symbol of the zero maintains all the order throughout the calligramme. Recall that its direct use is the powerful indicator of artificiality. But it is even more striking that the semantically arranged palindrome text is written by the code symbols within the code itself. It is as if the genetic code had "known" before its own origin how to code for the *Stop* and *Start* signs (as well as all other coding) in order to do inside itself the palindrome. It could only be possible if the genetic code was projected preliminarily. The standard *block-to-chain* balance and its decimal syntax are represented this time through the palindrome semantics in Fig. 12c. Both intact representations – symmetrical crossing and nucleon balance – show that after a long journey through time the palindrome and other parts of the message was delivered safely inside the code and that we have read it the message correctly enough.

The long *central* string, if taking into account its highest organization, should carry the most important part of the message. Any given protein gene is always located between the *Start* and *Stop* signs as shown in Fig. 12d; and vice versa there are nontranscribed intergenic spacers always located between the *Stop* and *Start* signs of two neighboring genes. So, the palindrome text could possibly be interpreted as "A symbol of nontranscribed intergenic spacer" with its possible meaning as "A final address of the message body and some structural details of intergenic spacers—a special accentuation of a Guanine base and AAA sequence." One can notice on the boundary between the degeneracy III and II or I and II sets the consensus sequence *GT-AG* that begin and end eukaryotic nuclear introns.

14   The Decimalism

Why are we so certain in speaking about the decimalism? As shown above, the genetic code is the result of numerous very precise summations. To be performed, these summations must privilege a certain number system over a variety of others. Therefore, that privileged numerical system should leave traces inside the code. Being nothing else but a syntactic rule, such a system indicates its own presence only through syntactic features of arithmetical written symbols. There are several phenomena relevant to the case.

First, a general and the most forcible argument: it has been found that the genetic code is governed directly by the arithmetical symbol of zero. This striking fact is verified simultaneously by several independent orderlinesses – logical, arithmetical, and semantical – in the previous section. Incidentally, such an acting zero alone might be sufficient to assume an artificial nature of the genetic code.

Second: it is obligatory that any place-value number system is preceded by revelation of the zero and that happened in the genetic code. Therefore, one should expect inside the code the traces of some privileged system; such system should have the positional principle and not a very big radix. On the other hand, the inevitability of arithmetical summation for the origin of the genetic code equilibria greatly strengthens this premise.

Third: there is a high number of couplings between balances and divisibility by *PN* 037 × 3 in the genetic code. Note that a balance is not the cause of certain divisibility and vice versa. Moreover, there are drastic differences among logical conditions that define balanced nucleons throughout the code. Even so, the strange couple – a balance and certain divisibility – remains the regular and immutable phenomenon. Recall that the same strangeness of coupling is inherent to linguistic symbols whose meaning and carrier are coupled for no apparent natural reason. Juxtaposed arithmetic and these couples have revealed a meaning of the couples. There is a complete set of information symbols utilizing the decimal syntax 111, 222, 333, 444, 555, 666, 777, 888, 999 in the genetic code. Each of these symbols consists uniformly of a carrier (balanced nucleons) and a meaning (the decimal syntax).

Fourth: there are no other systems in the neighborhood of the decimal one having similar criterions for three-digit numbers divisible by *PN* 037 × 3 (*sh*Cherbak, 2003). In this respect, the decimal system is a unique one as well, as unique is the above interpretation based on its syntax.

Fifth, indirect arguments: two systems with radixes four and seven precede the decimal system. These systems possess quaternary and septenary criterions for divisibility by *PN* 7 and *PN* 19, correspondingly. Both these criterions have affinity to the decimal one in Fig. 2. The quaternary criterion is useful for possible numerical calculations in DNA (see the final section). It seems that there is a syntactic symbol in the genetic code that unites the decimal world of amino acids and the quaternary world of bases. That is the final digital permutation 259 within the numerical symbol of the "Egyptian triangle" in Fig. 9. This number is the product of quaternary *PN* 7 and decimal *PN* 037. Incidentally, the canonical pair of the

Thymine and Adenine residues in DNA molecules has the same nucleon number 259 + "0"; the other pair of Guanine and Cytosine has 259 + "1" nucleons. One can speculate that this may be imposed by some standard in binary numbering for the four DNA bases.

Speaking about the septenary system, there are uniformly organized symbols in Fig. 7. These are two balanced sums 703 equal to the product of septenary *PN* 19 and decimal *PN* 37.

15 The Formula of the Genetic Code

It does not require sophisticated analytical methods to show to what extent the code is governed by arithmetic. Simple algebra could confirm this fact with scientific accuracy. For example, the symbol of the "Egyptian triangle" in Fig. 9 could be written down in the form of two Diophantine equations. One of these equations is *Gly + Ala + Ser + Pro + Val + Thr + Leu + Arg* = **333**. There are enough phenomena relevant to the case in the genetic code. Their records complete the algebraic system of Diophantine equations, which deal with integer variables only. Nucleon numbers of amino acid provide suitable system roots since they are never fractional. Actually, this algebraic system represents that very universal formula of the genetic code, which Gamow was looking for. This formula, however, no longer belongs to elementary arithmetic, and here again we provide a reference for those interested in a more detailed study of the issue (*sh*Cherbak, 2003).

The roots of the system are more than mere algebra. They show to what extent the code obeys the new ordering. In other words, the roots have to answer the question: could another arrangement of the canonical amino acids or other amino acid molecules with nucleon numbers different from the canonical ones produce the same ordering? The roots of an algebraic system may be unique or they may be contained within certain intervals depending on to what extent that system has been determined. The degree of determination becomes a reliable criterion to evaluate how the code obeys the new ordering. Computational studies of the system showed that it is a completely defined one.

It is obvious that a certain nucleon number is not always a unique symbol of a certain molecule out of many natural amino acids. However, the root of the variable *Gly* equal to 1 does not correspond to any side chain, but only to the amino acid glycine. The same is true, at least, for the variables *Ala*, and *Ser*. This list should be expanded over mid-size and large amino acids. Biochemical nomenclature denotes carbon atoms in α-amino acids alphabetically: C_α, C_β, C_γ, and so on (C_α atoms of standard blocks bond with C_β atoms of side chains). There is the nucleon sum **111** at the structural level of C_β atoms in Fig. 9 (see the degeneracy IV set) and the nucleon sum **333** at the structural levels of C_β and C_γ atoms in Fig. 10a. Note that this representation limits the possibility of natural amino acids to be canonical molecules. Thus, some roots are associated with certain molecules numerically assigned to them.

Jukes (1983) believes that the code "froze" too early, having not completed an optimal distribution of triplets and amino acids. For instance, amino acid arginine, which is rarely found in proteins, has six triplets, while lysine, having similar properties and being one of the most widely used amino acids, has only two. He called arginine an "evolutionary intruder." However, under new conditions it rather looks like a "captive." Its molecule, whose side chain has 100 nucleons, occupies precisely those triplets, which are intended to it by the system solution. The same is true for lysine with its 72 nucleons and for other canonical amino acids and syntactic signs. Of course, both the triplet (as the universal codon) and the regular degeneracy result from the formula.

16 Chemistry Obeying Arithmetic

It is well known that amino acids, triplet series, and series degeneracy are optimally distributed so that the effect of mutations of single bases is minimized. As a rule, possible consequences of the mutations such as an abrupt change of hydrophobicity are, to some extent, weakened (e.g. Sjöström and Wold, 1985; Figureau, 1987; Knight et al., 1999). It looks like some imperfect protection against mutations. However, Shulz and Schirmer (1979) raised the question: why is it that another property of amino acids – the geometrical size of side chain – has no comparable protection though size is no less important than hydrophobicity?

One can show that the cooperative symmetry does not allow the code to enhance protection against an abrupt change of size. Histograms of hydrophobicity (Lacey and Mullins, 1983) and molecular size (Jungck, 1978) are combined with a calligramme fragment in Fig. 13. Transversions are a kind of point mutations that alternate pyrimidine bases with purine bases and vice versa. All possible 3' base transversions of the largest and the most protected code set of degeneracy II are shown by arrows. Due to the cooperative symmetry of the fragment, the transversions are also centrosymmetrical in their arrangement. Recall that the triplet series are aligned as a row of monotonically increasing amino acid nucleon numbers.

The molecular size of an amino acid is generally proportional to its nucleon number. Therefore, the monotony excludes a centrosymmetrical shape of the size histogram. On the other hand, the hydrophobicity of an amino acid does not depend explicitly on its size and nucleon number. This allows the genetic code to form the centrosymmetrical shape of its hydrophobicity histogram. As a result, when amino acids are replaced through mutation the change of hydrophobicity is generally weak, while that of size is strong.

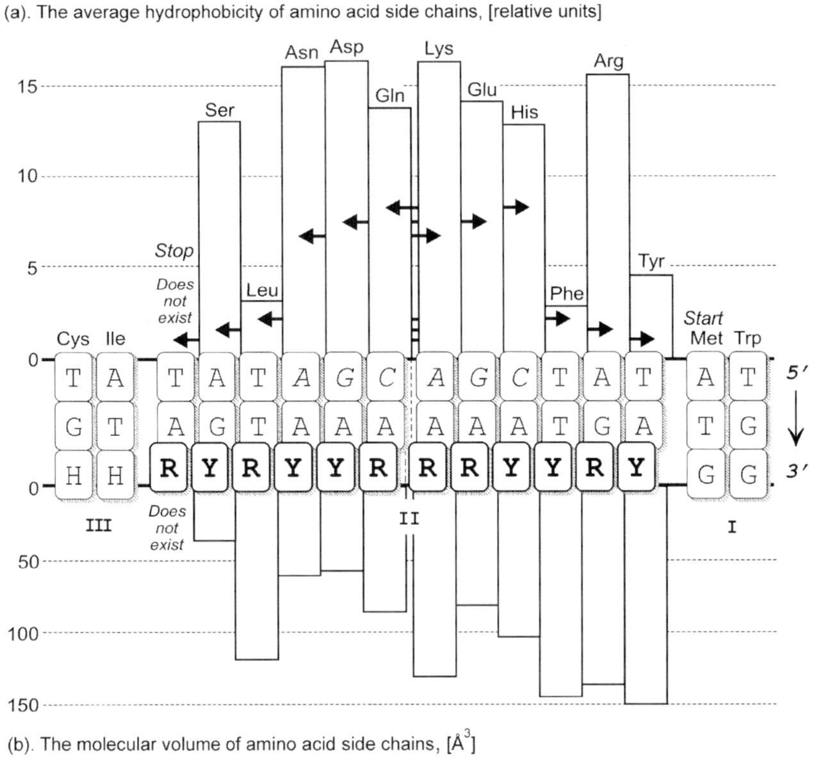

(a). The average hydrophobicity of amino acid side chains, [relative units]

(b). The molecular volume of amino acid side chains, [Å³]

Fig. 13 Physicochemical effect of the cooperative symmetry. If a point mutation occurs, the code amino acids are to a certain extent protected against an abrupt change of their hydrophobicity but not of their geometric size. Such limited protection results from the internal cooperative symmetry and its systematization principle. Arrows denote the transversion mutations of the triplet 3′ base positions and the corresponding alterations of the hydrophobicity and geometric size

Freeland and Hurst (1998) showed that, compared to defense capabilities of the universal genetic code against mutations, only one code version in every million randomly generated alternative codes is more efficient against abrupt changes in the hydrophobicity. Those alternative codes generated in computer simulation can hardly ever get some exact order at random. What is more, such codes lose the original arithmetic of the actual genetic code. However, necessity to have and save arithmetic rejects this and other statistical approaches including the "RNA world." Therefore, the genetic code is a unique one rather than one in a million in this respect.

The universal genetic code is arranged optimally (Di Giulio and Medugno, 2001). Only now after its concealed order had been disclosed, one can perceive a true scale of that optimality.

17   The *Gene Abacus*

A reliable prediction of artificiality is a relevant but secondary ability of arithmetic. Arithmetic is the only tool for producing information systems of extremely high efficiency. Life, being an information phenomenon, could use arithmetic for control, integrity, and precise alterations of its genetic texts. Gamow (1954) noted this possibility in his pioneering article on the genetic code. He wrote that a long number written in the quaternary system could characterize the hereditary properties of any given organism. In order to convert some DNA base sequence into Gamow's *long number* one should replace four bases T, C, A, and G with four quaternary digits 0, 1, 2, and 3.

After Gamow, various authors noted the digital nature of the genetic code (e.g. Eigen and Winkler, 1985; Yockey, 2000; Mac Dynaill, 2002; Gusev and Shulze-Makuch, 2004; Négadi, 2004; Rakočević, 2004; Gonzalez et al., 2006). Ordinary computers use the binary notation and a checksum even parity, i.e. divisibility by 2, as a data integrity control. Similarly, some genetic sequences could be arithmetically arranged using the number system(s). The number system inside the genetic code indicates – besides the code artificiality – a possible arithmetical power on the level of genetic sequences in genomic DNA. One can speculate that some regular arithmetical background may "underlie" genetic sequences without limitation to their biological context. Analyzing its own arithmetic, such background might check, restore, and alter superimposed biological context by means of calculations omitting translation. Is this the essence of the enigmatic meiosis prophase I? For instance, the meiosis prophase I lasts for maximum about 50 years for female humans (Bennet, 1977). Over that long period, homologous chromosomes remain conjugate without visible biochemical activity. It is not unlikely that information analysis – including arithmetical reckoning – prevails over biochemical activity at that time.

Thus, arithmetic residing inside the genetic code forces us to look for its possible analogue inside genomic DNA. We developed a computer code called a *Gene Abacus* (*sh*Cherbak, 2005). This program tool simulates a hypothetical molecular adding machine that slides along a DNA and performs appropriate summations over certain distances as shown in Fig. 14. Let a machine display the checksum **333** accepted as correct on the register in the form of the base triplet AAA. Homogenous digital structure of the register or its absence corresponds to the intact or damaged gene. A routine shape analysis could convert this digital notation into an analogous form and initiate certain biochemical reactions depending on the performed computations. We have used the new code arithmetic as an instruction to equip the *Gene Abacus* with suitable arithmetic abilities. The *Gene Abacus* should look for regularities of both arithmetical syntax and quantity in chromosomes. We fully realize that only the discovery of the same arithmetic in genomic DNA gives the strong key to the mystery of the origin of life.

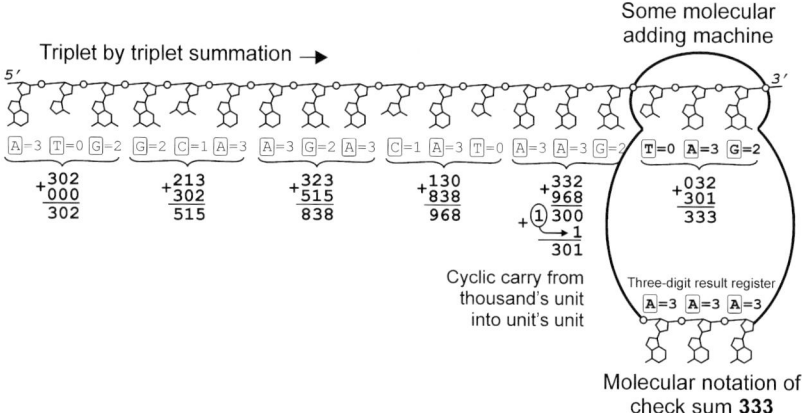

Fig. 14 A hypothetical molecular adding machine working in the quaternary system and a short gene invented for the sake of illustration. The quaternary criterion for divisibility by *PN* 7 is similar in every respect to the decimal one in Fig. 2 (*sh*Cherbak, 1993b, 1994). There are exactly 64 three-digit quaternary numbers in the range from 000 to 333 (decimal 63). The quaternary four digits and 64 three-digit numbers bear a close analogy with the four DNA bases and the 64 base triplets of the genetic code. There is one numbering T = 0, C = 1, G = 2, A=3 of 24 possible ones in the invented gene. The done reckoning shows checksum 333. As before, one needs only three-digit register to establish divisibility by quaternary *PN* 7 of any number, irrespectively of how large this number would be. Though there is the quaternary numbering, the decimal system was used in the reckoning for simplicity. The same summation performed in the quaternary system results in the same particular notation 333, but in another quantity

18   Conclusion

Almost half a century ago, the idea of chemical evolution determined a stochastic approach to the origin of the newly deciphered genetic code. The molecular machinery of genetic coding revealed however a baffling complexity. It is common practice to criticize the stochastic approach because the likelihood that this machinery could have been produced by chance is extremely low if not negligible. But, on the other hand, the basic premise of the stochastic approach is that billion years and countless natural events could realize, nevertheless, practically impossible things. This vicious circle exists because both these opposing opinions appeal to the same concept – to the assessment of chances of natural events. One needs therefore properties of the genetic code that eliminate any possibility of a dual interpretation or overlapping of the arguments. In these terms, only the facts whose essence cannot be reduced to natural events can solve the original problem. We believe that a part of these facts have been presented and discussed above.

Acknowledgements The work was financed by the Ministry of Science and Education of the Republic of Kazakhstan. Part of this study was made during my stay at Max-Planck-Institute für biophysikalische Chemie, Göttingen, Germany. I am greatly indebted to Professor Manfred

Eigen for his support and express special thanks to Ruthild Winkler-Oswatitsch for her valuable help. I would also extend my gratitude to Bakytzhan Zhumagulov and Alevtina Yevseyeva of National Engineering Academy of the Republic of Kazakhstan who promoted this chapter to see the light of the day.

References

Apollinaire, G., 1980. *Calligrammes*. University of California Press, Los Angeles/London:.

Barbieri, M., 2005. Life is "artifact-making". *J. Biosemiotics* 1, 113–142.

Bennet, M. D., 1977. The time and duration of meiosis. *Phil. Trans. R. Soc. Lond. B*. 277, 201–226.

Chiusano, M. L., Alvarez-Valin, F., Di Giulio, M., D'Onofrio, G., Ammirato, G., Colonna, G., and Bernardi, G., 2000. Second codon positions of genes and the secondary structures of proteins. Relationships and implications for the origin of the genetic code. *Gene* 261, 63–69.

Crick, F. H. C., 1966. Codon-anticodon pairing: the wobble hypothesis. *J. Mol. Biol.* 19, 548–555.

Crick, F. H. C., 1968. The origin of the genetic code. *J. Mol. Biol.* 38, 367–379.

Crick, F. H. C. and Orgel, L. E., 1973. Directed panspermia. *Icarus* 19, 341–346.

Dewachter, M., 1990. *Champollion, un scribe pour l'Egypte*. Coll. Découvertes, Gallimard, Paris.

Downes, A. M. and Richardson, B. J., 2002. Relationships between genomic base content and distribution of mass in coded proteins. *J. Mol. Evol.* 55, 476–490.

Di Giulio, M., 1989. Some aspects of the organization and evolution of the genetic code. *J. Mol. Evol.* 29, 191–201.

Di Giulio, M. and Medugno, M., 2001. The level and landscape of optimization in the origin of the genetic code. *J. Mol. Evol.* 52, 372–382.

Eigen, M. and Winkler, R., 1985. *Das Spiel*. Piper Verlag, München, Zürich, pp. 281–316.

Figureau, A., 1987. Information theory and the genetic code. *Origins Life* 17, 439–449.

Freeland, S. J. and Hurst, L. D., 1998. The genetic code is one in a million. *J. Mol. Evol.* 47, 238–248.

Gamow, G., 1954. Possible relation between deoxyribonucleic acid and protein structures. *Nature* 173, 318.

Gonzalez, D. L., Giannerini, S., and Rosa, R., 2006. Detecting structure in parity binary sequences: error correction and detection in DNA. *IEEE Eng. Med. Biol. Mag.*. Jan./Feb, 69–81.

Grimm, M., Brünen-Nieweler, C., Junker, V., Heckmann, K., and Beier, H., 1998. The hypotrichous ciliate *Euplotes octocarinatus* has only one type of $tRNA_{Cys}$ with GCA anticodon encoded on a single macromolecular DNA molecule. *Nucleic Acids Res.* 26, 4557–4565.

Gusev V. A. and Shulze-Makuch D., 2004. Genetic code: Lucky chance or fundamental law of nature? *Phy. Life Rev.* 1, 202–229.

Hasegawa, M. and Miyata, T., 1980. On the antisymmetry of the amino acid code table. *Origins Life* 10, 265–270.

Jukes, T. H., 1983. Evolution of the amino acid code: inferences from mitochondrial codes. *J. Mol. Evol.* 19, 219–225.

Jukes, T. H. and Osawa, S., 1990. The genetic code in mitochondria and chloroplasts. *Experientia* 46, Birkhäuser Verlag, CH-4010 Basel/Switzerland, 1117–1133.

Jungck, J. R., 1978. The genetic code as periodic table. *J. Mol. Evol.* 11, 211–224.

Kashkarov, V. V., Krassovitskiy, A. M., Mamleev, V. S., and *sh*Cherbak, V. I., 2002. Random sequences of proteins are exactly balanced like the canonical base pairs of DNA. In: *Proceedings of the 10th ISSOL Meeting and 13th International Conference on the Origin of Life*. Oaxaca City, Mexico, June 30–July 4.

Knight, R. D., Freeland, S. J., and Landweber, L. F., 1999. Selection, history and chemistry: the three faces of the genetic code. *Trends Biochem. Sci.* 24, 241–247.

Lacey, J. C. Jr. and Mullins, D. W. Jr., 1983. Experimental studies related to the origin of the genetic code and the process of protein synthesis—a review. *Origins Life* 13, 3–42.

Mac Dynaill, D. A., 2002. A parity code interpretation of nucleotide alphabet composition. *Chem. Commun.* 18, 2062–2063.

Marshal, R. E., Cascey, T. C., and Nirenberg, M., 1967. Fine structure of RNA codewords recognized by bacterial, amphibian and mammalian transfer RNA. *Science* 155, 820–825.

Marx, G., 1979. The message through time. *Acta astronaut.* 6, 221–226.

Meyer, F., Schmidt, H. I., Plümper, E., Hasilik, A., Mersmann, G., Meyer, H. E., Engström, A., and Heckmann, K., 1991. UGA is translated as cysteine in pheromone 3 of *Euplotes octocarinatus*. *Proc. Natl. Acad. Sci. U.S.A.* 88, 3758–3761.

Négadi, T., 2004. Symmetry groups for the Rumer-Konopel'chenko-*sh*Cherbak "bisections" of the genetic code and applications, *Internet Electron. J. Mol. Des.* 3, 247–270. Available at: http://www.biochempress.com

Nirenberg, M., Leder, P., Bernfield, M., Brimacombe, R., Trupin, J., Rottman, F., and O'Neal, C., 1965. RNA codewords and protein synthesis, VII. On the general nature of the RNA code. *Proc. Natl. Acad. Sci. U. S. A.* 53(5), 1161–1168.

Rakočević, M. M., 2004. A harmonic structure of the genetic code. *J. Theor. Biol.* 229, 221–234.

Rumer, Yu. B., 1966. About systematization of the codons of the genetic code. *Dokl. Acad. Nauk. SSSR* 167, 1393–1394.

Schutzenberger, M.-P., Gavaudan, P., and Besson, J., 1969. Sur l'existence d'une certaine correlation entre le poids moleculaire des acides amines et le nombre de triplets intervenan dans leur codage. *CR Acad. Sc. Paris, Serie D* 268, 1342–1344.

*sh*Cherbak, V. I., 1988. The co-operative symmetry of the genetic code. *J. Theor. Biol.* 132, 121–124.

*sh*Cherbak, V. I., 1989a. Rumer's rule and transformation in the context of the co-operative symmetry of the genetic code. *J. Theor. Biol.* 139, 271–276.

*sh*Cherbak, V. I., 1989b. Ways of wobble pairing are formalized with the co-operative symmetry of the genetic code. *J. Theor. Biol.* 139, 277–281.

*sh*Cherbak, V. I., 1989c. The "START" and "STOP" of the genetic code: why exactly ATG and TAG, TAA? *J. Theor. Biol.* 139, 283–286.

*sh*Cherbak, V. I., 1989d. The information artefact of the genetic code. In: Proceedings of the 6th ISSOL Meeting and 9th International Conference Origin of Life, Book of Abstract, Prague, July 3–8, Czechoslovakia.

*sh*Cherbak, V. I., 1993a. The symmetrical architecture of the genetic code systematization principle. *J. Theor. Biol.* 162, 395–398.

*sh*Cherbak, V. I., 1993b. Twenty canonical amino acids of the genetic code: the arithmetical regularities. Part I. *J. Theor. Biol.* 162, 399–401.

*sh*Cherbak, V. I., 1994. Sixty-four triplets and 20 canonical amino acids of the genetic code: the arithmetical regularities. Part II. *J. Theor. Biol.* 166, 475–477.

*sh*Cherbak, V. I., 1996. A new manifestation of the arithmetical regularity suggests the universal genetic code distinguishes the decimal system. In: *Proceedings of the 8th ISSOL Meeting of the 11th International Conference Origin of Life.* Book of Abstracts, Orleans July 5–12, France.

*sh*Cherbak, V. I., 1999. A new manifestation of the decimal system in the genetic code. In: *Proceedings of the 9th ISSOL Meeting and of the 12th International Conference Origin of Life.* Book of Abstracts, San-Diego, July 11–16, USA.

*sh*Cherbak, V. I., 2003. Arithmetic inside the universal genetic code. *BioSystems* 70, 187–209.

*sh*Cherbak, V. I., 2005. The origins of life and arithmetic zero. In: *Proceedings of the 11th ISSOL Meeting and 14th International Conference on the Origin of Life.* Book of Abstract, Beijing, June 6–11, People's Republic of China.

Shulz, G. E. and Schirmer, R. H., 1979. *Principles of Protein Structure.* Springer-Verlag, New York, Heidelberg, Berlin.

Sjöström, M. and Wold, S., 1985. A multivariate study of the relationship between the genetic code and the physical-chemical properties of amino acids. *J. Mol. Evol.* 22, 272–277.

Spengler, O., 1922. *Der Untergang des Abendlandes. Umrisse einer Morphologie der Weltgeschichte.* C.H. Beck Verlag, München.

Taylor F. J. R. and Coates D., 1989. The code within the codons. *BioSystems* 22, 117–187.

Yockey, H. P., 2000. Origin of life on earth and Shannon's theory of communication. *Comput. Chem.* 24, 105–123.

Part 3
Protein, Lipid, and Sugar Codes

Chapter 8

Protein Linguistics and the Modular Code of the Cytoskeleton

Mario Gimona

Abstract Protein assembly follows linguistic rules, and the combination of linguistic analysis and the use of modular units as building blocks are now beginning to allow first insights into the underlying parameters for the evolution of multidomain proteins. Filaments of the cytoskeleton systems, themselves assembled from modular protein units, display the hallmarks of self-replicating von Neumann automata. The actin cytoskeleton is a prototype for a molecular code that is generated by the assembly of identical subunits, and cells are able to respond to changes in the state of the cytoskeleton. Thus, cytoskeleton assembly generates signs whose purpose is to provide the necessary asymmetry for molecular interactions and that makes organic meaning accessible. The cytoskeleton acts as a code maker, and functions as an internalized, shared background knowledge of a historically evolved linguistic community. It is a modular, universal, and dynamic structure that employs adapters to interact with intracellular and extracellular entities, and it is a molecular machine that engages in biosemiosis via a continuous assembly and disassembly cycle. In this chapter I will discuss how the rules and parameters of protein linguistic assembly, and in particular the cytoskeleton code, reveal a cellular biosemiotic mechanism at work.

1 Introduction

The concept of cell language [42] suggests that: 'human language can be viewed as a transformation of cell language'. In consequence, a grammar that regulates verbal communication between humans might be a specific manifestation of a more universal biological grammar that dictates the basic conventions for all facets of communication, including that between molecules [83]. The recognition of syntactic patterns is the key to understanding any unknown language, and the features of

Consorzio Mario Negri Sud, Department of Cell Biology and Oncology,
Unit of Actin Cytoskeleton Regulation, Via Nazionale 8A, 66030 Santa Maria Imbaro, Italy,
e-mail: gimona@negrisud.it

M. Barbieri (ed.), *The Codes of Life: The Rules of Macroevolution.*
© Springer 2008

protein structure resemble conceptual expression patterns such as Chinese characters or Egyptian hieroglyphs. Portions of amino acid sequence that fold and function autonomously, and that maintain their inherent function also in the context of another protein are generally referred to as 'domains' or 'modules'. Such modules might delineate the authentic syntactic units in the protein language, not so much at the level of individual 'words', but rather as more complex 'phrases' or 'clauses', as these can better accommodate functional plasticity within structurally similar modules that have little conservation at the sequence level [29]. Within the framework of protein linguistics based on a modular protein syntax substantial modification of a protein's semantic value can be achieved without changing the length or structure of the syntactic unit (its size and basic fold). Modular proteins may thus represent a biological *Rosetta Stone* that can lead to the deciphering of the basic grammatical rules of protein assembly.

Modular proteins are most prominently identified in the cytoskeleton, in both the structural polymers, and the signalling and adapter components [54,55]. The conventions within a cell allow the recognition of individual phrases in the context of a complex, dynamic protein and for the most part preserve their semantic autonomy. As a consequence of the code-making process [7,8] the individual domains and modules represent *code messengers*, fractions of the code that convey a biological meaning. In this way the information that is contained in a protein interaction module is translated to make its semantic content accessible for the binding partners. The encoded information is read and interpreted by other proteins, lipids, or nucleic acid structures, themselves entities determined by the code-making process. Albeit not reproducing their own subunits by a direct copying process, the major cytoskeleton elements (actin, tubulin, and intermediate filament proteins) are capable of reproducing their major functional state, the filament. Both actin filaments and microtubules are produced from a template in a machine-like fashion, and treadmilling and dynamic instability govern the equilibrium dynamics of the polymers via molecular interactions. The recognition and interpretation of the semantic value of a protein module suggests that the elements of the cytoskeleton act as integrating interpreters that are accepted by objects for which a direct code relation exists (molecules that directly interact with the filaments).

In this chapter I will discuss how basic protein linguistic rules and the compositional semantics of modular protein domains in cytoskeleton components may allow us to take a first glimpse into a universal biological grammar, and how the modular cytoskeleton can be used to understand part of the protein codes that seem to govern cellular life.

2 Protein Linguistics

The need for protein linguistic analysis arose in the course of the annotation of genomic and proteomic data [12,20]. The prediction of function from a given nucleic acid or amino acid sequence requires precise and unambiguous knowledge of the functions that are attributed to similar sequences [27,72,74] or three-dimensional folds [21,41]. Currently, the ambiguity of annotations in databases is hindering progress in this direction. Protein linguistics deals with the potential categorization of the biological world to provide a robust framework for future test systems of protein functionality [13,57,60,68]. The aims range from (i) the definition of rules for protein assembly from modular protein domains, and (ii) help to formulate parameters for the prediction of protein function, to (iii) exploring the possible routes towards a deeper understanding of pragmatics (meaning) of molecular interactions beyond their apparent biochemical and physiological consequences. In light of a biosemiotic approach, one also have to consider that the biosemiotic process starts with evolution [6,44]. As a consequence, protein (and nucleic acid) linguistics can provide relevant insights for a definition of underlying rules and constraints that allow biosemiotics at the cellular and molecular level. It is these constraints of freedom or creativity (that go beyond those imposed upon molecules by the limitations of physics and thermodynamics) that provide the essential parameters for communication. As a first analogy to natural human language one may thus consider that in the biosemiotic world the rules of grammar restrict (syntactic) freedom and enable the use of signs (based on grammatical conventions) for communication.

The analogies between linguistics and biology [15] were first presented in detail in the work of David Searls [67,68,69]. In his model Searls compared nucleic acid and amino acid sequences with lexical entities, whereas structure would resemble syntax, and the function related to the molecular structure would give rise to what is called semantics in linguistics. A more complex issue is the relationship between biological role and linguistic pragmatics. Steven Benner's work [10] on the functional annotation of proteomic data has already foreseen a two-layered annotation of biological data: one in which the similarities in sequence, structure, and function are to be considered, and another, in which the differences in functional constraints and site-interdependence have a central role. Searls clearly demonstrated the existence of dependencies and even nested dependencies in nucleic acid molecules owing to the folding of the RNA, and identified striking similarities in linguistic complexity between nucleic acids and natural human languages. He also indicated the significance of information (or semantic value) that is contained within the linear or folded structure of macromolecules, and that the hierarchical order and underlying rules that govern macromolecular organization relate strongly to syntactic structures in natural human languages. Compositional semantics has a clear mathematical basis and in this framework the biological role of any given molecule can be described by the concerted actions and functions of its individual domains. The complexity of protein semantics is easily explained by the three- and four-dimensionality of protein function. Similar to the linear and nested dependencies in folded RNA molecules, proteins can (with an even higher complexity) adjoin modules not only in space, but also in time (by dimerization, folding, protein–protein and protein–lipid interactions, site-unmasking, or post-translational modifications) to alter the semantic value of the molecule [see, 78]. By accepting compositional semantics as a protein linguistic operation it becomes clear that not only the total composition of modules in a protein controls its semantic value, but also that the position of these modules along the amino acid chain needs to be considered.

In reviewing the current proteomic and genomic efforts it may be fair to say that the genome and proteome projects have not evolved much further than nineteenth and twentieth century lexicostatistics. Thus, a novel and considerably more radical view on proteomic data interpretation is needed for biology just as much as Chomsky's theories were vital to modern linguistics [18,19,25,37]. Notably, the work of another classical linguist, Mark C. Baker [2], offers promising new leads for a future linguistic annotation of biological data. In his book 'The Atoms of Language' Baker described 18 basic parameters that define the complete grammar for all known (about 6000) human languages. A consequence that follows from this work is that the much debated *Universal Grammar* (or *Faculty of Language*), as initially defined by Chomsky, can be encrypted in a minimal set of genetically-encoded structures (DNA, RNA, micro-RNA, or protein molecules) that can govern the initial identification of syntactic patterns in a language that is first being perceived by a developing (human) organism [13,20,48]. Baker also succeeded to accommodate the Chomsky hierarchy of language grammars by analogy to a three-dimensional periodic table of the classical Mendeleyev-type – a most astonishing comparison of the molecular and the cognitive world. With these parameters in mind one may have to ask if nucleic acid linguistics and protein linguistics, just as natural language grammar, are reflections of a more general principle of linguistic organization that may be applied to all domains of biology. Together with the Chomsky hierarchy of the levels of language grammar and the language parameters, as postulated by Baker, it indeed appears that the basis for all types of regulatory grammars is similar, if not identical in their basic principles. On this basis one can formulate a hypothesis in which human language is seen as a specific manifestation of a general biosemiotic process. A truly Universal Grammar (in a biosemiotic sense and therefore valid for all types of communication and interaction) can thus be encrypted in a basic set of context-dependent and multilayered rules (or parameters) that determine the communication and semantic validation between different parts of the biological world.

Knowledge derived from formal language grammar has been used in the past to analyze genome and proteome structures. In one approach, Sung Chul Ji [42,43] defined *Cell Language* as the underlying communication theory, and proposed that human language is ultimately founded on cell language: a provocative and much-debated theory, that requires consequent testing in the future. Albeit the isomorphism between cell and human language is far from rationalizing how general linguistics could explain biological processes, there is a most significant contribution in this work on cell language, namely the notion that 'human language can be viewed as a transformation of cell language'. What Ji addresses in his work is the difficulty of justifying a scenario where the human brain that created language and the underlying Universal Grammar (a natural endowment to the human organism, the basic elements of which are likely to be encoded in only a very small fraction of the genetic repertoire) would be anything different from what is possible in terms of physicochemical and biological constraints. The human mind cannot escape its boundaries and language grammar can thus not be anything else than a reflection of the Universal Grammar that we all carry without knowing. As such, Universal Grammar is dictated by biological information, and it may thus be possible that a grammar that regulates verbal communication between humans is a variation of the grammar that regulates all facets of communication, including that between molecules.

Brendel and Busse [14] applied automaton-driven fold recognition for genome annotation, and showed that indeed genome structure can be described by formal language grammar. The analysis of recursive protein domains by Theresa Przytycka and colleagues [61,62] led to the formulation of a folding grammar that is similar to the one used in human languages. A basic set of four grammatical rules dictates the folding of minimal regions of proteins, which governs all β–sheet protein domains. Notably, these studies led to the further suggestion that folding grammar is an expression of Nature per se. In the work from Eric Werner [85] in which *genome semantics* was probed for by in-silico modelling to elucidate the meaning of the genome, the problem of a suitable definition for the words that convey a semantic value (the *semantic unit*) in order to further embark on their combinatorial use for the generation of sentences became apparent. The work, however, recognized that a semantic code translates sequence into meaning, and *genome semantics* may thus serve to assign meaning to a regulatory code by way of its function.

3 Protein Modularity and the Syntactic Units of a Protein Linguistic Grammar

For protein linguistic considerations SYNTAX shall be seen as the branch of linguistics that defines how words (syntactic units) are combined to make phrases and sentences. The PHRASE is a vehicle for defining syntactic units that appear next to each other or stay together in the arrangement of a sentence, and that form a syntactic unit of equal or higher complexity [29]. Definition of the syntax is a necessary prerequisite for any further efforts, as one must have detailed knowledge of the syntactic units to which the new (protein linguistic) grammar can be applied. Three potential ways can be used to define protein linguistic syntax: (i) that of the individual amino acid; (ii) that of the primary structural elements (e.g. beta strands, loops, or alpha helices); and (iii) that of larger (30–100 residues in length) autonomously folding and stable modular structures that have persisted throughout evolution [1,36,77]. Clearly, the first scenario has little to offer, and computational analysis with the aim of structural and functional predictions from amino acid sequences wrestles with such parameters since a good two decades. The second scenario appears to define a domain in itself and may assist the efforts of the first (computational) scenario by providing shortcuts and helping to reduce complexity. For the third scenario there is support from attempts that appear to reside in between scenarios 2 and 3, namely the appealing '*n*-gram theory' developed over many years by Judith Klein-Seetharaman [28]. This work has defined short species-specific *n*-grams (three to four amino acid residues in length) that can be viewed as analogs to certain linguistic phrases, and are termed '*n*-gram phrases'. Such *n*-gram phrases may be closely to syntactic primitives that provide insight into the possible structure of a biological protein alphabet. Extrapolating from such a thought, syntactic structure would imply that *n*-gram motifs (and presumably also larger motifs, domains, or modules) become part of syntactic categories (multidomain proteins, multimolecular complexes, or entire signalling pathways) that constrain their combinatorial use and thus their biosemiotic value (or their 'interpretability').

The recent application of statistical coupling analysis (SCA) by Rama Ranganathan and colleagues [65,73] revealed first important insights into the evolutionary information that specifies protein folding in modular domains. The information that is required for specifying the tertiary structure of a protein is contained in its amino acid sequence, but the potential complexity of this information is enormous, thus preventing the definition of folding rules by computational energy minimization procedures. SCA uses only statistical information from multiple sequence alignments (without tertiary structure information) to define folding rules. Significantly, simple statistical energy functions that capture the co-evolution of amino acid residues were shown to be both necessary and sufficient to generate rules that, when obeyed, allow the synthetic construction of a natively folding and fully active protein module. As demonstrated for the proline-containing target sequence-binding WW domain, the conserved functional coupling of specific residues in a protein drives their co-evolution, and only small sets of positions co-evolve among the bulk of largely decoupled sites. Moreover, this work showed that the strongly co-evolving residues are organized into physically connected networks. Thus, evolutionary conservation together with a small set of coupling rules appears to dictates the complete set of positional constraints in WW domain sequences – the grammar of WW domain function. The view of modular protein domains as complex phrases has also been adopted in the recent work from Loose et al. [46] replacing the term 'pattern' with that of 'grammar'. Letting aside semantic variations, the successful application of a linguistic model to assist in the design of novel anti-microbial peptides is an impressive proof of concept.

About 75% of eukaryotic proteins have a modular architecture and contain at least two or more (mostly globular) domains that are combined [36]. Many protein modules have persisted in diverse species over a considerable period of evolution, and based on their predominantly structure-dependent classification and function, closely resemble rich conceptual linguistic expression patterns like Chinese characters or Egyptian hieroglyphs. The term modularity is defined as the standardized (recognized by conventions) and interchangeable use of components as structural units, but is also used to describe parts of a signal circuitry (with a clear input–output relationship) [53], or more versatile parts of a programme where the input-output relationship is defined by the context [35]. If such modules are used as the first manageable syntactic units of protein linguistics, then protein modules may reveal a first glimpse into the organization of protein language, just as the 'Rosetta Stone' provided the first access to understanding hieroglyphs. The validity for using protein modules is further underscored when the conservation of structure and/or function in a given module is weighed against the evolutionary pressure [11]. Like in a language, where the frequency of syntactic units, the persistence of phrases, and context relationships can be used to decipher even encrypted codes or languages, the persistence of modules and the thorough understanding of their function(s) and functional plasticity in the context of their diverse biological environment can help in unravelling the first grammatical pieces of protein linguistics.

4 The Cytoskeleton

Protein interaction network diagrams demonstrate the impressive wiring of pathways in cells. The potential interactions of components that are being revealed in the diverse '–omes' (like the proteome, genome, metabolome, interactome, transcriptome, etc.) is bewildering. Proteome-wide analysis of a single interaction module that is present, on average, in 100–200 individual molecules can result in over 10,000 potential interactions [5,52,84,88]. Protein–protein interactions consequently lead to transient or more stable protein complexes, which generate new signs by means of their assembly. The underlying codes and conventions are thus not easily accessible. However, rather simple biochemical operations, like homotypic dimerization (between two identical molecules) is similar to a basic linguistic operation that can potentially modulate the semantic content, and thus generate a new sign. Interestingly, each addition of another copy of the same molecule increases the linguistic complexity and hence the versatility of the sign. The assembly of actin filaments follows this scheme of homotypic multimerization of identical subunits. We may thus consider the actin cytoskeleton as a prototype for a molecular code generated by the assembly of identical subunits. The cytoskeleton may also be one of the oldest cellular conventions and all cell types rapidly respond to (or interpret) changes in the state of the cytoskeleton. This results in the generation of meaning [8] for the cell that requires coding. Therefore, I will take a closer look at the potential of the cytoskeleton to act as a code maker, as initially proposed by Barbieri [7].

One particular puzzle in biology is the evolution and function of the cytoskeleton, the functional plasticity of the molecules and adapters that regulate the dynamic assembly and disassembly of the individual cytoskeleton filament systems, and the temporally and spatially regulated interactions with cellular components and components of the environment of the cell. The cytoskeleton governs cell shape and force generation and notably, the majority of protein components in this system are built in a modular fashion. Luc Steels [75,76] noted that: 'evolution is the attempt to convert implicit meaning into explicit meaning, and to eliminate explicit meanings when they have become obsolete'. Indeed, the majority of new words in a given language that describe novel meanings are not constructed de novo, but rather by using existing words with an established meaning (that are accessible to existing conventions), and whose meaning can be used in an analogical or even metaphorical way to generate the new code for describing the new meaning. Here are some examples:

- Agri-tourism - the practice of touring agricultural areas to see farms and often to participate in farm activities
- Aqua-scape - an area having an aquatic feature
- Bio-diesel - a fuel similar to diesel, but commonly derived from vegetable sources
- Botox - the exotoxin derived from *Clostridium botulinum*
- Ring-tone - the sound made by a cell phone to signal an incoming call
- Spy-ware - unwanted, self-installing software that transmits information about the computer activities via the Internet
- Supersize - to increase considerably the size, amount, or extent of

(Source: Merriam-Webster online dictionary; available at: http://www.m-w.com/dictionary)

This principle is also observed in protein modularity throughout evolution as far as we can monitor the occurrence of protein species. Just as much as the use of an 'old' word or phrase to describe a certain meaning (even in the same language but from a different level of language evolution) can cause considerable confusion and even prevent transmission of the meaning as a whole (e.g. phone booth, cold war, rule of thumb [originally: an old English law declared that a man could not beat his wife with a stick any larger than the diameter of his thumb]; Available at http://www.rootsweb.com), the use of novel or old protein linguistic lexical entries (new proteins generated by in-frame gene fusion events, or viral components) can lead to substantial problems in cellular communication, and even to diseases. This principle appears also well-conserved in the primary modular structure of cells, the cytoskeleton.

Barbieri underscored that 'the synthesis of proteins is code making' [7]. This fundamental notion and the vigorous pursuit of its validation as an underlying principle is derived from the idea that it is indeed the generation of the sign that generates and drives the biosemiotic process of communication at the cellular/molecular level. It may, however, be useful to consider that in light of what I have discussed above (also with respect to the similarities between natural languages and modular protein assembly), the de novo synthesis of proteins is only one way to generate a sign. In cytoskeleton filament assembly the synthesis of filaments from subunits creates a new molecular species (filamentous (F)-actin) that is biochemically and semiotically different from the globular, monomeric (G)-actin subunits [16,70]. The repeated structure of subunits (one in the case of actin filaments, two subtly different subunits in the case of the alpha/beta tubulin building unit for microtubules) creates a different set of signs, and the code is read by a different set of adapters. In addition to the obvious differences between G- and F-actin (that have a clear basis, which can be described by mathematical equations that determine surface profiles and charge distributions, etc.) there are examples indicating that F-actin has varying qualities depending on its dynamic status and cellular context. Actin dynamics regulate cellular aging, and increasing cytoskeleton dynamics in yeast cells extends their life expectancy by 60% [32,33]. Actin and intermediate filaments inside the nucleus regulate gene transcription and chromosomal organization [38,40,50]. Cytoskeleton-mediated mechanotransduction in response to extracellular forces, and the levels of intracellular tension on the actin cytoskeleton regulate the isoform-specific incorporation of actin subunits [31].

Despite the findings of structurally related homologues of actin, tubulin, and intermediate filaments in bacteria [30], it is generally conceived that cytoskeletal systems are a hallmark feature of eukaryotic cells [59,63] that clearly distinguishes them from prokaryotic cells (of either bacterial or archaeal origin). From the evolutionary standpoint no prokaryote possesses true homologues of the major mammalian cytoskeleton proteins, such as actin or tubulin. However, proteins have been identified that exhibit not only limited sequence similarity with components of the actin, tubulin, and intermediate filament cytoskeleton, but that are also able to form filaments in vivo, namely MreB, a potential actin ortholog, FtsZ that closely resembles tubulin, and Crescentin that shows similarities to intermediate filament proteins [30,51,71]. Comparative genomic analysis revealed that the microtubule system of *Giardia lamblia* contains a complete array of microtubule cytoskeleton proteins, including all tubulin isoforms, as well as kinesin and dynein motor proteins. However, the actin cytoskeleton is lacking the majority of actin associated and regulatory proteins, or actin motors, other than actin itself. These results challenge the common assumption of a strong conservation of these proteins, and have intriguing implications for the evolution of the cytoskeleton, suggesting diversification of the microtubule cytoskeleton at an earlier time point than the microfilament system. [23,66].

The concept of the filamentous cytoskeleton is not anymore an invention of the eukaryotic cell – the principles are clearly visible in more ancestral cellular life forms, and if the cytoskeleton elements are indeed part of Nature's code-making plan, then this biosemiotic pathway was effective already in the earlier prokaryotes. But also the eukaryotic cytoskeleton elements are not variations of a theme or duplications of an archetypical pre-cytoskeleton – there are striking differences between the individual filament systems with respect to the mode that they employ to generate filaments. Actin filaments and microtubules follow the pattern of dynamic treadmilling [17] and instability [9,81,82], respectively. Both polymerization processes consume energy (ATP in the case of actin filaments, or GTP for microtubules), and the binding to nucleotide triphosphates is essential for the formation of filaments in both prokaryotic and eukaryotic actin and tubulin-based polymers. Although the structural basis for this may possibly be explained there is little rationale for such high-energy consumption other than maintaining a dynamic equilibrium.

Actin and tubulin polymerization are in contrast to the intermediate filament system [39,45]. There the formation of filamentous polymers is independent of the availability of tri phosphates in the GTP or ATP form [see also 59]. The assembly rather points towards a different evolutionary origin, as the assembly process is more akin to that of viral coat proteins, although genes for intermediate filaments appear to be a specific feature of eukaryotes as they are only poorly represented in bacteria. Intermediate filament turnover depends on the cell cycle machinery (a nuclear/cytoplasmic function) and not on the state of mitochondria, and intermediate filaments have the potential to remain stable even after the cellular life has ceased. Notably, the support of the fragile nuclear envelope is aligned and stabilized by nuclear intermediate filaments. So, while actin filaments and microtubules may be features that were contributed by the assimilation of bacterial ancestors, intermediate filaments may be remnants of viral infection or symbiosis.

Sometimes, cytoskeleton elements and their functions and evolutionary history are used inappropriately or incompletely to support surprising theories. In his 'cell body theory' Baluska includes the notion of microtubules as slaves of the cell body [4]. But while he discusses the actin filament and microtubule issue in his work, there is no explanation of intermediate filaments and a neglect of the bacterial ancestry of the cytoskeleton. Hence, the proposed 'explosive evolution of eukaryotic life' is poorly sustained by this theory. In other, more static theories the cytoskeleton is reduced to a purely mechanistic force generator. This might appear reasonable when the cytoskeleton is removed from its cellular context (see below), but clearly cytoskeleton elements are highly dynamic, and force comes from the polymerization cycles, the diverse and transient interaction with molecular motors, and the dynamic adhesion to the cell membrane.

The individual cytoskeleton systems are not variations on a theme, but separate entities that diversify the cytoskeleton code further – into an actin code, a microtubule code, and an intermediate filament code. Elements of the cytoskeleton also regulate and orchestrate transcriptional and translational events, and are a necessary prerequisite for the correct translation of the genetic code. Extrapolating from G. Witzany's notion, we can view the dynamic cytoskeleton as an 'internalized, shared background knowledge of a historically evolved linguistic community' [86,87]. Thus, the 'interpretation' of DNA as genetic material during the developmental cycle, is knowledge that is encrypted in the cytoskeleton of the egg cell. Barbieri substitutes the 'interpretation' via a mind with *code making* by the cell. This is important as the exchange of the genetic material in the nucleus by conventional genetic manipulations of cellular cloning alters the details of the process, but not the generation of codes per se. Extranuclear organelles (the Golgi apparatus, mitochondria) can influence heredity [49] and hence genetic determination is not exclusively governed by the genes in the nucleus. Extending such a view further it appears that following the deciphering of the genetic code, deciphering the cytoskeleton codes will indeed be a next major step forward in understanding biological structure and function, and life itself. Undoubtedly, understanding dynamic cytoskeleton assembly will provide useful new theories for further biosemiotic considerations.

5   The Cytoskeleton is a Self-Reproducing von Neumann Automaton

The von Neumann architecture is based upon the principle of one complex processor that sequentially performs a single complex task at any given moment. Space is divided into states that are either inactive or active (biologists use a similar terminology for the states of small signalling GTPases). At each time point a special set of rules is used to determine its next state as a function of its current state and the state of its immediate neighbour. Based on these local interactions an initially specified structure (a von Neumann automaton) constructs a duplicate of itself through a sequence of steps [47]. While the above definition stems from the description of computational attempts to generate self-replicating and self-healing programmes, it appears that similar parameters can be applied to cytoskeleton systems. The *concept* of cytoskeleton filament assembly (but not the individual elements of the cytoskeleton) may exist from the earliest cellular life forms onwards (clearly in bacteria, as shown above, and if the underlying concepts of protein assembly mechanisms are concerned, maybe even in viruses). From a protein linguistic/biosemiotic standpoint it is not too astonishing that bacterial cytoskeleton elements are poorly conserved at the amino acid sequence level. In unrelated languages, there is no phonetic resemblance for certain meanings, yet their semantic content is identical. Moreover, the underlying grammars are the same and based on the identical basic concept/parameters [2,19].

6 A Modular Code Encapsulated in the Cytoskeleton

The lack of an earliest common ancestor is generally a problem in evolutionary and comparative biology and some 'intermediate interpreters' or 'code makers' may have become lost during evolution. Yet, some of the early, basic, and essential rules that had functioned in the regulation of the first molecular biosemiotic inter-actions must still exist – and the cytoskeleton appears to be one of those very early biosemiotic constructs in Nature. Barbieri underlined that 'signs and meanings are mental entities when the code maker is the mind, but they are organic entities when the code maker is made of organic structures' (molecules, proteins, etc). According to our current knowledge of single cells there is no structure equivalent of a brain (as defined for cognitive sciences by the concerted action of a multicellular organ) to compute any of the signals/communication processes. Rather there may be a set of codes [7], and here the cytoskeleton can have an important role. Rules define code making and application. Information must be information for something, and is the result of a process. Hence, organic meaning is defined by its coding rules. Trifonov proposed that: 'a code is any pattern in a sequence which corresponds to one or another specific biological function' [79]. This definition can be applied to cytoskeleton filaments (particularly the actin filament system) in which the array of G-actin monomers in the filament generates diverse patterns within the sequence of monomers. Analogous to the regulation of the genetic code in coiled DNA [56], changes in actin filament structure (like twist and curvature) induced by adapter molecules can be transmitted over a substantial length of the filament and alter the pattern albeit no direct molecular interaction of adapter molecules with these sites occurs. These long-range effects can cause subtle changes in affinities of other adapters and thus modulate the biological meaning of a restricted part of the filament. As a consequence the sign stops being a sign for some adapters, but becomes a sign for others.

Physiology describes processes rather than components or structures, and has at its basis the concept that life is a dynamic process. The continuous polymerization and depolymerization cycle of cytoskeleton components is a reproducible, regulated, and error-correcting process. This process conveys information that goes beyond the presentation and occupancy of docking sites and thus fulfils Barbieri's criteria of code, code-making, and importantly also that of artefact making. The cytoskeleton-mediated cellular interactions depend on the dynamics of a non-living matter that cycles between a monomeric and a polymeric state that conveys different information in each state – and in addition has dual polarity. If the assembly of protein modules into higher order functional states indeed follows linguistic rules, then the proteome is also a linguistic community in which the stable and almost invariant cytoskeleton acts as background knowledge that is common to all cells, and is thus in the position to act as an general interpreter of/for code(s) – a versatile code maker in the Barbieri sense.

As discussed above, one cannot limit such considerations to the eukaryotic or multicellular (tissue/organism) world – the conventions must function also at the level of the single cell, with or without a nucleus. The physically and chemically determined parameters of polymerization of actin filaments are the building instructions, or grammar, which are present in each single G-actin subunit, thus functioning as an intrinsic molecular memory for the code maker. The process of polymerization clearly separates itself from the copy-making level as it regulates the complex dynamic structural rearrangement of itself (the G-actin) for the construction of something different than itself (the F-actin filament). Cytoskeleton dynamics reveal an astonishing diversity of the code-making principles. For one the act of polymerization consumes energy from nucleotide triphosphates. Secondly, mutational information is stored in the structure. Thirdly, the system is adaptive, it is self replicating, dynamic, and reversible. Finally, the system must posses the basic concept of self-reflection as actin expression is regulated by RNA–protein interactions, and actin mediates functions in the cytoplasm, in the nucleus, in the diverse cellular membrane compartments, and on cellular organelles. Hence, the cytoskeleton is a universal interpreter or code maker, and as such the cytoskeleton may be seen as an *organization principle*.

'Actin rocketing', the phrase that has been coined to describe the rapid propulsion of a number of pathogenic viruses and bacteria like *Shigella flexneri*, *Listeria monocytogenes*, or Vaccinia virus [26,34,58,64] was long thought to be a result of a miscommunication between the mammalian host cell and the invading pathogen. This view is documented by current textbook knowledge where pathogens 'highjack' the cellular actin polymerization machinery. Recent findings, however, have demonstrated that actin 'comet tails' also form inside the mammalian cell in the absence of viral or bacterial infection, and that the process of local asymmetric actin polymerization on solid fragments (membranous organelles, etc.) is used to rapidly transport cellular components, and to extend the surface of the plasma membrane for tissue and extracellular matrix invasion [3]. Notably, the core organizational and functional principles of the cytoskeleton remained conserved despite the possibility of genome fusions, as suggested by the symbiogenesis theory. If modern eukaryotes represent the endpoint of a continuous fusion process of DNA material from bacteria and viruses then the highjacking of cytoskeletal assembly processes by enteropathogenic bacteria is less astonishing, and the term rather inappropriate. Rather these pathogens appear to communicate perfectly well with the mammalian cell cytoskeleton. This potentially means that they contain the necessary elements for communication in their genetic material that provide knowledge of the meaning of the cytoskeleton.

7 Nature is Structured in a Language-like Fashion

The above considerations lead us back to the idea that molecular interactions are rule-governed and sign-mediated, and that cellular life is grounded on a linguistically structured and communication-oriented organization of life. Eigen suggested that: 'a prerequisite for the development of a language is a unambiguous symbol assignment' – that is, precise coding and code making [7,22]. For the molecular world this will be determined by physicochemical parameters regulating these interactions. However, it becomes increasingly clear today that for both molecular interactions and human languages an unambiguous assignment is not a prerequisite and that there is ambiguity in protein-protein interactions as much as there is in language. As few as 18 basic parameters can govern the entire grammatical repertoire of all known languages [2], and a similar scenario might be underlying all biological communication principles, including the arrangement and interaction of molecular components. Once again one arrives at the point where language as a sign combination matches the modularity of proteins. The transmission of signals or information is confined by the physical parameters like temperature, frequency, charge, folding (molecular recognition, molecular interaction, conformational modification, consumption or release of energy, etc). Signals are to be transmitted via physical interactions and cause a change in the energetic state of a molecular component, a protein in this case. Information about the signal process is stored in conformation (this is not new) but the question arises if there is also information (sign) in a process itself. For the actin cytoskeleton there appears to be no gain–loss relationship between the G- and F-conformation, but rather a qualitative change from state A to A'. As noted above, the process of polymerization of biological (cytoskeleton) components is not static but dynamic, and the dynamics per se appear to signal to other components in the cell conveying information about the state of the cell and the environment.

As with (mostly natural) languages the interpretation of the code/sign is context-dependent not at the grammatical level (in the Chomsky sense), but at the environmental level – this predetermines the 'semantic/semiotic expectation' in which the sign is received. And a context-dependent grammar can accommodate time-resolved plasticity. The current revival of more holistic views, reinvented and praised as

Systems Biology, raises the awareness that there is no context independence in biological processes just as much as there is no meaning in DNA sequence if it is not transcribed into something else, or engages in physical contact with other molecules. According to the definitions by Don Favareau that 'dynamics equals constant action' (the definition of the 'interpretant' in classical semiotics) dynamics as such then represents the latest point in the cascade of biosemiotic sign actions [24]. In the example of the cytoskeleton, the act of polymerization is viewed as an act of communication in which the context is precisely defined for every situation. In case of the actin filament it is not so much the mere production of the protein subunits (synthesis), but rather their subsequent 'grammatical trimming' by folding, assembly, post-translational modification, and proteolysis. The concept of grammatical trimming of actin filaments is a prerequisite for communicative anticipation or expectation. This communicative anticipation is seen in those cases in which the re-expression of a given cytoskeleton molecule following the initial knock down of the protein does not restore certain cellular pathways, despite the biochemical functionality of the molecule. Trimming restricts the possible interactions of the filament and thus acts as a truly grammatical parameter. As with most linguistic grammars the trimming does not reduce the total functions of the actin filaments but rather selects those required in a given situation.

8   Conclusions

Cognition is commonly portrayed as an abstract property of higher living organisms, and thus as a direct property of a brain. Cognitive research has focused on the capacities of abstraction, generalization, and specialization, and involves concepts like belief, knowledge, desire, preference, and intentions of intelligent individuals (or objects and systems). But cognition also means *the act of knowing* and can be interpreted in a social or cultural sense to describe the development of knowledge and concepts (modified from: Wikipedia, the free encyclopedia). *Mind* generally refers to the collective or combinatorial aspects of intellect, consciousness, and emotionality. As for our present understanding, the mind is a phenomenon of psychology and is thus tightly coupled to the human brain. It is clear, however, that at least certain aspects of biosemiotics that either do not involve humans or not even organisms run into difficulties with relating sign and meaning to molecular entities. Discussions about meaning commonly reduce the issue to human cognitive processes, but in the absence of a human brain (characterized as a specific multicellular organ with particular network functions) restrictions or specific considerations relating to the brain-dependent computation of signs cannot apply. Thus, if molecular biosemiotics is indeed a valid consideration, then the triad of sign, meaning and code maker must be put at a molecular level.

It is understood that protein linguistic attempts are in their infancy and that considerable progress must be demonstrated in the near future in order to show the usefulness of this route. Barbieri's 'Introduction to Biosemiotics' explains that organic meaning is the result of a coding process and that the presence of a code maker and of adapters is sufficient to bring a semiotic process to life. The cytoskeleton shows all the hallmarks of a semiotic mechanism: it is a modular, universal, and dynamic structure that consumes energy; it employs modular adapters to modulate its polymerization cycles; it interacts with all cellular compounds (proteins, nucleic acids, lipids, ions) and membrane-coated organelles; it transverses all cellular compartments, and interacts with the extracellular environment. But it also creates new active biosemiotic structures via a continuous and cyclic assembly process that itself has biosemiotic potential. It is an indispensable, cellular, molecular machine based on protein subunits that creates itself, and turns itself over – and it does so with high fidelity and precision. The cytoskeleton may contribute one of the most essential codes to life as polymerization works outside the cellular environment as well as it does within the context of membranous cellular compartment. Modulation of cytoskeleton assembly and stability by adapter molecules predicts and determines the future functional state of the cell (similar to the combination of histone modifications together with their adaptor molecules [80]), and allows the cell to anticipate the changes. Just like the genetic code the cytoskeleton is a true biosemiotic mechanism that fulfils all the necessary requirements – it generates signs whose purpose is to provide orientation or direction to molecular interactions and that enables organic meaning to become accessible. Language and code require rules in order to become tools for communication. A biosemiotic approach to understanding protein assemblies in cells and using the cytoskeleton as a model system appears a promising lead. Many of the theories that will derive from this endeavour will not be correct – but the critical discourse with biosemiotics will help us formulate significant questions that will allow us to gain a better understanding of the biological world, the living Nature, and of life as a whole.

Acknowledgements I am indebted to Marcello Barbieri, Don Favareau, and Guenther Witzany for making me acquainted with the Biosemiotic World.

References

1. Aasland R et al. (2002) Normalization of nomenclature for peptide motifs as ligands of modular protein domains. FEBS Lett 513:141–144
2. Baker MC (2001) The Atoms of Language. Basic Books
3. Baldassarre M, Ayala I, Beznoussenko G, Giacchetti G, Machesky, LM, Luini A, Buccione R (2006) Actin dynamics at sites of extracellular matrix degradation. Eur J Cell Biol 85:1217–1231
4. Baluska F, Volkmann D, Barlow PW (2004) Eukaryotic cells and their cell bodies: cell theory revised. Ann Bot 94:9–32
5. Barabasi A-L, Oltvai ZN (2004) Network Biology: understanding the cell's functional organization. Nature Rev Genetics 5:101–113
6. Barbieri M (1981) The ribotype theory of the origin of life. J Theor Biol 91:545–601
7. Barbieri M (2003) The organic codes: an introduction to semantic biology. Cambridge University Press, Cambridge

8. Barbieri M (2005) Life is artifact making. J Biosemiotics 1:107–134
9. Bayley PM, Martin SR (1991) Microtubule dynamic instability: some possible physical mechanisms and their implications. Biochem Soc Trans 19:1023–1028
10. Benner SA, Gaucher EA (2001) Evolution, language and analogy in functional genomics. Trends Gene 17:414–418
11. Benner SA, Sismour M (2005) Synthetic biology. Nature Rev Genetics 6:533–543
12. Bogusky MS (1999) Biosequence exegesis. Science 286:453–455
13. Botstein D, Cherry JM (1997) Molecular Linguistics: extracting information from gene and protein sequences. Proc Natl Acad Sci USA 94:5506–5507
14. Brendel V, Busse HG (1984) Genome structure described by formal languages. Nucleic Acid Res 12:2561–2568
15. Brendel V, Beckman JS, Trifonov EN (1986) Linguistics of nucleotide sequences: morphology and comparison of vocabularies. J Biomol Struct Dyn 4:11–21
16. Brenner SL, Korn ED (1983) On the mechanism of actin monomer-polymer subunit exchange at steady state. J Biol Chem 258:5013–5020
17. Carlier MF (1991) Nucleotide hydrolysis in cytoskeletal assembly. Curr Opin Cell Biol 3:12–17
18. Chomsky N (1964) The logical basis of linguistic theory. In: Proceedings of the 9th International Congress of Linguistics, Cambridge, Massachusetts/The Hague, pp. 914–978
19. Chomsky N (2005) Universals of human nature. Psychother Psychosom 74:263–268
20. Doerfler W (1982) In search of more complex genetic codes – can linguistics be a guide? Med. Hypotheses 9:563–579
21. (2002) Editorial Folding as grammar. Nature Struct Biol 9:713
22. Eigen M, Winkler R (1975) Das Spiel - Naturgesetze steuern den Zufall. Piper Verlag Muenchen
23. Eriksson HP (2001) Evolution in bacteria. Nature 413:30–31
24. Favareau D (2006) The evolutionary history of biosemiotics. In: Barbieri M (ed.) Introduction to biosemiotics. Springer life sciences. ISBN:978-1-4020-4813-5
25. Fitch WT, Hauser MD, Chomsky N (2005) The evolution of the language faculty: clarifications and implications. Cognition 97:179–210
26. Frischknecht F, Way M (2001) Surfing pathogens and the lessons learned for actin polymerization. Trends Cell Biol 11:30–38
27. Galzitskaya OV, Melnik BS (2003) Prediction of protein domain boundaries from sequence alone. Prot Sci 12:696–701
28. Ganapathiraju M, Manoharan V, Klein-Seetharaman J (2004) BLMT: statistical sequence analysis using N-grams. Applied Bioinformatics 3:193–200
29. Gimona M (2006) Protein linguistics – a grammar for modular protein assembly? Nature Rev Mol Cell Biol 7:68–73
30. Gitai Z (2005) The new bacterial cell biology: moving parts and subcellular architecture. Cell 120:577–586
31. Goffin JM, Pittet P, Csucs G, Lussi JW, Meister JJ, Hinz B (2006) Focal adhesion size controls tension-dependent recruitment of alpha-smooth muscle actin to stress fibers. J Cell Biol 172:259–268
32. Gourlay CW, Ayscough KR (2005) The actin cytoskeleton: a key regulator of apoptosis and ageing? Nature Rev Mol Cell Biol 6:583–589
33. Gourlay CW, Carpp LN, Timpson P, Winder SJ, Ayscough KR (2004) A role for the actin cytoskeleton in cell death and aging in yeast. J Cell Biol 164:803–809
34. Greber UF, Way M (2006) A superhighway to virus infection. Cell 124:741–754
35. Han J-D et al. (2004) Evidence for dynamically organized modularity in the yeast protein-protein interaction network. Nature 430:88–93
36. Hartwell LH, Hopfield JJ, Leibler S, Murray AW (1999) From molecular to modular cell biology. Nature 402:C47–C52
37. Hauser MD, Chomsky N, Fitch WT (2002) The faculty of language: what is it, who has it, and how did it evolve? Science 298:1569–1579

38. Herrmann H, Foisner R (2003) Intermediate filaments: novel assembly models and exciting new functions for nuclear lamins. Cell Mol Life Sci 60:1607–1612
39. Herrmann H, Aebi U (2004) Intermediate filaments: molecular structure, assembly mechanism, and integration into functionally distinct intracellular *Scaffolds*. Annu Rev Biochem 73:749–89
40. Huang S, Chen L, Libina N, Janes J, Martin GM, Campisi J, Oshima J (2005) Correction of cellular phenotypes of Hutchinson-Gilford Progeria cells by RNA interference. Hum Genet 118:444–450
41. Huynen MA, Snel B, Mering C, Bork P (2003) Function prediction and protein networks. Curr Opin Cell Biol 15:191–198
42. Ji S (1997) Isomorphism between cell and human languages: molecular biological, bioinformatic and linguistic implications. Biosynthesis 44:17–39
43. Ji S, Ciobanu G (2002) Conformon-driven biopolymer shape changes in cell modelling. Biosystems 70:165–181
44. Koonin EV, Wolf YI, Karev GP (2002) The structure of the protein universe and genome evolution. Nature 420:218–223
45. Kreplak L, Aebi U, Herrmann H (2004) Molecular mechanisms underlying the assembly of intermediate filaments. Exp Cell Res 301:77–83
46. Loose C, Jensen K, Rigoutsos I, Stephanopoulos (2006) A linguistic model for the rational design of antimicrobial peptides. Nature 443:867–869
47. Mange D, Madon D, Stauffer A, Tempesti G (1997) Von Neumann revisited: a turing machine with self-repair and self-reproduction properties. Robotics Autonom Syst 22:35–58
48. Mantegna RN, Buldyrev SV, Goldberger AL, Havlin S, Peng CK, Simons M, Stanley HE (1994) Linguistic features of noncoding DNA sequences. Phys Rev Lett 73:3169–3172
49. Margulies L, Dolan MF, Guerrero R (2000) The chimeric eucaryote: origin of the nucleus from karyomastigont in amitochondriate protists. Proc Natl Acad Sci USA 97:6954–6959
50. Mattout A, Dechat T, Adam SA, Goldman RD, Gruenbaum Y (2006) Nuclear lamins, diseases and aging. Curr Opin Cell Biol 18:335–341
51. Moller-Jensen J, Lowe J (2005) Increasing complexity of the bacterial cytoskeleton. Curr Opin Cell Biol 17:75–81
52. Papin JA, Hunter T, Palsson BO, Subramaniam S (2005) Reconstruction of cellular signalling networks and analysis of their properties. Nature Rev Mol Cell Biol 6:99–111
53. Park S-H, Zarrinpar A, Lim WA (2003) Rewiring MAP kinase pathways using alternative scaffold assembly mechanisms. Science 299:1061–1064
54. Pawson T (1995) Protein modules and signalling networks. Nature 373:573–580
55. Pawson T (2001) Specificity in signal transduction: from phosphotyrosine-SH2 domain. Trends Genetics 17:414–418
56. Pearson H (2006) Codes and enigmas. Nature 444:259–261
57. Pesole G, Attimonelli M, Saccone C (1994) Linguistic approaches to the analysis of sequence information. Trends Biotechnol 12:401–408
58. Ploubidou A, Way M (2001) Viral transport and the cytoskeleton. Curr Opin Cell Biol 13:97–105
59. Pollard TD (2003) The cytoskeleton, cellular motility and the reductionist agenda. Nature 422:741–745
60. Popov O, Segal DM, Trifonov EN (1996) Linguistic complexity of protein sequences as compared to texts of human languages. Biosystems 38:65–74
61. Przytycka T, Aurora R, Rose GD (1999) A protein taxonomy based on secondary structure. Nature Struct Biol 6:672–682
62. Przytycka T, Srinivasan R, Rose GD (2002) Recursive domains in proteins. Prot Sci 11:409–417
63. Rivero F, Cvrcková F Origins and Evolution of the Actin Cytoskeleton. In: Gáspár J. (ed.) Origins and Evolution of Eukaryotic Endomembranes and Cytoskeleton. ISBN: 1-58706-204-6

64. Rottner K, Stradal TE, Wehland J (2005) Bacteria-host-cell interactions at the plasma membrane: stories on actin cytoskeleton subversion. Dev Cell 9:3–17
65. Russ WP, Lowery DM, Mishra P, Yaffe MB, Ranganathan R (2005) Natural-like function in artificial WW domains. Nature 437:579–583
66. Sagolla MS, Dawson SC, Mancuso JJ, Cande WZ (2006) Three-dimensional analysis of mitosis and cytokinesis in the binucleate parasite Giardia intestinalis. J Cell Sci 119:4889–4900
67. Searls DB (2001) Reading the book of life. Bioinformatics. 17:579–580
68. Searls DB (2002) The language of genes. Nature 420:211–217
69. Searls DB (2003) Trees of life and of language. Nature 426:391–392
70. Sheterline P, Handel SE, Molloy C, Hendry KA (1991) The nature and regulation of actin filament turnover in cells. Acta Histochem. Suppl 41:303–309
71. Shih YL, Rothfield L (2006) The bacterial cytoskeleton. Microbiol Mol Biol Rev 70:729–574
72. Sim J, Kim SY, Lee J (2005) PPRODO: prediction of protein domain boundaries using neural networks. Proteins 59:627–632
73. Socolich M, Lockless SW, Russ WP, Lee H, Gardner KH, Ranganathan R (2005) Evolutionary information for specifying a protein fold. Nature 437:512–518
74. Sonnhammer ELL, Kahn D (1994) Modular arrangement of proteins as inferred from analysis of homology. Protein Sci 3:482–492
75. Steels L, Oudeyer P-Y (2000) The cultural evolution of syntactic constraints in phonology. In: Artificial Life VII: Proceedings of the seventh International Conference on Artificial Life. The MIT Press, Cambridge, Massachusetts, pp. 382–394
76. Steels L (2000) The puzzle of language evolution. Kognitionswissenschaft 8:143–150
77. Sudol M (1998) From src homology modules to other signalling domains: proposal of the 'Protein Recognition Code'. Oncogene 17:1469–1474
78. Sudol M, Recinos CC, Abraczinskas J, Humbert J, Farooq A (2005) WW or WoW: the WW domains in a union of bliss. IUBMB Life 57:773–778
79. Trifonov EN (2000) Earliest pages in bioinformatics. Bioinformatics 16:5–9
80. Turner BM (2007) Defining an epigenetic code. Nature Cell Biol 9:2–6
81. Tuszynski JA, Trpisova B, Sept D, Sataric MV (1997) The enigma of microtubules and their self-organizing behavior in the cytoskeleton. Biosystems 42:153–175
82. Van Buren V, Cassimeris L, Odde DJ (2005) Mechanochemical model of microtubule structure and self-assembly kinetics. Biophys J 89:2911–2926
83. Vendramini D (2005) Noncoding DNA and the teem theory of inheritance, emotions and innate behaviour. Medical Hypotheses 64:512–519
84. Vidal M (2005) Interactome modelling. FEBS Lett 579:1834–1838
85. Werner E (2005) Genome semantics, in silico multicellular systems and the Central Dogma. FEBS Lett 579:1779–1782
86. Witzany G (2006) From Umwelt to Mitwelt: natural laws versus rule-governed sign mediated interactions (rsi's). Semiotica 158:425–438
87. Witzany G (2006) Serial Endosymbiotic Theory (set): the biosemiotic update. Acta Biotheor 54:103–117
88. Wuchty S, Oltvai ZN, Barabasi A-L (2003) Evolutionary conservation of motif constituents in the yeast interaction network. Nature Genetics 35:176–179

Chapter 9

A Lipid-based Code in Nuclear Signalling

Nadir M. Maraldi

Abstract Cell signalling in eukaryotes requires mechanisms more complex than in prokaryotes, because the genome is segregated within the nucleus. This is not merely due to the physical gap between the receptor and the genome, owing to the presence of the nuclear envelope, but because of the major complexity of the transcriptional and translational mechanisms in eukaryotes.

This chapter reviews the main evidence of a multiple localization of the inositol lipid signalling system in the cell, i.e. plasma membrane, cytoskeleton, and nucleus. This results in a variety of functions, which depend on the intracellular localization (contestual modulation).

In the nucleus, the elements of the inositol lipid signalling system are located at nuclear domains involved in pre-mRNA processing and in the modulation of the chromatin arrangement. The nuclear signalling system presents the characteristics of an organic code; furthermore, it does not represent a redundancy of the system located at the plasma membrane, but the result of an evolutionary process.

1 Introduction

Signal transduction is required to allow the genome to selectively express a gene set in response to environmental changes. Specific chemical interactions between the environmental signals, *first messengers*, and the genome are not required in signal transduction, since a limited series of *adaptors* are capable of modifying genome expression. At the present stage of evolution, gene availability can be modulated by proteins belonging to the class of *transcription factors*. The interaction between the first messengers and the transcription factors is regulated by 'conventions', mediated by a limited number of *second messengers*. Convention means that there

Laboratory of Cell Biology and Electron Microscopy, IOR, Via di Barbiano, 1/10,
40126 Bologna, Italy, e-mail: maraldi@area.bo.cnr.it

M. Barbieri (ed.), *The Codes of Life: The Rules of Macroevolution.*   207
© Springer 2008

is no univocal correspondence in the chain first messenger–second messenger transcription factor. In fact, the correspondence depends on the context, i.e. cell type, differentiation stage, etc. This implies that signal transduction is a mechanism in which meaning evolves, without information changes, as occurs for natural codes. In signal transduction, indeed, there is a connection between two independent systems, the environment and the genome; the information exchanged can evolve by modifying its meaning according to the context; the connections and their variations are subjected to rules that promote the biological specificity. The rules that cells utilize to regulate the effects that extracellular signals have on genes, independently of the energy and the information that they carry, constitute a *signal transduction code*, as implied by the organic code theory (Barbieri, 1998). Both the signal transduction and the splicing codes, however, differ with respect to the genetic code. In fact, whilst the genetic code is practically universal, in the other cases, there are different sets of rules depending on the cell context, and we are therefore in the presence of a plurality of codes. This implies that the information carried by signals undergoes two transformations before reaching the genome. Firstly, it is transformed, according to the rules of the signal transduction code, into a series of intracellular signals, and then these signals are given contextual information by being channelled along three-dimensional complex circuits (Barbieri, 2003).

Signal transduction is a multistep mechanism. In prokaryotes two steps are involved: environmental signals, such as osmolarity, pH, and specific ion concentration, are sensed by a component of the transduction pathway, which acts as a *sensor*. As a rule, sensors are represented by a transmembrane protein, which, by phosphorylation, activates a further component, which acts as a *transcriptional activator* (Albright et al., 1989). In eukaryotes the system is more complex due to the genome segregation into the nucleus. This is not due to the presence of the nuclear envelope; in fact, also liposoluble agonists, which freely reach the nucleoplasm, require receptor proteins acting as *adaptors*, capable of recognizing specific DNA sequences through protein motifs.

Receptor proteins for non-liposoluble agonists are either proteins with seven transmembrane domains associated to G proteins, or proteins with a single transmembrane coiled domain with a direct kinase activity, or associated with protein kinases. Receptor proteins are able to recognize the agonists and to originate either directly, or more frequently through an adaptor molecule, the second messengers. Whilst the number of agonists is very high, second messengers are relatively few. The firstly identified second messenger was cyclic AMP (cAMP); all receptors that act via cAMP are coupled with a stimulatory G protein (G_s), which activates the enzyme adenyl cyclase, thereby increasing cAMP concentration. cAMP, in turn, exerts its effects mainly by activating cAMP-dependent protein kinase A (PKA) which phosphorylates, upregulating their activity, target proteins.

A further group of second messengers is constituted by the phosphoinositides, derived from the inositol phosphorylation at the D-3, -4 and -5 positions, yielding a family of seven distinct signalling molecules identified to date (Toker, 2002). The most important, although the least abundant of the inositol lipids in the cell, is the phosphatidylinositol 4,5-bisphosphate ($PI(4,5)P_2$), a membrane phospholipid, which is hydrolyzed to inositol 1,4,-trisphosphate (IP_3) and diacylglycerol (DG) by the plasma membrane-bounded enzyme phospholipase C-β (PLC-β), activated by a G-protein linked receptor. IP_3 and DG act as second messengers, by releasing Ca^{2+} from intra-cellular stores, and activating protein kinase C (PKC), respectively. Other phosphoi-nositides, but also PI(4,5)P_2 can act directly as second messengers (Irvine, 2003).

While an agonist is recognized by a specific receptor, the effect induced by the elicited signalling pathway is different depending on the cell type, cell cycle phase, or developmental stage. This depends on the variety of the adaptor molecules, which results into a release of different second messengers, or on the different intracellular localization of the signal transduction pathway.

In the following sections we will focus on a peculiar aspect of signal transduction based on phosphoinositides. In particular, we will consider that this pathway is not restricted, as other signalling pathways, to the plasma membrane, but localized also at the cytoskeleton and nuclear level. This experimental finding provided evidence that in the cell, independent signalling processes exist, although based on the same second messengers. Furthermore, these independent signalling pathways play different functions depending on the cellular context. The different intracellular localization of these signal transduction complexes depend on the formation of molecular interactions among the lipid substrates, the enzymes involved in their phosphorylation or hydrolysis, as well as the molecular targets of the released second messengers. The precise intracellular location of these complexes depends on the presence of specific components which are different at the plasma membrane, at the microfilament cytoskeleton and inside the nucleus.

2   Multiple Role of Inositides in Signal Transduction

In recent years, PI(4,5)P_2 has been shown to be not only the source of second messengers at the plasma membrane, but also to be directly involved in regulating a daunting number of cellular processes, including assembly and regulation of the actin cytoskeleton, and of the intracellular trafficking of exo and endocytotic vesicles (Czech, 2002). PI(4,5)P_2 is localized in distinct subdomains of the plasma membrane, as well as at the Golgi stacks and endoplasmic reticulum. Its bisphos-phorylated isomers PI(3,4)P_2 and PI(3,5)P_2 are involved in the trafficking of multivesicular bodies, or in anion superoxide production after phagocytosis, respectively (Cooke, 2002; Cullen et al., 2001). Also PI(3,4,5)P_3 regulates vesicular trafficking, actin dynamics, and cell proliferation (Toker, 2002), whilst PI(3)P modulates endosomal trafficking, ion channel regulation, and cell survival (Osborne et al., 2001). The soluble head-group IP_3 is also the precursor to a family of inositol polyphosphates involved in the intracellular calcium homeostasis, ion channel physiology, and membrane dynamics (Irvine and Schell, 2001).

The multiplicity of phosphoinositide functions also depends on their ability to be segregated at different intracellular compartments, characterized by a specific set of interacting proteins (Fig. 1). In fact, phosphoinositides can act directly as

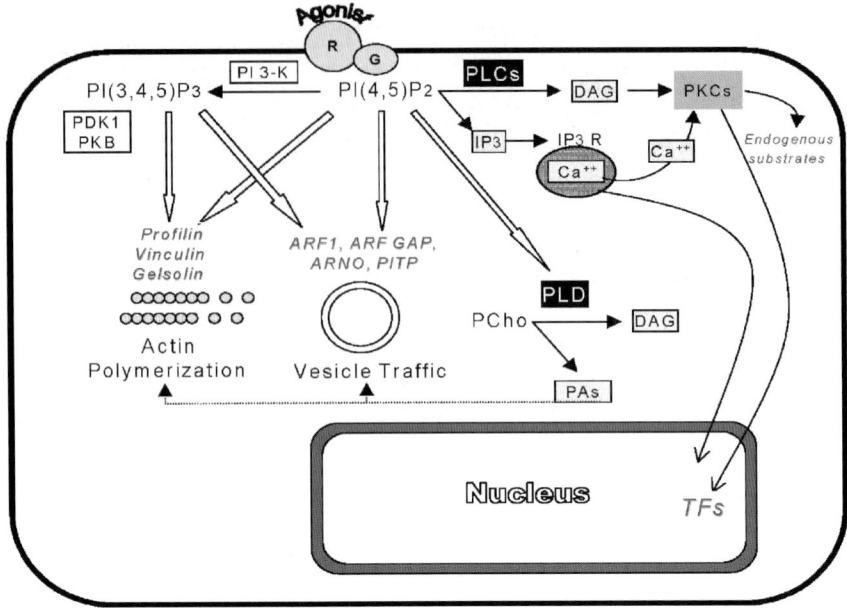

Fig. 1 Multiple roles of inositides in signal transduction. PI(4,5)P$_2$ is indicated both as precursor of second messengers (*black arrows*) and as a direct signalling molecule (*open arrows*). Through interaction with protein modules (*grey italics*) it affects vesicle trafficking and actin cytoskeleton assembly. In the upper part of the drawing is represented the recruitment, at the membrane-associated receptor, of the elements of the signalling cascade. PKC isoforms are the target of DAG and Ca^{2+} and provide phosphorylation of cytoplasmic transcription factors. PKC isoforms and Ca^{2+} can also be translocated to the nucleus (*bent arrows*). Graphic outlines used in this and other drawings: *black square*, inositol PLC isoforms; *open square*, lipid kinases; *gray square*, inositol-derived second messengers; *grey italics*, molecular targets

signals, because their polar heads recognize protein domains, such as plekstrin homology (PH) or src oncogene homology (SH) domain. Since the affinity of phosphoinositides to proteins can greatly vary, the localization of a phosphoinositide at a given site induces a lipid-dependent segregation of the related enzymes, including lipid kinases and lipases (Anderson et al., 1999). Many of the actin-regulating proteins, such as profilin, gelsolin, and vinculin, present protein affinity domains to PI(4,5)P$_2$. The presence of this lipid substrate at the cytoskeleton level is able to segregate at the same site some lipid kinases and phospholipases. In this sense, actin microfilaments behave as a platform to coordinate the different constituents of the signalling system. For example, it is known that PLCγ recognizes a SH3 domain of profilin; because PI(4,5)P$_2$ associated to profilin is the elective substrate of PLCγ, this results in the formation of actin-profilin complexes, that prevent actin polymerization. An inverse effect is caused by PI(3,4,5)P$_3$ (Toker, 2002).

The complexity of the system is further increased by the observed translocation of some phosphoinositide-related enzymes in response to agonists. Generally, indeed, cytosolic enzymes are translocated to membrane receptors upon agonist stimulation. However, since lipid substrates capable of sequestering enzymes are located at other cellular sites, translocations involve also other districts. For example, PI-3 kinase, depending on the agonist, translocates to the plasma membrane, the microtubule organizing centre, intracellular vesicles, cytoskeleton, or to the nucleus.

3 Lipid Signal Transduction at the Nucleus

A short history will resume the main experimental evidence accumulated in the past twenty years on the presence, precise localization, and functional significance of nuclear phosphoinositides involved in a signalling pathway completely independent of the cytoplasmic counterpart.

The presence of lipids and mainly phospholipids in the nucleus has been known since the 1970s; in particular, Manzoli and coworkers defined the composition of nuclear phospholipids and outlined a possible relationship among nuclear phospholipid species and amounts and DNA replication, mRNA transcription, and chromatin organization (Manzoli et al., 1977, 1979; Maraldi et al., 1984). The first direct evidence on the existence of a lipid signalling system in the nucleus was the identification, in mouse erythroleukemia cells, of separate nuclear pools of PIP and $PI(4,5)P_2$ that were metabolized differently from lipids in the cytoplasm (Cocco et al., 1987). Subsequently, a distinct polyphosphoinositol lipid metabolism was shown to be regulated by insulin-like growth factor-1; the increased amount of diacylglycerol due to the $PI(4,5)P_2$ hydrolysis by PI-PLCβ1 caused a translocation to the nucleus of PKC, where it phosphorylates nuclear targets (Martelli et al., 1992). There is now solid and wide evidence that phosphoinositides are a key component of nuclei and that their physical characteristics, regulation of synthesis, and functions are all entirely different from their counterparts elsewhere in the cell, and that they constitute a signalling system at the nuclear level capable of modulating many cell activities (Irvine, 2003; Martelli et al., 2004; Hammond et al., 2004; Raben, 2006).

4 Clues for the Nuclear Localization of the Inositol Lipid Signalling System

A remarkable feature of the inositol lipid cycle present in the nucleus with respect to that found in the cytoplasm is the fact that it is located mainly outside of membrane bilayers. The second and possibly related distinctive feature is its operational distinctiveness from the cycle present at the plasma membrane. This implies that many agonists that stimulate the membrane cycle do not activate the nuclear system and vice versa, or that, when they are both respondent, they do so in a temporally distinct manner, or through different regulation mechanisms (Irvine, 2003).

The identification of the nuclear localization sites of inositol lipids and of the enzymes involved in their metabolism was a crucial issue for the interpretation of the functional significance of a signal transduction system located at the nuclear interior. In fact, compartmentalization is a potential solution to account for the different responses given by signal transduction pathways that share their molecular constituents.

The enzymes involved in inositol lipid metabolism were found not associated with the nuclear membrane but with intranuclear components. Cell fractionation methods did not ensure an absolute exclusion of contaminating membrane-derived lipids in the preparation of nuclei. Therefore, the presence of a given lipid substrate or of a related enzyme within the nucleus has to be demonstrated also by morphological techniques in whole cells not subjected to fractionation (Maraldi et al., 1992). Moreover, the identification of the intranuclear sites where the inositol lipid signalling systems are located could clarify which nuclear domain mediates the transduction of signals to the genome (Fig. 2).

The availability of a reliable antibody allowed the identification of the nuclear sites of localization of PI(4,5)P$_2$ (Mazzotti et al., 1995). The sites of PI(4,5)P$_2$

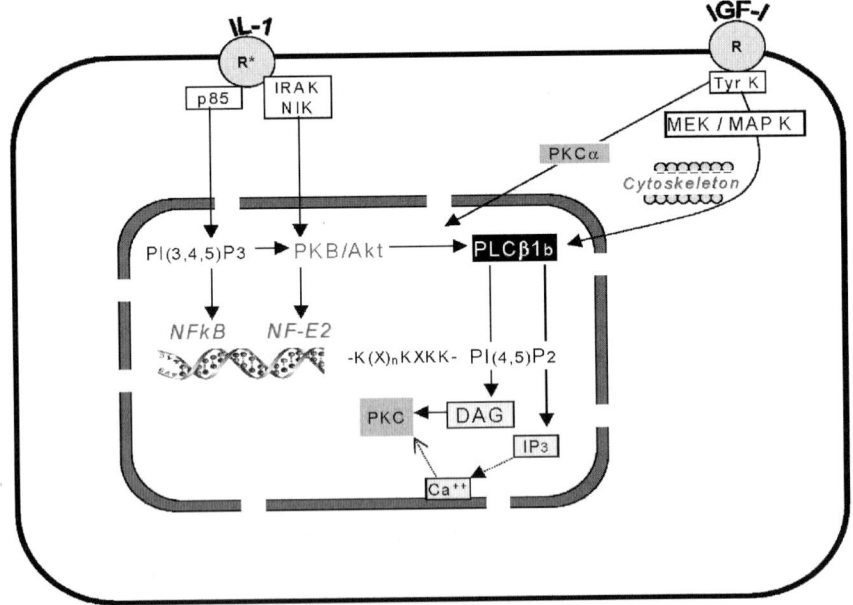

Fig. 2 Phosphoinositide signalling at the nucleus. PI(4,5)P$_2$ is recruited at the cell nucleus through the PI-binding consensus sequence –K(X)$_n$KHKK- present in many chromatin and nuclear matrix proteins. Phospholipases and inositol kinases are recruited at the same sites by a lipid substrate-dependent compartmentalization. Agonists at the cell membrane can trigger the nuclear inositide signalling system either through translocation of the lipid kinase PI 3-kinase, or of protein kinases, such as PKB/Akt or PKCα, which activates nuclear PLCβ1. This activation requires the translocation of MAP kinase, mediated by the cytoskeleton (*bent arrow*)

localization are multiple, due to the presence of a PI-binding consensus sequence in several nuclear proteins. The stability of PI(4,5)P$_2$ association with nuclear proteins is more or less strong; in fact, part of the PI(4,5)P$_2$ pool associated with heterochromatin is lost during nuclei isolation in mild ionic conditions (Mazzotti et al., 1995; Maraldi et al., 1995). The other part of PI(4,5)P$_2$ pool is localized at the perichromatin fibrils and at the interchromatin granule clusters (Maraldi et al., 1995); these interactions are more stable and are maintained also in chromatin-depleted nuclear matrix preparations (Mazzotti et al., 1995). This suggests that inositol lipids form lipoprotein complexes in which the lipids are not oriented as in membranes (Maraldi et al., 1992).

PI(4,5)P$_2$ is a minor phospholipid (0.3% of total phospholipids); nevertheless, the amount of the nuclear PI(4,5)P$_2$, estimated about 0.5 mmol/mg nuclear protein, represents more than 20% of the PI(4,5)P$_2$ detectable in the cell (Watt et al., 2002). This finding is particularly significant because it has been obtained with a method completely different with respect to that used in the first demonstration of a presence of PI(4,5)P$_2$ within the nucleus (Mazzotti et al., 1995). In fact, instead of using a post-embedding immunocytochemical detection of the anti-PI(4,5)P$_2$ antibody, the method utilizes the PH domain of PLCδ in thawed cryosectioned material (Watt et al., 2002). In any case, the quantitative evaluations of the nuclear PI(4,5)P$_2$ amount are quite similar, as well as the sites where the lipid is located within the nucleus (electron-dense patches of heterochromatin). A similar localization has been obtained at lower resolution by confocal microscopy, using PI(4,5)P$_2$ fluorescently tagged with a green NBD fluorophore, using Texas Red-labelled histone H1 as carrier (Ozaki et al., 2000).

To address the problem of the functional role of nuclear polyphosphoinositides, the detection and quantitative evaluation of the enzymes involved in their synthesis, phosphorylation and hydrolysis has been widely investigated. An inverse proportion has been reported between the PI(4,5)P$_2$ amount detectable within the nucleus by immunogold electron microscopy and the nuclear amount and activity of nuclear PI-phospholipase PLCβ1 (Maraldi et al., 1995).

PI(4,5)P$_2$ is also present at other intranuclear sites, in association with the other elements of the inosito signalling system, such as PIP kinases, PI 3-kinase (Boronenkov et al., 1998), PLC isoforms, and PKC isoforms (Maraldi et al., 1999). The speckles detectable by fluorescence microscopy correspond to clusters of interchromatin granules that represent sites of accumulation of splicing factors and hnRNPs (Boronenkov et al., 1998). The members of the inositide signalling system have been localized at the nuclear speckles by fluorescence microscopy, and at the interchromatin granule clusters, as well as at the perichromatin fibrils associated with the periphery of heterochromatin (Maraldi et al., 1999). The inositol signalling elements, found at these sites, can be essential for the activation of protein kinases and phosphatases to induce the re-location of splicing factors (Misteli and Spector, 1997).

Hydrolysis of nuclear PI(4,5)P$_2$ by PLC might provide a localized release of IP3 that, in addition to the regulation of Ca^{2+} from cell stores, could serve as precursor of inositol polyphosphates, which are essential for RNA transport (Odom et al., 2000; Osborne et al., 2001).

Finally, PI(4,5)P$_2$ mediated actin-dependent mechanisms have been demonstrated to be involved in chromatin remodelling (Zhao et al., 1998). PI(4,5)P$_2$ but not PIP enhances the binding of the BAF complex to the nuclear matrix (Zhao et al., 1998); PI(4,5)P$_2$ binds to BAF at one molecule per complex, which, in turn, increases actin polymerisation in a PI(4,5)P$_2$-sensitive manner, suggesting that PI(4,5)P$_2$ can induce the uncapping of actin, leading to nucleation or filament assembly (Rando, 2002).

5 Nuclear Domains Involved in Inositide Signalling

The identification of elements of inositide signalling at nuclear sites has been achieved by immunocytochemical techniques at confocal and electron microscopy level (Maraldi et al., 1999). Each method presents advantages and drawbacks; however, the results obtained are generally consistent and complementary. The main advantage of fluorescent-labelled probes at the confocal microscope is the possibility of examining the whole cell maintaining its three-dimensional organization, while gold-labelled probes at the electron microscope ensure very high resolution and quantitative evaluations. The availability of GFP-tagged probe technique overcomes the limits of traditional cytochemistry based on fixed specimens, and allows one the use of in vivo localization assay. The results till now obtained with this technique confirm those previously obtained with conventional immunocyto-chemistry (Bavelloni et al., 1999).

The use of in situ nuclear matrix preparations allowed one to identify the localization sites of the insoluble versus soluble pools of some elements of the signalling system. The PI(4,5)P$_2$ soluble pool is quite diffused within the nucleus; in fact it is localized at both heterochromatin and interchromatin domains, in agreement with the presence of a large number of nuclear proteins presenting PI-binding motifs. As a consequence of the interactions between polyphosphoinositides and nuclear proteins such as histones, polymerases, and BAF complexes, PI(4,5)P$_2$ localization sites have been detected at the heterochromatin, mainly at its periphery, at some nucleolar components, and at ribonucleoproteins in the interchromatin domains. The insoluble pool, on the other hand, is restricted to interchromatin granules. Other elements of the signalling system were detected to colocalize with PI(4,5)P$_2$ at the same sites (Maraldi et al., 1999). The molecular events occurring at these districts, that is transcription of hnRNA, and pre-mRNA processing, have been widely investigated. Nascent transcripts correspond to perichromatin fibrils, while interchromatin granule clusters, detectable as fluorescent speckles at confocal microscope, contain many of the splicing factors. Splicing can occur both co- and post-transcriptionally, that is at the perichromatin fibres as well as at the interchromatin granules. Activation of splicing factors involves phosphorylation that modulates their shuttling from transcriptional to post-transcriptional sites (Misteli and Spector,

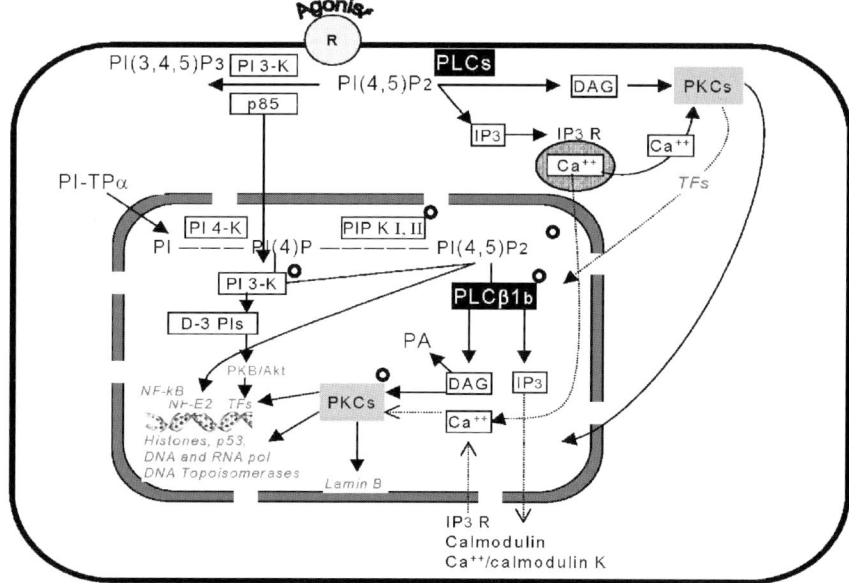

Fig. 3 Nuclear inositide signalling pathways and nuclear targets. The elements of the signalling system identified by immunocytochemistry within the nucleus are indicated (o). Some of them are constitutively associated to nuclear proteins, such as PI(4,5)P$_2$ and PLCβ1b, or translocated upon stimuli, such as PI 3-kinase and some PKC isoforms. The targets of lipid second messengers, including PKC isoforms and PKB/Akt, are able to phosphorylate specific transcription factors, or chromatin-associated proteins, or nuclear lamins (*grey italics*)

1997). Therefore, the presence of a signalling system at these sites, colocalized with the protein kinases involved in the phosphorylation of splicing factors (Maraldi et al., 1999), strengths the possibility of a functional involvement of inositides into crucial steps of nuclear metabolism (Fig. 3).

6 Evolution of the Inositide Signalling System

The contention of a signal transduction system located within the nucleus, separated and autonomous with respect to the canonical signal transduction system localized at the cell membrane, has been initially regarded as a non-necessary redundancy. The growing experimental evidence on this topic in the last years has not only confirmed the actual role of the nuclear system, but also stressed its evolutionary significance. First of all, in the same cell there is not a mere replication of the transduction system at the membrane and nuclear level. Let us consider, as an example, the case of IGF-1 responding 3T3 cells. 3T3 cells present a clone-dependent response to IGF-1. In fact, although all clones present IGF-1 receptors, a clone does not respond mitogenically to the agonist and does not show changes in nuclear inositide metabolism, whilst another clone responds mitogenically and shows an exclusive activation of the nuclear PLC β_1 isoform (Martelli et al., 1992). This suggested a direct link between compartmentalization of PICs, nuclear PI(4,5)P$_2$ hydrolysis, and the onset of DNA synthesis.

The different PLCs (β, γ, δ) are not only regulated by covalent modifications mediated by intrinsic protein-tyrosine kinase activity of the receptors, by non-receptor tyrosine kinase, or by GTP-binding proteins, but, being constituted by isoforms with specific subcellular localization, they could present a multiplicity of responses to the same agonist in different moments of the cell cycle.

Thus, the quite non-expected response to the binding of IGF-1 at a plasma membrane tyrosine kinase receptor which, instead of the activation of the cytoplasmic PLC γ by phosphorylation of tyrosine residues, causes the activation of the nuclear PLC β isoform, can be regarded as a further potential of the complex pathway involved in transmitting the signals from the environment to the genome. On the other hand, the IGF-1-like agonist bombesin induces the activation of PLC γ at the cytoplasm, but does not induce either PKC translocation to the nucleus, or DNA replication, confirming the presence of distinct inositide cycles at the cell membrane and at the nucleus (Divecha et al., 1993).

Therefore, the inositol signal transduction system at the nucleus appears to utilize the same but not identical enzymatic principles of the system at the cell membrane. The nuclear system plays a crucial role in the control of the cell cycle, modulating the translocation and activation of nuclear PKCs, thus controlling nuclear envelope assembly, triggering the onset of DNA replication, and modulating the expression of transcription factors. This evidence strongly suggests that such a system which carries out crucial physiological role and is localized at the heart of the cell, is not a by-product of the system localized at the cell membrane. On the contrary, it seems much more reasonably that the nucleus has been the site at which PI cycle evolved firstly, being its function crucial for the regulation of DNA replication and gene expression by environmental messages. It is then conceivable that, for cytosolic signalling purposes, the system, initially evolved in the nucleus, has been later duplicated at the plasma membrane (Divecha et al., 1993). Traces of this evolution have been left, which are represented by the presence of a cytoplasmic inositol signalling system, which is structurally and functionally bound to the cytoskeleton. IGF-1-dependent activation of nuclear PLCβ1 in 3T3 cells is prevented when the integrity of the cytoskeleton is affected; indeed, the translocation to the nucleus of MAP kinase, possibly involved in PLCβ1 phosphorylation, appears mediated by the cytoskeleton (Martelli et al., 1999). It must be considered that cytoskeleton modifications in the course of the cell cycle are related with those occurring at the nuclear level and which, in combination, determine the assumption of the cell morphotype. Cell morphology depends mainly on the cytoskeletal arrangement, and the relationships between the skeletal components of the inner nuclear matrix and those of the cytoplasm are very close, in spite of the presence of the nuclear membranes. The same occurs for the interactions between the cytokeleton and the extracellular matrix, in spite of the presence of the cell membrane. Therefore, the inositol lipid signalling systems bound to the inner nuclear matrix and to the cytoskeletal components can be considered as steps of the evolution of a complex system that, eventually, has been extended to the cell membrane, in order to allow a cross-talk between the extracellular signals and the genome.

7 Towards the Deciphering of the Nuclear Inositol Lipid Signal Transduction Code

Once established that the nuclear inositol lipid signal transduction presents some of the crucial characteristics of an organic code, the further step should imply the deciphering of the code by which the system works. We know for sure that the understanding of other organic codes, such as splicing code and histone code, is still extraordinarily difficult; nevertheless we have some indications that the bases for understanding the inositol lipid signal transduction code have been posed.

Phosphoinositides are known to transduce signals essentially via two events: their modification and their interaction with specific proteins. In the nucleus it has been demonstrated that some phosphoinositide species are specifically represented, and nuclear proteins have been identified capable of interacting with these phosphoinositides. The main nuclear phosphoinositides are represented by PIPs, $PI(4,5)P_2$ and inositol phosphates, whose relative amounts vary in response to environmental conditions. Protein domains that interact with phosphoinositides include PH, ENTH, FYVE, PHOX domains, PHD finger and lysine/arginine rich patches. Many of the cellular proteins that display these domains are nuclear, being chromatin-associated or lamina-associated proteins, capable of modulating chromatin remodelling and gene domain positioning. It has been largely demonstrated that extracellular signals that induce processes such as cell proliferation, differentiation, and stress adaptation/apoptosis, lead to temporal and spatial changing in nuclear inositide profiles. This results in specific relative proportions of the different nuclear inositide pool. The combinatorial changes in nuclear phosphoinositides could be decoded through their interaction with specific phosphoinositide-binding proteins to elicit differential outputs (Fig. 4).

These may include changes in chromatin structure to regulate gene expression, DNA replication or repair. For example, signal inputs that increase the level of nuclear PI(5)P within the chromatin lead to the translocation of ING2 that, in turn, may regulate p53 function and histone acetylation to coordinate the apoptotic response (Jones and Divecha, 2004). Moreover, intranuclear changes of the $PI(4,5)P_2$ pool have been demonstrated to modulate chromatin remodelling complex activity (Rando et al., 2002) which are essential to induce gene transcription. These experimental findings strongly suggest that the transductional complexes constituted by phosphoinosites-proteins, located at different nuclear domains behave as adaptors, suggesting a key towards the deciphering of the nuclear inositide signal transduction code.

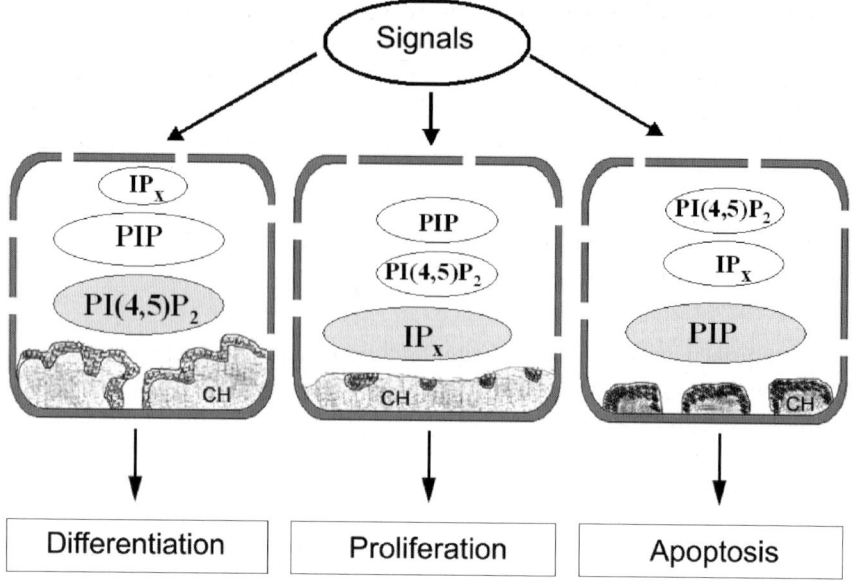

Fig. 4 In the drawing the size of the oval reflects the relative amount of the nuclear inositide pool. The interaction of the more represented nuclear inositide with chromatin-associated proteins (*shaded*) leads, in response to extracellular signals, to differential processes, such as cell differentiation, proliferation, or apoptosis

8   Conclusions

Tissue-specific patterns of gene transcription are established during development. However, because identical signalling pathways are known to produce different developmental outcomes, cells are likely to respond to the signals depending on their developmental history. This, in turn, may reflect either the mixture of activator and repressor factors present in each cell type, or the epigenetic modifications of the genome that render particular gene sets available or restricted for transcription. Epigenetic regulation occurs mainly at the level of chromatin organization, but we must also consider that the local contest of the whole nucleus may represent a formidable multidimensional element of epigenetic regulation. In mammals, the whole arrangement of the nucleus is essential to establish, maintain, or change, a tissue-specific pattern of gene expression, by targeting given chromatin domains to nuclear territories characterized by three-dimensional arrays of regulatory elements. Thus it has been established that the hallmarks of constitutive heterochromatin, containing highly repetitive DNA, such as satellite centromeric sequences with a very reduced amount of genomic sequences, include the absence of methylation at H3 Lys4, the tri-methylation of H4 Lys20, and the tri-methylation of H3 Lys9. Therefore, the establishment and maintenance of heterochromatin assembly is a multistep process, which requires a concerted action between DNA methyltransferases, histone methyltransferases, siRNAs, TFs, and repressors.

Besides changes at specific loci of the repressed chromatin, during differentiation, changes also take place on a genome-wide scale that involve the whole nuclear organization. These changes of the nuclear phenotype represent key epigenetic markers that could contribute to cell memory by "indexing" the genome during tissue-specific cell differentiation.

The process of the dynamic transition of silenced loci to nuclear regions that allow their transition to the active state has been reported to require a global reorganization of the nucleus, of the chromatin arrangement and of protein factor localization.

It is quite likely that, in different cell types where similar TFs are expressed, distinct differentiation patterns might be achieved by regulating the accessibility of promoters through local changes in the chromatin structure; these changes constitute heritable epigenetic markers for fixing patterns of tissue-specific gene expression. The presence of a phosphoinositide nuclear pool which responds to external cues modifying its interaction with chromatin-associated proteins, would provide a formidable mechanism for controlling gene positioning shown to be crucial in regulating gene silencing or expression (Maraldi et al., 2003; Maraldi and Lattanzi, 2005).

References

Albright, L. M., Hualu, E., and Ausubel, F. M. (1989), Prokaryotic signal transduction mediated by sensor and regulatory protein pairs. *Annu. Rev. Genet.* 23: 311–336.

Anderson, R. A., Boronenkov, I. V., Dougham, S. D., Kunz, J., and Loijens, J. C. (1999), Phosphatidylinositol phosphate kinases, a multifaceted family of signaling enzymes. *J. Biol. Chem.* 274: 9907–9910.

Barbieri, M. (1998), The organic codes: the basic mechanism of macroevolution. *Rivista di Biologia-Biology Forum,* 91: 481–514.

Barbieri, M. (2003), *The Organic Codes.* Cambridge Univesity Press, Cambridge.

Bavelloni, A., Santi, S., Sirri, A., Riccio, M., Faenza, I., Zini, N., Cecchi, S., Ferri, A., Auron, P. F., Maraldi, N. M., and Marmiroli, S. (1999), Phosphatidylinositol 3-kinase translocation to the nucleus is induced by interleukin 1 and prevented by mutation of interleukin 1 receptor in human osteosarcoma Saos-2 cells. *J. Cell Sci.* 112: 631–640.

Boronenkov, I. V., Loijens, J. C., Umeda, M., and Anderson, R. A. (1998), Phosphoinositide signaling pathways in nuclei are associated with nuclear speckles containing pre-mRNA processing factors. *Mol. Biol. Cell* 9: 3547–3560.

Cocco, L., Gilmour R. S., Ognibene, A., Letcher A. J., Manzoli, F. A., and Irvine, R. F. (1987), Synthesis of polyphosphoinositides in nuclei of Friend cells. Evidence for polyphosphoinositide metabolism inside the nucleus which changes with cell differentiation. *Biochem. J.* 248: 765–770.

Cooke, F. T. (2002), Phosphatidylinositol 3,5-bisphosphate: metabolism and function. *Arch. Biochem. Biophys.* 407: 143–151.

Cullen, P. I., Cozier, G. E., Banting, G., and Mellor, H. (2001), Modular phosphoinositide-binding domains-their role in signalling and membrane trafficking. *Curr. Biol.* 11: R882–893.

Czech, M. P. (2002), Dynamics of phosphoinositides in membrane retrieval and insertion. *Annu. Rev. Physiol,* 65: 33.1–33.25.

Divecha, N., Banfic, H., and Irvine, R. F. (1993), Inositides and the nucleus and inositides in the nucleus. *Cell* 74: 405–407.

Hammond, G., Thomas, C. L., and Schiavo, G. (2004), Nuclear phosphoinositides and their functions. *Curr. Top. Microbiol. Immunol.* 282: 177–206.

Irvine R. F. (2003), Nuclear lipid signalling. *Nat. Rev. Mol. Cell Biol.* 4:1–12.

Irvine, R. F. and Schell, M. J. (2001), Back in the water: the return of the inositol pphosphates. *Nat. Rev. Mol. Cell Biol.* 2: 327–338.

Jones, D. R. and Divecha, N. (2004), Linking lipids to chromatin. *Curr. Opin. Genet. Develop.* 14: 196–202.

Manzoli, F. A., Capitani, S., Maraldi, N. M., Cocco, L., and Barnabei, O. (1979), Chromatin lipids and their possibile role in gene expression. A study in normal and neoplastic cells. *Advan. Enzyme Regul.* 17: 175–194.

Manzoli, F. A., Maraldi, N. M., Cocco, L., Capitani, S., and Facchini, A. (1977), Chromatin phospholipids in normal and chronic lymphocytic leukemia lymphocytes. *Cancer Res.* 37: 843–849.

Maraldi, N. M. and Capitani, S. (2003), The topology of nuclear lipids, in: Cocco, L. and Martelli, A. M. (eds) Nuclear lipid metabolism and signalling. Research Signpost, Kerala, India, pp. 101–121.

Maraldi, N. M., Capitani, S., Caramelli, E., Cocco, L., Barnabei, O., and Manzoli, F. A. (1984), Conformational changes of nuclear chromatin related to phospholipid-induced modifications of the template availability. *Advan. Enzyme Regul.* 22: 447–464.

Maraldi, N. M. and Lattanzi, G. (2005), Linkage of lamins to fidelity of gene transcription. Crit. Rev. Eukar. *Gene Express.* 15: 277–293.

Maraldi, N. M., Zini, N., Ognibene, A., Martelli, A. M., Barbieri, M., Mazzotti, G., and Manzoli, F. A. (1995), Immunocytochemical detection of the intranuclear variations of phosphatidylinositol 4,5-bisphosphate amount associated with changes of activity and amount of phospholipase Cβ1 in cells exposed to mitogenic or differentiating agonists. *Biol. Cell* 83: 201–210.

Maraldi, N. M., Zini, N., Santi, S., and Manzoli, F. A. (1999), Topology of inositol lipid signal transduction in the nucleus. *J. Cell. Physiol.* 181: 203–217.

Maraldi, N. M., Zini, N., Squarzoni, S., Del Coco, R., Sabatelli, O., and Manzoli, F. A. (1992), Intranuclear localization of phospholipids by ultrastructural cytochemistry. *J. Histochem. Cytochem.* 40: 1383–1392.

Martelli, A. M., Gilmour, R. S., Bertagnolo, V., Neri, L. M., Manzoli, L., and Cocco, L. (1992), Nuclear localization and signalling activity of phospholipase Cβ in Swiss 3T3 cells. *Nature* 358: 242–245.

Martelli, A. M., Cocco, L., Bareggi, R., Tabelloni, G., Rizzoli, R., Ghibellini, M. D., and Narducci, P. (1999), Insulin-like growth factor-I-dependent stimulation of nuclear phospholipase C-β1 activity in Swiss 3T3 cells requires an intact cytoskeleton and is paralleled by increased phosphorylation of the phospholipase. *J. Cell Biochem.* 72: 339–348.

Martelli, A. M., Manzoli, L., and Cocco, L. (2004), Nuclear inositides: facts and perspectives. *Pharmacol. Therap.* 101: 47–64.

Mazzotti, G., Zini, N., Rizzi, E., Rizzoli, R., Galanzi, A., Ognibene, A., Santi, S., Matteucci, S., Martelli, A. M., and Maraldi, N. M. (1995), Immunocytochemical detection of phosphatidylinositol 4,5-bisphosphate localization sites within the nucleus. *J. Histochem. Cytochem.* 43: 181–191.

Misteli, T. and Spector, D. L. (1997), Protein phosphorylation and the nuclear organization of pre-mRNA splicing. *Trends Cell Biol.* 7: 135–138.

Odom, A. R., Stahlberg, A., Wente, S. R., and York, J. D. (2000), A role for nuclear inositol 1,4,5-trisphosphate kinase in transcriptional control. *Science* 287: 2026–2029.

Osborne, S. L., Meunier, F. A., and Schiavo, G. (2001), Phosphoinositides as key regulators of synaptic functions. *Neuron* 32: 9–12.

Osborne, S. L., Thomas, C. L., Gschmeissner, S., and Schiavo, G. (2001), Nuclear PtdIns(4,5)P$_2$ assembles in a mitotically regulated particle involved in pre-mRNA splicing. *J. Cell Sci.* 114: 2501–2511.

Ozaki, S., DeWald, D. B., Shope, J. C., Chen, J., and Prestwich, G. D. (2000), Intracellular delivery of phosphoinositides and inositol phosphates using polyamine carriers. *Proc. Natl. Acad. Sci. USA* 97: 11286–11291.

Raben, D. M. (2006), Lipid signaling in the nucleus. *Biochem. Biophys. Acta* 1761: 503–504.

Rando, O. J., Zhao, K., Janmey, P., and Crabtree, G. R. (2002), Phosphatidylinositol-dependent actin filament binding by the SWI/SNF-like BAF chromatin remodeling complex. *Proc. Natl. Acad. Sci. USA* 99: 2824–2829.

Toker, A. (2002), Phosphoinositides and signal transduction. *Cell. Mol. Life Sci.* 59: 761–779.

Watt, S. A., Kular, G., Fleming, I. N., Downes, C. P., and Lucocq, J. M. (2002), Subcellular localization of phosphatidylinositol 4,5-bisphosphate using the pleckstrin homology domain of phospholipase Cδ1. *Biochem. J.* 363: 657–666.

Zhao, K., Wang, W., Rando, O. J., Xue, Y., Swiderek, K., Kuo, A., and Crabtree, G. R. (1998), Rapid and phosphoinositol-dependent binding of the SWI/SNF-like BAF complex to chromatin after T lymphocyte receptor signaling. *Cell* 95: 625–636.

Chapter 10

Biological Information Transfer Beyond the Genetic Code: The Sugar Code*

Hans-Joachim Gabius

Abstract In the era of genetic engineering, cloning, and genome sequencing, the focus of research on the genetic code has received an even further accentuation in the public eye. When, however, aspiring to understand intra- and intercellular recognition processes comprehensively, the two biochemical dimensions established by nucleic acids and proteins are not sufficient to satisfactorily explain all molecular events in, e.g. cell adhesion or routing. To bridge this gap consideration of further code systems is essential. A third biochemical alphabet forming code words with an information storage capacity second to no other substance class in rather small units (words, sentences) is established by monosaccharides (letters). As hardware oligosaccharides surpass peptides by more than seven orders of magnitude in the theoretical ability to build isomers, then the total of conceivable hexamers is calculated. Beyond the sequence complexity application of nuclear magnetic resonance (NMR) spectroscopy and molecular modeling have been instrumental to discover that even small glycans can often reside in not only one but several distinct low-energy conformations (keys). Intriguingly, conformers can display notably different capacities to fit snugly into the binding site of nonhomologous receptors (locks). This process, experimentally verified for two classes of lectins, is termed "differential conformer selection." It adds potential for shifts of the conformer equilibrium to modulate ligand properties dynamically and reversibly to the well-known changes of sequence (including anomeric positioning and linkage points) and of pattern of substitution, e.g. by sulfation. In the intimate interplay with sugar receptors (lectins, enzymes, and antibodies) the message of coding units of the sugar code is deciphered. This communication will trigger postbinding signaling and the intended biological response. Knowledge about the driving forces for the molecular rendezvous, that is, contributions of bidentate or cooperative hydrogen bonds, dispersion forces, stacking and solvent rearrangement, will enable the design

Reprinted with permission from Naturwissenschaften, Vol. 87 (2000), pp. 108–121.

Institut für Physiologische Chemie, Tierärztliche Fakultät, Ludwig-Maximilians-Universität München, Veterinärstr. 13, D-80539 München, Germany,
e-mail: gabius@tiph.vetmed.uni-muenchen.de

of high-affinity ligands or mimetics thereof. They embody clinical applications reaching from receptor localization in diagnostic pathology to cell-type-selective targeting of drugs and inhibition of undesired cell adhesion in bacterial/viral infections, inflammation, or metastasis.

1 Introduction

Basic biochemical knowledge assigns nucleic acids and proteins the decisive role in information flow in biosystems. Connected by the genetic code the transcribed portions of the genome govern the expression of a complex set of messages on the level of polypeptides. To meet the requirement of flexible regulation of product availability, the synthesis and degradation of proteins can be intimately modulated. Moreover, posttranslational modifications, phosphorylation taking a prominent place in textbooks, assure rapid and reversible fine-tuning of enzyme and receptor activities. With genome sequencing becoming routine practice, the allurement to view biological information as being exclusively epitomized by the genetic code becomes nearly irresistible. To be mindful of substance classes, which would otherwise be unfairly and incorrectly treated as "second class citizens" (von der Lieth et al., 1997b), this review furnishes information to underscore that the current judgment to place primary emphasis in research on the genetic code is unlikely to be final. After introducing first the concept of the sugar code on the level of sequence and conformation and then documenting presence of sophisticated decoding devices (among them endogenous lectins), the versatility of the sugar code will be exemplified leading to the description of perspectives to turn these discoveries into biomedical applications.

2 The Sugar Code: Basic Principles

In order to succeed as hardware for information transfer any substance class must offer the potential for specific coding. The message will have to be deciphered with sufficient biochemical affinity and low probability for ambiguities and misinterpretation. A high-density coding capacity is beneficial to keep the size of the active sections of biomolecules small, thereby reducing the energetic expenses during synthesis. Moreover, spatially easy accessibility and the potential for rapid structural modulations by reversible variations of the chain length and/or introduction of small but decisive substituents are eminent factors in the design of an efficient code system. This set of conditions describes the frame in which the quality of biological coding is to be rated. By performing such calculations on the theoretical storage capacity expressed as the total number of isomers without preconceptions it takes no effort of persuasion to convincingly show that nucleotides and amino acids are surpassed, by far, by another class of biomolecules.

Currently, carbohydrates have their main place in textbooks in chapters on energy metabolism and cell wall composition. The regular repetitive arrangement of monosaccharides in plant, insect, fungal, or bacterial cell walls or coats seduces to underestimate the other inherent talents of carbohydrates. Amazingly, they are readily discernible when closely looking at a simple structural representation (Fig. 1). Each monosaccharide offers various hydroxyl groups for oligomer formation by glycosidic bonds including the anomeric C1-position. In contrast to nucleic acids and proteins branching of chains is a common feature of the glycan part of cellular glycoconjugates (glycoproteins, glycolipids). Taking stock of the peculiarities of monosaccharide structure the total number of isomer permutation for a hexamer with an alphabet of 20 letters (monosaccharides) reaches the staggering number of 1.44×10^{15} (Laine, 1997). Under the same conditions only 6.4×10^7 (20^6) structures can be devised from 20 amino acids, the four nucleotides just yielding 4096 (4^6) hexanucleotides. Allowing two different substitutions in a hexasaccharide, occurring in Nature, e.g. as sulfation in glycosaminoglycan chains, further increases

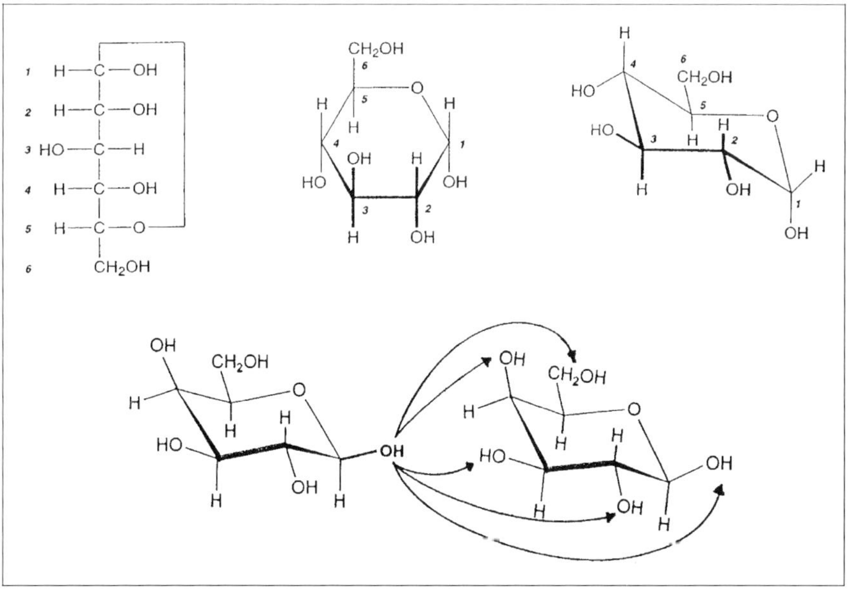

Fig. 1 The different graphic representations of the structure of a hexapyranose using α-D-glucose (Glc) as example (*top*). Commonly, the Haworth formula (*middle*) with the ring being placed perpendicular to the plane is given preference to the traditional Fischer projection of the hemiacetal (*left*). The relative positioning of the axial and equatorial substituents can readily be visualized by drawing the relatively rigid and energetically privileged chair conformation (*right*). For the formation of an acetal (disaccharide) by a glycosidic bond using D-galactose (Gal), the 4′-epimer of glucose as example, the anomeric hydroxyl group of the left monosaccharide can theoretically react with any of the five acceptors present on a second hexopyranose yielding 11 isomers with full consideration of the two anomeric positions (*bottom*). The structure of the β1–3-linked digalactoside is drawn in Fig. 3

the number of isomers by more than two orders of magnitude (Laine, 1997). In the prophetic words of Winterburn and Phelps, "carbohydrates are ideal for generating compact units with explicit informational properties, since the permutations on linkages are larger than can be achieved by amino acids, and, uniquely in biological polymers, branching is possible" (Winterburn and Phelps, 1972).

It is no treading on thin ice to follow the authors to their conclusion that "the significance of the glycosyl residues is to impart a discrete recognitional role on the protein" (Winterburn and Phelps, 1972), and it is not surprising that at least 1.0% of the translated genome in animals is devoted to the generation of code words with as many as 70% of proteins harboring the tripeptide sequon for N-glycosylation (Reuter and Gabius, 1999; Varki and Marth, 1995; Wormald and Dwek, 1999). The core region and complex extensions of this ubiquitous type of protein glycosylation in eukaryotes are shown in Fig. 2. It gives a graphic example how branching sets in and how to read the sugar code. Each linkage is characterized by the anomeric configuration and the positions of the two linkage points, such as β1-4 as opposed to α1-4 or α1-3. Since nucleotide sugars are employed as donors by the glycosyltransferases (Brockhausen and Schachter, 1997; Sears and Wong, 1998), chain growth generally involves the anomeric position restricting the range of products by enzymatic synthesis in relation to all theoretically possible isomers. Nonetheless, the presented staggering complexity of glycan structures has already placed severe obstacles in the way to go beyond merely acknowledging the enormous potential for structural variability towards precise structure determination.

These problems have mainly been solved by the development of sophisticated isolation and analysis methods combining the power of liquid chromatography, capillary zone electrophoresis, mass spectrometry, and NMR spectrometry with that of biochemical reagents such as endo- and exoglycosidases and sugar receptors (Cummings, 1997; Geyer and Geyer, 1998; Hounsell, 1997; Reuter and Gabius, 1999). Application of these techniques has revealed that subtle variations and

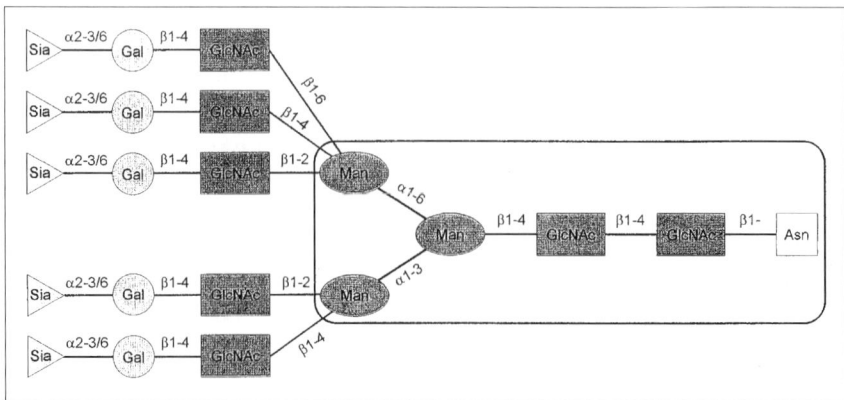

Fig. 2 Structure of the core pentasaccharide of N-glycans given within the frame and the additional branching yielding a penta-antennary complex-type sugar structure (*left*)

modifications are especially frequent in the terminal, spatially accessible sections of the sugar antennae. The strategic placement of distinctive substitutions is expected for a role in information transfer. They are marked by introduction of small substituents (sulfate and *O*-acetyl groups, etc.) into sugar moieties such as *N*-acetylgalactosamine or *N*-acetylneuraminic acid, comparable to the formation of an umlaut in the German language, or by directing a synthetic intermediate to various end products by mutually exclusive refinements, e.g. α1-3 fucosylation, α2-3/6 sialylation, and 4-sulfation (Hooper et al., 1997; Reuter and Gabius, 1996, 1999; Reutter et al., 1997; Sharon and Lis, 1997; Varki, 1996). Intercellular and temporal flexibility turns the available letter repertoire into an array of alternative structures (biosignals). Indeed, the observations that the profile of glycans is not genetically strictly coded but influenced by the presence and relative positioning of the set of enzymes in the assembly line and the actual availability of activated substrates such as nucleotide donors argues in favor of purpose versus randomness (Abeijon et al., 1997; Pavelka, 1997; Varki, 1998). Thus, the prerequisite for rapid and multifarious modulation mentioned in the introductory paragraph is adequately fulfilled in the sugar code.

In view of the assumed importance for maintaining diversity, a multicellular organism with lack of presence of one of the mentioned pathways will allow to probe into the question whether this deficit is accompanied by any remodeling in the overall glycosylation system or not. Assisted by genome sequencing, it can indeed be proposed that absence of sialylation in the nematode *Caenorhabditis elegans* might be compensated by elaboration of another part of the enzymatic machinery. The discovery of 18 different genes for putative fucosyltransferases in the genome of this nematode argues in favor of this notion (Oriol et al., 1999). In these authors' own words, "for some unknown reasons, these nematodes have favored through evolution fucosylation instead of sialylation of their terminal nonreducing oligosaccharide epitopes or glycotopes and since sialic acid and fucose are usually in competition for the same acceptors, the lack of all forms of sialic acid in *C. elegans* fits well with a large expression of different fucosyl-transferase genes, making this animal an ideal model for evolutionary studies of fucosyltransferases" (Oriol et al., 1999). All these reactions in glycosylation result in a typical pattern of glycan chains on the level of cells and organs. It is as characteristic as a fingerprint or a signature. Yeast cells, for example, pro-duce mannose-rich surface glycans, while multicellular organisms prominently put histo-blood group epitope-rich complex-type glycans on display. Enzymes for these extensions at the end of antennae (Fig. 2) typically reside in the medial- and *trans*-Golgi regions. Since the number of activities operating upon these sections has especially expanded in the animal kingdom, it is rather unrea-sonable to assume these refinements to have survived fortuitously. Driving this evolutionary process can be attributed to functions of the glycans ranging from purely physical aspects such as solubility or protection of surface against prote-olytic attack to any involvement in recognition (Drickamer and Taylor, 1998; Gagneux and Varki, 1999; Reuter and Gabius, 1999; Sharon and Lis, 1997; Varki, 1996).

A principal comment is warranted on the surmised evolutionary mechanisms of selection of letters for the alphabet of this code system. As insightfully discussed by Hirabayashi (1996), elementary hexose synthesis under prebiotic conditions was most probably facilitated by the following cascade. It started with formol condensation, yielding basic trioses known from glycolysis. The next step is the aldol condensation to 3,4-*trans*-ketoses and a conversion of D-fructose to D-glucose and D-mannose via an enediol-intermediate and the keto-enol tautomerism (Lobry de Bruyn rearrangement). Notably, D-glucose harbors no 1,3-diaxial interactions involving a hydroxyl group (Fig. 1), and the favored "tridymite" water structure is maintained in the presence of equatorial hydroxyl groups (Uedaira and Uedaira, 1985). In mannose as in galactose, a biochemical derivative obtained by the NAD⁺-dependent epimerization of glucose, only one hydroxyl group is axial, keeping unfavorable 1,3-diaxial interactions and perturbation of solvent structure minimal. In contrast to the 2′- and 4′-epimers, the 3′-epimer has two 1,3-axial interactions. Origin from synthesis under prebiotic conditions and energetic consequences entail the organization of the initial hardware of the sugar code. From them, further letters of the alphabet comprising also the *N*-acetyl derivatives of the 2′-amines of glucose and galactose, L-fucose, D-xylose, and *N*-acetylneuraminic acid are biosynthetically produced. Interestingly, the core section of *N*-glycans (Fig. 2) is composed of basic units derived from a presumably prebiotic origin. This fact invites to speculate on a relationship of evolutionary pathways on the levels of eukaryotic organisms and of glycan complexity. Setting this aspect which is further discussed elsewhere (Drickamer and Taylor, 1998; Gagneux and Varki, 1999; Hirabayashi, 1996; Oriol et al., 1999) aside in this context, it can at least be reliably concluded at this stage that oligosaccharides by their inherent potential for ample sequence permutations including variations in the anomeric position and the linkage groups for a glycosidic bond deserve attention as coding units. Remarkably, recent work extends the capacity for information storage from two dimensions of linear and branched oligosaccharide chains to the third dimension.

3   The Sugar Code: The Third Dimension

The shape of a glycan will be determined by the conformation of the furanose/pyranose rings and the relative positioning of the rings in the chain. Based on x-ray crystallography, neutron diffraction, and homonuclear coupling constant data the $^4C_1$ chair conformer ($^1C_4$ for L-sugars) is the energetically preferred pyranose ring structure (Abeygunawardana and Bush, 1993; Brown and Levy, 1965). In rare cases, for example, for L-iduronic acid as constituent of heparan and dermatan sulfates, and to accommodate mechanical stress, conformational flexibility and elasticity of a pyranose can be generated by chair-boat transitions, which allow L-iduronic acid to acquire the skew-boat form ^2S_o (Casu et al., 1988; Marszalek et al., 1998). Yet the main contribution to define a glycan's shape will generally originate not

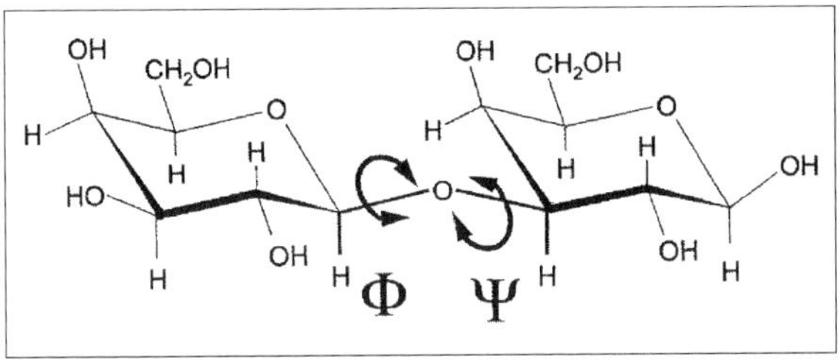

Fig. 3 Depiction of the main source of conformational flexibility of the disaccharide Galβ1-3Gal (see Fig. 1) by independent rotations about the two dihedral angles φ and ψ of the glycosidic bond

from this source. In contrast, it will arise from changes of the two dihedral angles φ and ψ of each glycosidic bond (Fig. 3).

By letting the thumbs of each hand touch, independent variations of these two parameters by movements of the hands can swiftly be visualized. Since the pyranose rings linked by the glycosidic bond and their exocyclic substituents are rather bulky, their size will impose topological restraints to the intramolecular movements of the oligomer. Compared to oligopeptides with small side chains, the conformational space accessible to the molecule at room temperature will thus be relatively restricted. That this spatial factor limits the range of interchangeable conformations has been inferred by computer-assisted molecular mechanics and dynamics calculations and convincingly documented by experimental evidence primarily from sophisticated NMR-spectroscopy (Bush et al., 1999; Imberty, 1997; Siebert et al., 1999; von der Lieth et al., 1997a, 1998; Woods, 1998). Exploring the actual position(s) of each oligosaccharide on the scale between high flexibility with an ensemble of conformers and almost complete rigidity will definitely have salient implications to predict its role as coding unit. In this respect, it is also worth pointing out that a notable level of intramolecular flexibility is not a favorable factor for crystallization. Indeed, such an extent of unrestrained conformational entropy can contribute to explain the frequently frustrating experience in respective attempts in carbohydrate chemistry. If on the other hand the level of conformational entropy is confined to only very few stable conformers (keys), the presented shape distribution is not only a function of the sequence but also of external factors affecting the actual status of the equilibrium. In this context it should not escape notice that environmental parameters with impact on presentation of the glycan in glycoconjugates might shift the dynamic equilibrium of shape interconversions between attainable positions without requirement to alter the primary structure. Sugar receptors as probes for distinguishing bioactive or bioinert glycan presentation modes on proteins have already given the hypothesis experimental credit (Mann and Waterman, 1998; Noorman et al., 1998; Solís et al., 1987; White et al., 1997). This support brings to view an attractive means to modify shape, which warrants contriving further appropriate experiments to underpin its actual operation beyond any doubt.

As implied by referring to a code system, information stored as sequence and shape will have to be grasped. Translating and transmitting it into intended responses is the task of decoding devices. They should specifically recognize coding units established by glycans. Thus, in addition to physicochemically serving roles to control folding, oligomerization and access of proteolytic enzymes, as already mentioned (Drickamer and Taylor, 1998; Gagneux and Varki, 1999; Reuter and Gabius, 1999; Sharon and Lis, 1997; Varki, 1996), oligosaccharides in glycan chains can be likened to the postal code in an address to convey distinct messages read by suitable receptors. These carbohydrate-binding proteins are classified into enzymes responsible to assemble, modify, and degrade sugar structures, immunoglobulins homing in on carbohydrates as antigens, and, last but not least, lectins. Evidently, the third class encompasses all carbohydrate-binding proteins, which are neither antibodies nor are they enzymes which couple ligand recognition with catalytic activity to process the target (Barondes, 1988; Gabius, 1994). That lectin/glycan recognition has been assigned pivotal duties in an organism can at best be rendered perceptible by aberrations causing diseases. Knowledge accrued from the study of the biochemical basis of human diseases (e.g. mucolipidosis II or leukocyte adhesion deficiency (LAD) type II syndrome) underscores how trafficking of lysosomal enzymes or leukocytes can go awry owing to a lack of generation of the essential carbohydrate signal (Brockhausen et al., 1998; Lee and Lee, 1996; Paulson, 1996; Reuter and Gabius, 1999; Schachter, 1999; von Figura, 1990).

3.1   Lectins: Translators of the Sugar Code

The concept of a recognitive interplay between a sugar ligand and a lectin readily receives support, when the assumed ligand properties can be ascertained. As compiled in Table 1, various experimental approaches exploit the lectin's binding specificity in assays for their detection and characterization. The success in establishing these techniques and the power of affinity chromatography together with expression cloning and homology searches have spurred the transition from the early phase to categorize lectins according to their monosaccharide specificity and requirement for cations to the era to draw genealogical trees of lectin families. Having its roots in the structural definition of the folding pattern and architecture of the carbohydrate recognition domain, the classification scheme is currently agreed upon with five distinct families of animal lectins, i.e. C-type lectins, galectins, I-type lectin, P-type lectins, and pentraxins (Drickamer, 1988, 1993; Gabius, 1997a; Powell and Varki, 1995; Rini and Lobsanov, 1999). That this compilation is unlikely to be final is implied by the description of lectin sequences lacking invariant characteristics of any of the five classes (e.g. the chaperones calnexin and calreticulin mentioned in Table 2) and the detection of new folding arrangements (e.g. the five-bladed β-propeller in the invertebrate lectin tachylectin-2 (Beisel et al., 1999)).

Table 1 Methods used in the search for lectins. (Modified from Gabius, 1997a.)

Tools	Parameter
Multivalent glycans and (neo)glyco-conjugates or defined cell populations	Carbohydrate-dependent inhibition of lectin-mediated glycan precipitation or cell agglutination
Labelled (neo)glycoconjugates and matrix-immobilized extract fractions or purified proteins	Signal intensity
Cell populations	Labeling intensity
Tissue sections	Staining intensity
Animal (neo)glycoconjugatedrug chimera and cell populations	Biodistribution of signal intensity cellular responses (cell viability etc.)
Matrix-immobilized (neo)glycoconju-gates and cell populations	Carbohydrate-inhibitable cell adhesion
Cell extracts	Carbohydrate-elutable proteins
Homology searches with computer programs (e.g. Gene-finder or Blast), expressed sequence tags and knowledge of key structural aspects of carbohydrate recognition domains	Homology score in sequence alignment or knowledge-based modeling
Lectin motif-reactive probe (antibody, primer sets)	Extent of cross-reactivity

In each lectin family sequence alignments and homology searches have so far been conducive to unravel the divergent pathway from an ancestral gene to the current diversity. The intrafamily genealogy of mammalian C-type lectins has elegantly been traced back in a dendrogram to common ancestors for the seven subfamilies (Drickamer, 1993). To illustrate that such domains, often a part of modular arrangements, are no rare peculiarity in animal genomes, it is telling to add that a current database lists 389 C-type lectin-like sequences in animals (Sonnhammer et al., 1998). Yeast lacks this module in its domain collection. In the nematode *C. elegans*, whose elaborate enzymatic system for fucosylation has already been referred to (Oriol et al., 1999), this domain is ranked on the seventh place in frequency of occurrence, excelling for example the abundance of the EGF-like motif (The *C. elegans* Sequencing Consortium, 1998). At present, 183 C-type lectin-like domains have been traced in 125 proteins (Drickamer and Dodd, 1999). However, it is presently unclear how many of these proteins will be actually operative in Ca^{2+}-dependent sugar (or peptide) binding (Drickamer, 1999). Also, at least eight functional galectin genes and a tentative total of 28 candidate galectin genes among the approximately 20,000 genetic reading frames (current number predicted: 19,099) in its genome were identified in the nematode (Hirabayashi et al., 1997; Cooper and Barondes, 1999). These new insights into lectin abundance further increase the percentage of the coding genome devoted to glycan production and recognition.

Table 2 Functions of animal lectins

Activity	Example of Lectin
Ligand-selective molecular chaperones in endoplasmic reticulum	Calnexin, calreticulin
Intracellular routing of glycoproteins and vesicles	ERGIC-53, VIP-36, P-type lectins, comitin
Intracellular transport and extracellular assembly	Non-integrin 67 kDa elastin/laminin-binding protein
Cell type-specific endocytosis	Hepatic asialoglycoprotein receptor, macrophage C-type lectins, hepatic endothelial cell receptor for GalNAc-4-SO$_4$-bearing glycoproteins
Recognition of foreign glycans (β1,3-glucans, LPS)	CR3 (CD11b/CD18), *Limulus* coagulation factors C and G
Recognition of foreign or aberrant glycosignatures on cells (incl. endocytosis or initiation of opsonization or complement activation)	Collectins, C-type macrophage receptors, pentraxins (CRP, limulin), L-ficolin, tachylectins
Targeting of enzymatic activity in multimodular proteins	Acrosin, *Limulus* coagulation factor C
Bridging of molecules	Homodimeric and tandem-repeat galectins, cytokines (e.g. IL-2:IL-2R and CD3 of TCR), cerebellar soluble lectin
Effector release (H$_2$O$_2$, cytokines etc.)	Galectins, selectins, CD23
Cell growth control and apoptosis	Galectins, C-type lectins, amphoterin-like protein, cerebellar soluble lectin
Cell routing	Selectins, I-type lectins, galectins
Cell–cell interactions	Selectins and other C-type lectins, galectins, I-type lectins
Cell–matrix interactions	Galectins, heparin- and hyaluronic acid-binding lectins
Matrix network assembly	Proteoglycan core proteins (C-type CRD), galectins, non-integrin 67 kDa elastin/laminin-binding protein

For further information, see Gabius (1997a), Gabius and Gabius (1993, 1997), Kaltner and Stierstorfer (1998), Kishore et al. (1997), Vasta et al. (1999), Zanetta (1998) for recent reviews.

Equaling the strides being taken in the structural research on lectins, elucidation of their in vivo significance has steadily moved forward in the last decade. Summarized in Table 2, our present status of knowledge bears witness to the versatility to ply glycan recognition for a variety of purposes. In addition to mediating a physical contact between molecules and cells their initial recognition can trigger postbinding signaling with impact, for example, on growth regulation (Villalobo and Gabius, 1998). With focus on the homodimeric galectin-1 its mediation of downregulation of cell growth of responsive human neuroblastoma cells and of T-cell apoptosis to alleviate collagen-induced arthritis depicts representative examples with potential clinical relevance (Kopitz et al., 1998; Rabinovich et al., 1999).

Albeit necessarily centered in basic science, such cases illustrate the conceivable future potential to turn an endogenous lectin into a pharmaceutical.

Having already moved closer to applied science, the participation of lectins and glycoconjugates in cell adhesion has prompted attempts to rationally interfere with the molecular rendezvous, conceptually visualized as anti-adhesion therapy in Fig. 4. This approach mimics the natural strategy for success achieved with the complex cocktail of milk glycoconjugates.

They are protective by blocking docking of pathogens such as enteropathogenic and hemorrhagic *Escherichia coli, Campylobacter jejuni,* or rotavirus (Newburg, 1999). Although realization of this approach can prove tedious, because the pattern of recognition pairs is often not restricted to very few lectins (*Helicobacter pylori* with at least ten different carbohydrate-binding activities compared to the single type of influenza sialidase whose inhibition will noticeably affect virus propagation (Karlsson, 1999; Lingwood, 1998; von Itzstein and Thomson, 1997)), the custom-made design of tools, drawn as symbols in the strategy-outlining Fig. 4, justifies efforts to first localize binding partners and then to interfere with their activity aimed at therapy.

Notably, the first method can be used independently, e.g. in diagnostic procedures to characterize cell features. The visualization of carbohydrate-specific activities is commonly performed with carrier-immobilized sugar structures. Covalent attachment of a suitable derivative furnishes the versatility to produce neoglycoconjugates tailored to the experimental requirements (Bovin and Gabius, 1995; Lee Y.C. and Lee, 1994). Compared to a single carbohydrate unit the affinity of the multivalent ligand "is often beyond that expected from the increase in sugar concentration due to the presence of multiple residues on the protein (or polymeric backbone). Such an affinity enhancement is termed the glycoside cluster effect" (Lee R.T. and Lee, 1994). The geometrical increase in affinity with a numerical increase in valence for

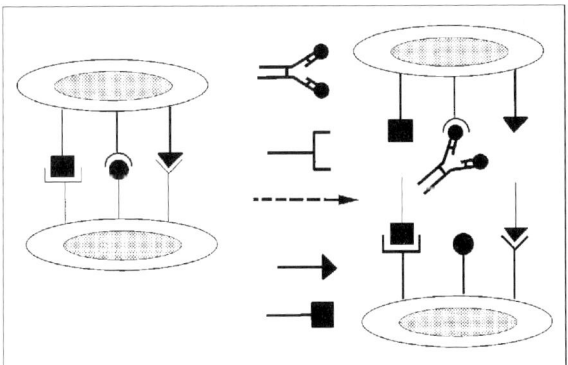

Fig. 4 Interference in lectin-mediated cell contact formation or recognition processes in general with target-specific blocking reagents, i.e. antibodies, sugar receptors, and oligosaccharides or mimetics thereof. (Kindly provided by Priv.-Doz. Dr. H. Kaltner, Munich. Details on the current status of anti-adhesion therapy are given by Cornejo et al., 1997; Gabius, 1997b; Gabius and Gabius, 1997; Karlsson, 1998; Simon, 1996; Zopf and Roth, 1996.)

mono-, bi-, and trivalent Gal-terminated oligosaccharides and mammalian asialoglycoprotein receptor, a C-type lectin, has been attributed to the topological complementarity between multiple ligand and receptor sites (Lee and Lee, 1997). Membrane solubilization by detergent treatment will in this case disrupt the essential spatial arrangement. An important caveat for approaches to detect the cluster effect concerns the use of agglutination assays. In contrast to affinity measurements in direct binding assays, the ongoing aggregation of multivalent receptors and ligands in solution can lead to erroneous conclusions. Indeed, under these circumstances isothermal titration calorimetry failed to record enhancements of Gibbs' free energy of binding but measured an endothermic, entropically favored process, its extent correlating with the inhibitory potency (IC_{50}-values) of tetra- and hexavalent ligands (Dimick et al., 1999).

Adding a label to the neoglycoconjugates renders them serviceable for detection of ligand-specific sites in cells and tissues, as listed in Table 1 with special practical emphasis being currently placed in tumor diagnosis (Gabius et al. 1995; Danguy et al., 1998; Gabius et al., 1998; Kayser and Gabius, 1999). In view of common lectin histochemistry with plant agglutinins, this method has been designated as "reverse lectin histochemistry" (Gabius et al., 1993). Following the description of a relevant clinical correlation, e.g. binding of histo-blood group A- and H-trisaccharides to lung cancer cells and survival of patients (Kayser et al., 1994), further work will aim to define the tissue target and to refine the ligand for optimal selectivity and specificity (Mammen et al., 1998) en route to running assays to unveil, if possible, therapeutic benefit in lectin-directed anti-adhesion therapy (see references given in legend for Fig. 4) and drug targeting (Gabius, 1989, 1997b). To attain this objective, it is indispensable to comprehend the how and why of protein–carbohydrate recognition. Thus, it is inostructive to proceed with a brief outline of these principles relevant for drug design.

4 Principles of Protein–Carbohydrate Recognition

Basically, typical contributions to the Gibbs' free energy of ligand binding originate from hydrogen bonding, van der Waals forces and the consequences of the hydrophobic effect. Factors to be reckoned with to predict the affinity of a ligand further include any alterations of the geometry and motional dynamics of the receptor and/or the ligand and/or the solvent molecules. As experimentally readily accessible parameters by calorimetric techniques, the determination of the reaction enthalpy and entropy delineates the global driving force towards complex formation. These parameters have, for example, been measured for an array of mono- and disaccharides in the cases of a plant and an animal lectin sharing specificity to D-galactose (Bharadwaj et al., 1999), and the plot of the data (Fig. 5) according to the equation:

$$-\Delta H = -\Delta G - T\Delta S$$

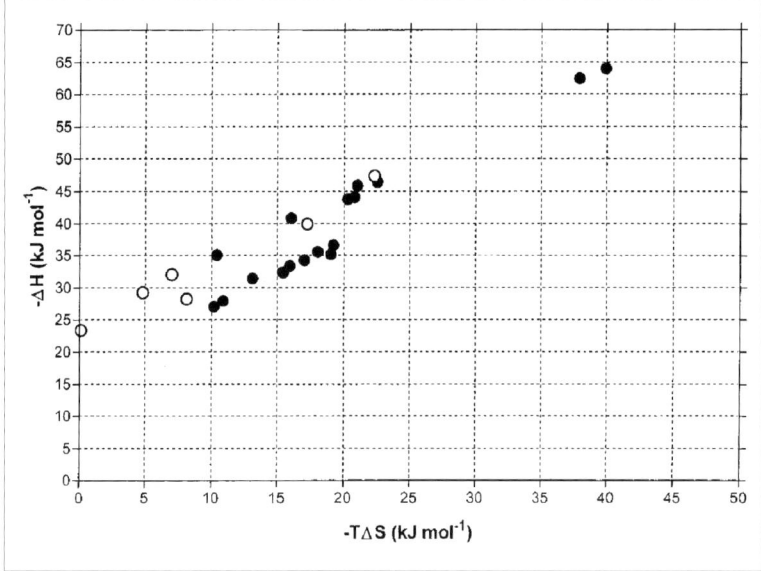

Fig. 5 Enthalpy–entropy compensation plot for the binding of a panel of mono- and disaccharides to the galactoside-specific mistletoe lectin (•) and the galectin from adult chicken liver (o)

reveals a slope near unity and intercepts of −16.45 kJ/mol (plant lectin) and −23.12 kJ/mol (animal lectin).

This figure conveys a fundamental message on the relation between enthalpic and entropic factors attributed to the participation of weak intermolecular forces. An increase in enthalpy for ligand binding is inherently balanced by an entropic penalty (or vice versa), an obvious example of common enthalpy–entropy compensation (Dunitz, 1995; Gilli et al., 1994; Lumry and Rajender, 1970). Its illustration automatically poses an ambitious question. The major challenge is to assign events on the level of the molecules in the course of association to the global enthalpic and entropic factors to bridge the gap between the demand for rules to optimize shape recognition and the thermodynamics. With this knowledge in hand, it might be feasible to intentionally shift the specificity and selectivity of derivatives. As the controversial discussion on the positive or negative role of water molecules for the enthalpy of complexation illuminates (Gabius, 1998; García-Hernández and Hernández-Arana, 1999; Lemieux, 1996; Toone, 1994), it will be essential to scrutinize the behavior of each participant of the molecular rendezvous in detail. Consequently, quick complete answers should not be expected but stepwise advances by the combination of computer-assisted calculations, spectroscopic techniques in solution, chemical tinkering with the ligand structure towards potent mimetics, and x-ray crystallography. An impression into the practical implementation of this interdisciplinary approach is given in the next paragraph.

5   How to Define Potent Ligand Mimetics

Taking the meaning of the word "carbohydrate" (C $(H_2O)_n$) literally, the abundant display of hydroxyl groups with their sp^3-hybridized oxygen atoms acting as acceptors with two lone electron pairs and the protons as donors nourishes the view that hydrogen bonds will dominate the spectrum of binding forces. When the spacing between two hydroxyl groups or the axial 4'-hydroxyl group and the ring oxygen atom matches that of an amino acid side chain (amide or carboxylate), two neighboring sites on the ligand can well be engaged in bidentate hydrogen bonds. The necessity for topological complementarity to yield the intricate network, schematically shown in Fig. 6, may not only be a source for enthalpy but also for selectivity, distinguishing anomers such D-Gal versus D-Man/D-Glc. It can thus be expected that the axial 4'-position for recognition of D-Gal and the equatorial 3',4'-positions for binding D-Man/D-Glc play decisive roles. This assumption is strikingly verified by x-ray crystallography and in solution by chemically engineered ligand derivatives (Rini, 1995; Solís et al., 1996; Weis and Drickamer, 1996; Gabius, 1997a, Solís and Díaz-Mauriño, 1997; Gabius, 1998; Lis and Sharon, 1998; Loris et al., 1998; Rüdiger et al., 1999;). With this structural explanation it becomes obvious why the change of the position of one hydroxyl group to form an epimer discussed during the presentation of the individual members of the monosaccharide alphabet unmistakably has the effect of creating a new letter. By the way, the same principle holds true for the characteristic formation of two coordination

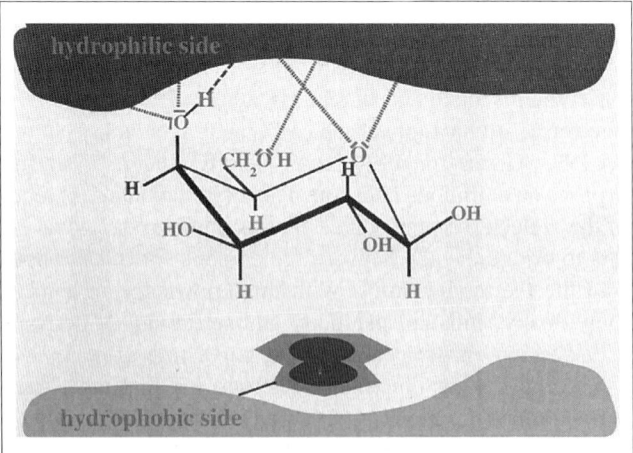

Fig. 6 The potential of D-galactose (see Fig. 1, bottom, and Fig. 3) for establishing interactions with constituents of the binding pocket of a sugar receptor. While the rather polar upper side can be engaged in frequent hydrogen bonds exploiting lone electron pairs of oxygen atoms as acceptors and the protons of appropriately positioned hydroxyl groups as donors (and also coordination bonds with a Ca^{2+}-ion in the case of C-type lectins), C-H/π-electron interactions and entropically favorable stacking can be engendered by an intimate contact of an aromatic (here: indolyl) amino acid side chain and the sugar's less polar bottom section. (Kindly provided by Dr. C.-W. von der Lieth, Heidelberg.)

bonds with the central Ca^{2+}-ion in the mentioned C-type lectins. Thereby, any wrong combination for the two adjacent hydroxyl groups involved in contacting the metal ion is excluded and sugar specificity is assured, unless the access-restricting impediment by a constraining loop close to the metal ion is lifted (Weis and Drickamer, 1996; Gabius, 1997a; Lis and Sharon, 1998; Loukas et al., 1999).

Inspecting Fig. 6 more closely, another important feature to drive ligand binding can be discovered. While the upper side of D-Gal is rather polar, the B-face exhibits a hydrophobic character. Stacking to the bulky aromatic amino acid side chain in the binding pocket removes both nonpolar surfaces from solvent accessibility, although the two rings may not be perfectly aligned in parallel. In fact, their positioning can tolerate distortions with angles between 17° and 52° in lectins (Weis and Drickamer, 1996). Nonetheless, this alignment will still contribute to complex stability and also to ligand selection despite its lower degree of directionality relative to hydrogen bonds (Quiocho, 1988; Vyas, 1991). The ensuing shielding of the indolyl side chain by the ligand is reflected for galectins in molecular dynamics calculations as well as differential UV, fluorescence, and laser photo chemically induced dynamic nuclear polarization (CIDNP) spectra (Levi and Teichberg, 1981; Siebert et al., 1997). Beyond this impact on solvent molecules by reducing the apolar surface area the π-electron cloud of the aromatic ring is likely to interact with the aliphatic D-Gal protons which harbor a net positive charge (Dougherty, 1996; Nishio et al., 1995; Weis and Drickamer, 1996). That the ensuing hydrophobic effect and van der Waals interactions do not deserve to be underestimated for impinging on the overall Gibb's free energy gain is underscored by the analysis of dominant forces in tight ligand binding for a variety of cases, where these factors can even surpass by far the contribution of hydrogen bonds (Davis and Teague, 1999; Kuntz et al., 1999).

This observation illustrates the complexity of the question how to account for the global enthalpic and entropic parameters on the level of molecules. For that galectin, whose data set from isothermal titration calorimetry is given in Fig. 5, it has recently been described by crystallographical work that six structural water molecules occupy the binding site in the ligand-free state stabilizing its topology and yielding a not yet precisely quantitated contribution to the Gibbs' free energy change upon displacement (Varela et al., 1999). In the case of a related galectin from the conger eel one additional water molecule even takes place of D-Gal's B-face substituting stacking by forming a π-electron hydrogen bond with a distance of 3.36 Å and an angle of 6.5° between the vector of the weight center of the five-membered section of the indole ring to the water molecule and the vector perpendicular to the ring plane (Shirai et al., 1999). The total exchange of the water molecules with the ligand will not only directly affect these solvent molecules but may also have a bearing on the proteins' intramolecular motions in solution. Remarkably, also the impact of ligand binding on protein flexibility is to be reckoned with. An increase in its vibrational entropy (14.6 kJ/mol for binding of one water molecule to bovine pancreatic trypsin inhibitor as model (Fischer and Verma, 1999)) can offset a substantial portion of the entropic penalty of the immobilization. The extent of this factor will certainly depend on the inherent mobility dynamics of the carbohydrate ligand free in solution. This parameter has already been inferred above to be often restricted due to spatial interference of the rather bulky rings and substituents. Graphically drawing on E. Fischer's (1894) classical "lock and key" paradigm, the metaphor has tentatively been introduced for this ligand type to view certain oligosaccharides as "bunch of keys" moving in solution through a limited set of shapes (Hardy, 1997). Only one of them may be selected by a receptor.

With a digalactoside (Galβ1-2Gal) as model, the formation of two "keys" from the same sequence is displayed in Fig. 7. Based on the φ, ψ, E-plot, shown in its left section, molecular dynamics calculations and nuclear Overhauser effect (NOE) NMR-spectroscopy (Siebert et al., 1996, 1999; von der Lieth et al., 1998), two distinct conformations are present in solution, each molecule rapidly fluctuating between these two topological constellations (Fig. 7, right side). Due to the inability to acquire spectroscopic snapshots with a resolution in the ps range, spectroscopical monitoring will be subject of time and ensemble averaging (Carver, 1991; Jardetzky, 1980). Since the term "key" implies its accurate fit into an appropriate lock, monitoring of transferred NOE signals, reflecting through space dipolar interactions between two protons in the bound ligand in double-resonance experiments, will resolve the gripping question as to which ligand topology will be accommodated in the binding pocket (Gabius, 1998; Jiménez-Barbero et al., 1999; Peters and Pinto, 1996; Poveda and Jiménez-Barbero, 1998; Rüdiger et al., 1999; von der Lieth et al., 1998).

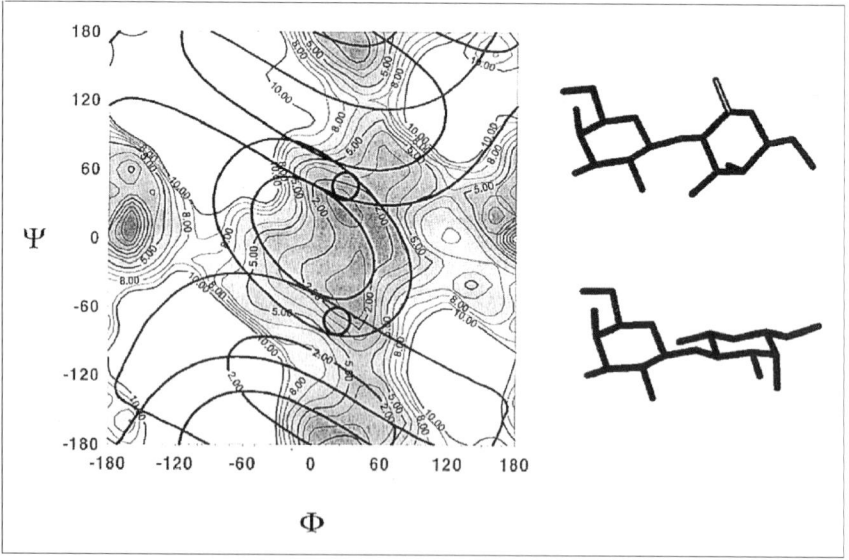

Fig. 7 Illustration of the principle of differential conformer selection. Based on NMR-spectroscopical analysis and molecular mechanics/dynamics calculations the disaccharide Galβ1-2Gal can adopt two distinct conformations in solution, which reside in energetically preferred regions of the φ, ψ, E-plot, symbolized by circles (*left*). Keeping the topological positioning of the nonreducing Gal-unit constant, the two sets of φ, ψ-values are readily visualized to translate into two significantly different conformers (*right*) which harbor disparate ligand properties. (Kindly provided by Priv.-Doz. Dr. H.-C. Siebert, Munich, and Dr. C.-W. von der Lieth, Heidelberg.)

These experiments provide two captivating answers for the studied case of lectins. Firstly, a lectin can actually select a distinct conformer, as seen for galactoside-binding lectins and selectins (Asensio et al., 1999; Espinosa et al., 1996; Gilleron et al., 1998; Harris et al., 1999; Poppe et al., 1997; Siebert et al., 1996; von der Lieth et al., 1998). Despite the same sequence the shape of other conformers renders them unsuitable for binding. Of course, a wrong key will not open a nonadaptable (rigid) lock designed for a different shape. Secondly, different receptors even with the same saccharide specificity harbor the capacity to bind different conformers. Thus, freezing a distinct conformation should have a dramatic impact on receptor binding as alluded to above. This principle is referred to as "differential conformer selection." It is visualized in Fig. 7 by noting that the conformer defined by the upper ϕ, ψ-combination is exclusively bound by a plant (mistletoe) agglutinin, while the tested galectin homes in on the second conformer (Gabius, 1998; Gilleron et al., 1998; Siebert et al., 1996; von der Lieth et al., 1998). Thus, not only the hydrogen-bonding patterns of these lectins toward D-Gal differ, as delineated by chemical mapping with deoxy and fluoro derivatives (Lee et al., 1992; Rüdiger et al., 1999; Solís et al., 1996), but also the pair of ϕ, ψ-torsion angles of β-Gal-terminated disaccharides. Because the importance of the intramolecular flexibility of the free ligand and conformer selection is only gradually explored as factor to be rationally manipulated, this result together with insights into favorable energetic interactions between the binding partners including solvent molecules warrants consideration for the design of mimetics. Thereby, they can eventually meet the high expectations for potency expressed in Fig. 4. When the geometry of crucial groups is maintained or even improved, the obtained substances do not even need to belong to the class of carbohydrates. To grant adequate heed to mimetics is probably a means to open a wide field for rational drug design, currently, for example, exploited for the influenza A/B neuraminidase and selectins (Sears and Wong, 1999; Simanek et al., 1998; von Itzstein and Thomson, 1997). As caveats to caution against prematurely advocating clinical effectiveness of anti-adhesion therapy in inflammation or of sugar-based drugs in epidemic flu, detrimental long-term effects in an animal model mimicking both acute and chronic intestinal inflammation has been reported (McCafferty et al., 1999). Similarly, stress has been laid upon the necessity to prove clinical benefit for an elegantly invented but costly antiflu drug in terms of an obvious impact on mortality beyond that of common, less expensive medications including vaccination (Cox and Hughes, 1999; Institut für Arzneimittelinformation, 1999; Yamey, 1999).

6   Conclusions

Elucidation of the structural basis of the genetic code and its translation into peptide sequences with milestones set by J.D. Watson and F. Crick (1953) and M.W. Nirenberg and J.H. Matthei (1961) has paved the way for medical applications more than three decades after the pioneering work in basic science. Today, nearly 20% of the new drugs tested in final phases are based on the technology of genetic engineering, up from 12% last year. To fathom the intricacy of the sugar code and transfer this knowledge on how sugar words are formed and these messages are decoded by receptors to applied science should therefore not be anticipated to be a matter of only a few years. The versatility of exploiting oligosaccharides as carriers of biological information presented by nature should first be thoroughly unraveled. Building this solid basis of experimental data will most likely entail to be able to venture into newly defined areas of glycan functionality and then to launch further projects of interdisciplinary research leading from basic to applied science. In view of the current focus on genomics, the presented evidence and reasoning offer well-grounded arguments to shake the conviction that as source for understanding of recognitive and regulatory processes in normal and disease states fruitful work will solely be confined to handling the data bank of the human genome.

References

Abeijon C, Mandon EC, Hirschberg CB (1997) Transporters of nucleotide sugars, nucleotide sulfate and ATP in the Golgi apparatus. Trends Biochem Sci 22:203–207

Abeygunawardana C, Bush CA (1993) Determination of the chemical structure of complex polysaccharides by heteronuclear NMR spectroscopy. Adv Biophys Chem 3:199–249

Asensio JL, Espinosa JF, Dietrich H, Cañada FJ, Schmidt RR, Martín-Lomas M, André S, Gabius H-J, Jiménez-Barbero J (1999) Bovine heart galectin-1 selects a distinct *(syn)* conformation of C-lactose, a flexible lactose analogue. J Am Chem Soc 121:8995–9000

Barondes SH (1988) Bifunctional properties of lectins: lectins redefined. Trends Biochem Sci 13:480–482

Beisel H-G, Kawabata S-i, Iwanaga S, Huber R, Bode W (1999) Tachylectin-2: crystal structure of a specific GlcNAc/GalNAc-binding lectin involved in the innate immunity host defense of the Japanese horseshoe crab *Tachypleus tridentatus*. EMBO J 18:2313–2322

Bharadwaj S, Kaltner H, Korchagina EY, Bovin NV, Gabius H-J, Surolia A (1999) Microcalorimetric indications for ligand binding as a function of the protein for galactoside-specific plant and avian lectins. Biochim Biophys Acta 1472:191–196

Bovin NV, Gabius H-J (1995) Polymer-immobilized carbohydrate ligands: versatile chemical tools for biochemistry and medical sciences. Chem Soc Rev 24:413–421

Brockhausen I, Schachter H (1997) Glycosyltransferases involved in N- and O-glycan biosynthesis. In: Gabius H-J, Gabius S (eds) Glycosciences: Status and Perspectives, Chapman & Hall, London/Weinheim, pp 79–113

Brockhausen I, Schutzbach J, Kuhns W (1998) Glycoproteins and their relationship to human disease. Acta Anat 161:36–78

Brown GM, Levy HA (1965) a-D-glucose: precise determination of crystal and molecular structure by neutron-diffraction analysis. Science 147:1038–1039

Bush CA, Martin-Pastor M, Imberty A (1999) Structure and conformation of complex carbohydrates of glycoproteins, glycolipids, and bacterial polysaccharides. Annu Rev Biophys Biomol Struct 28:269–293

Carver JP (1991) Experimental structure determination of oligosaccharides. Curr Opin Struct Biol 1:716–720

Casu B, Petitou M, Provasoli M, Sinay P (1988) Conformational flexibility: a new concept for explaining binding and biological properties of iduronic acid-containing glycosaminoglycans. Trends Biochem Sci 13:221–225

Cooper DNW, Barondes SH (1999) God must love galectins; He made so many of them. Glycobiology 9:979–984

Cornejo CJ, Winn RK, Harlan JM (1997) Anti-adhesion therapy. Adv Pharmacol 39:99–142

Cox NJ, Hughes JM (1999) New options for the prevention of influenza. N Engl J Med 341:1387–1388

Cummings RD (1997) Lectins as tools for glycoconjugate purification and characterization. In: Gabius H-J, Gabius S (eds) Glycosciences: Status and Perspectives, Chapman & Hall, London/ Weinheim, pp 191–199

Danguy A, Decaestecker C, Genten F, Salmon I, Kiss R (1998) Application of lectins and neogly-coconjugates in histology and pathology. Acta Anat 161:206–218

Davis AM, Teague SJ (1999) Die Bedeutung der Balance von Wasserstoffbrückenbindungen und hydrophoben Wechselwirkungen im Wirkstoff-Rezeptor-Komplex. Angew Chem 111:778–792

Dimick SM, Powell SC, McMahon SA, Moothoo DM, Naismith JH, Toone EJ (1999) On the meaning of affinity: cluster glycoside effects and concanavalin A. J Am Chem Soc 121:10286–10296

Dougherty DA (1996) Cation-p interactions in chemistry and biology: a new view of benzene, Phe, Tyr, and Trp. Science 271:163–168

Drickamer K (1988) Two distinct classes of carbohydrate recognition domains in animal lectins. J Biol Chem 263:9557–9560

Drickamer K (1993) Evolution of Ca^{2+}-dependent animal lectins. Progr Nucl Acid Res Mol Biol 45:207–233

Drickamer K (1999) C-type lectin-like domains. Curr Opin Struct Biol 9:585–590

Drickamer K, Dodd RB (1999) C-type lectin-like domains in Caenorhabditis elegans: predictions form the complete genome sequence. Glycobiology 9:1357–1369

Drickamer K, Taylor ME (1998) Evolving views of protein glycosylation. Trends Biochem Sci 23:321–324

Dunitz JD (1995) Win some, lose some: enthalpy-entropy compensation in weak intermolecular interactions. Chem Biol 2:709–712

Espinosa JF, Cañada FJ, Asensio JL, Martin-Pastor M, Dietrich H, Martin-Lomas M, Schmidt RR, Jiménez-Barbero J (1996) Experimental evidence of conformational differences between C-glycosides and O-glycosides in solution and in the protein-bound state: the C-lactose/O-lactose case. J Am Chem Soc 118:10862–10871

Fischer E (1894) Einfluß der Configuration auf die Wirkung der Enzyme. Ber Dt Chem Ges 27:2985–2993

Fischer S, Verma CS (1999) Binding of buried structural water increases the flexibility of proteins. Proc Natl Acad Sci USA 96:9613–9615

Gabius H-J (1989) Endogene Lektine in Tumoren und ihre mögliche Bedeutung für Diagnose und Therapie von Krebserkrankungen. Onkologie 12:175–181

Gabius H-J (1994) Non-carbohydrate binding partners/domains of animal lectins. Int J Biochem 26:469–477

Gabius H-J (1997a) Animal lectins. Eur J Biochem 243:543–576

Gabius H-J (1997b) Concepts of tumor lectinology. Cancer Investig 15:454–464

Gabius H-J (1998) The how and why of protein-carbohydrate interaction: a primer to the theoretical concept and a guide to application in drug design. Pharm Res 15:23–30

Gabius H-J, Gabius S (eds) (1993) Lectins and Glycobiology. Springer Verlag, Berlin/ New York

Gabius H-J, Gabius S (eds) (1997) Glycosciences: Status and Perspectives. Chapman & Hall, London/Weinheim

Gabius H-J, Gabius S, Zemlyanukhina TV, Bovin NV, Brinck U, Danguy A, Joshi SS, Kayser K, Schottelius J, Sinowatz F, Tietze LF, Vidal-Vanaclocha F, Zanetta J-P (1993) Reverse lectin histochemistry: design and application of glycoligands for detection of cell and tissue lectins. Histol Histopathol 8:369–383

Gabius H-J, Kayser K, Gabius S (1995) Protein-Zucker-Erkennung. Grundlagen und medizinische Anwendung am Beispiel der Tumorlektinologie. Naturwissenschaften 82:533–543

Gabius H-J, Unverzagt C, Kayser K (1998) Beyond plant lectin histochemistry: preparation and application of markers to visualize the cellular capacity for protein-carbohydrate recognition. Biotech Histochem 73:263–277

Gagneux P, Varki A (1999) Evolutionary considerations in relating oligosaccharide diversity to biological function. Glycobiology 9:747–755

García-Hernández E, Hernández-Arana A (1999) Structural bases of lectin-carbohydrate affinities: comparison with protein-folding energetics. Protein Sci 8:1075–1086

Geyer H, Geyer R (1998) Strategies for glycoconjugate analysis. Acta Anat 161:18–35

Gilleron M, Siebert H-C, Kaltner H, von der Lieth C-W, Kozár T, Halkes KM, Korchagina EY, Bovin NV, Gabius H-J, Vliegenthart JFG (1998) Conformer selection and differential restriction of ligand mobility by a plant lectin. Conformational behavior of Galb1-3GlcNAcb1-R, Galb1-3GalNAcb1-R and Galb1-2Galb1-R' in the free state and complexed with mistletoe lectin as revealed by random walk and conformational clustering molecular mechanics calculations, molecular dynamics simulations and nuclear Overhauser experiments. Eur J Biochem 252:416–427

Gilli P, Ferretti V, Gilli G, Borea PA (1994) Enthalpy-entropy compensation in drug-receptor binding. J Phys Chem 98:1515–1518

Hardy BJ (1997) The glycosidic linkage flexibility and time-scale similarity hypotheses. J Mol Struct 395–396:187–200

Harris R, Kiddle GR, Field RA, Milton MJ, Ernst B, Magnani JL, Homans SW (1999) Stable-isotope-assisted NMR studies on ^{13}C-enriched sialyl Lewis^x in solution and bound to E-selectin. J Am Chem Soc 121:2546–2551

Hirabayashi J (1996) On the origin of elementary hexoses. Quart Rev Biol 71:365–380

Hirabayashi J, Arata Y, Kasai K-i (1997) Galectins from the nematode Caenorhabditis elegans and the genome project. Trends Glycosci Glycotechnol 9:113–122

Hooper LV, Manzella SM, Baenziger JU (1997) The biology of sulfated oligosaccharides. In: Gabius H-J, Gabius S (eds) Glycosciences: Status and Perspectives, Chapman & Hall, London/Weinheim, pp. 261–276

Hounsell EF (1997) Methods of glycoconjugate analysis. In: Gabius H-J, Gabius S (eds) Glycosciences: Status and Perspectives, Chapman & Hall, London/Weinheim, pp. 15–29

Imberty A (1997) Oligosaccharide structures: theory versus experiment. Curr Opin Struct Biol 7:617–623

Institut für Arzneimittelinformation (eds) (1999) Neue Konzepte gegen Virusgrippe: vielversprechend? arznei-telegramm 30:98–100

Jardetzky O (1980) On the nature of molecular conformations inferred from high-resolution NMR. Biochim Biophys Acta 621:227–232

Jiménez-Barbero J, Asensio JL, Cañada FJ, Poveda A (1999) Free and protein-bound carbohydrate structures. Curr Opin Struct Biol 9:549–555

Kaltner H, Stierstorfer B (1998) Animal lectins as cell adhesion molecules. Acta Anat 161:162–179

Karlsson K-A (1998) Meaning and therapeutic potential of microbial recognition of host glycoconjugates. Mol Microbiol 29:1–11

Karlsson K-A (1999) Bacterium-host protein-carbohydrate interactions and pathogenicity. Biochem Soc Transact 27:471–474

Kayser K, Gabius H-J (1999) The application of thermodynamic principles to histochemical and morphometric tissue research: principles and practical outline with focus on glycosciences. Cell Tissue Res 296:443–455

Kayser K, Bovin NV, Korchagina EY, Zeilinger C, Zeng F-Y, Gabius H-J (1994) Correlation of expression of binding sites for synthetic blood group A-, B-, and H-trisaccharides and for sarcolectin with survival of patients with bronchial carcinoma. Eur J Cancer 30A:653–657

Kishore U, Eggleton P, Reid KBM (1997) Modular organization of carbohydrate recognition domains in animal lectins. Matrix Biol 15:583–592

Kopitz J, von Reitzenstein C, Burchert M, Cantz M, Gabius H-J (1998) Galectin-1 is a major receptor for ganglioside GM$_1$, a product of the growth-controlling activity of a cell surface ganglioside sialidase, on human neuroblastoma cells in culture. J Biol Chem 273:11205–11211

Kuntz ID, Chen K, Sharp KA, Kollman PA (1999) The maximal affinity of ligands. Proc Natl Acad Sci USA 96:9997–10002

Laine RA (1997) The information-storing potential of the sugar code. In: Gabius H-J, Gabius S (eds) Glycosciences: Status and Perspectives, Chapman & Hall, London/Weinheim, pp. 1–14

Lee RT, Lee YC (1994) Enhanced biochemical affinities of multivalent neoglycoconjugates. In: Lee YC, Lee RT (eds) Neoglycoconjugates. Preparation and Applications, Academic Press, San Diego, pp. 23–50

Lee YC, Lee RT (eds) (1994) Neoglycoconjugates. Preparation and Applications. Academic Press, San Diego

Lee YC, Lee RT (1996) Glycobiology in medicine. J Biomed Sci 3:221–237

Lee RT, Lee YC (1997) Neoglycoconjugates. In: Gabius H-J, Gabius S (eds) Glycosciences: Status and Perspectives, Chapman & Hall, London/Weinheim, pp. 55–77

Lee RT, Gabius H-J, Lee YC (1992) Ligand-binding characteristics of the major mistletoe lectin. J Biol Chem 267:23722–23727

Lemieux RU (1996) How water provides the impetus for molecular recognition in aqueous solution. Acc Chem Res 29:373–380

Levi G, Teichberg VI (1981) Isolation and physicochemical characterization of electrolectin, a b-D-galactoside-binding lectin from the electric organ of *Electrophorus electricus*. J Biol Chem 256:5735–5740

Lingwood CD (1998) Oligosaccharide receptors for bacteria: a view to kill. Curr Opin Chem Biol 2:695–700

Lis H, Sharon N (1998) Lectins: carbohydrate-specific proteins that mediate cellular recognition. Chem Rev 98:637–674

Loris R, Hamelryck T, Bouckaert J, Wyns L (1998) Legume lectin structure. Biochim Biophys Acta 1383:9–36

Loukas A, Mullin NP, Tetteh KKA, Moens L, Maizels RM (1999) A novel C-type lectin secreted by a tissue-dwelling parasitic nematode. Curr Biol 9:825–828

Lumry R, Rajender S (1970) Enthalpy-entropy compensation phenomena in water solutions of proteins and small molecules: an ubiquitous property of water. Biopolymers 9:1125–1227

Mammen M, Choi S-K, Whitesides GM (1998) Polyvalente Wechselwirkungn in biologischen Systemen: Auswirkungen auf das Design und die Verwendung multivalenter Liganden und Inhibitoren. Angew Chem 110:2908–2953

Mann PL, Waterman RE (1998) Glycocoding as an information management system in embryonic development. Acta Anat 161:153–161

Marszalek PE, Oberhauser AF, Pang Y-P, Fernandez JM (1998) Polysaccharide elasticity governed by chair-boat transitions of the glucopyranose ring. Nature 396:661–664

McCafferty D-M, Smith CW, Granger DN, Kubes P (1999) Intestinal inflammation in adhesion molecule-deficient mice: an assessment of P-selectin alone and in combination with ICAM-1 or E-selectin. J Leukoc Biol 66:67–74

Newburg DS (1999) Human milk glycoconjugates that inhibit pathogens. Curr Med Chem 6:117–127

Nishio M, Umezawa Y, Hirota M, Takeuchi Y (1995) The CH/p interaction: significance in molecular recognition. Tetrahedron 51:8665–8701

Noorman F, Barrett-Bergshoeff MM, Rijken DC (1998) Role of carbohydrate and protein in the binding of tissue-type plasminogen activator to the human mannose receptor. Eur J Biochem 251:107–113

Oriol R, Mollicone R, Cailleau A, Balanzino L, Breton C (1999) Divergent evolution of fucosyl-transferase genes from vertebrates, invertebrates and bacteria. Glycobiology 9:323–334

Paulson JC (1996) Leukocyte adhesion deficiency type II. In: Montreuil J, Vliegenthart JFG, Schachter H (eds) Glycoproteins and Disease, Elsevier, Amsterdam, pp. 405–411

Pavelka M (1997) Topology of glycosylation - a histochemist's view. In: Gabius H-J, Gabius S (eds) Glycosciences: Status and Perspectives, Chapman & Hall, London/Weinheim, pp. 115–120

Peters T, Pinto BM (1996) Structure and dynamics of oligosaccharides: NMR and modeling studies. Curr Opin Struct Biol 6:710–720

Poppe L, Brown GS, Philo JS, Nikrad PV, Shah BH (1997) Conformation of sLe^x tetrasaccharide, free in solution and bound to E-, P-, and L-selectin. J Am Chem Soc 119:1727–1736

Poveda A, Jiménez-Barbero J (1998) NMR studies of carbohydrate-protein interaction in solution. Chem Soc Rev 27:133–143

Powell LD, Varki A (1995) I-type lectins. J Biol Chem 270:14243–14246

Quiocho FA (1988) Molecular features and basic understanding of protein-carbohydrate interactions: the arabinose-binding protein-sugar complex. Curr Top Microbiol Immunol 139:135–148

Rabinovich GA, Daly G, Dreja H, Tailor H, Riera CM, Hirabayashi J, Chernajovsky Y (1999) Recombinant galectin-1 and its genetic delivery suppress collagen-induced arthritis via T cell apoptosis. J Exp Med 190:385–397

Reuter G, Gabius H-J (1996) Sialic acids. Structure, analysis, metabolism, and recognition. Biol Chem Hoppe-Seyler 377:325–342

Reuter G, Gabius H-J (1999) Eukaryotic glycosylation - whim of nature or multipurpose tool? Cell Mol Life Sci 55:368–422

Reutter W, Stäsche R, Stehling P, Baum O (1997) The biology of sialic acids: insights into their structure, metabolism and function in particular during viral infection. In: Gabius H-J, Gabius S (eds) Glycosciences: Status and Perspectives, Chapman & Hall, London/Weinheim, pp. 245–259

Rini JM (1995) Lectin structure. Annu Rev Biophys Biomol Struct 24:551–577

Rini JM, Lobsanov YD (1999) New animal lectin structures. Curr Opin Struct Biol 9:578–584

Rüdiger H, Siebert H-C, Solís D, Jiménez-Barbero J, Romero A, von der Lieth C-W, Díaz-Mauriño T, Gabius H-J (1999) Medicinal chemistry based on the sugar code: fundamentals of lectinology and experimental strategies with lectins as targets. Curr Med Chem (in press)

Schachter H (ed) (1999) Molecular basis of glycoconjugate disease. Biochim Biophys Acta 1455:61–418

Sears P, Wong C-H (1998) Enzyme action in glycoprotein synthesis. Cell Mol Life Sci 54: 223–252

Sears P, Wong C-H (1999) Kohlenhydratmimetika: ein neuer Lösungsansatz für das Problem der kohlenhydratvermittelten biologischen Erkennung. Angew Chem 111:2446–2471

Sharon N, Lis H (1997) Glycoproteins: structure and function. In: Gabius H-J, Gabius S (eds) Glycosciences: Status and Perspectives, Chapman & Hall, London/Weinheim, pp. 133–162

Shirai T, Mitsuyama C, Niwa Y, Matsui Y, Hotta H, Yamane T, Kamiya H, Ishii C, Ogawa T, Muramoto K (1999) High-resolution structure of the conger eel galectin, congerin I, in lactose-liganded and ligand-free forms: emergence of a new structure class by accelerated evolution. Structure 7:1223–1233

Siebert H-C, Adar R, Arango R, Burchert M, Kaltner H, Kayser G, Tajkhorshid E, von der Lieth C-W, Kaptein R, Sharon N, Vliegenthart JFG, Gabius H-J (1997) Involvement of laser photo CIDNP (chemically induced dynamic nuclear polarization)-reactive amino acid side chains in ligand binding by galactoside-specific lectins in solution. Similarities in the role of tryptophan/tyrosine residues for ligand binding between a plant agglutinin and mammalian/avian galectins and the detection of an influence of single-site mutagenesis on surface presentation of spatially separated residues. Eur J Biochem 249:27–38

Siebert H-C, Gilleron M, Kaltner H, von der Lieth C-W, Kozár T, Bovin NV, Korchagina EY, Vliegenthart JFG, Gabius H-J (1996) NMR-based, molecular dynamics- and random walk molecular mechanics-supported study of conformational aspects of a carbohydrate ligand (Galb1-2Galb1-R) for an animal galectin in the free and in the bound state. Biochem Biophys Res Commun 219:205–212

Siebert H-C, von der Lieth C-W, Gabius H-J (1999) Der Zuckerkode - Oligosaccharide als Träger biologischer Information. Dtsch Apotheker Ztg 139:272–282

Simanek EE, McGarvey GJ, Jablonowski JA, Wong C-H (1998) Selectin-carbohydrate interactions: from natural ligands to designed mimics. Chem Rev 98:833–862

Simon PM (1996) Pharmaceutical oligosaccharides. Drug Discovery Today 1:522–528

Solís D, Díaz-Mauriño T (1997) Analysis of protein-carbohydrate interaction by engineered ligands. In: Gabius H-J, Gabius S (eds) Glycosciences: Status and Perspectives, Chapman & Hall, London/Weinheim, pp. 345–354

Solís D, Estremera D, Usobiaga P, Díaz-Mauriño T (1987) Differential binding of mannose-specific lectins to the carbohydrate chains of fibrinogen domains D and E. Eur J Biochem 165:131–138

Solís D, Romero A, Kaltner H, Gabius H-J, Díaz-Mauriño T (1996) Different architecture of the combining sites of two chicken galectins revealed by chemical-mapping studies with synthetic ligand derivatives. J Biol Chem 271:12744–12748

Sonnhammer EL, Eddy SR, Birney E, Bateman A, Durbin R (1998) Pfam: multiple sequence alignment and HMM-profiles of protein domains. Nucl Acids Res 26:320–322

The C. elegans Sequencing Consortium (1998) Genome sequence of the nematode C. elegans: a platform for investigating biology. Science 282:2012–2018

Toone EJ (1994) Structure and energetics of protein-carbohydrate complexes. Curr Opin Struct Biol 4:719–728

Uedaira H, Uedaira H (1985) The relationship between partial molecular heat capacities and the number of equatorial hydroxyl groups of saccharides. J Sol Chem 14:27–34

Varela PF, Solís D, Díaz-Mauriño T, Kaltner H, Gabius H-J, Romero A (1999) The 2.15 Å crystal structure of CG-16, the developmentally regulated homodimeric chicken galectin. J Mol Biol 294:537–549

Varki A (1996) "Unusual" modifications and variations of vertebrate oligosaccharides: are we missing the flowers for the trees? Glycobiology 6:707–710

Varki A (1998) Factors controlling the glycosylation potential of the Golgi apparatus. Trends Cell Biol 8:34–40

Varki A, Marth J (1995) Oligosaccharides in vertebrate development. Sem Develop Biol 6:127–138

Vasta GR, Quesenberry M, Ahmed H, O'Leary N (1999) C-type lectins and galectins mediate innate and adaptive immune functions: their roles in the complement activation pathway. Dev Comp Immunol 23:401–420

Villalobo A, Gabius H-J (1998) Signaling pathways for transduction of the initial message of the glycocode into cellular responses. Acta Anat 161:110–129

von der Lieth C-W, Kozár T, Hull WE (1997a) A (critical) survey of modeling protocols used to explore the conformational space of oligosaccharides. J Mol Struct 395-396:225–244

von der Lieth C-W, Lang E, Kozár T (1997b) Carbohydrates: second-class citizens in biomedicine and in bioinformatics? In: Hofestädt R, Lengauer T, Löffler M, Schomburg D (eds) Bioinformatics, Springer, Berlin/New York, pp. 147–155

von der Lieth C-W, Siebert H-C, Kozár T, Burchert M, Frank M, Gilleron M, Kaltner H, Kayser G, Tajkhorshid E, Bovin NV, Vliegenthart JFG, Gabius H-J (1998) Lectin ligands: new insights into their conformations and their dynamic behavior and the discovery of conformer selection by lectins. Acta Anat 161:91–109

von Figura K (1990) Mannose-6-phosphat-Rezeptoren: ihre Rolle beim Transport lysosomaler Proteine. Naturwissenschaften 77:116–122

von Itzstein M, Thomson RJ (1997) Sialic acids and sialic acid-recognising proteins: drug discovery targets and potential glycopharmaceuticals. Curr Med Chem 4:185–210

Vyas NK (1991) Atomic features of protein-carbohydrate interactions. Curr Opin Struct Biol 1:732–740

Weis WI, Drickamer K (1996) Structural basis of lectin-carbohydrate recognition. Annu Rev Biochem 65:441–473

White KD, Cummings RD, Waxman FJ (1997) Ig N-glycan orientation can influence interactions with the complement system. J Immunol 158:426–435

Winterburn PJ, Phelps CF (1972) The significance of glycosylated proteins. Nature 236:147–151

Woods RJ (1998) Computational carbohydrate chemistry: what theroretical methods can tell us. Glycoconjugate J 15:209–216

Wormald MR, Dwek RA (1999) Glycoproteins: glycan presentation and protein-fold stability. Structure 7:R155–R160

Yamey G (1999) Anti-flu drug may not reduce death rate. Br Med J 319:659

Zanetta J-P (1998) Structure and functions of lectins in the central and peripheral nervous system. Acta Anat 161:180–195

Zopf D, Roth S (1996) Oligosaccharide anti-infective agents. Lancet 347:1017–1021

Chapter 11

The Immune Self Code: From Correspondence to Complexity

Yair Neuman

Abstract Codes are conventions of correspondence between realms. These conventions may be considered emergent products of complex microlevel interactions. In this chapter, I illustrate the complexity of coding by discussing the immune self. I argue that our evolving understanding of the immune self bears a striking resemblance to the way our understanding of signs evolved. According to the thesis presented in this chapter, the immune system is a meaning-making system and therefore a biosemiotic analysis of the immune self may shed new light on a variety of unsolved theoretical questions in immunology.

1 Introduction: Codes of Complexity

A watershed differentiating between living and nonliving matter is that living matter is *mediated*. In fact, we can even argue that the qualitative difference between the two is evident whenever mediation enters the picture. This argument is far from clear or trivial, and its implications are even less trivial. What do we mean when we say that living matter is "mediated"? In a nutshell, the idea is that what characterizes living systems as a unique category of matter is that their behavior cannot be reduced to simple interactions between dyads. There is always a third party – a boundary condition, to use Michael Polanyi's term – whose description cannot be reduced to the language of the dyads. Whereas interactions between atoms are unmediated and occur through the direct exchange of electrons or through weak forces, the interactions that characterize living systems are always mediated by a third party. For example, the synthesis of proteins based on genes is not direct; mediation by RNA is a must.

The concept of mediation becomes more comprehensible if we explain it in terms of *codes*. A code is a set of rules of correspondence between two independent worlds (see Barbieri's editorial), and as such it prevails in both the organic and the mental realms. Codes are conventional in the sense that the correspondence between the realms is not established through rigid, universal, and transcendental laws but through

Ben-Gurion University of the Negev, Beer-Sheva, Israel, e-mail: yneuman@bgu.ac.il

M. Barbieri (ed.), *The Codes of Life: The Rules of Macroevolution.*
© Springer 2008

convention (Barbieri, 2003): biological, psychological, or sociological. The "rules" that govern the codes of life are "rules" of coordination rather than the laws of logic or physics. Understanding living systems means understanding codes, from the organic codes that constitute the organism to the codes that constitute its mental life. It should be noted that the study of codes does not imply the study of simple rules of correspondence between different realms. Living organisms are paragons of complexity and studying biological complexity obliges us to move beyond simple models of coding to the way mediation constitutes the human organism.

The aim of this chapter is to illustrate this paradigmatic shift by dealing with a specific case – the immune self. My aim is to show how our understanding of the immune self can shift from simple rules of correspondence to the complexity of a meaning-making perspective on the immune self. This will be done in terms of the interrelations of two fields – semiotics and immunology – that have been following a similar trajectory from correspondence to complexity.

2   The Immune Self

Immunology has been described as the science of self and nonself discrimination (Klein, 1982). Wikipedia, the highly popular Internet encyclopedia, explains: "The immune system defends the body by recognizing agents that represent *self* and those that represent *nonself*, and launching attacks against harmful members of the latter group." Understanding that certain agents "represent" self means understanding the rules of correspondence between the "self" as an abstract concept and its corresponding agents at the molecular level.

Wikipedia also explains: "Distinguishing between self and nonself and between harmful nonself and harmless nonself is a difficult problem" (Available at: http://en.wikipedia.org/wiki/Immune_system). Indeed, the meaning of the "immune self" is disputed. What is the immune self and why was the concept of the self introduced to immunology in the first place?

The concept of self is traditionally associated with disciplines such as philosophy and psychology. Indeed, there is a whole branch of psychology known as "self psychology" and in philosophy the concept of the self has been discussed at length with regard to the issue of personal identity. My aim, however, is not to discuss the self as a property of human beings, or to answer the questions "Who am I?" and "What is the stable essence of my identity?" My aim is to discuss the "self" of the immune system. Are those two different selves? Is the self a concept that is applicable both to philosophy/psychology and immunology? Is the self a metaphor that was imposed on the immune system or is it a crucial organizing concept for theoretical immunology?

As suggested by Howes (1998, p. 1), there are numbers of "fascinating parallels that might be drawn between theoretical developments concerning self in philosophy and in immunology." These parallels cannot be denied. The concept of "self" was introduced in immunology by Sir Frank Macfarlane Burnet after it had been intensively elaborated in philosophy and psychology (Tauber, 1996, 2002). As a metaphor imported from other fields of inquiry, the concept of the self was inevitably loaded with associations and connotations, and these have clearly influenced the idea of the immune self. Although interesting parallels exist between the concept of the self in the humanities and in immunology, these parallels should not obscure the significant differences between the concept as it is used in these distinct fields of inquiry.

In this chapter, I do not present a complete survey of the literature dealing with the immune self or review its various senses or historical and philosophical origins, a task carried out by Tauber (1996, 1997, 1998, 1999, 2002). Instead, I present the argument that the problem of understanding the meaning of the immune self is analogous to the problem of finding the meaning of signs in semiotics, and that in both cases we should move from naïve "rules" of correspondence to complex patterns of coding. In this context, I present the idea that the immune system is a meaning-making system (Neuman, 2004) and provide a novel conceptualization of the immune self that integrates several ideas from immunology and semiotics.

Before I delve into the issue of the immune self, let me make a comment – a trivial one, but one nevertheless worth making. The immune self is our way of conceptualizing processes of the immune system or using a heuristic for approaching problems, research questions, or findings concerning the immune system. The immune self is not a "real" entity in the same sense that lymphocytes, cytokines, and the thymus are real entities. It is an abstraction of both our minds and the immune system. Identifying the immune self with certain components of the immune system, a step taken by some scholars (Tauber, 1998), is an error of logical typing: *Pars pro toto*! The immune self is embodied by these components; however, it is still a concept used by us as outside observers, and as such deserves a critical analysis.

In this context, it is worth asking what the meaning of the immune self is and whether it is a meaningful concept that is really crucial for immunology. In other words, do we really need the concept in order to understand the immune system? Surprisingly, this fundamental question is still being debated in immunology.

3   The Reductionist Perspective

How do we know to differentiate between self and nonself? The genetic reductionist approach suggests that there is a single genetic criterion for identification of the self: a simple code, which is a genetic fingerprint that allows the immune system to differentiate between the self and the other (nonself). In other words, the reductionist perspective offers a simple correspondence between the immune self and a molecular component of the cell.

The MHC is the set of gene loci specifying major histocompatibility antigens. It first looked as though this genetic marker might provide us with the ultimate criterion for self–nonself differentiation. Rolston (1996, quoted in Howes, 1998, p. 3) expresses this idea clearly:

> Recognition of the nonself is signaled by the molecules of the major histocompatibility complex (MHC). There are class I molecules on every nucleated cell in the body to identify the self. It is also important to determine which cells to kill, and this is done by T cells, using class II molecules located in macrophages, B cells and some T cells.

It should be noted that according to this suggestion *the nonself is an empty slot.* There is only a self, which is identified through the genetic marker of the MHC. Whereas the immune self corresponds to the MHC, the nonself corresponds to nothing. An entity is recognized as "nonself" if it lacks the sign of the self. In other words, the foreignness of the antigen is implicit in not having a self marker.

Identity markers such as the MHC or fingerprints may be very helpful in identifying a specific self. However, the advancement of our understanding of identity markers does not solve a problem inherent in identity markers: *The map is not the territory and the identity marker should not be confused with the self it supposedly signifies.* No simple coding exists between an emerging phenomenon on one scale of analysis (i.e. the immune self) and another phenomenon on a lower scale of analysis. For example, although my right thumbprint may serve to identify me, my right thumb is not my self. Similarly, the MHC may signify my immune self but it is not *the* self. The fallacy of identifying the identity marker with the self it signifies should always be avoided, but in practice some people make the mistake of identifying the sign with the signified, the identity marker with the self. The genetic-reductionist approach is a fertile ground for this fallacy since it clearly adheres to the *referential* theories of meaning. In semiotics this is taken to mean that the meaning of a sign is explained through a simple correspondence with a reference. If we adopt this perspective, then the MHC seems to be a sign that clearly corresponds to the "self," whatever that is.

Following the work of the philosopher Gottlob Frege, we should differentiate between "sense" (the semantic content of the sign) and its reference (what the sign refers to). The MHC does not seem to have a simple and concrete reference. It does not point to a concrete object. Nevertheless, it has a sense. It is a sign of the self. What is this self? Here we come to a blind alley. The genetic-reductionist approach does not explain what the immune self is. Our complete genome? The codable part of our genome? It just points to the MHC as a sign corresponding to the self.

The genetic-reductionist approach is not totally wrong, just as the referential theories of meaning are not totally wrong. Meaning can be interpreted to a certain extent and in certain contexts in terms of a correspondence between a sign and a signified. When a child learns to use language by pointing to a bird and saying "bird," a direct correspondence is established between the sign and the signified. The problem is that the meaning of a sign or the "immune self" cannot be exhausted by using simple means of direct correspondence between a sign and a signified. As summarized with regard to immunology: "Although an understanding of such immune behavior canonically begins with the major histocompatibility complex (MHC), *its complete characterization* appears to reside at levels of biological organization beyond the gene" (Tauber, 1998: p. 458, emphasis mine).

Let me support Tauber's argument by pointing out the theoretical and empirical difficulties with the genetic-reductionist approach to the immune self. The first theoretical problem is that, in nature, the "self" is a dynamic object. One does not have to be an orthodox Darwinist to recognize this fact. If the self is recognized by a single, strict genetic criterion, how can we explain the changes in organisms' identity through evolution? This question remains unanswered if we adopt a simplistic reductionist approach to the immune self. The same problem is evident in semiotics. Indeed, the sign "cat" may correspond to a member of the feline family. But the sign "cat" may also signify a jazz musician or may be used as a slang word for a sexy woman. The correspondence between a sign and a signified is not simple, static, and permanent.

When we examine the empirical evidence, things become even more problematic. To see why, let us consider *tolerance* and *autoimmunity*, two phenomena central to understanding the problems of the genetic-reductionist approach.

Autoimmunity is a process in which the immune system turns against constituents of the host that it is supposed to defend, that is: against the self. Autoimmunity is usually associated with disease; the body's attack on its own self is described as a kind of a pathological deviation. For example, lupus is an autoimmune disease in which antibodies identify host tissues as nonself and may cause arthritis and kidney damage. Autoimmunity is usually associated with disease. However, it has been found that autoimmunity is not necessarily a pathological process. For example, Schwartz et al. (1999, p. 295) argue that "Autoimmune T cells that are specific for a component of myelin can protect CNS neurons from the catastrophic secondary degeneration, which extends traumatic lesions to adjacent CNS areas that did not suffer direct damage."

In other words, autoimmunity is not necessarily a problem in self–nonself discrimination, and the meaning of the self turns out to be less simple than we might have thought. The implication of the above argument for the genetic-reductionist conception of the self is clear. The self is not a stable, well-defined entity protected from the nonself by the immune system but a *contextual* construct. In a certain context, the immune system may turn against host constituents, against the self, as a normative function of bodily maintenance. As Cohen (2000a,b, p. 215) argues with respect to inflammation: "The difference between autoimmune protection and autoimmune disease, it appears, is a matter of intensity and the timing of the autoimmune inflammation." The idea that the immune self is a contextual construct will be discussed below. For now, we should realize that no simple correspondence exists in the realm of the living, and that any correspondence is subject to regulation by environmental factors, i.e. context.

Tolerance is the other side of autoimmunity. It concerns the immune system's ability to ignore its own constituents. Interestingly, these constituents may be ignored *even if they do not bear the genetic identity marker of the self.* My example concerns the bacterium *Escherichia coli* When this bacterium is found in high concentrations in food, it is indicative of poor hygienic standards in the restaurant. However, this bacterium resides peacefully in our colon and mouth, without being aggressively attacked by our immune system. It is clearly not a part of the self as defined by the genetic-reductionist approach. How can we explain this tolerance?

Tolerance of parasites is not unusual in nature; *E. coli* is just a specific instance. Sometimes, like talented imposters, intruders in the host self develop a unique mechanism for hiding their identity. However, in many other cases they are simply tolerated by the host. Organisms, human beings, for example, host a variety of parasites that live in perfect symbiosis with them. These parasites are not part of the self in the genetic sense. However, during the evolution of mammals, a symbiotic relationship was established with *E. coli*, for instance, which produces vitamins B_{12} and K and aids in the digestion process. In sum, the immune system tolerates the presence of *E. coli*, a fact which the genetic-reductionist approach to the immune self may find difficult to explain.

Another simple example of immune tolerance to cells that clearly do not have the genetic marker of the self is that of a woman having sexual relations with a man. The woman hosts his sperm cells in her womb. Unless her partner is her twin brother, his sperm cells clearly do not carry the marker of her self. How is it that the host immune system does not destroy the sperm as nonself? And when the fertilized egg develops into a fetus, how does the immune system tolerate the fetus?

Medawar (quoted in Choudhury and Knapp, 2000) suggests that the fetus represents an immunologically foreign graft that is tolerated by the mother's body during pregnancy. He proposes three hypotheses to explain this (Mellor and Munn, 2000): (1) physical separation of mother and fetus; (2) antigenic immaturity of the fetus; and (3) immunological inertness of the mother. However, it is clear that no single mechanism resolves the quandary.

How the mother's immune system tolerates the fertilized egg is an interesting and largely unanswered question in the biology of reproduction. Another interesting finding should be mentioned in this context. According to a news article entitled "Sex Is Good for You" (Buckland, 2002), recreational sex – sex with no procreational purpose – can have a positive impact on pregnancy. But an important qualification should be added: *sex with the same partner*. Sex, early, often, and with the intended father, may help overcome the reluctance of the mother's immune system to accept a fetus that is producing foreign proteins from the father's genes. That is, the more accustomed the woman's immune system is to the man's sperm, the more *habitual* or *conventional* the encounter, the less likely her body will be to reject the fetus. From this study we learn two things. First, the Catholic Church was again found to misunderstand the nature of living organisms. If pregnancy should be encouraged, then sex with the same partner with no procreational purpose should be encouraged too. The second lesson is that the *somatic* aspect of the immune system is crucial for understanding a variety of immunological phenomena. Not everything can be reduced to genes and delivered to us by our parents through germline cells. There are things we have to learn by ourselves, such as conventions. Learning is built into every intelligent system, and the immune system is just a particular case of an intelligent system that learns conventions.

Another example along the same lines: Sperm proteins arise in the testes after the development of neonatal immune tolerance, that is, after immune tolerance has basically been established. It is known that crude sperm proteins are highly immunogenic in all species (McLachlan, 2002). How is it that these nonself cells are tolerated by the host? Why does not it usually attack them as nonself? In fact, some infertility problems in men are caused by the immune system's identifying the sperm cells as nonself by means of antibodies (i.e. sperm antibodies). However, tolerance to the emerging sperm cells is the norm. We do not really know how this tolerance is established, and although a lack of knowledge is usually not a proof, in this case the lack of knowledge concerning sperm tolerance or autoimmunity is an indication that the genetic-reductionist approach to the immune self is overly simplistic and cannot provide us with the answers we are looking for. I now move on to another perspective on the immune self.

4   Putting Complexity into the Picture

The genetic-reductionist approach suggests that there is only a self signified by a genetic marker. This theoretical position implies that the nonself is not an actual entity but a synonym for a genetic foreigner. The opposite perspective was presented by Burnet in his clonal selection theory (CST). He suggested that lymphocytes with reactivity to host constituents are destroyed during development, and only those lymphocytes that are nonreactive are left to engage the antigens of the outside world. The foreign object is destroyed by the immune cells and their products, whereas the normal constituents of the organism are ignored. That is, the immune system recognizes only the nonself – and the self is an empty term. Burnet's CST explains from a very simple evolutionary perspective why we tolerate ourselves. We tolerate ourselves because those who were unable to tolerate themselves (i.e. differentiate between self and nonself) did not survive.

There are major difficulties with Burnet's conception of the immune self. One is the fact that self-recognition is clearly evident in the immune system (Cohen, 1994). I will point out these difficulties in the following sections, but for the present phase of our analysis, I would like to note some similarities between Burnet's conception of the self and Saussure's conception of the sign.

Burnet's concept of the self is purely differential and negative. The self exists only as the background for identifying the foreign object, the nonself (Cohen, 1994). In a certain sense, this position is similar to the one presented by Ferdinand de Saussure in his classical text *Course in General Linguistics*. According to Saussure (1972, p. 118), "*In the language itself, there are only differences.* Even more important than that is the fact that, although in general a difference presupposes positive terms between which the difference holds, in language there are only differences and no *positive terms.*"

What does he mean when he says that in language there are "only differences"? For Saussure, language as an abstract system (*la langue*) is "a system of distinct signs corresponding to distinct ideas" (1972, p. 26). That is, in itself a sign means nothing. It exists solely by being differentiated. According to this interpretation, the sign "cat" has no intrinsic meaning. The "catness" of the cat is not embedded either in the way "cat" is pronounced or in concept of a cat. The same is true of our self.

During our lifetime, our self changes significantly: cells die and are replaced by new ones, our mental content changes during our development, and so on. There is nothing intrinsic to our self that can define us as the same person over the years. According to this line of reasoning, our identity is primarily and negatively established by differentiation from others. To use an analogy from mathematics, a pair of points consists of units that are indistinguishable in isolation. Each unit in the pair is distinct only in that its position is different from the other.

Saussure's statement is applied to the sign as an isolated unit that is "purely differential and negative" (1972, p. 118) as a phonetic or a conceptual unit. Bear in mind that for Saussure the meaning of a word is the "counterpart of a sound pattern" (p. 112). In this sense the meaning of the sign "cat" is the corresponding concept of a cat. Saussure suggests that *meaning* should be distinguished from *value*, which is important for understanding the abstract nature of any system of signs.

A value involves "(1) something *dissimilar* which can be exchanged for the item whose value is under consideration, and (2) *similar* things which can be *compared* with the item whose value is under consideration" (Saussure, 1972, p. 113). For example, money is an abstract system of signs/values. In this system, as in the linguistic system, a dollar bill has no intrinsic meaning. The meaning of a dollar bill can be determined only in a closed system of values. To determine the value of a dollar bill, we should know that it can be exchanged for something else (e.g. ice cream) and that its value can be compared to another value within the same system of currency (e.g. it can be exchanged for euros). The linguistic system is a system of pure values whose function is to combine the two orders of difference – phonic and conceptual – in the making of signs.

Turning to immunology, the similarities are clear: the immune self has no intrinsic meaning. The immune self is only negatively established through the existence of the other, nonself. However, at the point where Burnet stops his analysis, Saussure presents a system-oriented approach, moving away from the isolated sign in language as an abstract system and pointing to the social semiotic dynamics that flesh out this abstract system of values in practice. Surprisingly, Saussure's theory of language as a *social network of signs* is highly relevant for understanding self and nonself discrimination.

As suggested by another semiotician without any reference to immunology (Thibault, 2005, p. 4), "Meaning is an embodied relation between self and nonself on the basis of the individual's entraining into the higher-order and transindividual structures and relations of langue." In simple words, this means that only by going beyond the individual level of analysis and entering the semiotic network can the relation between self and nonself be clarified. As will be shown in the Section 5, this statement has clear relevance for studying the immune self.

5   Where is the Self?

I think there is now a need for a novel and fundamental idea that may give a new look to immunological theory. (Jerne, 1974, p. 380)

An interesting alternative to Burnet's concept of the self was suggested by Jerne in his network theory of the immune system (Jerne, 1974). This theory clearly corresponds to the Saussurian theory and pushes it to its limits within immunological theory.

Jerne suggests that the "progress of ideas" in immunology follows a path from application (i.e. vaccine) to description (e.g. of antibodies) and mechanisms (e.g. selection clones), to systems analysis of network cooperation and suppression of immune agents. He places his theory in the final phase of this progression and approaches the immune system with a network metaphor. Before presenting the gist of his theory, let us clarify some of his terms.

An *antigenic determinant* is a single antigenic site or epitope on a complex antigenic molecule or particle. Jerne replaces the term *antigenic determinant* with *epitope*. He also replaces the term *antibody combining site* with *paratope*. In this sense, the paratope complements the epitope. Next he introduces the terms *allotype* and *idiotope*. Allotypes are "antigenic determinants that are present in allelic (alternate genetic) forms. When used in association with immunoglobulin, allotypes describe allelic variants of immunoglobulins detected by antibodies raised between members of the same species." Idiotypes are "the combined antigenic determinants found on antibodies of an individual that are directed at a particular antigen; such antigenic determinants are found only in the variable region." In other words, an idiotope is a set of epitopes (Jerne, 1974, p. 380). The term *repertoire* is used for the repertoire of antibody combining sites or the total number of different paratopes in the immune system.

Using this terminology, Jerne suggests that "[the] immune system is an enormous and complex network of paratopes that recognize sets of idiotopes, and of idiotopes that are recognized by sets of paratopes" (Jerne, 1974, p. 381).

According to this suggestion, antibody molecules can recognize as well as be recognized. This situation raises a question: What happens to a lymphocyte when its idiotopes are recognized by paratopes (e.g. of another cell)? Jerne suggests that the lymphocyte is then repressed. Stressing the importance of suppression, he suggests that the *essence of the immune system is the repression of its lymphocytes* (Jerne, 1974, p. 382).

This is a radical statement since it suggests that the immune system is a closed system oriented towards itself rather than towards the destruction of foreign invaders. In other words, the system is "complete onto itself" (Bersini, 2003). The idea of a system "complete unto itself" is a natural outcome of avoidance of a direct encounter with the relation between a sign and a signified. If we cannot explain the relation between a sign (e.g. an antigen) and a signified (e.g. nonself), there is a dangerous tendency to deny the existence of a signified (i.e. the immune self/nonself) while assuring the autonomous realm of a sign system. It is a way of ignoring coding!

Tauber (1997, p. 424) describes Jerne as the "true author of the cognitive immune model," meaning the notion that the immune system is designed to know itself. In this context, antigens are interpreted as stimuli that cause *perturbations* in the network. There is no nonself and therefore not even a self, just a "source" of perturbation that causes the network to reorganize itself in order to restore its lost equilibrium. As summarized by Tauber (2002): "In the Jernian network, 'foreign' is defined as perturbation of the system above a certain threshold. Only as observers do we designate 'self' and 'non-self'. From the immune system perspective it only knows itself."

Elsewhere he explains this perspective further: "antigenicity is only a question of degree, where 'self' evokes one kind of response, and the 'foreign' another, based not on its intrinsic foreignness but, rather, because the immune system sees that foreign antigen in the context of invasion or degeneracy" (Tauber, 1997, p. 425).

Readers familiar with the work of Humberto Maturana and Francisco Varela may immediately recognize the similarity between their theory and Jerne's perspective. Both were inspired by the system metaphor and both promote the notion of an autonomous system that is "closed" and subject to perturbations only. Indeed, as Vaz and Varela (1978, p. 231) argue, "all immune events are understood as a form of self-recognition, and whatever falls outside this domain, shaped by genetics and ontogeny, is simply nonsensical."

The problem with the network theory, originated by Jerne and further developed by others, is that it suffers from conceptual obscurity regarding the way in which meaning is established in a closed system. The key term for understanding this difficulty is the *hall of mirrors*. Let us first read Jerne and then explain this difficulty:

> The immune system (like the brain) reflects first ourselves, then produces a reflection of this reflection, and that subsequently it reflects the outside world: *a hall of mirrors*. The second mirror images (i.e., stable anti-idiotypic elements) may well be more complex than the first images (i.e., antiself). Both give rise to distortions (e.g., mutations, gene rearrangements) permitting the recognition of nonself. The mirror images of the outside world, however, do not have permanency in the genome. Every individual must start with self. (Jerne, 1984, p. 5, emphasis mine)

Jerne's use of the term *hall of mirrors* is not an intellectual whim. It corresponds to an established position concerning the relation between a sign and a signified. Rosen (2004, pp. 24–25) explains this position as follows:

> Structuralist semioticians like Saussure still sought to preserve the invariance of the link between the given signifier and what it signifies. The problem is that, once classical signification is surpassed by signifying the signifier, the door is opened to an infinite regress. For now, it seems that no signifier is exempted from mutation into that which is signified. A new signifier is presumably needed to signify what *had* been the signifier, but this new signifier is subject to signification by a still newer signifier, and so on *ad infinitum*. And each time the tacit operation of the signifier is undermined by being explicitly signified, the functioning of what had been signified by that signifier is also affected. Ultimately then, we have in this "hall of mirrors" neither signifier nor signified in any stable, abidingly meaningful form.

This position is attributed by Rosen to Derrida, but one may also find its more sophisticated and constructive aspects in the semiotic theory of Peirce. For Peirce a sign is "a Medium for the communication of a Form" (MS 793 [On Signs], n.d., p. 1). In this sense it is a member of a triad and holds a mediating position between an *object* (i.e. anything we can think or talk about (MS 966 [Reflections on Real and Unreal Objects], n.d. Available at: http://www.helsinki.fi/science/commens/ dictionary.html) and an "interpretant" (an agent, not necessarily human, that is the source of the interpretation process). This triad of object, sign, and interpretant is the indivisible unit of *semiosis* – an action or influence that cannot be reduced to a direct encounter between pairs such as an agent and an object. In other words, any sign-mediated activity is semiosis. In Peirce's words: "This tri-relative influence not being in any way resolvable into actions between pairs" (EP 2, p. 411). This process of semiosis is irreducible but ever-expanding since the interpretant itself exists as long it is a part of a dynamic process of semiosis.

Let me explain this point. According to Peirce, "meaning" is that which a sign conveys. "In fact, it is nothing but the representation itself conceived as stripped of irrelevant clothing. But this clothing never can be completely stripped off; it is only enacted for something more diaphanous. So there is an infinite regression here" (CP 1, p. 339).

The meaning of the interpretant-self is therefore "nothing but another representation" (CP 1, p. 339). In other words, "like the signs in general, the self manifests a trinary character. Every self, in collaboration with its signs, addresses itself to some other" (CP 5, p.252) (quoted in Merrell, 1997, p. 56). The self is mediated and inferred and like all signs must be related to otherness.

The relevance to self–nonself discrimination in immunology is implicit: "That is, the self, upon inferring itself into existence, sets itself apart from everything else in order that there may be a distinction between something and something else." (Merrell, 1997, p. 57). In sum, the self is *reasoned out* by a process of semiosis. It is not a construct that is given a priori.

Jerne's network theory clearly corresponds to Peirce's theory of semiosis. However, Peirce's theory may shed light on Jerne's ideas and add depth to his conceptualization of the network. For example, in the Peircian sense, a "perturbation" of the system is a break in a *habit*, where *habit* is used in the sense of regularity. This perturbation in the process of semiosis results in an effort to reorganize the system and to restore the lost equilibrium.

According to this interpretation, the immune network is not absolutely autonomous. It is context-sensitive and attunes itself to perturbations, violations of habits/regularity that we may define post hoc as "nonself." In other words, the *self may be considered the regularity of relations and interactions that constitute the systemic closure of the organism.* Any disturbance to this regularity (local or global), whether it emerges from within or outside the system, may be responded to by the immune system and defined as "nonself." This interpretation preserves the flexible, dynamic, and commonsensical notion of the self, while also explaining the case sensitivity and the contextual nature of immune activity. This interpretation of the Jernian network brings it closer to the contextualist approach proposed by Irun Cohen. Section 6 presents the contextualist approach.

6   Codes and Context

Another response to Burnet comes from Irun Cohen's contextual theory of immunology. Cohen (1994, p. 11) disagrees with Burnet's conception of self–nonself discrimination in which the foreign object is considered the figure/subject and the self is the background: "According to clonal selection, only the picture of nonself has substance; the picture of the self must be virtual. The immunological self can exist legitimately only as that which bounds the foreign."

Cohen argues that this conception of the immunological self is wrong because the immune system can recognize the self: "Healthy immune systems are replete with T and B cells that recognize self-antigens" (Cohen, 1994, p. 11). While genetic reductionists suggest that only the self really exists and Burnet suggests that only the nonself exists, Cohen suggests that the self and the nonself are complementary. He discusses this idea by means of four concepts: (1) substance; (2) essence; (3) origins; and (4) harmony.

"Substance" has to do with the fact that "self-antigens and foreign antigens are made of similar chemicals and are apprehended by the same receptor machinery" (Cohen, 1994, p. 12). There is no *substantial* difference between self and nonself, and the "selfness and foreignness of an antigen depends on the interpretation given it by the immune system." No *essential* difference exists between self and nonself. The concept of *origins* suggests that experience is crucial to our ability to differentiate self from nonself. Two sources of experience help the immune system differentiate between self and nonself: the genetic and the somatic. Evolution has endowed organisms with inherited mechanisms for handling infection through inflammation. Bacterial and viral products are identified by germline-encoded elements, and objects identified in this context (i.e. antigens) are interpreted as "nonself." In other words, it is the *context* of infection/inflammation that serves as the background for identifying foreignness. This idea is also been advanced by Janeway (1992), who argues that the immune system evolved to discriminate infectious nonself from noninfectious self. This suggestion does not solve the conceptual difficulties associated with the self concept; it merely explains the context that supports self–nonself discrimination.

Interpretation must assume basic familiarity not only with the "text" but with *context*, too. The somatic experience is the actual organization of the immune network in each individual. Somatic experience is no less important than evolution. We are all born with general templates for recognizing something foreign. However, the actual experience is indispensable for recognizing the threatening foreign object. The idea of somatic selection can be explained by an evolutionary perspective. It should be remembered (Langman and Cohn, 2000) that mammals like human beings have a relatively slower rate of evolution (e.g. mutation) than do bacterial and viral pathogens. Therefore, a germline selection might have been disastrous for them in the "arms race" with the pathogens. In other words, relying on genetic reshuffling of antibodies would have been a poor evolutionary strategy. In contrast, somatic selection can respond more flexibly to the higher rates of mutation among the possible pathogens.

Harmony is "the concern of the immune system: recognition of the right self-antigens and the right foreign antigens, interpretation of the context of recognition and a suitable response" (Cohen, 1994, p. 16). It means that in contrast to the simple idea of self–nonself discrimination, the immune system is a highly orchestrated, contextual system of interpretation that transcends the simple dichotomy of self and nonself.

Surprising evidence supporting Cohen's thesis comes from the immunology of reproduction. Among the risk factors for sperm antibodies (SpAb) is testicular trauma. McLachlan (2002) suggests that even minor and/or repetitive sporting testicular trauma may be sufficient for the production of SpAb. This factor is explainable by Cohen's thesis. Testicular trauma, such as a kick in the groin during a soccer game, may cause infection in the damaged tissues. This may result in the identification of the sperm cells as foreign and the production of SpAb to fight them. However, the idea of infectious context has its critics. Anderson and Matzinger (2000) argue that the "infectious hypothesis" is inconsistent with the rejection of transplants by the host body even when no infectious agents are evident. This critique is a serious challenge to the contextualist approach. Moreover, why does the immune system reject some tumors when they are not accompanied by infection? These are open questions that must be answered by a contextualist theory of the immune self.

In response to this challenge, we may suggest that the immune system responds to the perturbation of regularity and that regularity consists of the *embedded contexts of relations that constitute a member of a given category*. For example, the reproductive system of mammals evolved in such a way that the fetus develops in the uterus. This form of reproduction is regular and is therefore a context in which the fertilized zygote is tolerated. (Further theoretical elaborations will be presented below to contend with the difficulties of the contextualist approach.)

Meanwhile, we should add another tier to our discussion by introducing the idea of the immune system as a complex system. Cohen positions his contextual perspective within the complex-systems perspective: "The immune system is a paragon of complexity and needs the tools of complex systems research to understand it" (Efroni and Cohen, 2002, p. 24). Indeed, achieving harmony is not a simple task. According to this suggestion, the observed properties of the immune system, such as self–nonself discrimination, are *emergent properties* that result from microlevel interactions between the heterogeneous constituents of the system. Moreover, Cohen (Efroni and Cohen, 2003) does not consider the immune system merely a biodestructive system but also a regulatory system responsible for a certain portion of body maintenance. Wound healing, tissue repair, and cell regeneration are just some of the maintenance processes in which the immune system is involved. Rather than using the metaphor of the immune system as a warrior defending his castle against invaders, Cohen portrays the system in more prosaic language, comparing it to the superintendent of the apartment building known as the organism. This role is much less heroic but much more complex. In this context, the simplicity of self–nonself discrimination is replaced by the complexity of *meaning-making* (Neuman, 2004, 2005a). Certain molecules are identified as antigens not because they are a signs of a nonself but because in a certain context they do not integrate with the local maintenance activity of the organism and this disharmony is interpreted as an immune response. This suggestion brings up the question of the meaning of a context in which biological agents are judged and responded to; below I attempt to answer this question by invoking Valentine Volosinov and his contextual theory of meaning.

Cohen's contextualist approach to immunology clearly corresponds to the contextualist approach to semiotics. Let us dwell a bit on this approach by contemplating a wonderful example by Valentine Volosinov. Consider the following scenario (Volosinov, 1926, in Shukman, 1983):

> A couple is sitting in a room. They are silent. One says, "Well!" The other says nothing in reply. For us who were not present in the room at the time of the exchange, this "conversation" is completely inexplicable. Taken in isolation the utterance "well" is void and quite meaningless. Nevertheless the couple's peculiar exchange, consisting of only one word, though one to be sure which is expressively inflected, is full of meaning and significance and quite complete.

Understanding the sign "Well" in the above example and extracting the information it conveys is a *meaning-making* process that relies heavily on contextual cues and inferences. The meaning of "Well" is not encapsulated in the sign. *The meaning is inferred from contextual cues.* What are these contextual cues? Volosinov suggests that we examine the "nonverbal context," which is formed from (1) the *spatial purview* shared by the interlocutors (the totality of what is visible—the room, the window and so on), i.e. their phenomenological field; (2) the couple's *common knowledge and understanding of the circumstances – the result of years of being involved in patterns of interactions –* and finally, (3) their *common evaluation* of these circumstances (Volosinov, 1926, pp. 10–11, in Shukman, 1983), what Gregory Bateson describes as "belief."

According to this suggestion, the sign "Well" is totally devoid of meaning in itself. If, however, we find that the two people are sitting in front of a window and see snow falling outside, and if it is winter where snow usually falls, the "Well" makes sense. *Meaning-making is therefore our response to an indeterminate sign* (Neuman, 2007). It is a response within a local context. An analogous thesis may be advanced concerning self–nonself discrimination. Certain entities may be identified as self or nonself only in context. The same agent may be ignored in a healthy tissue and attacked in a damaged tissue. Meaning, whether in semiotics or immunology, emerges in context.

7   Codes of Complexity

What general conclusions may we draw from the analysis so far? The first conclusion is that the immune self is not a platonic, autonomous, and monolithic entity that corresponds through rigid rules to a certain molecular entity, but a context-dependent construct. There is no Self with a capital S. In other words, the question of what is self or nonself cannot be answered by reference to a specific entity. Being self or nonself depends on the response of the immune system in a given context, and this context, although governed by regularity (e.g. the context of infection), is always a local one, as Volosinov suggests.

If we adopt this perspective, then the self turns out to be a highly contextual and fuzzy concept that is *actively inferred* from raw data rather than *passively conferred* by our genes. This perspective can be illustrated by tolerance of sperm in the testes and tolerance of malignant tumors.

The components of a context suggested by Volosinov may be easily applied to the case of the testes. The *spatial purview* shared by the agents is the totality of biological objects in the local functional organ or complex. Spermatozoa in the testes are in an immunologically legitimate spatial position. The *common knowledge and understanding of the circumstances – the result of years of being involved in patterns of interactions* – is the established pattern of relations between the objects. It is a biological convention and a regularity (or habit, to use Peirce's term) among male mammals for sperm cells to be produced in the testes. Transferring sperm cells to another biological site would be a violation of this habit/regularity and would elicit a response. Finally, the "*common evaluation* of these circumstances" is produced by the immunological agents' complex process of communicating with and responding to each other. Hormones that signal the production of sperm cells and macrophages that sense the state of a tissue are just a few of the agents that provide input for evaluating the circumstances. When one gets a kick in the groin, sperm antibodies might be produced because the evaluation of the circumstances has changed.

The idea that the system's response to a given entity is what defines the meaning of the entity is not new either in semiotics (Volosinov, 1986) or in immunology (Cohen, 2000a,b). A genuine contextualist always insists that meaning is encapsulated not in the message, which is in itself devoid of meaning, but in the response to the message. In this context the immune system is no exception; immune tolerance in the testes is just a concrete example of this logic.

What is the major implication of considering the immune self to be a contextual construct? Identifying the objects involved in the immune response is a relatively easy task that has been carried out successfully. However, mapping the relations between this polyphony of agents is a demanding and complex task, and understanding the correspondence between the objects involved in the immune response and the abstract, dynamic pattern of relations that organize their behavior is currently beyond our grasp. As the late Ray Paton (2002, p. 63) argued, "From a biological system's point-of-view there is a lack of tools of thought for dealing with integrative issues [such as this]." However, we are now in a better position to understand the immune self. We realize that the meaning of the immune self, like the meaning of any other sign, is inferred from the system's response to a given signal and is not encapsulated in the signal itself. There is no positive definition of the immune self as the genetic-reductionist approach says there is; there is no negative definition of the self as Burnet would have it; and there is no postmodernist hall of mirrors with which the immune system narcissistically occupies itself. The immune self is defined post hoc as those objects to which the system responds with tolerance. It is defined in terms of the system's *response* as revealed by the semiotic equivalent of a contextual analysis.

How can a simple convention of correspondence turn into a complex coding system? How is it that as children we learn that the word "cat" corresponds to the feline mammal and as adults we understand it as a context-dependent sign that can correspond, for example, to a jazz player? We are still far from answering this question. Identifying the dynamics that turn simple rules of correspondence into complex patterns of meaning is a challenge facing those interested in living systems. Inquiring into the complexity of codes may provide us with insight into immune memory (Neuman, 2007), the irreducibility of biological systems (Neuman, 2005b), the mystery of cryptobiosis (Neuman, 2006), and many other biological phenomena that the current paradigm fails to address.

Acknowledgements A previous version of this manuscript was published in *S.E.E.D. Journal*. I would like to thank Irun Cohen and Peter Harries-Jones for their constructive comments and support and Marcello Barbieri for his editorial assistance.

References

Anderson CC, Matzinger P 2000 Danger: the view from the bottom of the cliff. Seminars in Immunology 12:231–238

Barbieri M 2003 The Organic Codes. Cambridge University Press, Cambridge

Bersini H 2003 Revising idiotypic immune networks. Lecture Notes in Computer Science 2801:164–174

Buckland J 2002 Sex is good for you! Nature Reviews Immunology 2:148

Choudhury SRC, Knapp L 2000 Human reproductive failure I: immunological factors. Human Reproduction Update 7:113–134

Cohen IR 1994 Kadishman's tree, Escher's angels, and the immunological homunculus. In: Coutinho A, Kazatchkine MD (eds) Autoimmunity: Physiology and Disease. Wiley-Liss, New York, pp. 7–18

Cohen IR 2000a Discrimination and dialogue in the immune system. Seminar in Immunology 12:21–219

Cohen IR 2000b Tending Adam's Garden. Academic Press, New York

Efroni S, Cohen I 2002 Simplicity belies a complex system: a response to the minimal model of immunity of Langman and Cohn. Cellular Immunology 216:23–30

Efroni S, Cohen IR 2003 The heuristics of biologic theory: the case of self-noself discrimination. Cellular Immunology 223:87–89

Howes M 1998 The self of philosophy and the self of immunology. Perspectives in Biology and Medicine 42:118–130

Janeway CA Jr. 1992 The immune system evolved to discriminate infectious nonself from noninfectious self. Immunology Today 13:11–16

Jerne NK 1974 Toward a network theory of the immune system. Ann. Immunol. (Inst. Pasteur) 125c:373–389

Jerne NK 1984 Idiotypic networks and other preconceived ideas. Immunological Review 79:5–24

Klein J 1982 Immunology: The science of self nonself discrimination. Wiley, London

Langman RE, Cohn M 2000 A minimal model for the self-nonself discrimination: a return to the basics. Seminars in Immunology 12:180–195

McLachlan RI 2002 Basis, diagnosis and treatment of immunological infertility in men. Journal of Reproductive Immunology 57:35–45

Matzinger P 2002 The danger model: A renewed sense of self. Science 296:301–305

Mellor AL, Munn DH 2000 Immunology at the maternal-fetal interface: lessons for the T cell tolerance and suppression. Annual Review of Immunology 18:367–391

Merrell F 1997 Peirce, Signs and Meaning. Toronto University Press, Toronto

Neuman Y 2004 Meaning making in the immune system. Perspectives in Biology and Medicine 47:317–328

Neuman Y 2005a Meaning making in language and biology. Perspectives in Biology and Medicine 48:317–327

Neuman Y 2005b Why do we need signs in biology? Rivisita di Biologia/Biology Forum 98:497–513

Neuman Y 2006 Cryptobiosis: a new theoretical perspective. Progress in Biophysics and Molecular Biology 92:258–267

Neuman Y (2007) Immune memory, immune oblivion: a lesson from Funes the memorious. *Progress in Biophysics and Molecular Biology* (submitted)

Paton R 2002 Process, structure and context in relation to integrative biology. BioSystems 64:63–72

Rosen SM 2004 What is a radical recursion? *S.E.E.D.* 4(1):38–57

Saussure F de 1972 Course in General Linguistics. Trans. Harris R Duckworth, London

Schwartz M et al. 1999 Innate and adaptive immune responses can be beneficial for CNS repair. Trends in Neuroscience 22:295–299

Shukman A (ed.) 1983 Bakhtin School Papers. RPT Publications, Oxford

Tauber AI 1996 The Immune Self: Theory or metaphor? Cambridge University Press, Cambridge

Tauber AI 1997 Historical and philosophical perspectives concerning immune cognition. Journal of the History of Biology 30:419–440

Tauber AI 1998 Conceptual shifts in immunology: comments on the "two-way" paradigm. Theoretical Medicine and Bioethics 19:457–473

Tauber AI 1999 The elusive immune self: a case of category errors. Perspectives in Biology and Medicine 42:459–474

Tauber AI 2002 The biological notion of self and non-self. Stanford Encyclopedia of Philosophy. Available at: http://plato.stanford.edu/entries/biology-self

Thibault PJ 2005 Saussure and beyond. Lecture one: Speech and writing: two distinct systems of sign. Available at: www.chass.utoronto.ca/epc/srb/cyber/thi1.html

Vaz NM, Varela FJ 1978 Self and non-sense: an organism-centered approach to immunology. Medical Hypotheses 4:231–267

Volosinov VN 1986 Maximum and the Philosophy of Language. Harvard University Press, Cambridge, MA.

Abbreviations of Peirce's writings:

Peirce, C. S.

MS xxx (number) = Peirce manuscripts

CP x.xxx (volume.paragraph) = Collected Papers of Charles Sanders Peirce, 8 volumes. Vols. 1–6, Charles Hartshorne and Paul Weiss (ed.), vols. 7–8, Arthur W. Burks (ed.) Harvard University Press, Cambridge, Massachusetts, pp. 1931–1958

EP x:xxx (volume:page number) = The Essential Peirce. Selected Philosophical Writings. Vol. 1 (1867–1893), Nathan Houser & Christian Kloesel (eds), 1992, vol. 2 (1893–1913), Peirce Edition Project (ed.), 1998. Indiana University Press, Bloomington/Indianapolis

Chapter 12

Signal Transduction Codes and Cell Fate

Marcella Faria

Abstract In cells in general, regardless of their identity and functional status, the mediators of signal transduction (ST), the classic second messengers, are highly conserved: calcium, cAMP, nitric oxide, phosphorylation cascades, etc. At the same time, they are significantly less numerous than the extracellular signals (or first messengers) they represent, suggesting that this universal conversion of signals into second messengers follows the conventional rules of an organic code. Nevertheless, the way these second messengers are integrated and the consequences they trigger change dramatically according to cell organization – its structure and function. Here we examine ST beyond the generation of second messengers, and more as the ability of a cell in its different configurations to assign meaning to signals through discrimination of their context. In metabolism, cell cycle, differentiation, neuronal, and immune function the circuitry operating at cell level will proceed by the creation of conventional links between an increasing number of physiological activities, that is, changes in environment are progressively coupled to: transcription patterns; transcription and replication patterns; transcription, replication, and differentiation patterns; and transcription, replication, differentiation, and functional patterns.

   The categorial framework [1] consisting of CELL/SELF/SENSE has been previously proposed [2] as an attempt to classify the levels of organization adopted by living systems. Our working hypothesis is that these categories reflect: (i) an improved comprehension of self-organization and the convergent gain of complexity that are crucial traits of biological systems, (ii) the possibility of a research agenda, which aims to identify organic codes at the transitions between levels. In the present work we shall use the CELL/SELF/SENSE categories to analyze the progressive complexity of cell fate control through evolution and through development, showing how it is related to switches in ST codes. The notion of a physical attractor [3] will be introduced to reframe the role of classical ST pathways in cell function. The notions of degeneracy and polisemy will also

Laboratory of History of Science, Instituto Butantan, Av Vital Brazil,
1500, São Paulo 05503-900, SP, Brazil
e-mail: Marcella@butantan.gv.br

M. Barbieri (ed.), *The Codes of Life: The Rules of Macroevolution.*   265
© Springer 2008

be examined as possible defining resources for the convergent gain of complexity taking place in biological systems.

1   Signal Transduction as a Recognition Science

In 2004, the Journal of Biological Chemistry celebrated its centenary with a series of commissioned papers called reflections. There, the eminent neuroscientist Gerald Edelman wrote a contribution entitled "Biochemistry and the Sciences of Recognition" [4] in which he uses the term "recognition" to emphasize some of the crucial features that evolution, embryology, immunology, and the neurobiology of complex brains display in common. Biochemical rules, he claims, have their roots in the precision of organic chemistry and the generality of thermodynamics, but at the same time are constrained by the flexible organization of life's forms and behaviors across many hierarchical levels. It is only when embedded within the complexity of cells, organs, and organisms that biochemical processes acquire their significance. The emergence of biochemical rules arise by selection acting over time on variable populations of molecules, cells, and organisms and it is precisely these two notions, i.e. variation and selection as the substrate for biological interaction, that are fully expressed in the four sciences of recognition. Selective processes guide the interaction among variable molecules, cells, and individuals. In each case we can see the deterministic rules of biochemistry being constrained by higher order principles.

In fact, whenever variation is a substrate for selection rather than a source of noise that corrupts proper function, one is certainly dealing with a particular kind of complexity, namely, that of biological systems. That is what is insightful about Edelman's categories; they unify four formulations to the same general question as to how living systems become selective rather than instructive when dealing with choices, an essential question that certainly fits all the proposed arenas: evolution, development, immunology, and neurobiology. It is easy to see that mutation, competition, and differential reproduction in evolution, cell–cell interaction in morphogenesis, antigen recognition in immune response, and network connectivity in neurobiology, are all selection-driven recognition processes. Nevertheless, these properties seem too indistinguishable of life itself to have their origin in cell populations (organisms, embryos, immune, or nervous systems) rather than in single cells.

Strongly motivated by this hypothesis (that selective recognition has its most basic expression in single cell behavior) we will turn back to signal transduction (ST), the process by which single cells endow environmental change with contextual meaning. This process was recently defined as "the ability to sense changing environmental conditions and then implement appropriate responses" [5]. In its original context the quotation talks specifically about how prokaryotic and eukaryotic cells react to their environment. This same definition if applied to cell collectives, to species and/or individuals is strikingly similar to the minimal features defining the "recognition sciences" proposed by Edelman. This similarity only illustrates that ST in cells does in fact qualify as a selection-driven recognition phenomenon.

Besides variation and selection, Edelman emphasizes that the semantic dimension of biochemical processes is only realized by virtue of their embeddedness in biological systems and their multilevel hierarchical organization. Here, we will approach the subject of ST in cells from an evolutionary perspective fully inspired by Edelman's synthetic efforts; a theoretical background that will consider any biological recognition as a selective process and will take into consideration the organization of the system at the focal level of the interactions described. It is important to note that a large body of work exists that considers the cell as a semiotic structure and ST as a "meaning-making" process and they will be brought into our discussion where suitable [6,7,8]. The contribution of the present work is in line with these previous efforts, but goes further by introducing categories for considering how living systems increase their complexity through evolution, development, and function. Most importantly, by dissecting the semiotic structure, biological function, and evolved form associated with each of the categories we hope to illuminate some general principles of their organization in biological systems through layers of complexity.

As we have mentioned in the Abstract, the scope of ST that we will explore goes beyond the traditional textbook definition, which is the conversion of the so-called first messengers into second messengers. It has been proposed by Marcello Barbieri that this conversion clearly follows an organic code [7] that allows for the transformation of a multitude of agents (the various sorts of growth factors, hormones, cytokines, neurofactors, etc.) into a many fewer number of charged molecules (calcium, cAMP, nitric oxide, phosphorylation cascades), the second messengers. But as stated by Barbieri himself there is more to ST codes than that "The effects that external signals have on cells (…) do not depend on the energy and information they carry, but on the meanings that cells give them with rules that can be called ST codes."

2 A Census of Cell Senses

Models are key elements guiding and reflecting the pace of our scientific knowledge at a given time. The way a community chooses to represent a natural phenomenon does necessarily change to accommodate new data and changing theoretical perspectives. In the case of ST, it is interesting to examine some of the representations that mainstream sciences have generated in later years. By doing so, we shall be able to understand the recent developments made by experimentalists and, hopefully, to start building up new theoretical syntheses in the field.

Any effort to understand how a cell responds to its changing environment begins with a description. Since the advent of recombinant DNA technology about 20 years ago, a lot of data have been produced by the artificial elimination of specific gene products and assessment of their contribution to the adaptive cell behavior under investigation. The approach of choice has been to depict this information in graphical form. The wiring diagrams representing ST pathways activated under specific conditions are wall decorations in every laboratory interested in cell biology worldwide and indeed a large-scale approach to the question of how the cell senses its environment. These maps look very much like the wiring diagrams of electronic devices, but their building blocks are genes, proteins, and metabolites hooked together by chemical reactions, rather than transistors and resistors connected by wires. The accuracy in molecular details provided by some of these descriptions is stunning; chemotaxis in bacteria [9], cell cycle in yeast [10], cell differentiation during fruit fly development [11], and mammalian cell transformation [12]. But as rich as these descriptions might be, it seems appropriate to quote Gregory Bateson declaring that "the map is not the territory" [13], because there are no simple rules to predict a cell response to a given signal based only on the information gathered in the maps. As opposed to metabolic maps and electronic device wiring diagrams in which wires and arrows represent an actual flow of matter and/ or energy, the nature of "signs" that should be represented in ST maps is much more evasive, having to do with information rather than directly with matter and energy. The following attempts of representing ST are strongly motivated by the search for informational unities in these maps, minimal components that are responsible for the specificity of the biological response that the cell attributes to a given sign.

At first, molecular biologists and especially those working with microorganisms tried to identify conserved multicomponent regulatory systems [14,15], switching from the large-scale level of wiring diagrams to the small-scale level of molecular devices. One interesting example is that of two-component systems, the prototypical signaling machinery of prokaryotic cells. These modular proteins are organized as follows: their first component is typically formed by a sensor kinase autophosphorylated on its transmitter domain by a response to environmental signs, with the second component composed of a response regulator that will be the acceptor of the previously generated phosphoryl groups. The response regulator activity varies according to the phosphorylation status of its receiver domain, which in turn regulates the activity of an output domain. Most of the response regulators are in fact transcriptional regulators (for review, see [5]). These highly conserved molecular devices were analyzed and classified according to the nature of their modular domains even before genome sequencing was available, and what were informal surveys at that time have acquired the status of systematic census in our postgenomic era. Later analysis revealed the existence of simpler devices such as one-component systems, they have also dissected the great diversity in sensor kinases and response regulators, suggesting that the combinatorial potential among modular domains could account for the great adaptability of prokaryotic organisms, despite the economic (frugal) organization of their genomes. This line of reasoning implies that the phosphorylation of one receiver domain could control the activity of all identified receiver domains, assuming that in small genomes everything that is coded tends to be functional. There are almost 45 receiver domain types according to a new survey including most sequenced prokaryote genomes to date [16]. Interestingly, although two-component systems are a satisfactory description for prokaryotic ST networks [17], they do not account for the complexity of signaling identified in nucleated cells.

Eukaryotic transcriptional organization is completely different from that of prokaryotes; chromatin compaction, chromosome localization, promoter organization, RNA splicing, and processing are only some of the new potential targets of ST that exist in the context of nuclear transcription. Thus a survey of domain architecture in signaling proteins is not likely to provide much insight to the actual mechanisms of eukaryotic ST. But also in the case of highly complex ST networks of mammalian cells, understanding their functional organization starts with mapping the interactions that can exist in a given cell type; in short, the wiring diagrams. Transcription-regulatory (TR) networks reflect the potential interactions of transcription regulators by mapping their connectivity. Preliminary studies of such diagrams have shown that they have the defining properties of scale-free networks, in which typically a reduced number of hub nodes are connected to a great number of other nodes. The analogy to the elements of two-component devices is direct, but here we are dealing with the topology of interaction patterns rather than that of protein structure. Therefore the search for elementary unities of a greater order of magnitude than signaling proteins will better fit this level of the scale. A possible methodological approach for TRs is establishing which particular interaction patterns (or in more precise jargon, "regulatory motifs") are more frequent than would be expected by chance [18,19], implying that their stable position in the network under analysis was obtained by selective processes, either evolutionary, functional, or both. A seminal work in systems biology established that two of these motifs are overrepresented in transcriptional networks of two rather diverse organisms, *Escherichia coli* and *Saccharomyces cerevisae*, and also in the neuronal networks of the worm *Caenorhabditis elegans* [19]. The connectivity paradigms called "feed forward loop" and "Bi-fan" may be the origin of the synergistic and/or cooperative response triggered by some sign combinations. Nevertheless, other motifs of known biological relevance, such as negative feedback loops, were not found by this kind of rationale, implying that the motifs playing a biological role in ST may not always have been selected by means of increasing their abundance. This should not come as a surprise, since it is most probable in biology that rare events and structures will have regulatory potential as compared to the abundant events and structures that most often play structural roles.

"Statistical inventories" of conserved regulatory motifs are small-scale studies. A simple zoom-out can show that the position of motifs within condition-specific subnetworks and their aggregation into larger topological modules [20] may dramatically change their behavior. This realization, in line with the observation that most of the hubs and motifs activated in differential experimental conditions (growth, oxidative stress, starvation, or DNA damage) are transient, emphasizes the need to complement large-scale studies (the topology of TRs wiring diagrams) and small-scale studies (the identification of abundant components and motifs in TRs) with medium-scale studies – the identification of subnetworks that have their dynamic behavior affected by specific conditions. As a matter of fact, the search for dynamic regularities in transcriptional networks has been systematically pursued by the scientific community, with at least two books and hundreds of papers devoted to the subject in the last 10 years [21,22]. The central idea in this kind of approach is: on the one hand, to use cell physiology parameters as higher order constraints on the dynamic behavior of transcriptional networks; on the other hand, choosing simple dynamic interactions that can be ascribed to a defined subnetwork inside the system so that they can generate more complex behaviors when the greater topology is considered. From this perspective, regulatory motifs that were not significantly abundant (as negative and positive feedback loops) may become important as they connect the previously mentioned overrepresented motifs (such as bi-fan and feedforward) to generate more complex behaviors – oscillations and switches – provided a suitable network topology. A beautiful work on the matter was the fruit of collaborative efforts between the mathematician John Tyson and cell biologist Bella Novak [23], starting with two simple examples of protein dynamics; synthesis/degradation and phosphorylation/dephosphorylation and their mathematical representations by nonlinear differential equations. They show how it is possible to obtain various kinds of signal–response curves only by implementing the subnetworks with precise feedback and feedforward connections, and they present us with the mathematical representation of such curves; one-parameter bifurcation diagrams. This approach has generated a general classification of signal–response curves that is worth mentioning, for it will be used in our further discussions: there is a *graded and reversible mode* which includes the simple linear and hyperbolic curves, as well as sigmoid and adapted curves of a more complex kind (in each case the responses depend on signal state and do not take into account the signal's dynamic properties); and there is *the discontinuous* (the response changes abruptly as signal magnitude comes to a critical level) *and irreversible* (the response may remain high even if the signal strength decreases) *mode* which includes switch-like, homeostatic and oscillatory responses. Interestingly, this kind of classification allows the tracing of the response mode to a subnetwork topology; oscillatory responses, for example, can be produced by three alternative signaling modes: negative feedback, activator inhibitor, and substrate depletion. Conversely, two kinds of switch-like response can be produced by different positive feedback network topologies: a one-way switch and a toggle switch. It is not within the scope of the present work to dissect the mathematical details underlying each signaling mode, but it is important to note that these representations of transcriptional networks that use formal mathematical tools allow an assessment of the regularities in the dynamics of subnetworks that was not possible by the previously presented approaches.

We propose (see Fig. 1) that the different strategies chosen by scientists to represent signal transduction in cells reflect various levels of a multiscale phenomenon. We have the wiring diagrams representing all the potential interactions taking place within a cell at a given experimental condition; a large-scale description unable to provide mechanistic keys per se. Then we have the identification of multicomponent systems and regulatory motifs, small-scale representations which may shed light into the mechanisms of ST but only to a certain extent; two-component systems can explain cell responses only for the transcriptional organization of

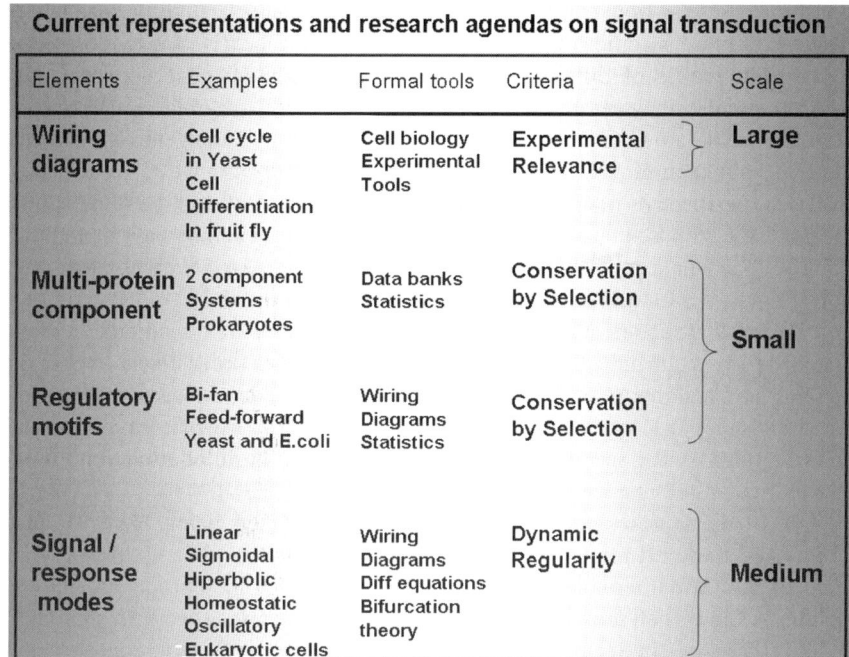

Fig. 1 The various strategies chosen by the scientific community to represent Signal Transduction in cells

prokaryotes and the identification of motifs biased by their relative abundance may neglect the really meaningful patterns of connection. The need for medium-scale representations of the signal-transduction problem is emphasized by the limited scope of both describing a phenomenon too generally by its higher order organization, and reducing it too drastically to its lower level components. The identification of signaling modes (low-frequency motifs that control the behavior of high-frequency ones to generate a response mode) and response modes (regularities in the dynamics of cell behavior) are medium-scale approaches. They explain the mechanism by connecting transcriptional networks to cell physiology, using mathematics as a formal tool, such that subnetworks may be extracted from a whole map not based on their frequency but rather on their functional relevance.

In fact, medium-scale approaches are becoming increasingly popular among the "systems biologists." By deriving differential equations and bifurcation curves from subnetworks clearly involved in specific cell physiology transitions, they produce signal–response curves that display the dynamic behavior of the system and the number of stable configurations it may adopt under specific conditions. The immunologist and theoretical biologist Irun Cohen defines an attractor as follows: "The long-term stable state of a system is called attractor. The attractor, which is a solution to the dynamics of the system, exists a priori, even if it surprises the modeler who formulated the model."[24]. This definition is general enough to fit any complex system operating far from equilibrium. Alternatively, attractors can be defined taking into account more specific contexts; in a recent paper the term is used to refer to the transcription profile of differentiated cells; "Large, random gene networks can, given a particular network architecture, produce "ordered" dynamics, i.e. have relatively few stable attractor states in which a large fraction of the genes remain stationary despite their global interdependence" [3]. In either case, we can draw a new theoretical perspective for ST by incorporating the notion of a "high dimensional attractor." We propose that a cell changes its behavior by switching basins of attraction, from one stable configuration to another. This notion has extremely important implications for ST studies, as signs become a source of instability to the cell, forcing it into new attractors at repeated cycles. The attractors are already there, interpolated by evolution into the cell's physiology as higher order constraints to its potential fates and functions. From this perspective, ST codes are particular subnetwork topologies that create conventional links between a specific environmental condition (sign) and a specific transition into a different basin of attraction implying a particular cell response (meaning).

ST has been represented in various ways in the last few years. Here we have summarized these efforts to show that experimentalists have been realizing – by means of the formal tools and conceptual framework that they produced, that the question of "how a cell senses its environment by implementing specific responses?" is a multiscale one. In Section 3 we shall introduce three previously proposed categories (CELL/SELF/SENSE) reflecting stable configurations of increasing complexity adopted by living systems [2] into this discussion, and investigate how medium-scale parameters behave in each category as means to access the contextual development of ST codes.

3   Levels of Organization and Signal Transduction Codes

The topology of transcriptional networks, as we have seen, has a decisive role in the potential fates and functions a cell can adopt. Here we propose that the three stable configurations adopted by living systems – CELL, SELF, SENSE – increase their own complexity by changing TR topology in a precise manner: by interpolating new attractors and increasing the connectivity between signaling and response modes at each level (see Fig. 2 for a schematic representation). Let us illustrate this general idea by dissecting each level's structure.

The CELL level has two prototypical configurations: prokaryotes and eukaryotes. In both cases semiosis occurs to maintain the system's autonomy; it is by means of simple feedback loops that the CELL adjusts its own behavior. The main task to be performed by ST codes at the CELL level is to *couple environmental changes with the subnetworks controlling cell growth*. In fact, the way transcription is organized in prokaryotes assures that transcription is modulated by environmental signs; two-component regulatory systems work like adaptors coupling the presence of an specific energy source in the environment (e.g. glucose) to the transcription of

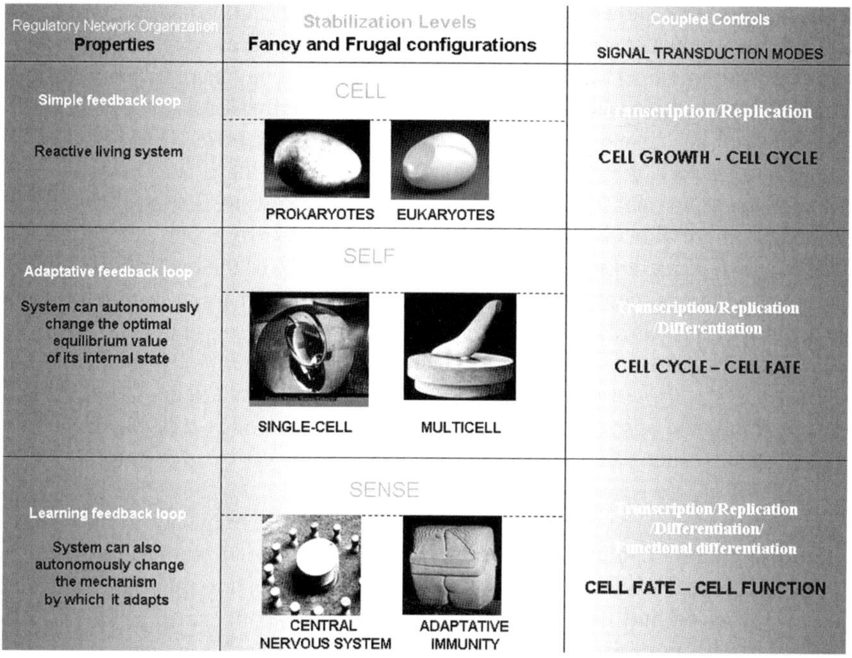

Regulatory Network Organization Properties	Stabilization Levels Fancy and Frugal configurations		Coupled Controls SIGNAL TRANSDUCTION MODES
Simple feedback loop Reactive living system	CELL PROKARYOTES	EUKARYOTES	Transcription/Replication CELL GROWTH - CELL CYCLE
Adaptive feedback loop System can autonomously change the optimal equilibrium value of its internal state	SELF SINGLE-CELL	MULTICELL	Transcription/Replication /Differentiation CELL CYCLE – CELL FATE
Learning feedback loop System can also autonomously change the mechanism by which it adapts	SENSE CENTRAL NERVOUS SYSTEM	ADAPTIVE IMMUNITY	Transcription/Replication /Differentiation/ Functional differentiation CELL FATE – CELL FUNCTION

Fig. 2 Signal transduction modes and network organization properties displayed by living systems in their different stable configuration

enzymes involved in its metabolic processing. Transcriptional profiles are adjusted to keep the cell growing as conditioned by available resources, and the response modes are graded and reversible, including very simple signal–response curves of the linear and hyperbolic kinds and slightly more complex ones like sniffers and buzzers [23]. Irreversible and discontinuous modules, however, do not function in prokaryotic cell growth control. Nevertheless, in prokaryotes DNA replication and cell division are uncoupled from the transcription/metabolism subnetwork, an example of a discontinuous response mode. It has been proposed that in such cells when a reason (number of replication origins/cell mass) reaches a certain threshold, DNA replication and cell division are triggered, regardless of transcriptional patterns and metabolic needs of the cell [25]. DNA replication and cell division seem to be controlled by what qualifies as a one-way switch.

In eukaryotes – the more sophisticated configuration of CELL level – ST codes assure that *environmental changes* are coupled to *transcriptional controls*, but in addition, nucleated cells couple these changes to *replication and division controls*. With the advent of the nucleus, the increased genome size and its organization in chromosomes with independent origins of replication for various DNA segments, eukaryotic cells require coordination between cell growth, DNA replication, and cell division. In this context the organization of a cell cycle, with checkpoints that assure DNA integrity before, during and after its replication, assumes utmost importance. This particular kind of organization favors autonomy; eukaryotes depend only on the presence of growth factors in their environment during the initial hours of the GAP 1 (G1) cell cycle phase. At the same time, cell division is no longer a simple one-way switch transition dependent on one threshold transition (as in prokaryotes); in eukaryotes cell growth and cell division form a unified program, the cell cycle, with at least three critical points (the G1/S, G2/M, and M/G1 transitions). In fact, the molecular apparatus controlling these transitions are proteins called cyclins and the kinases they recruit and activate. The cyclins have an oscillatory expression pattern and a modular regulation by means of association with cyclin-dependent kinases (Cdks). It is the topology of the cyclin/Cdk network that enables environmental changes to be coupled to transcription, which thus integrates metabolic controls with DNA replication and cell division controls. The cell cycle control system has been described in the following terms by Tyson and collaborators [23]: "Progress through the cell cycle is viewed as a sequence of bifurcations. A small new born cell is attracted to the stable G1 steady-state. As it grows, it eventually passes the saddle-node bifurcation where the G1 steady state disappears, and the cell makes an irreversible transition into S/G2. It stays in S/G2 until it grows so large that the S/G2 steady state disappears, giving way to an infinite-period oscillation. Cyclin-B-dependent kinase soars, driving the cell into mitosis. The drop in Cdk1- cyclin B activity is the signal for the cell to divide, causing cell size to be halved and the control system is returned to its starting point, in the domain of attraction of G1 steady state." This is a strikingly detailed example of how ST codes can match categories very different from molecular devices in their logical typing as building blocks. The transcription topology associated with the eukaryotic cell cycle connects regulatory elements (cyclin/Cdk subnetwork interactions) with higher order motifs (by introducing positive and negative feedback loops) to create (or to select) stable response modes (toggle switches and oscillators associated with G1/S and G2/M, and M/G1, respectively). Response mode modules work together to coordinate otherwise separated cell physiology transitions; cell growth, DNA replication, and cell division.

To put it simply, in prokaryotes cell growth is metabolism-driven and cell division is cell growth-driven, whereas in eukaryotes the cell cycle (which includes cell growth and cell division) is metabolism driven. This is only possible because prokaryotes and eukaryotes have different ST codes.

When cells have their dynamic behavior stabilized at the level of SELF configurations, new properties are developed due to increasing complexity in system organization. Besides the simple feedback loops employed by the semiotic structures at the previous level, SELF-configurations are able to autonomously adjust the optimal equilibrium values of their internal states by adaptive feedback loops. At such a level eukaryotic cells living as single-cell or multicell organisms will develop a differentiation skill, such that other attractors are interpolated onto the basic cell cycle structure. We suppose that this bifurcation assures, very much like the cell cycle checkpoints, that a specific cell state is to be coupled to a specific phenotypic transition. In single-lived eukaryotes this kind of organization results in *life cycles with alternative forms*, in multicell organisms this will reflect in *embryogenesis with alternative cell fates*.

Protozoans, the single-cell eukaryotes, may alternate between haploid and diploid life cycles, such as happens with budding yeast for example. Alternatively, as does the amoeba Dictyostelium discoideum, can switch from unicellular to a multicellular stage during their life cycle [26]. In any case, as ST codes are concerned, it seems that new attractors are interpolated into the previous cell cycle organization. In yeast, the G2/M module is replaced by a "G2/2M" module, opening the way for a choice between mitosis and meiosis after G2 steady-state disruption during cell cycle. In Dictyostelium, new attractors are interpolated to the cell cycle signal transduction system after the M/G1 module, so that newborn cells will be presented with the choice between continuous division or division arrest plus aggregation. If cells enter the aggregation mode they will be further presented with the choice between two alternative cell fates to generate fruiting bodies [27]. By integration of new attractors and increasing the connectivity between regulatory motifs, evolution transforms cell cycle transduction codes into cell fate transduction codes; starvation becomes a sign for aggregation, and oscillatory cAMP patterns become a sign for differentiation.

Typically, in order to differentiate, metazoan eukaryotic cells have to commit, migrate, and adhere to their target organ and perform specific functions [27]. An interesting idea that will require experimental corroboration is that each of these transitions (commitment, migration, contact) function as higher dimensional attractors to select cell populations during embryogenesis. In fact, the work of Huang and collaborators suggests that, at least in the case of neutrophil differentiation, final differentiation functions as an attractor by the stabilization of transcriptional profiles. Of course the level of detail available for cell differentiation networks is not comparable to that of cell cycle data, but solving cell differentiation codes as they emerge, constrained by cell cycle codes, would improve the overall quality of available explanation.

We turn now to consider the SENSE level, wherein canonical frugal and fancy configurations lie: the central nervous system; and the adaptive cell immune system of vertebrates. SENSE configurations have a basic semiotic structure with feedback loops, adaptive feedback loops, and learning feedback loops incorporated as regulatory modes. Besides the properties shared by the previous levels, they can autonomously change the mechanism by which the system adapts to changes. Here, the logic of interpolating new transitions to previously existing programs remains the same; in this case, environmental changes previously coupled to transcription, replication, and differentiation controls are additionally coupled to functional differentiation controls. The similarities between nervous and immune system have been emphasized over time, as both systems use specific molecular recognition events between single cells, cell–cell adhesion patterns, positional stability, and directed secretion to fulfill their respective functions. They have also both evolved highly sophisticated forms of information storage, providing the systems with memory capabilities. Their prototypical cells, neurons, and lymphocytes, are excitable cells which can be somatically induced to perform specific functions in a threshold dependent manner, which can be considered to be a form of *somatic differentiation*.

By supplementing the previous signaling transduction codes with new oscillatory modules with very specific spatiotemporal patterns, the system increases its own flexibility to unprecedented levels. During immune and nervous system function, entire cells acquire the status of regulatory elements able to create, propagate, or extinguish a sign, or sign construct, and the cell network topologies (hard wired in the nervous system and dynamically assembled during immune response) acquire the status of signaling and response modes. At the same time, neurons and lymphocytes are still organized according to lower level programs, i.e. cell growth, cell cycle, and cell fate controls. In neurons, particular patterns of chemical triggers are coupled to the propagation of electric waves through cell membrane, which are converted again into chemical triggers. In the immune system, different cell types will react to particular chemical triggers coupling them to a recognition process that will include active remodeling of genetic material, membrane protein repertoire, and cell fate. In both cases, between the stable states predicted by the previous ST codes, another potential transition appears that allows these to switch between active and resting states. Synapses, despite clear structural differences between their neural and immunological forms [28], are functionally organized cell to cell communication structures. They provide otherwise nonspecific soluble agents (neuromediators and cytokines) with specificity through their confinement to precise spatiotemporal patterns. From this perspective, it is interesting to think about neural and immune function in terms of their constitutive cell-signaling network topology, trying to identify subnetworks important for somatic function.

We have seen that signs are selected by ST codes operating at the different cell stabilization levels. We can envisage different mechanisms implied by the emergence of ST codes. Firstly, different aspects of a changing environment can become relevant and instructive to the system as new codes emerge. In this framework, properties of increasingly complex logical types, as quantities, interactions in time, and arrangements in space, would progressively acquire a signaling potential. They would become signs only by the emergence of suitable codes. If we examine the ST codes matching the CELL/SELF/SENSE categories for these sign/code relationships, the description goes as follows: quantities are signs to be integrated at CELL level through cell growth, and cell division codes; interactions in time are signs to be integrated at CELL level by cell cycle codes and at SELF level by cell fate codes; arrangements in space are signs to be integrated at the SELF level by cell fate codes and at the SENSE level by cell function codes.

Another mechanism can be superimposed upon the first, rather than being an alternative to it. Through organizational levels, by increasing the subnetworks' connectivity, and the number of its stable configurations (attractors), the system selects oscillatory behaviors of various timescales as potential operative modes. In Fig. 3 we show that a same-sign modality – intracellular oscillations – can control diverse cell responses: cell growth codes provide glycolitic oscillations with a metabolic meaning, a dynamic measure of energetic resources in microorganisms [29]; cell fate codes integrate cAMP oscillations as a differentiation trigger; cyclin/Cdk oscillation are given various meanings by cell cycle codes, controlling cell growth, DNA replication, and cell division; calcium oscillations are converted into differentiation triggers by

BIOCHEMICAL OSCILLATIONS AND CELL BEHAVIOR		
Agent	**Period**	**Regulated Transition**
glycolitic	Minutes	Cell function
cAMP	Hours	Life cycle
CYCLINS/CdK	Hours	Cell cycle
Ca^{2+}	Seconds → Minutes	Cell fate Cell function
Membrane potential	Seconds	Cell function

Fig. 3 Biochemical oscillations with different time scales regulate different cell behaviour

cell fate codes and hormone secretion triggers by cell function codes in hepatic cells [29]; and neuronal function codes provide oscillating membrane potentials with a very precise biological meaning, switching cells from resting to active states and triggering neuronal firing.

Biochemical oscillations in biological systems are conserved through various hierarchical levels and have been the object of experimental studies [30,31] and theoretical analyses [32,33]. To fully understand the role of biochemical oscillations in cellular rhythms one must go back to nonequilibrium thermodynamics and some of the ideas coined by chemist Ilya Prigogine. In his Nobel lecture [34], he states that nonequilibrium may be a source of order by generating a precise kind of organization (self-organization) illustrated by dissipative structures. But for this to occur, a precise relation is to be established between the system's function (as expressed by chemical equations), its space-time structure (necessarily unstable), and fluctuations triggering such instabilities. It is the interplay between these three aspects that leads to most unexpected phenomena, including "order through fluctuations." The subject is so vast and full of implications that we could devote a whole section to this discussion. Nevertheless, for our present purposes it is sufficient to consider that every ST code we have described requires an oscillator and that this oscillator will conduct the cell among the different attractors implied by the system's organization at the code operating level.

4   Polysemic Signs, Degenerated Codes, Selected Meanings

In the present work we have seen that ST, defined as the way cells sense changes in the environment to implement proper responses, can be viewed as a recognition process. It implies variation and selection at the various levels of a multiscale hierarchical organization. Previously, we had also defined a cell as a semiotic system that converts linear signs into dimensional meanings, through conventional organic codes, increasing its own complexity. In this context, an interesting challenge is to examine the semiotic status of the generally applicable biological concepts of variation and selection.

In semiotic terms, there are two conserved sources of variation: the polysemy of signs and the degeneracy of codes. All signs are polysemous – with a range of potential meanings in different contexts [35], and all codes are degenerate – with a range of potential signs to produce a given meaning [36]. We believe that these two properties are intrinsically related to the selective nature of semiotic processes, in general, and of biological recognition, in particular. For a meaning to be specific to, but not materially determined by, its triggering sign (which is equivalent to the process of "meaning-making" being selective rather than instructive), some extent of indetermination has to be maintained through the whole codification process. Polysemy assures such indetermination at the level of signs and degeneracy keeps the flexibility at code level. Nevertheless, specific (contextual) meanings can be extracted by such higher order parameters as the sign's position and connections in the code(s) network. Let us illustrate this point using the genetic code as an example: every gene has the potential for pleiotropic effects and, at the same time, multiple gene products contribute to almost any phenotypic function. If there were a one-to-one correspondence between every gene product (sign) and a particular cell response (meaning), there would be no way to produce a conditional cell response by taking the context into account. As it is, a range of potential meanings, assured by the very nature of signs and codes, is potentially maintained until an interaction-in-context determines a precise output (a selective cell response).

The process of ST in cells can be considered an example of a broader phenomenon, the "differentiation" of polysemic signs into specific meanings. From this perspective, like all others adopted here, it is important to acknowledge the hierarchical and multilevel organization displayed by cells as semiotic systems. A number of biosemiotiscians have discussed the semiotic dimensions of biological information in hierarchical terms [37,6,8]. It is not in the scope of the present work to review this great amount of work, however, we will stress some common features emphasized by different authors as they matter to our discussion.

An important point is that any system of signs has an abstract nature. Rather than being based in discrete unities of matter and energy, they use differences and values to build contextual meanings [38,13]. Difference and value were first defined as informational dimensions by Bateson and interestingly, in a recent work Neuman [35] describes them both in a semiotic system represented by a signaling network. In such context, he claims, differences are expressed as precise regulatory elements and motifs (i.e. nodes and attractors), and value, on the other hand, is expressed by the connectivity and relative positioning of elementary differences. It follows almost naturally to correlate a difference to a digital mode of communication based on binary and discontinuous choices, and a value to the analogue mode based on comparisons and qualitative choices.

Any attempt to describe ST effectively has been trying to coordinate these two communication modes. Bruni [6] has proposed that "One way to look at how elementary differences build up and are sensed up and down the biological hierarchy, and how biological systems … obtain relevant information out of otherwise ubiquitous differences is by considering a communication pattern I have referred to as digital-analogical consensus. … As the mediatory action of codes which are formed at different hierarchical levels out of an indefinite number of "lock and key" interactions, that by their simultaneous occurrence give rise to emergent specificities and triadic relations." This is an interesting notion as it accounts for the contextual nature of organic codes across hierarchic levels. The alternation between digital and analogue modes occurs when thresholds are reached, summing a number of events of one or each kind. Nevertheless, the relative simplicity of this alternating dynamic seems suitable to describe ST only at the small-scale of second messengers, signaling components, and regulatory elements, but its transposition to higher hierarchical levels may well prove to be too laborious and ineffective in explanatory terms.

The way digital and analogue differences combine as means to create the whole of a semiotic system may vary according to the complexity of the operating system. As we have seen, there are two sign–response curve modes also at the medium scale of TR subnetworks; a graded and reversible mode and a discontinuous and irreversible one. These two modes can be easily described in digital/analogue terms, but it is only through their aggregation to build oscillatory behaviors that lead a cell between stable states, that we can understand the logic of ST codes. In this sense, before describing any code modality, it seems important to identify the potential stable configurations that semiotic system shall adopt through time.

We propose that in the particular case of ST codes, analogue and digital modes combine to produce an increasing number of potential cell fates (here broadly understood as stable states) as the system organization becomes more complex (following the CELL/SELF/SENSE categories). In Fig. 4 we illustrate alternative cell fates presented as binary choices for cells at each organization level (a prokaryotic cell will choose between growth and arrest, an eukaryotic cell will choose further between clonal and differential cell division, a multicell eukaryote will choose further between migrate and adhere, and finally a fully differentiated neuron or lymphocyte will choose between activity and rest). At every level the "here and now" choice that the cell has to face follows a digital code: provided a threshold of cell size the prokaryotic cell will necessarily divide, eukaryotic cell division is organized through checkpoints as "points of no return"; eukaryotic cell differentiate by becoming irreversibly committed, by migrating and by tissue targeting; a neuron will necessarily fire provided a certain membrane potential threshold; and a lymphocyte will be activated provided a particular biochemical pattern is achieved. Nevertheless, coexisting with the digital codes underling function at each level, are greater signaling

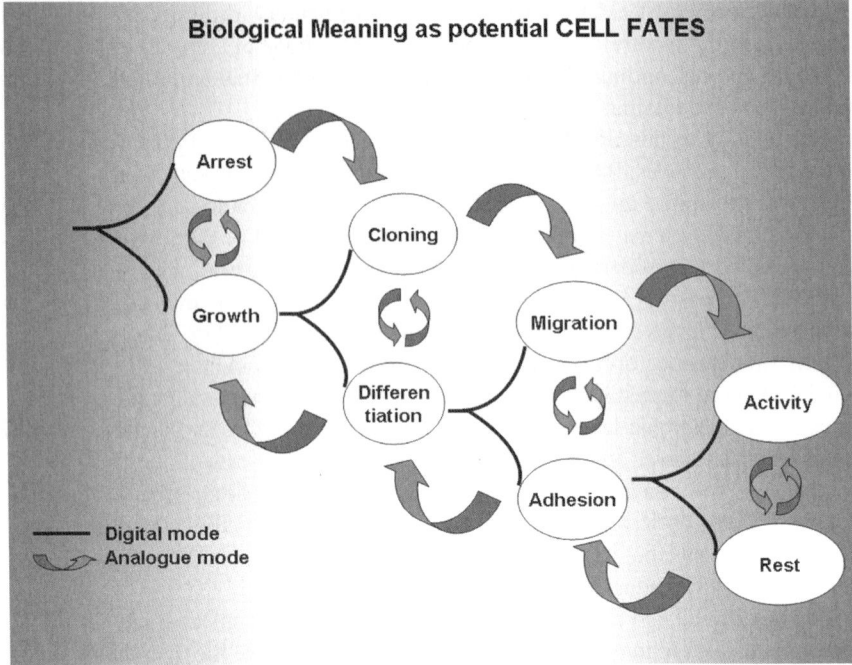

Fig. 4 Alternative cell fates as binary choices according to cell organization

programs based on oscillations and basins of attraction, which are instead analogue in nature. We have seen that different ST codes control cell growth, cell cycle, cell fate, and cell function, and in each of these codes, stabilities are reached by virtue of the dynamic properties of the code network which vary by degree. There is a precise interplay between digital and analogue communication modes inside and between each of these codes, and most importantly, each of these codes can be ascribed to a precise functional property or selected biological meaning.

At the medium scale of signal–response modes, variation and selection will manifest as oscillatory behaviors and high-dimensional attractors, respectively. The specificity of biological meaning expressed as cell states is achieved by the dynamic interplay between polysemous signs and degenerate codes as constrained by signaling system's organization. The computational skills displayed by each organization level in relation to environmental changes are very different, ranging from simple reaction (CELL), to adaptation (SELF), and to learning (SENSE).

It seems that there is a direct correlation between degeneracy and complexity [36]. In the particular case of our medium scale analysis, the interpolation of attractors and increased connectivity of regulatory networks will produce structures of higher complexity. The potential destinies a cell may embrace become more numerous through levels of organization. At the same time, before the system faces the alternative choices presented to each level, it has to remain undifferentiated, lifted above choices. We have seen that in semiotic terms this indetermination can be produced by two sources: polysemy of signs and degeneracy of codes. Therefore, we could affirm that as far as ST by cells is concerned, the more signs are polysemous, and the more codes are degenerate, the more complex will be the organization adopted by a semiotic system. This is equivalent to stating that variation is the source for complexification by natural selection in the most basic level of ST.

As a final consideration, we propose that at the large-scale of wiring diagrams we could describe ST in thermodynamic terms as the coupling between chemical instability (environmental signs), biochemical oscillations (signaling motifs), and periodic cell behaviors (response modes). To maintain the organizational pattern required by dissipative structures, there must be a way of: assuring instabilities; propagating fluctuations; and selecting functions. Because these couplings between processes with no necessary physical connection are arbitrary, they must have been produced by a combination of natural selection and natural convention [7]. This is equivalent to assuming that variation is the fundamental source for the convergent gain of complexity biological systems have acquired through evolution by the unique combination of two mechanisms; natural selection and natural convention.

In natural languages there are many instances of this kind of variation. The very existence of figurative language attests to the powerful role of equivalent structures in conveying meaning and/or of disparate meanings to be traced to a unique structure. Metaphor, simile, and irony are examples of signs that, by a slight change in their structure, and because the language is a degenerate code, have their meaning radically altered. On the other hand, ambiguity and jargon are components of language denoting the role of multifunctionality at the sign level, with a same sign producing imprecise or multiple meanings. In both cases variation is at work with the creative potential of natural languages. It is important to remark that the significance of figurative language can only be understood through some social perspective – as poetry, politeness, or power.

Biological systems, like natural languages, are highly complex and as we have seen, are therefore degenerated. We believe that these characteristics are intrinsically related to another property they share: being creative. In cell biology, like in semiotics, elementary sign constructs can only be understood through some integrative perspective – as cell cycle, cell differentiation, and cell function. In this sense, it will be interesting to turn back to our initial working hypothesis, namely that of recognition having its origin in single-cells rather than in cell collectives. We have seen that to fully understand ST in terms of a precise orchestration between variation and selection, we must consider the cells from an integrative perspective. In this sense there are no single-cells, but only "social-cells" with different perspectives to sense and respond to changes according to their own organization.

The deeper understanding of how complex systems – based in the interplay of polysemic signs, degenerate codes and selected meanings, become linked and synchronized across levels is a major challenge for all the subjects contemplated by recognition sciences.

References

1. Körner S (1984) Metaphysics. University Press, Cambridge
2. Faria M (2006). RNA as Code Makers: A Biosemiotic view of Rnai and Cell Immunity, in Introduction to Biosemiotics. In: Marcello Barbieri (ed) Springer, Dordrecht, The Nethelands
3. Huang S, Eichler G, Bar-Yam Y, and Ingber, DE (2005) Cell fates as high-dimensional attractor states of a complex gene regulatory network. Phy Rev Lett 94 (12):128701–128704
4. Edelman G (2004) Biochemistry and the sciences of recognition. J Biol Chem 279(9):7361–7369
5. Bourret R (2006) Census of prokaryotic senses. J Bacteriol 188(12):4165–4168
6. Bruni LE. (2006).Cellular semiotics and signal transduction. In: Barbieri M (ed) Introduction to Biosemiotics. Springer, Dordrecht, The Netherlands
7. Barbieri M (2003) The Organic Codes. University Press, Cambridge
8. Emmeche C (1999) The Sakar challenge to biosemiotics: is there any information in a cell? Semiotica, 127(1/4):273–293
9. Falke JJ et al. (1997) The two-component signaling pathway of bacterial chemotaxis: a molecular view of signal transduction by receptors, kinases and adaptor enzymes. Annu Rev Cell Dev Biol 13:457–512
10. Mendenhall MD and Hodge AE (1998) Regulation of Cdc28 cyclin-dependent protein kinase activity during the cell cycle of the yeast Saccharomyces cerevisae. Microbiol Mol Biol Rev 62:1191–1243
11. DiNardo S, Heemskerk J, Dougan S, O'Farrell PH (1994) The making of maggot:patterning the Drosophila embryonic epidermis. Curr Opin Genet Dev 4:529–534
12. Hanahan D, Weinberg RA. (2000). The hallmarks of cancer. Cell 100:57–70
13. Bateson G (1979) Mind and Nature. A Necessary Unity. Bantam Books, New York
14. Nixon BT, Ronson CW, Ausubel FM (1986) Two-component regulatory systems responsive to environmental stimuli share strongly conserved domains with the nitrogen assimilation regulatory genes *ntr*B. Proc Natl Acad Sci USA 83:7850–7854
15. Stock A, Koshland D, Stock J (1985) Homologies between the Salmonella typhimurium Che Y protein and proteins involved in the regulation of chemotaxis, membrane protein synthesis, and sporulation. Proc Natl Acad Sci USA 82:7989–7993
16. Galperin M (2006) Response regulators encoded in bacterial and archael genomes. Available at: http://www.ncbi.nlm.nih.gov/Complete-Genomes/RRcensus.html
17. Bijlsma J, Grisman E (2003) Making informed decisions: regulatory interactions between two-componentsystems. Trends Microbiol 11:359–364
18. Shen-Orr S, Milo R, Mangan S, Alon U (2002) Network motifs in the transcription regulation network of Escherichia coli. Nature Genetics 31:64–69
19. Milo R, Shen-Orr S, Itzkovitz S, Kashtan N, Alon U (2002) Network Motifs: simple building blocks of complex networks. Science 298:824–829
20. Dobrin R, Beg Q, Barabasi A, Oltvai Z (2004) Aggregation of topological motifs in the E. coli transcriptional regulatory network. BMC Bioinformat 5:10–13
21. Keener J, Sneyd J (1998) Mathematical Physiology. Springer Books, Berlin
22. Fall C, Marland E, Wagner J, Tyson J (2002) Computational Cell Biology. Springer Books, Berlin
23. Tyson JJ, Chen KC, Novak B (2003) Sniffers, buzzers, toggles and blinkers: dynamics of regulatory and signaling pathways in cells. Current Opin. Cell Biol. 15:221–231
24. Cohen IR (2004) Tending Adam's Garden. Elsevier, London
25. Hidetsugu K, Yota M (2005) Transcription factors and DNA replication origin selection. BioEssays 27:1107–1116
26. Raper, KB (1940) The communal nature of the fruiting process in the acrasiae. Am. J. Bot 27:436–448
27. Newell PC (1982) Cell surface binding of adenosine to Dictyostelium and inhibition of pulsatile signaling. FEMS Microbiol Lett 13:417–421

28. Dustin M, Colman D (2002) Neural and immunoological synaptic relations. Science 298:785–789
29. Goldbeter A (1996) Biochemical Oscillations and Cellular Rhythms. University Press, Cambridge
30. Lahav G (2004) The strength of Indecisiveness: oscillatory behavior for better cell fate determination. Science's STKE 264:55–57
31. Nelson DE et al. (2004) Oscillations in NF-kB signaling control the dynamics of gene expression. Science 306:704–708
32. Tyson J, Csikasz-Nagy A, Novak B (2002) The dynamics of cell cycle regulation. Bioessays 24:1095–1109
33. Goldbeter A (2002) Computational approaches to cellular rhythms. Nature 420:238–245
34. Prigogine I (1977) Time, Structure and Fluctuations. Nobel Lecture
35. Neuman Y (2006) The Polysemy of the Sign: From Quantum Computing to the Garden of Forking Paths. Proceedings of the Gathering in Biosemiotics 6
36. Edelman G, Gally J (2001) Degeneracy and complexity in biological systems. Proc Natl Acad Sci USA 98(24):13763–13768
37. Artmann S (2006) Biological information. To appear in: Sahotra Sarkar, Anya Plutynski (eds) A Companion to the Philosophy of Biology, (in press)
38. Saussure F (1962) Cours de Linguistique Générale. Payot, Paris

Part 4
Neural, Mental, and Cultural Codes

Chapter 13

Towards an Understanding of Language Origins

Eörs Szathmáry

Abstract Understanding of language is difficult since we do not know how it is being processed in the brain. Many areas of the human brain are involved in the relevant activities, even for syntactic operations. Aspects of the language faculty have significant heritability. There seems to have been positive selection for enhanced linguistic ability in our evolutionary past, even if most implied genes are unlikely to affect only the language faculty. It seems that complex theory of mind, teaching, complex cooperation and language together form an adaptive suite in the human race. It is plausible that genes changed in evolution so as to render the human brain more proficient in linguistic processing. An Evolutionary Neurogenetic Algorithm (ENGA) is also reviewed that holds promise that we shall ultimately understand how genes can rig the development of cognitively specialised neuronal networks.

1 Introduction

The 'original resources' available to humans has puzzled the founder of modern neuroscience, Santiago Ramon y Cajal. In his autobiography he writes (Cajal, 1917, pp. 345–350):

> At that time, the generally accepted idea that the differences between the brain of [non-human] mammals (cat, dog, monkey, etc.) and that of man are only quantitative, seemed to me unlikely and even a little offensive to human dignity … but do not articulate language, the capability of abstraction, the ability to create concepts, and, finally, the art of inventing ingenious instruments … seem to indicate (even admitting fundamental structural correspondences with the animals) the existence of original resources, of something qualitatively new which justifies the psychological nobility of *Homo sapiens*? (Cited by De Felipe et al., 2002, p. 299)

Collegium Budapest (Institute for Advanced Study), 2 Szentháromság utca, H-1014 Budapest, Hungary, Research Group for Theoretical Biology and Ecology, Institute of Biology, Eötvös University, Budapest, Parmenides Center for the Study of Thinking, München, Germany,
e-mail: szathmary@colbud.hu

M. Barbieri (ed.), *The Codes of Life: The Rules of Macroevolution.*
© Springer 2008

Natural language is a unique communication and cultural inheritance system. In its practically unlimited hereditary potential it is similar to the genetic and the immune systems. The underlying principle is also similar in that all these systems are generative: they achieve unlimited capacity by the combination of limited primitives. The origin of natural language is the last of the major evolutionary transitions [Maynard Smith and Szathmáry, 1995]. Although later in society important transitions did happen in the way of storing, transmitting, and using inherited information, they were not made possible or accompanied by relevant genetic changes in the biology of our species. In contrast, language has a genetic background, but it is an open question how a set of genes affects our language faculty. It is fair to say that with respect to their capacity to deal with the complexity of language, even 'linguistically trained' animals are very far from us.

Understanding language origins and change is difficult because it involves three interwoven timescales and processes: individual learning, cultural transmission, and biological evolution. These cannot be neatly separated from one another (Christiansen and Kirby, 2003). The fact that a population uses some useful language that is culturally transmitted changes the fitness landscape of the population genetic processes.

Language has certain design features, such as symbolic reference, compositionality, recursion, and cultural transmission (Hockett, 1960). Theories of language and language evolution can be divided into two sets of hypotheses: a nativist versus empiricist account and a non-adaptationist versus adaptationist account, respectively (Smith, 2003).

The *nativist paradigm* argues that language capacity is a collection of domain-specific cognitive skills that is unique to humans and is somehow encoded into our genome. Perhaps the most famous proponent of this approach is Noam Chomsky, who coined the term 'language organ' and argued in favour of the uniqueness and the innateness of human linguistic skills (Chomsky, 1986). Different scholars agree with Chomsky on this issue (Pinker and Bloom, 1990; Jackendoff, 1992; Pinker, 1994; Maynard Smith and Szathmáry, 1995; Pinker and Jackendoff, 2004). The *empiricist paradigm,* however, argues that linguistic performance by humans can be explained with domain-general learning techniques (Sampson, 1997).

Fisher and Marcus (2006, p. 13) are right in stating that 'In short, language is a rich computational system that simultaneously coordinates syntactic, semantic, phonological and pragmatic representations with each other, motor and sensory systems, and both the speaker's and listener's knowledge of the world. As such, tracing the genetic origins of language will require an understanding of a great number of sensory, motor and cognitive systems, of how they have changed individually, and of how the interactions between them have evolved'. The study of language origins is however hampered by the fact that there is a critical lack of detailed understanding at all levels, including the linguistic one. There is no general agreement among linguists how language should be described: widely different approaches do exist and their proponents can have very tense scientific and other relationships. As a biologist I would maintain that symbolic reference combined with complicated syntax (including the capacity of recursion) is a least common denominator in this debate. Within this broad characterisation I just call attention to two approaches that have, perhaps surprisingly, a strongly chemical flavour. One is the minimalist programme of Chomksy (1995) where the crucial operator is *merge*, the action of which triggers certain rearrangements of the representation of a sentence. There is a broad similarity between this proposal and chemical reactions (Maynard Smith and Szathmáry, 1999). An even closer analogy between chemistry and linguistics can be detected in Steel's Fluid Construction Grammar (Steels, 2004; Steels and De Beule, 2006), in which semantic and syntactical 'valences' have to be filled for correct sentence construction and parsing. We should note (see also chapters on the genetic code) that the roots of genetic inheritance are of course in chemistry, and that even at the phenomenological level Mendelian genetics was a stoichiometric paradigm, influenced by contemporary chemical understanding (elementary units that can be combined in certain *fixed proportions* give rise to new qualities). Chemical reactions can be characterised by rewrite rules. It will take in-depth study to consider how deep this analogy goes. The deeper it goes, the more benefit one can hope from taking the analogy seriously.

Non-adaptationist accounts of language evolution rely heavily on 'spandrels' (Gould and Lewontin, 1979). The idea is that language or linguistic skills evolved not because it gave fitness advantage to its users; rather it evolved as a side effect of other skills as spandrels are side effect of architectural constraints. Chomsky again has a prominent role in this debate as the protagonist of the non-adaptationist approach. In the latest reworking of the theory (Hauser et al., 2002), Chomsky and colleagues distinguish between the 'Faculty of Language in the Broad Sense' (FLB) and 'Faculty of Language in the Narrow Sense' (FLN). They argue that FLB consists of skills that evolved in other animals as well as in humans, whereas FLN consists of only one skill (merge), which evolved in a different (unspecified) context and was then co-opted for linguistic use. However, the finding that European starlings appear able to recognise context-free grammatical structures (i.e. hierarchical syntax (Gentner et al., 2006) is somewhat contrary to Chomsky's position given that it shows that precursor of the skill they have assigned to FLN (i.e. merge) may have independently evolved in other animals too; although a strict proof of appropriate parsing of these structures by starlings is lacking (M. Corballis, personal communication, 2007, email].

The first *adaptationist* account of human language was by Darwin (1971), later defended by Pinker and Bloom (1990) in their influential paper about the Darwinian account of language. More specifically, these authors argued that language, as any complex adaptations, could only be explained by means of natural selection. This paper catalysed many linguists and biologists to study language and language evolution from the perspective of evolutionary biology and was followed by many influential publications (Jackendoff, 1992; Maynard Smith and Szathmáry, 1995; Knight et al., 2000; Christiansen and Kirby, 2003). Recently, Pinker and Jackendoff (2005) made a forceful defence of the adaptationist paradigm in response to Chomsky and colleagues (Hauser et al., 2002).

Language needs certain prerequisites. There are some obvious prerequisites of language that are not especially relevant to our approach. For example, apes do not have a descended larynx or cortical control of their vocalisations. Undoubtedly, these traits must have evolved in the human lineage, but we do not think that they are indispensable for language as such. One could have a functional language with a smaller number of phonemes, and sign language (Senghas et al., 2004) does not need either vocalisation or auditory analysis. Thus, we are mostly concerned with the *neuronal implementation* of linguistic operations, irrespective of the modality. It seems difficult to imagine the origin of language without capacities for teaching (which differs from learning), imitation, and some theory of mind (Premack, 2004). Apes are limited in all these capacities. It is fair to assume that these traits have undergone significant evolution because they were evolving together with language in the hominine lineage. To this one should add, not as a prerequisite, but as a significant human adaptation the ability to cooperate in large non-kin groups (Maynard Smith and Szathmáry, 1995). These traits together form an *adaptive suite*, specific to humans. I suggest that in any selective scenario, capacities for teaching, imitation, some theory of mind, and complex cooperation must be rewarded, because an innate capacity for these renders language emergence more likely.

On the neurobiological side I must call attention to the fact that some textbooks, [e.g. Kandel et al., 2000] still give a distorted image of the neurobiological basis of language. It would be very simple to have the Wernicke and Broca areas of the left hemisphere for semantics and syntax, respectively. But the localisation of language components in the brain is extremely plastic, both between and within individuals (Neville and Bavelier, 1998; Müller et al., 1999). Surprisingly, if a removal of the left hemisphere happens early enough, the patient can nearly completely retain his/her capacity to acquire language. This is of course in sharp contrast to the idea of anatomical modularity. It also puts severe limitation on the idea that it is only the afferent channels that changed in the evolution of the human brain: modality independence, and the enormous brain plasticity in the localisation of language favour the idea that whatever has changed in the brain that has rendered it capable of linguistic processing must be a very widespread property of the neuronal networks (Szathmáry, 2001). Components of language get localised somewhere in any particular brain in the most functionally 'convenient' parts available. Language is just a certain activity pattern of the brain that finds its habitat like an amoeba in a medium. The metaphor 'language amoeba' expresses the plasticity of language but it also calls attention to the fact that a large part of the human brain is apparently a potential habitat for it, but no such habitat seems to exist in non-human ape brains (Szathmáry, 2001).

A dogma concerning the histological uniformity of homologous brain areas in different primate species has also been around for some time. Recent investigations do not support such a claim (DeFelipe et al., 2002). In fact the primary visual cortex shows marked cytoarchitectonic variation (Preuss, 2000), even between chimps and man. It is therefore not at all excluded that some of the species-specific differences in brain networks are genetically determined, and that some of these are crucial for our language capacity. But, as discussed above, these language-critical features must be a rather widespread network property. Genes affect language through the development of the brain. One could thus say that the origin of language is to a large extent an exercise in the linguistically relevant developmental genetics of the human brain (Szathmáry, 2001).

The close genetic similarity between humans and chimps strongly suggests that the majority of changes relevant to the human condition are likely to have resulted from changes in gene regulation rather than from widespread changes of downstream structural genes. Recent genetic and genomic evidence corroborates this view. In contrast to other organs, genes expressed in the human brain seem almost always upregulated relative to the homologous genes in chimp brains (Caceres et al., 2003). The functional consequences of this consistent pattern await further analysis.

We know something about genetic changes more directly relevant to language. The FOXP2 gene was discovered to have mutated in an English-speaking family (Gopnik, 1990; Gopnik, 1999). It has a pleiotropic effect: it causes orofacial dyspraxia, but it also affects the morphology of language: affected patients must learn or form the past tense of verbs or the plurals of nouns case by case, and even after practice they do so differently from unaffected humans (see Marcus and Fisher, (2003) for review). The gene has been under positive selection (Enard et al., 2002) in the past, which shows that there are genetically influenced important traits of language other than recursion (Pinker and Jackendoff, 2005), contrary to some opinions (Hauser et al., 2002). There is a known human language, apparently with no recursion (Everett, 2005). It would be good to know how these particular people (speaking the Pirahã language in the Amazon) manage recursion in other domains, such as object manipulation. Apes are very bad at recursion both in the theory of mind or 'action grammar' (Greenfield, 1991).

It does seem that the capacity to handle recursion is different from species to species. Although the relevant experiment must be conducted with chimps as well, it has been demonstrated that tamarin monkeys are insensitive to auditory patterns defined by more general phrase structure grammar, whereas they discover violations of input conforming to finite state grammar (Fitch and Hauser, 2004). Human adults are sensitive to both violations. Needless to say it would be very interesting to know the relevant sensitivities in apes and human children (preferably before they can talk fluently). It will be interesting to see what kind of experiment can produce consistent patterns in such a capacity in evolving neuronal networks, and then reverse engineer proficient networks to discover evolved mechanisms for this capacity.

I share the view that language is a complex, genetically influenced system for communication that has been under positive selection in the human lineage (Pinker and Jackendoff, 2005). The task of the modeller is then to try to model intermediate stages of a hypothetical scenario and, ultimately, to re-enact critical steps of the transition from protolanguage (Bickerton, 1990) to language. It cannot be denied that language is also a means for representation. This is probably most obvious for abstract concepts, for which the generative properties of language may lead to the emergence of a clear concept itself. This is well demonstrated for arithmetics: for instance, an Amazonian indigenous group lacks words for numbers greater than 5; hence they are unable to perform exact calculations in the range of larger numbers, but they have approximate arithmetics (Pica et al., 2004).

I mentioned before that the fact that language changes while the genetic background also changes (which must have been true especially for the initial phases of language evolution) the processes and timescales are interwoven. This opens up the possibility for genetic assimilation (the Baldwin effect). Some changes that each individual must learn at first can become hard-wired in the brain later. Some have endorsed (Pinker and Bloom, 1990), while others have doubted (Deacon, 1997) the importance of this mechanism in language evolution. Deacon's argument against it was that linguistic structures change so fast that there is no chance for the genetic system to assimilate any grammatical rule. This is likely to be true but not very important. There are linguistic operations, performed by neuronal computations, related to compositionality and recursion that must have appeared sometime in evolution. Whatever the explicit grammatical rules are, such operations must be executed.

Hence a much more likely scenario for the importance of genetic assimilation proposes that many operations must have first been learned, and those individuals whose brain was genetically preconditioned to a better (faster, more accurate) performance of these operations had a selective advantage (Szathmáry, 2001). Learning was important in rendering the fitness landscape more climbable (Hinton and Nolan, 1987). This view is consonant with Rapoport's (1990) view of brain evolution. This thesis is also open for experimental test.

The origin of language is an unsolved problem; some have even called it the 'hardest problem of science' (Christiansen and Kirby, 2003). It is very hard because physiological and genetic experimentation on humans and even apes is very limited. The uniqueness of language prohibits, strictly speaking, application of the comparative method, so infinitely useful in other branches of biology. Fortunately, some elements of language lend itself to a comparative approach, as we shall see in relation to bird song. Nevertheless, limitation of the approaches calls for other types of investigation. I believe that simulations of various kinds are indispensable elements of a successful research programme. Yet, a vast range of computational approaches has brought less than spectacular success (Elman et al., 1996). This is attributable, I think, to the utterly artificial nature of many of the systems involved, such as connectionist networks using back-propagation (e.g. see Marcus (1998) for a detailed criticism). In Section 6, I present an alternative, potentially rewarding, modelling approach.

2   Genetic Background of Language

Information about the human and the chimp genome (The Chimpanzee Sequencing and Analysis Consortium, 2005) is now 'complete', and one can ask how far previous optimism seems justified in the light of comparative studies using this information. It is clear that a lot of work lies ahead: knowing all the genes of chimps and humans is not the whole thing: one should know how genotype is mapped to phenotype, and this is a formidable problem. Genes are expressed in specific ways, under the influence of other genes and the environment. Interaction between genes is not the exception but the rule. One gene can affect several traits (pleiotropy) and actions of different genes do not affect traits (including fitness) independently (epistasis). It is the network of interactions that one should know, and one must not forget that there are networks at different levels, from genetic regulatory networks through protein interaction networks and signal transduction pathways to the immune system or neuronal networks. The question is how the effect of genes percolates upwards. Genes act on expressed molecules (proteins and RNA) that do their job in their context. There is something amazing about the fact that hereditary action on such primitive molecules percolates upwards resulting in heritability of complex cognitive processes, including language.

The chimp and human genomes are indeed similar, but one should understand clearly what this means (Fisher and Marcus, 2005). Substitutions make-up for 1.23% of difference between the two genomes, which translates into 35 million altered sites in the single-copy regions of the genomes! Insertions and deletions yield a further 3% genomic difference. It is convenient to distinguish between altered structural and regulatory genes. The first code for altered enzymes or structural proteins, the latter code for altered transcription factors, for example. Both kinds of changes did happen since the humans diverged from chimps, and both can affect language in critical ways.

There seems to have been acceleration in the changes of neural gene expression patterns in human evolution, although this should be evaluated against the background that liver and heart expression patterns have diverged a lot more between chimps and humans. The usual interpretation is that neural tissue is under stronger stabilising selection. Another observed tendency is the upregulation of human neural gene expression relative to the chimp, but the functional significance of this finding is unclear (it may be a more or less direct consequence of recent genomic region duplications).

It is not yet clear what is the gene expression difference behind the cytoarchitectonic differences among the Brodman areas: most known gene expression differences between chimps and humans are common to all cortical regions. This view has been refined very recently. Oldham et al. (2006) analyzed gene coexpression patterns in humans and chimps. They could identify network modules that correspond to gross anatomical structures including the cerebellum, caudate nucleus, anterior cingulated cortex, and cortex. The similarity of network connectivity between the respective human and chimp areas decreased in that order; consistent with the radical evolutionary expansion of the cortex in humans. It is intriguing that in the cortical module there is a strong coexpressive link between genes of energy metabolism, cytoskeletal remodelling, and synaptic plasticity.

There are genetic changes that probably did boost language evolution but in a general, aspecific way. Genes influencing brain size are likely to have been important in this sense. A note is in order, however. Genes involved in primary microcephaly seem to have been under positive selection in the past, but children with this syndrome have rather normal neuroanatomical structures despite the fact that their overall brain size can be reduced to a mere one third of the normal. They show mild to moderate mental retardation and pass several developmental stages. Fisher and Marcus (2005, p. 13) conclude: 'In our view the honing of traits such as language probably depended not just on increased "raw materials" in the form of a more ample cortex, but also on more specific modification of particular neural pathways.'

Perhaps the most revealing recent finding concerning genetic brain evolution is the identification of an RNA gene that underwent rapid change in the human lineage (Pollard et al., 2006). It is expressed in the Cajal-Retzius cells of the developing cortex from 7 to 19 gestational weeks. It is coexpressed with reelin, a product of the same cells, that is important in specifying the six-layer structure of the human cortex.

Even if some of our linguistic endowment is innate, there may not be genetic variation for the trait in normal people, just as normal people have ten fingers. In contrast, our linguistic capacity may be like height: whereas all people have height, there are quantitative differences in normal people. To be sure, children as well as adults differ in their linguistic skills; the question is what part of this variation genes account for.

Surveying many studies Stromswold (2001) concluded that twin concordance rates have been significantly higher for monozygotic twins than for dizygotic twins. Twins are concordant for a trait if both express the trait or neither expresses it. Twins are discordant for a trait if one exhibits the trait and the other does not. If the concordance rate for language disorders is significantly greater for monozygotic than dizygotic twins, this suggests that genetic factors play a role in language disorders such as dyslexia and specific language impairment (SLI). The concordance rates for written and spoken language disorders are similar. For both written-language and spoken-language disorders, mean and overall concordance rates were approximately 30% higher for monozygotic twins than for dizygotic twins, with genetic factors accounting for between one half and two thirds of the written- and spoken-language abilities of language-impaired people. In studies of normal twins, depending on the aspect of language being tested, between one quarter and one half of the variance in linguistic performance was attributable to genetic factors. People have been tested on phonological short-term memory, articulation, vocabulary, and morphosyntactic tasks. It seems that different genes may be responsible for the variance in different components of language and that some genetic effects may be language specific.

The sum of all genetic effects is usually not much greater than 50% for various aspects of cognition (Stromswold, 2001). Most individual genes are expected to have small effects. Candidate genes affect functions including the cholinergic receptor, episodic memory, dopamine degradation, forebrain development, axonal growth cone guidance, and the serotonin receptor. It is a great problem that cognitive skills are likely to have been at least in part inadequately parsed, thus so-called intermediate phenotypes with a clearer genetic background should be sought. By this token schizophrenia as such does not exist; rather, different genes may go wrong and the symptoms such as hallucinations are emergent outcomes (Goldberg and Weinberger, 2004). The situation may be similar to that of geotaxis in *Drosophila*, where the individual involvement of different genes that collectively determine this capacity is counter-intuitive (Toma et al., 2002).

It is worth calling attention to the fact that the genetics of human cognitive skills is a notoriously difficult problem. One common reason is that usually the clinical characterisations are not sufficient as descriptions of phenotypes (Flint, 1999). A consensus seems to emerge that the genes involved are 'liability genes' that, when present in the right allelic form, significantly enhance the probability of developing the respective cognitive skills.

Perhaps the most important neurodevelopmental syndrome for our topic is SLI, where there is significant difference between verbal and non-verbal skills. Several candidate chromosomal regions have been identified (SLI consortium, 2002); importantly, they involve the gene USP10 (an ubiquitin-specific protease), which encodes a protein involved in synaptic growth, and shows an increased copy umber in the human lineage (Fortna et al., 2004). In *Drosophila*, its overexpression of its homolog results in increased synaptic branching and altered synaptic function (DiAntonio et al., 2001).

A by now famous gene is FOXP2, that was first called attention to by Gopnik (1990). In a certain English-speaking family there is a dominant allele that causes the syndrome, formerly grouped under SLI, but recently termed developmental verbal dyspraxia (DVD). There is no disagreement that SLI is real. It is more contested how closely it is limited to, or rooted in, a specific grammatical impairment. The Gopnik (1990, 1999) case has been very stimulating because of its characterisation as 'feature-blind' dysphasia and its obvious genetic background (a single dominant allele). Whether other cognitive skills are also, or even primarily, affected, has been debated ever since (Vargha-Kadem et al., 1998). More evidence with other linguistic groups is accumulating (Dalalakis, 1999; Rose and Royle, 1999; Tomblin and Pandich, 1999). A study (Van der Lely et al., 1998), sadly without genetics, claims to demonstrate that grammatically limited SLI does exist in 'children' (although only one child is analysed in the paper!).

The FOXP2 protein is an old transcription factor present in vertebrates, and there is evidence that it has been under positive selection in the human lineage. It seems to affect development of distributed neural networks across the cortex, striatum, thalamus, and cerebellum. The DVD condition makes the phenomenon different from SLI, but it is important that speech and language deficits are *always* present, even in otherwise normal children. In other affected individual general intelligence is also impaired. It is also important that the grammar deficits (difficulty with morphological features such as the suffix –s for plural or –(e)d for past tense) occur in written language as well. The selective sweep that affected this gene in the human lineage occurred within the last 200,000 years (Enard et al., 2002; Zhang et al., 2002).

Analysis of the expression patterns of FOXP2 in other species suggests that this gene has been involved in the development of neural circuitry processing sensory-motor integration and coordinated movements, lending support to the notion that language has its roots in motor control (e.g. Lieberman, 2007), which makes the involvement of basal ganglia in speech and language less than surprising.

Recent studies (reported by White et al., 2006) demonstrate that FOXP2, although without accelerated evolution, plays a crucial role in the development and seasonal activation of relevant brain areas in songbirds. Interestingly, although the avian FOXP2 is very similar to the human version, neither of the human-specific mutations is found in the former. Also of interest in the fact that the ganglia involved in bird song learning seem to be analogous to the basal ganglia involved in human vocal learning [Scharff and Haesler, 2005].

Researchers have called attention to the fact that not only FOXP2, but also FOXP1 is expressed in functionally similar brain regions in songbirds and humans that are involved in sensorimotor integration and skilled motor control (Teramitsu et al., 2004). Moreover, differential expression of FOXP2 in avian vocal learners is correlated with vocal plasticity (Haesler et al., 2004). Mice, like man, have also two copies of the FOXP2 gene. If only one of them is affected in mice, the pups are severely affected in the ultrasonic vocalisation upon separation from their mother. This suggests a role of this gene is social communication across different species. The Purkinje cells in the cerebellum are affected in the pups (Shu et al., 2005). Determination of the expression pattern in the developing mouse and human brain is consonant with these investigations: regions include the cortical plate, basal ganglia, thalamus, inferior olives, and cerebellum. Impairments in sequencing of movement and procedural learning thus may be behind the linguistic symptoms in humans (Lai et al., 2003).

We think a key issue is the biologically motivated dissection of the language faculty. Put differently, what are the intermediate phenotypes composing language? This question cannot be answered, we believe, without an appropriate formulation of aspects of language. Thus linguistic theories must ultimately be biologically constrained. A good start in this direction may be Luc Steels' fluid construction grammar (Steels and De Beule, 2006). There is no coupling yet between details of linguistic theories and those of brain mechanisms.

3   Brain and Language

Analyses of neural activity during the performance of cognitive tasks have become a growth industry. Their sensitivity has increased over the years. These methods are increasingly applied to the recording of brain activity during linguistic performance.

The recognition that neural localisation of language can be plastic is now widely known (Nobre and Plunkett, 1997; Neville and Bavelier, 1998; Musso et al., 1999). Studies of brain injury revealed that damage to the left hemisphere before a critical period is not for life: the right hemisphere can take over the necessary functions (Müller et al., 1999). This does not contradict the finding that in normal people Broca's area does seem specialised for syntax (Embick et al., 2000). It seems that the common left hemisphere localisation of language is just the most likely outcome when there is no genetic or epigenetic disturbance. What is more, cortical and subcortical areas both contribute to processing of language; reward systems and motor control provided by basal ganglia and the cerebellum seem to be critical components of our language faculty (Lieberman, 2002, 2007). PET studies have revealed a truly shocking feature of language development: the localisation of linguistic processing is shifting during normal ontogenesis. The outcome in 'normal' people is also highly variable.

Analysis of a particularly interesting genetic syndrome, called Williams syndrome, also reveals surprisingly dynamical manifestations during ontogenesis. Whereas affected children seem to be bad at language and good with numbers, adults perform the other way round (Paterson et al., 1999). The classical characterisation of the disease was based on adult performance.

The conclusions that we can draw from brain studies are the following:

- Localisation of language is not fully genetically determined: even large injuries can be tolerated before a critical period.
- Language localisation to certain brain areas is a highly plastic process, both in its development and its end result.
- It does seem that a surprisingly large part of the brain can sustain language: there are (traditionally recognised) areas that seem to be most commonly associated with language, but by no means are they exclusive, either at the individual or the population level, during either normal or impaired ontogenesis.
- Whereas a large part of the human brain can sustain language, no such region exists in apes.
- Language processing has a distributed character.

It is instructive to look at the evolutionary patterns of the sensory neocortex in mammals (Krubitzer and Kaas, 2005). Auditory, somatosensory, and visual fields (continuous brain tissue regions) have changed in location and size in different species. Fields can change in absolute and relative size, and in number. Connections of cortical fields can also change. Such alterations can be elicited by manipulation of either the peripheral morphology or activity, or that of the expression level of certain genes. Phenotypic within-species variation can be extremely broad; little is known about the relative magnitude of the genetic part of this variation, however. A good example of genetic influence is the variation in the cortical area map of inbred mice, reflecting strain identity (Airey et al., 2005).

Evolution of the vertebrate brain has produced increase in cortical size, and elaboration of the cortical circuit diagram (Hill and Walsh, 2005). Importantly, cortical layers II and IIIb, IIIc of the chimp differ from layers IIa, IIb, and IIIa, IIIb, and IIIc, respectively, in humans. A tentative conclusion, based on 'rewired' ferrets and three-eyed frogs, is that layers form independently of patterned input, but also that instructive electrical signals play a crucial role in fine network development, also affecting intracortical connections (Sur and Learney, 2001).

Genetically determined patterning of parts of the brain follows mechanisms well-known from conventional developmental studies. For example, during the formation of the retinotopic map, axons from the retinal ganglion cells find their targets in the tectum as a result of matching between two receptor/ligand pairs (Schmitt et al., 2006), both expressed according to (altogether four) gradients, (two in the eye and two in the tectum).

Several people, including Greenfield (1991) have suggested the involvement of tool making in the evolution of language; for example in the form of 'action grammar' that can be recursive when agents use the 'subassembly' strategy in the 'nesting cups' experiment. The idea is that selection for efficient tool use could have aided language evolution and vice versa. Stout and Chaminade (2007) investigated the neurobiological bases in modern naïve humans of 2.5 million-year-old Oldowan practice of tool making by brain imaging. (Incidentally, in that material culture we see evidence for the uniquely human practice to use a tool to make another tool.) Premotor cortex was activated in the task but not the prefrontal executive cortex (involved in planning), nor the inferior parietal cortex. The activation of caudal Broca's area in this task underlines the possible link between language and tool making, and is consonant with views of the importance of 'mirror neurons' in language evolution (Rizzolatti and Arbib, 1998), although one should not forget that the latter are by no means sufficient for language, as many animals have it. As we learn words by imitation, and tool-making requires an 'action grammar', it is unlikely to be accidental that human Broca's area evolved from structures that are involved in these capacities beyond (and prior to) language. Caplan (2006) makes the suggestion that Broca's area is involved in syntactic processing not merely because it is evolutionarily related to the dorsolateral prefrontal cortex or its original involvement in sensory-motor functions, but because of its intrinsic neural organization – but this leaves the very essence of the suggested neural organization obscure.

4 Brain Epigenesis and Gene-language Co-evolution

It has to be admitted that on the whole we do not understand how the brain works. Nevertheless, some crucial elements seem to emerge. One is that development of the normal brain is enormously plastic, even though the power of genetic factors is obvious. One classic example is that in the same brain areas of identical twins the two hemispheres of the same individual resemble each other more closely than the same hemispheres in the two people (Changeux, 1983).

Another insight is that a tremendous amount of variation and selection is going on during brain ontogenesis. This is a Darwinian-type process, no doubt. As the psychologist William James recognised a long time ago, natural selection of heritable variation is the only known force that can lead to adaptations, so let us apply it to brain ontogenesis and problem-solving as well (James thought that even learning is a result of selection of variation within the brain). There are several expositions that all regard the brain, one way or the other, as a 'Darwin machine' (Calvin and Bickerton, 2000). Here I stick to the formulation by Changeux (1983), because we think this is the most relevant to the language problem. According to this view, the functional microanatomy of the adult cortex is the result of the vastly surplus initial stock of synapses and their selective elimination according to functional criteria (performance).

We have just learnt in Section 3 that a very large part of the human brain can process linguistic information, including syntactical operations. This means that there is no fixed macro-anatomical structure that is exclusively dedicated to language, but some functional micro-anatomical structure *must* be appropriate, otherwise it could not sustain language. This further suggests that there is some *statistical connectivity feature* of a large part of the human brain that renders is suitable for linguistic processing. From the selectionist perspective there are three options: the initial variation in synaptic connectivity is novel; the means of selection on functional criteria is novel; or both. Maybe both component processes are different in the relevant human brain areas, and I do not dare to speculate about their relative importance.

This idea must be seen in close connection to the one presented by Rapoport (1990) about the co-evolution of brain and cognition (within a population of humans). The traditional view is the so-called bottom-up mechanism: that a genetic change of some neural structure is subjected to selection and, based on its performance, it either does spread or it does not. There is, however, a so-called top-down mechanism, which could have more significantly contributed to the evolution of human cognitive skills, including, especially we argue, language. The crucial idea is as follows:

- Due to the plasticity in brain development, enhanced demands on a certain brain region lead to less synaptic pruning (a known mechanism).
- Less synaptic pruning is assumed to lead to more elaborate (and more adaptive) performance.
- Any genetic change contributing to the growth of the brain area thus affected will be favoured by natural selection.

There are two important connections that must be pointed out. First (observed by Rapoport himself), the top-down mechanism is a more detailed exposition of the late Allan Wilson's idea (Wyles et al., 1983). Thus an larger brain, due to its more complex performance, alters the selective environment (in social animals composed of conspecifics to a great extent), which selects for an even larger brain, and so on. Second, and perhaps more important, this mechanism is also a neat example of a Baldwin effect (or genetic assimilation), when 'learning guides evolution'. As Deacon (1997) pointed out, it is trickier to apply the idea of genetic assimilation to language than usually thought. The reason for this is that the performed behaviour must be sufficiently long lasting and uniform in the population. It is thus hard to imagine how specific grammatical rules, for example, could have been genetically assimilated. This point is well taken, but here we speak of a different thing: the genetic assimilation of a general processing mechanism that is performed by virtue of the connectivity of the underlying neural structures.

Our claim is that the most important, and largely novel, faculty selected for was the ability of the networks to process syntactical operations on symbols that are part of a semantically interwoven network. The specific hypothesis is that linguistically competent areas of the human brain have a statistical connectivity pattern that renders them especially suitable for syntactical operations. In conclusion, we think:

- The origin of human language required genetic changes in the mechanism of the epigenesis in large parts of the brain.
- This change affected statistical connectivity patterns and dynamical development of the neural networks involved.
- Due to the selectionist plasticity of brain epigenesis, co-evolution of language and the brain resulted in the genetic assimilation of syntactical processing ability as such.

An intriguing possible example of gene-culture co-evolution has recently been raised by Bufill and Carbonell (2004). They call attention to a number of facts. First, human brain size did not increase in the last 150,000 years, and it did even decrease somewhat in the last 35,000 years. Second, a new allele of the gene for apolipoprotein E originated sometime between 220,000 and 150,000 years ago. This allele improves synaptic repair (Teter et al., 2002). The original form entails a greater risk of Alzheimer disease and a more rapid, age-related decline in general (Raber et al., 2000). More importantly, ApoE4 impairs hippocampal plasticity and interferes with environmental stimulation of synaptogenesis and memory in transgenic mice (Levi et al., 2003). Interestingly, the ancestral allele decreases fertility in men (Gerdes et al., 1996). The facts taken together indicate, but do not prove, a role in enhanced synaptogenesis in a period when syntactically complex language is thought to have originated. More evidence like this would be welcome in the future, since one such case can at best be suggestive.

Various people (e.g. Premack, 2004) have called attention to the fact that besides language, efficient teaching (which differs from learning), imitation, and a developed theory of mind are also uniquely human resources. I also stress the trait of human cooperation (Maynard Smith and Szathmáry, 1995), which is remarkable because we can cooperate even in large non-kin groups. My proposal is that these traits are not by accident together. They form an adaptive suite, and presumably they have co-evolved in the last 5 million years in a synergistic fashion. The relevant image is a *co-evolutionary wheel* (Fig. 1): evolution along any of the radial

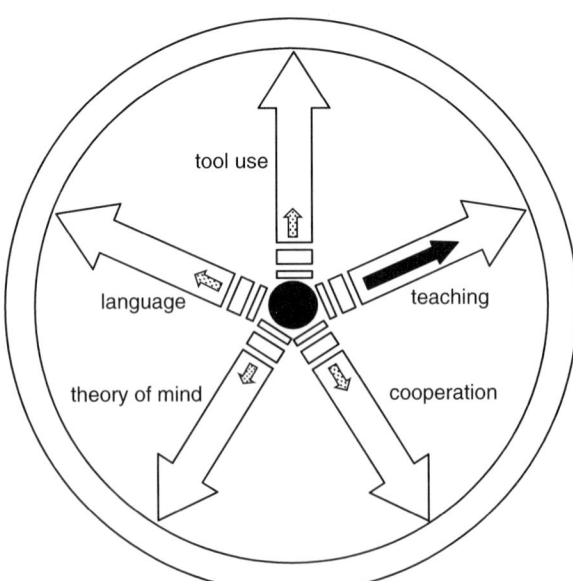

Fig. 1 The coevolutionary wheel and the human adaptive suite. In this example, direct selection on genetic variation is on teaching/docility (*black arrow*), but this gives some improvement, to varying degrees, in other dimensions (*patterned arrows*) as well

spokes presumably gave a mileage to all the other capacities, even if the focus of selection may have changed spokes several times.

This hypothesis is testable; and there is evidence in its favour already. Take the case of autism, for example. Affected people have a problem with the theory of mind, communication, and they can be seriously challenged in the strictly linguistic domain as well (Fisher and Marcus, 2005). The prediction is that there will be several to many genes found, that will have pleiotropic effects on more than one spoke of the wheel in Fig. 1.

5   Selective Scenarios for the Origin of Language

The issue of the origin of human language has provided fertile ground for speculation and alternative theories have been proposed (Box 1).

Most of the theories that suggest a given context for the evolution of human language attempt to account for its functional role. Given that, functionally, all of these theories are more or less plausible, it is almost impossible to decide on their usefulness based only on this criterion. However, recent game theoretical research can help us evaluate various contexts. These criteria concern the interest of communicating parties and the cost of equilibrium signals.

The central issue is whether early linguistic communication was honest. If signal cost is the same for all signallers, then honest cost-free signalling can be evolutionarily stable only if there is no conflict of interest between the participants (Maynard Smith, 1991). If the cost of signals varies with the quality of the signaller, then the situation is more complicated. In this case, it is possible to construct cost functions that give an arbitrarily low cost at equilibrium even if there is a conflict of interest (Hurd, 1995; Számadó, 1999; Lachmann et al., 2001) (Box 2).

In case of human language, the most obvious way to construct such a cost function is to punish dishonest signallers (Lachmann et al., 2001). However, this solution assumes that dishonest signallers can, on average, be detected (i.e. signals can be cross-checked); it also assumes that dishonest signallers are punished (which is a non-trivial assumption). Thus, one can conclude that, ' "conventional" signals will be used when communicating about (i) coincident interest or (ii) verifiable aspects of conflicting interest; "costly" signals will be used otherwise' (Lachmann et al., 2001). Although theory so far says nothing about the evolution of such systems of communication, there are a few computer simulations that suggest that honest cost-free communication evolves only if there is shared interest between the participants (Bullock, 1998; Noble, 2000; Harris and Bullock, 2002).

What does this tell us about the emergence of human language? The production cost of speech or gesturing appears to be low, thus human language consists of cost-free or low-cost signals at equilibrium (not counting time constraints). Thus, based on the above criteria, one should favour either those theories that propose a context with no conflict of interest (e.g. hunting, tool making, motherese, grooming, or the group bonding and/or ritual theory) or a context in which there might be a conflict

Box 1 Alternatives theories to explain language evolution

Gossip: menstrual ritual can be a costly signal of commitment; hence participating in such rituals can create female groups of shared interest in which sharing information about the social life of others (i.e. gossiping) can be beneficial (Power, 1998).

Grooming hypothesis: language evolved as a substitution for physical grooming (Dunbar, 1998). The need for this substitution derived from the increasing size of the early hominid groups.

Group bonding and/or ritual: language evolved in the context of intergroup rituals, which first occurred as a kind of 'strike action' against non-provisioning males. Once such rituals were established, a 'safe' environment was created for further language evolution (Knight, 1998).

Hunting theories: 'our intellect, interests, emotions, and basic social life – all are evolutionary products of the success of the hunting adaptation.' (Washburn and Lancaster, 1968). Later, Hewes (1973) in his paper about the gestural origins of language takes up the idea and argues that the probable first use of language was to coordinate the hunting effort of the group.

Language as a mental tool: language evolved primarily for the function of thinking and was only later coopted for the purpose of communication (Burling, 1993).

Mating contract and/or pair bonding: increasing size of the early hominid groups and the need for male provisioning also necessitated 'social contract' between males and females (Deacon, 1997).

Motherese: language evolved in the context of mother–child communication. Mothers had to put down their babies to collect food efficiently, and their only option to calm down babies was to use some form of vocal communication (Falk, 2004).

Sexual selection: language is a costly ornament that enables females to assess the fitness of a male. According to this theory, language is more elaborate than a pure survival function would require (Miller, 2001).

Song hypothesis: language evolved rapidly and only recently by a process of cultural evolution. The theory assumes two important sets of pre-adaptations; one is the ability to sing; the other is better representation abilities (i.e. thinking and mental syntax) (Vaneechoutte and Skoyles, 1998).

Status for information: language evolved in the context of an 'asymmetric cooperation', where information (that was beneficial to the group) was traded for status (Desalles, 1998).

Tool making: assumes a double homology: 'a homologous neural substrate for early ontogeny of the hierarchical organisations shared by two domains – language and manual object combination – and a homologous neural substrate and behavioural organisation shared by human and non-human primates in phylogeny.' (Greenfield, 1991).

Box 2 The problem of honesty

There is an ongoing debate about the honesty of animal communication, centred on the proposition that signals need to be costly to be honest (Zahavi, 1975). Although some models appear to provide support for this statement (Grafen, 1990; Godfray, 1991; Maynard Smith, 1991), there are exceptions. First, cost-free signals can be evolutionarily stable provided that there is no conflict of interest between the communicating parties (Maynard Smith, 1991). Second, even where a conflict of interest exists, cost-free signals can be evolutionarily stable provided that the cost of signals is a function of the quality being signalled (Hurd, 1995; Számadó, 1999; Lachmann et al., 2001). The most general case is when the fitness of the signaller depends on its state and the fitness of both players is influenced by the survival of the other. Assuming a discrete model with two states, two signals and two responses, the conditions of evolutionarily stable honest signalling are as follows (Eqs I–VI) (Számadó, 1999):

$$W_h + rV_h > 0 \tag{Eq. I}$$

$$W_l + rV_l > 0 \tag{Eq. II}$$

$$V_h + rW_h > C_h \tag{Eq. III}$$

$$V_l + rW_l > C_l \tag{Eq. IV}$$

$$V_h + rW_h > 0 \tag{Eq. V}$$

$$V_l + rW_l > 0 \tag{Eq. VI}$$

where W, V, and C denotes the fitness of the receiver and signaller, and the cost of signalling, respectively. l and h denote the quality of the signaller ('high' and 'low', respectively). The fitness of each player can be influenced by the survival of the other player (r). Equations I and II describe the conditions of the receiver for honest signalling; Eqs III and IV reflect are the condition of the signaller; and Eqs V and VI describe conflicts of interest. Reversing the inequality in Eq. VI would mean that there is no conflict of interest between signaller and receiver. If $r = 1$, then there can be no conflict of interest, assuming that signalling is beneficial for the receiver (because $V_l + rW_l = W_l + rV_l$, Eq. VI cannot be fulfilled). This implies that, in this case, C_l can equally zero or less than zero (given that the left-hand side of Eq. IV need not be greater than zero). However, if $0 < r < 1$ then at least C_l should be greater than zero (Eq. IV). That is, signalling for low-quality individuals must be costly in case of a conflict of interest. Signalling for high-quality individuals need not be costly even in this case (Eq. III). Given that at the honest equilibrium only high-quality individuals signal, the observed cost can be zero. If, however, $C_l = C_h$, then both costs should be greater than zero for honest signalling to be stable. Figure 2 depicts the regions of honest signalling in case of conflict of interest. The same logic also applies and, thus, the same results hold, for continuous models (Lachmann et al., 2001).

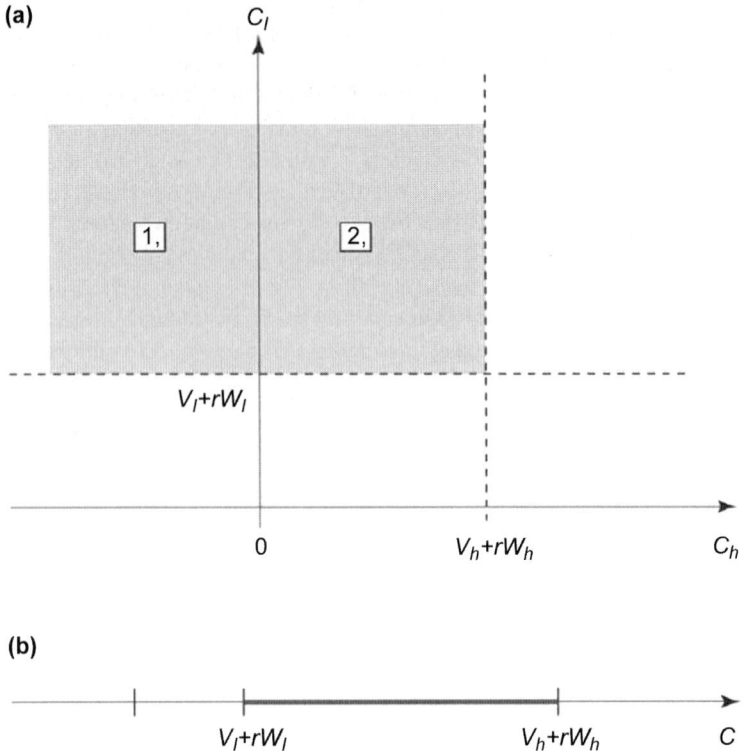

Fig. 2 The relationship between signal cost and evolutionarily stable honest signalling. (a) Regions of honest evolutionarily stable signalling: 1st, 2nd (green shading) when $C_h \neq C_l$, $V_h \neq V_l$ and $r > 0$. C_l must be greater than zero, whereas C_h can equal zero or even less than zero. (b) Regions of honest evolutionarily stable signalling when $C_h = C_l$ (thick line); both C_h and C_l must be greater than zero. In cases of shared interest ($r = 1$), both C_h and C_l can equal zero or even less than zero (see Eqs IV and VI) (Számadó, 1999)

of interest but signals can be easily cross-checked. None of the theories fit the second context: for example, mating contract and gossiping both assume a context in which conflict of interest exists and signals cannot be easily cross-checked.

Explaining the evolution of human language is likely to remain a challenge for the coming decade. There is no single theory that could sufficiently answer all the questions about honesty and groundedness, power of generalisation, and uniqueness. Table 1 gives a summary of these criteria (Számadó and Szathmáry, 2006). As one can see most of the theories fail to answer the majority of the questions. Perhaps the easiest criterion to fulfil is shared interest, as there are a number of social situations, which assume shared interest between communicating parties (such as hunting or contact calls). There are only two theories – 'tool making' (Greenfield, 1991) and 'hunting' (Washburn and Lancaster, 1968) – that do significantly better than the others as they can answer three out of the four questions asked of them (Table 1).

Table 1 The properties and the explanatory power of the various theories. We ask the following questions to evaluate the various alternative theories: (1) Honesty: Can the theory account for the honesty of early language, that is, is there a shared interest between the proposed communicating parties? (2) Groundedness: Can the concepts proposed by the theory grounded in reality? (3) Power of generalization: Can the theory account for the power of generalisation unique to human language? (4) Uniqueness: Can the theory account for the uniqueness of human language? As one can see most of the theories can answer only one or two questions, some none at all; only the tool making and hunting theories can answer three questions out of four. ('?': no information available. Mod.: modality; T: thought, V: vocalisation, G: gestures). (From Számadó and Szathmáry, 2006.)

	Mod.	First words	Topic	#1	#2	#3	#4
Language as a mental tool (Burling, 1993)	T	?	?	Yes	No	Yes	No
Grooming hypothesis (Dunbar, 1998)	V	?	?	Yes	No	No	No
Gossip (Power, 1998)	V	'Faithful', 'Philander'	Social life	No	No	Yes	No
Tool making (Greenfield, 1998)	?	?	?	Yes	Yes	Yes	No
Mating contract (Deacon, 1997)	?	?	Social contract	No	No	No	No
Sexual selection (Miller, 2001)	?	?	Anything	No	No	No	No
Status for information (Desalles, 1998)	?	?	Valuable information	No	No	Yes	No
Song hypothesis (Vaneechoutte and Skoyles, 1998)	V	?	?	No	No	No	No
Group bonding/ritual (Knight, 1998)	?/V	?	?	Yes	No	No	No
Motherese (Falk, 2004)	V	'Mama'	Contact call	Yes	Yes	No	No
Hunting theories (Hewes, 1973; Washburn and Lancaster, 1968)	G/V	Prey animals	Coordination of the hunt	Yes	Yes	Yes	No

Thus, it might be tempting to say that some combination of the two could provide a series of selective scenarios that would fit all of our criteria. The most notable conclusion, however, is that all the theories fail to explain the uniqueness of human language. Thus, even though indirect evidence strongly suggests that the evolution of human language was selection limited, it remains difficult to envisage a scenario that would show why.

Although the different scenarios suggest all kinds of selective forces, none of these scenarios has been consistently implemented in a family of models. Given the limitations on experimentation on humans and chimps, researchers should consider implementing the different scenarios in various model-based settings. Ultimately, researchers should be able to re-enact the emergence of language in artificial worlds, many of which will probably involve robots (e.g. available at: http://ecagents.istc.cnr.it/). The use of robots offers a unique and probably indispensable way of symbol grounding (basic words, via concepts, should be linked to physical reality (Steels, 2003)) and somatosensory feedback (actions, or results of actions, on behalf of the agent feed back into its own cognitive system via sensory channels (Nolfi and Floreano, 2002)).

Some Major Transitions in evolution (such as the origin of multicellular organisms or that of social animals) occurred a number of times, whereas others (the origin of the genetic code, or language) seem to have been unique events (Maynard Smith and Szathmáry, 1995). One must be cautious with the word 'unique', however. Due to a lack of the 'true' phylogeny of all extinct and extant organisms, one can give it only an operational definition (Szathmáry, 2003). If all the extant and fossil species, which possess traits due to a particular transition, share a last common ancestor after that transition, then the transition is said to be unique. Obviously, it is quite possible that there have been independent 'trials', as it were, but we do not have comparative or fossil evidence for them. What factors can lead to 'true' uniqueness of a transition? (A) The transition is variation-limited. This means that the set of requisite genetic alterations has a very low probability. 'Constraints' operate here in a broad sense. (B) The transition is selection-limited. This means that there is something special in the selective environment that can favour the fixation of otherwise not really rare variants. Abiotic and biotic factors can both contribute to this limitation. For example (Maynard Smith, 1998), a single mutation in the haemoglobin gene can confer on the coded protein a greater affinity for oxygen: yet such a mutation got fixed in some animals living at high altitudes only (such as the lama or the barred goose, the latter migrating over the Himalayas at an altitude of 9000 m).

There are interesting subcases for both types of limitation. For (A), one can always enquire about the time-scale. 'Not enough time' means that given a short evolutionary time horizon, the requisite variations have a very low probability indeed, but this could change with a widened horizon. An interesting subcase of (B) is 'preemption', meaning that the traits resulting from the transitions act via a selective overkill, and sweep through the biota so quickly that further evolutionary trials are competitively suppressed. The genetic code could be a case in point.

It is hard to assess at the moment why language is unique. Even the 'not enough time' case could apply, which would be amusing. But preemption, due to the subsequent cultural evolution that language has triggered, may render further trials very difficult indeed. Let us point out, however, yet another consideration that indicates that language could be variation-limited in a deeper sense. The habitat of the language amoeba is a large, appropriately connected neural network: most of the information processing within the network elaborates on information coming from other parts of the network. There is a special type of processing likely to be required: that of hierarchically embedded syntactic structures. It is far from obvious how this can be achieved in a network full of cycles. One must be able to show how a stack memory (last in, first out) can be neurobiologically implemented.

6 A Possible Modelling Approach

Motivated by the surveyed observation, the modeller also would like to get a handle on the language problem. Clearly, purely linguistic modelling or the application of unnatural neural algorithms is not enlightening for a biologist. Experimentation is fine, except that there are severe (and understandable) practical and ethical constraints on physiological and genetic experiments of primates, including humans. Hence *in vivo* experiments and field observations should be complemented by an *in silico* approach. Such an approach should ideally be based on the distillation of available biological knowledge, as presented above. The modelling framework must be flexible enough to accommodate the necessary genetic and neural details; with the complication that 'necessary' depends on the actual tasks and cannot always be set in advance. Such an approach cannot be based on an elegant but limited analytic model: rather, a flexible simulation platform is needed, which will be presented in this section (Szathmáry et al., 2007).

A crucial difficulty of such a research programme is that we do not know how far one can go with contemporary understanding of the nervous system. With the biochemistry of the early twentieth century one had zero chance even to conceptualise aptly the problem of origin of life, let alone to solve it. By the same token crucial elements of the understanding of the brain may be a serious obstacle in our understanding of language origins from a biological point of view. This objection is serious. My response to it is that unless we try, we shall never know whether we have sufficient basic knowledge of neurobiology to tackle the language problem. A complete understanding of all the details of the brain is unlikely to be necessary. Also, crucial components of the language faculty (e.g. symbolic reference) may be understood in neurobiological terms without an understanding of other components (e.g. syntax).

6.1   Evolutionary Neurogenetic Algorithm

We have developed a software framework called Evolutionary Neurogenetic Algorithm (ENGA), which offers researchers a fine control over biological detail in their simulations. Our original intent was to create software with much potential for variability. That is, we wanted a piece of software which is general enough to allow for a wide range of experimentation but appears as a coherent system and does not fall apart into a loosest of unrelated pieces of code. This required careful specification and design; especially in partitioning it into modules and the specification of interfaces in a programme that has grown to about 90,000 lines of C++ code. In such a short communication it is impossible to acknowledge all researchers of all important input fields to this paper. We have been especially influenced by evolutionary robotics, such as the work by Baldassarre et al. (2003), and by the evolutionary approach to neuronal networks with indirect encoding by Rolls and Stringer (2000). Our model is a recombinant of these approaches, with some key new elements, such as topographical network architecture.

The software is organised into packages that are built upon each other, i.e. there is a dependency hierarchy between them. This gives the architecture a layered nature so that lower modules do not know about the existence of higher modules. The most important packages and their dependencies are shown in Fig. 3.

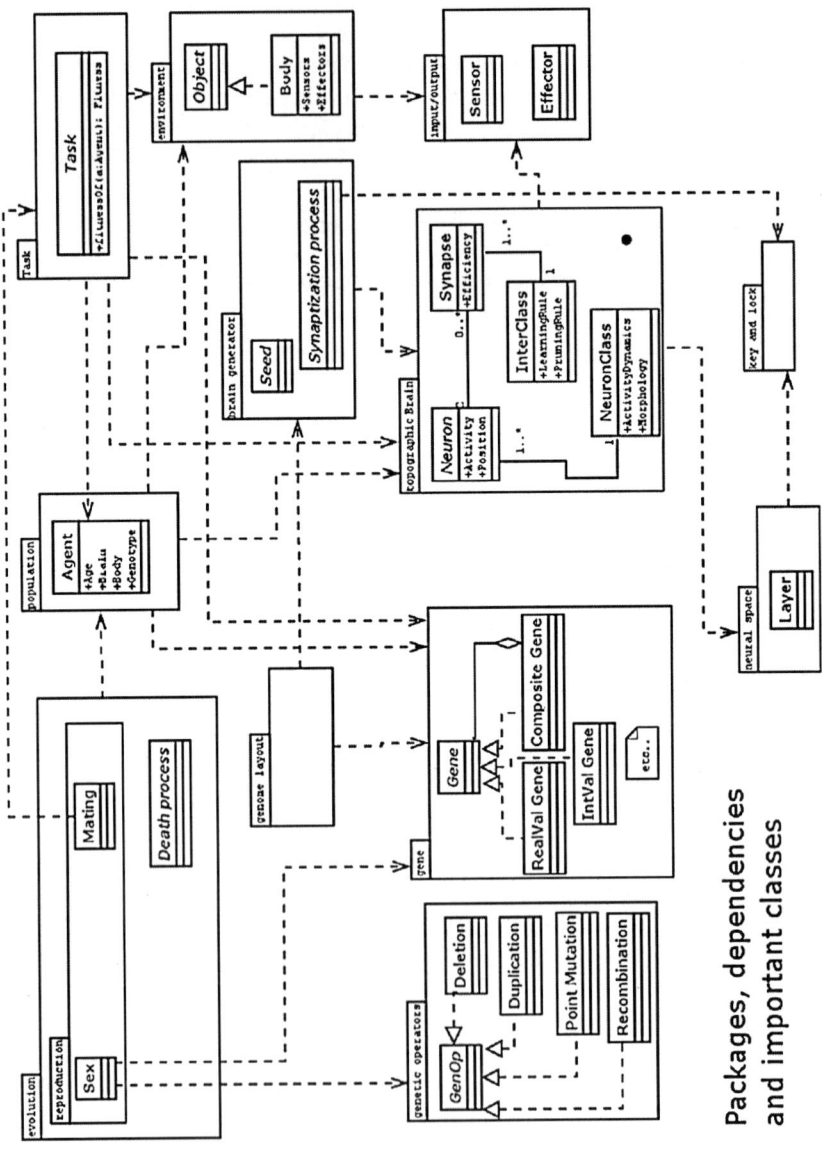

Fig. 3 Software design for the ENGA platform (Szathmáry et al., 2007). Modularity of the different components is apparent

Layered design allows easy modifiability of higher levels without the need to modify lower levels. Moreover, each layer exposes an interface that can be used by any client, even those deviating from the original purpose of simulating evolution of embodied communicating agents. The genetic module, for example, can be used in any evolutionary computation, not only those evolving artificial neural networks. We may as well talk about a multilevel software framework consisting of several modules that can be used individually or in combination with others to produce various kinds of evolutionary and neural computation related simulations. In the following sections individual packages are described in more detail.

The most important feature of the model is that it is deliberately *biomimetic:* within the constraints of computation, we intended to model real life, rather than artifical life. The most important element is *indirect genetic control* of the evolving agents: few genes specify a potentially very large neuronal network. This is very different from the merely engineering approach where each neuron and connection is affected by a dedicated gene. I refer for most details to Szathmáry et al. (2007). Here I highlight the 'developmental neurobiology' of the model to illustrate its potential.

6.2   Simulation of Brain Development

Development of the brain is controlled by the genotype in a highly indirect manner. By indirect we mean that no part of the genotype corresponds to individual neurons and synapses, and that only gross statistical and topographical properties of the brain are encoded in the genes. Individual brains are sampled from this statistical description. The neurogenesis phase goes on as follows. Neurons are situated in a layered topographic neural space mimicking real cortical layers but low-level biological mechanisms shaping the cortex such as concentration gradient dependence in neurogenesis are not present in the simulation. Instead, neuron classes define a probability density function over the neural space from which individual neuron soma positions are sampled. Neurons possess morphologies, i.e. there is some function over neural space describing their dendritic arborisation. This morphology is applied to each neuron relative to its sampled soma position. Synaptogenesis exploits two mechanisms, just as in biology: a long-range one (called projections) and a subsequent short-range one (lock-and-key mechanism). Each neuron class has an associated list of projections. Projections are probability density functions over neural space, used in the following way. They can be defined either in absolute coordinates or in coordinates relative to a neuron's soma. When a neuron's efferents are to be determined, putative synapse locations are sampled from its projections (determined by its neuron class). Neurons having dendritic arborisation near these putative locations become candidate efferents. Then the short-range mechanism selects from competing candidates at each synapse location. The short-range lock-and-key mechanisms mimic the receptor/ligand-based binding mechanisms present in

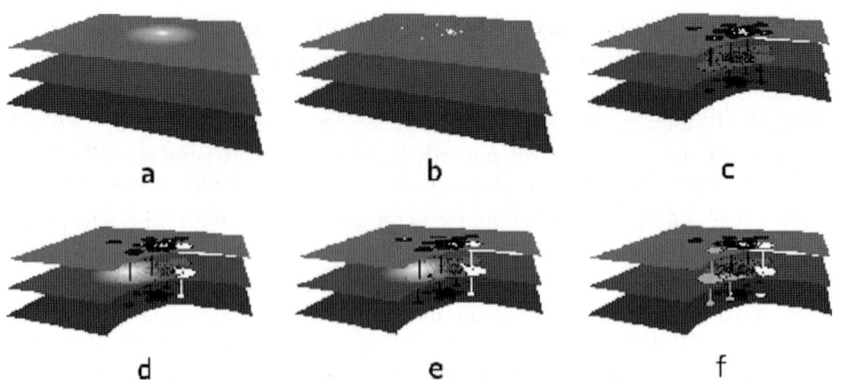

Fig. 4 Neurogenesis and long-range mechanism of synaptogenesis. (a) Probability distribution of positions. (b) Sampled soma positions. (c) Morphology added to soma. (d) Projection (onto layer 2) from a presynaptic cell, marked with white. (e) Putative synapse location (X mark) is sampled from projection. (f) Candidate postsynaptic cells are determined (Szathmáry et al., 2007)

real synaptogenesis. Locks and keys are strings of 30bits. Every candidate postsynaptic neuron's lock is matched to the presynaptic neuron's key. Binding probability is then a decreasing function of the Hamming distance between the key and the lock. The complete ontogenetic algorithm is depicted in Fig. 4.

Unlike the majority of approaches to evolve neural networks we maintain the full topographic information of our networks, i.e. neurons are situated in a layered neural space. Topographical information in a neural network can have a number of advantages. First, the interpretation of the structure can be easier. Second, developmental processes can model biological processes of neuron and synapse growth more accurately. Models, which acknowledge spatial information in biological systems yield various scale-free and small-world network attributes like the ones that are common in brain structure (Sporns et al., 2004).

6.3   Benchmars Tasks: Game Theory

We have used simple 2 × 2 matrix games with one-way pre-play communication as test tasks: a coordination and a task-allocation game. There are two reasons we have chosen these games. First, because of their simplicity these scenarios are easy to implement. Second, these games are well studied in the economics literature, and it is shown both theoretically and empirically that (one-way) cost-free, pre-play communication (so called 'cheap talk') can resolve coordination problems in both coordination and task-allocation games (Cooper et al., 1992; Hurkens and Schlag, 1999). Thus it is reasonable to expect that our agents will evolve the use of signals under these scenarios.

We have introduced two environments, which have an influence on the pay-off matrices. Accordingly we have three different scenarios:

1. Coordination game. Agents have to pick the same response, corresponding to the given environment to get high fitness. A pair of agents plays a coordination game in which they have to choose a given response conditional on the environment. (i) If both agents pick the correct response they both get a high reward; (ii) if only one of them picks the correct response then one gets a higher reward than the other; (iii) if both of them pick the wrong response then they both get low rewards.
2. Task-allocation game. Agents have to pick complementary responses to get high fitness. A pair of agents plays a task-allocation game in which they have to choose complementary responses to get high fitness. If they pick the same response they get low reward regardless of the environment.
3. Coordination or task-allocation depending on the environment. A pair of agents plays a coordination or task-allocation game depending on the state of the environment. In case of E(1) they play a task-allocation game, in case of E(−1) they play a coordination.

Each game has the same plan:

1. A random partner is picked to play with the focal agent.
2. One of the agents out of the pair is picked as the signaller the other is the listener.
3. 'Mother Nature' picks the state of the environment (E(1) or E(−1)) with equal probability.
4. The signaller can see the state of the environment and gives signal that can have three states (S(1), S(0), or S(−1)).
5. The listener receives the signal.
6. Both agents make a decision (D(1), D(0), or D(−1)) independently from each other (i.e. they do not know what decision the other agent makes). Whether the listener takes the signal into account when making its decision depends on its brain activity.
7. Fitness of the agents determined by the state of the environment, and by their decisions according to the type of game they play.

Agents have five input/output neurons and they start with 10–30 interneurons (chosen from an uniform distribution) in this task. The input/output neurons are as follows:

1. Vision: this is the input neuron for the state of the environment, provided that the agent plays the role of the signaller. The listener receives a zero input on this neuron.
2. AudioIn: this is the input for the signal. The listener gets the value of the AudioOut neuron of the signaller. The signaller gets a zero input on this neuron.
3. AudioOut: this is the output neuron for the signal. The signaller receives the value of this neuron on the AudioIn input channel.
4. Decision: this is the output neuron for the decision. The value of this neuron determines the fitness gain of the given agent according to the pay-off matrix of the given game (see Appendix).

The connections between the input/output neurons and the interneurons are determined only indirectly by the genes and they are subject to all the developmental rules mentioned above. The genes coding for the number of interneurons are allowed to mutate during the simulations; as a result the number of interneurons, and of course their number of connections to the input/output neurons can vary from generation to generation.

6.4   Outlook

The results of the games analysed (Szathmáry et al., 2007) suggest that the basic idea – the biomimetic approach – behind our neuro-genetic system is correct, and that the highlighted design principles of the system are sufficient at least in this simple task. The other goal of the test was to prove that the developed software is working properly and is useful for scientific experimentation. The core of our neuro-genetic system is the stochastic ontogenetic process and the indirect tree-based coding behind it. Evolution of agents in these scenarios found the right solution in a relatively short period, and the length of the search time was proportional to the complexity of a particular game. The topology and size of the networks both changed in the simulations. It is important to emphasise that we have not taken advantage of any grossly unnatural (but sometimes very effective) algorithms such as back-propagation.

Our results show that indirect encoding and a genetically controlled but stochastic ontogenetic process together can provide an appropriate framework for evolving communicating agents in simple scenarios. Our next goal is to introduce our neuronal networks to more and more complicated selective scenarios, by which we hope to obtain better understanding of the evolution of, among other things, symbolic communication.

Such ambitious research projects would have associated difficulties too numerous to mention here; however, the tasks that the agents (e.g. robots) would be subjected to imply a complicated fitness landscape that is similar to climbing a staircase rather than a hill: a good capacity for imitation would probably co-evolve with a capacity to learn symbols (words), which then opens up the possibility to climb to the first level ('stair') of syntax (Jackendoff, 1992). Successful modelling could answer the burning question of how genes, under the influence of natural selection, could rig a neuronal system so that it becomes able to handle linguistic input and output at the level of symbolic reference combined with complex syntax (Szathmáry et al., 2007).

The only process that appears to have solved the 'language problem' is evolution by natural selection. But there is no guarantee that just any kind of selection scenario, even if implemented in silico, would lead to the origin of such a faculty, partly owing to the results of the analysis presented here, partly because of what is known as the 'no free lunches theorem' (Wolpert and Macready, 1997), which states that the efficiency of an evolutionary search process is dependent on the problem. Putting constraints on the selective scenarios might constrain the search space to such an extent that simulated evolution will be able to re-enact the fascinating evolutionary transition of language evolution.

Acknowledgements I thank Mauro Santos, Luc Steels, Michael Corballis, Peter Gardenfors, and Chrisantha Fernando for useful discussions. Partial support of this work has generously been provided by the National Office for Research and Technology (NAP 2005/KCKHA005) and the ECAGENTS project (Available at: http://ecagents.istc.cnr.it/). The ECAGENTS project is funded by the Future and Emerging Technologies programme (IST-FET) of the European Community under EU R&D contract IST-2003-1940. The information provided is the sole responsibility of the authors and does not reflect the Community's opinion. The Community is not responsible for any use that may be made of data appearing in this publication.

References

Airey DC, Robbins AI, Enzinger KM, Wu F, Collins CE (2005) Variation in the cortical map of C57BL/6J and DBA/2J inbred mice predicts strain identity. BMC Neurosci 6:18

Baldassarre G, Nolfi S, Parisi D (2003) Evolving mobile robots able to display collective behavior. Artif Life 9:255–267

Bickerton D (1990) Language and Species. The University of Chicago Press, Chicago

Bufill E, Carbonell E, Are symbolic behaviour and neuroplasticity an example of gene-culture coevolution (in Spanish)? Rev Neurol 39:48–55

Bullock S (1998) A continuous evolutionary simulation model of the attainability of honest signalling equilibria. In: Adami C, Belew RK, Kitano H, Taylor CE (eds) Artificial Life VI. Proceedings of the sixth International Conference on Artificial Life, Los Angeles, 26–29 June. MIT Press/Bradford Books, Cambridge, Massachusetts, pp. 339–348

Burling R (1993) Primate calls, human language, and nonverbal communication. Current Anthropology 34:25–53

Caceres M, Lachuer J, Zapala MA, Redmond JC, Kudo L, Geschwind DH, Lockhart DJ, Preuss TM, Barlow C (2003) Elevated gene expression levels distinguish human from non-human primate brains. Proc Natl Acad Sci USA 100:13030–13035

Calvin WH, Bickerton D (2000) Lingua ex Machina: reconciling Darwin and Chomsky with the Human Brain. MIT Press, Cambridge, Massachusetts

Caplan D (2006) Why is Broca's area involved in syntax? Cortex 42:469–471

Changeux J-P (1983) L'Homme Neuronal. Librairie Arthème Fayard, Paris

Chomsky N (1986) Knowledge of Language: Its Nature, Origin and Use. Praeger, New York

Chomsky N (1995) The Minimalist Program. MIT Press, Camridge, MA

Christiansen M, Kirby S (2003) Language evolution: the hardest problem in science? In: Christiansen M, Kirby S (eds) Language Evolution. Oxford University Press, Oxford, pp. 1–15

Christiansen M, Kirby S (eds) (2003) Language Evolution. Oxford University Press, Oxford

Christiansen MH, Kirby S (2003) Language evolution: consensus and controversies. Trends Cogn Sci 7:300–307

Cooper R, DeJong DV, Forshyte R, Ross TW (1992) Communication in Coordination Games. Quart J Econ 107:739–771

Dalalakis JE (1999) Morphological representation in specific language impairment: evidence from Greek word formation. Folia Phoniatr. Logop. 51:20–35

Darwin CR (1871) The Descent of Man and Selection in Relation to Sex. John Murray, London

De Waal F (1982) Chimpanzee Politics. Harper & Row, New York

Deacon T (197) The Symbolic Species. The Coevolution of Language and the Brain. W.W. Norton, New York

DeFelipe J, Alonso-Nanclares L, Arellano JI (2002) Microstructure of the neocortex: comparative aspects. J Neurocytol 31:299–316

Dessalles J (1998) Altruism, status and the origin of relevance. In: Hurford J, Studdert-Kenedy M, Knight C (eds) Approaches to the Evolution of Language. Cambridge University Press, Cambridge, pp 130–147

DiAntonio A, Haghighi AP, Portman SL, Lee JD, Amaranto AM et al. (2001) Ubiquitination-dependent mechanisms regulate synaptic growth and function. Nature 412:449–452

Dunbar R (1998) Theory of mind and the evolution of language. In: Hurford J, Studdert-Kenedy M, Knight C (eds) Approaches to the Evolution of Language. Cambridge University Press, Cambridge, pp. 92–110

Elman JL, Bates E, Johnson MH, Karmiloff-Smith A, Parisi D, Plunkett K (1996) Rethinking Innateness. MIT Press, Cambridge, Massachusetts

Embick D, Marantz A, Miyashita Y, O'Neil W, Sakai KL (2000) A syntactic specialization for Broca's area. Proc Natl Acad Sci USA 97:6150–6154

Enard W, Przeworski M, Fisher SE, Lai CSL, Victor Wiebe V, Kitano T, Monaco AP, Paabo S (2002) Molecular evolution of FOXP2, a gene involved in speech and language. Nature 418:869–872

Everett D (2005) Cultural constraints on Pirahã grammar. Curr Anthropol 46:621–646

Falk D (2004) Prelinguistic evolution in early hominins: Whence motherese? Behavior Brain Sci 27:491–503

Fisher SE, Marcus GF (2006) The eloquent ape: genes, brains and the evolution of language. Nat Rev Genet 7:9–20

Fitch WT, Hauser MD (2004) Computational constraints on syntactic processing in nonhuman primates. Science 303:377–380

Flint J (1999) The genetic basis of cognition. Brain 122:2015–2031

Fortna A et al. (2004) Lineage-specific gene duplication and loss in human and great ape evolution. PloS Biol 2:e207

Gentner TQ et al. (2006) Recursive syntactic learning by songbirds. *Nature* 440:1204–1207

Gerdes LU, Gerdes C, Hansen PS, Klausen IC, Faergman O (1996) Are men carrying the apolipoprotein ε4- or ε2 allele less fertile than ε3ε3 genotypes? Hum Genet 98:239–242

Godfray HCJ (1991) Signalling of need by offspring to their parents. Nature 352:328–330

Goldberg TE, Weinberger DR (2004) Genes and the parsing of cognitive processes. Trends Cogn Sci 8:325–335

Gopnik M (1990) Feature-blind grammar and dysphasia. Nature 344:715

Gopnik M (1999) Familial language impairment: more English evidence. Folia Phoniatr Logop 51:5–19

Gould SJ, Lewontin RC (1979) The Spandrels of San Marco and the Panglossian paradigm: a critique of the adaptationist programme. Proc Roy Soc London B 205:581–598

Grafen A (1990) Biological signals as handicaps. J Theor Biol 144:217–246

Greenfield PM (1991) Language, tools and brain: the ontogeny and phylogeny of hierarchically organized sequential behaviour. Behavior. Brain Sci. 14:531–595

Haesler S, Wada K, Morrisey EE, Lints T, Jarvis ED, Scharff C (2004) FoxP2 expression in avian vocal learners and non-learners. J Neurosci 24:3164–3175

Harris D, Bullock S (2002) Enhancing game theory with coevolutionary simulation models of honest signalling. In: Fogel D (ed.) Congress on Evolutionary Computation, 12–17 May, IEEE Press, pp. 1594–1599

Hauser MD, Chomsky N, Fitch WT (2002) The faculty of language: what is it, who has it, and how did it evolve? Science 298:1569–1579

Hewes G (1973) Primate communication and the gestural origin of language. Curr Anthropol 14:5–25

Hill RS, Walsh CA (2005) Molecular insights into human brain evolution. Nature 437:64–67

Hinton GE, Nowlan SJ (1987) How learning can guide evolution. Comp Syst 1:495–502

Hockett CF (1960) The origin of speech. Sci Am 203:88–111

Hurd PL (1995) Communication in discrete action-response games. J Theor Biol 174:217–222

Hurkens S, Schlag KH (1999) Communication, Coordination, and Efficiency in Evolutionary One-Population Models. Universitat Pompeu Fabra Department of Economics, Working Paper No. 387
Jackendoff RS (1992) Languages of the Mind. MIT Press
Johnston MV (2001) Developmental disorders of activity dependent neuronal plasticity. Ind J Pediat 68:423–426
Kandel ER, Schwartz JH, Jessell TM (2000) Principles of Neural Science, 4th edn. McGraw-Hill, New York
Khaitovich P, Muetzel B, She X, Lachmann M, Hellmann I, Dietzsch J, Steigele S, Do HH, Weiss G, Enard W, Heissig F, Arendt T, Nieselt-Struwe K, Eichler EE, Paabo S (2004) Regional patterns of gene expression in human and chimpanzee brains. Genome Res 14:1462–1473
Knight C (1998) Ritual/speech coevolution: a solution to the problem of deception. In: Hurford, J, Studdert-Kenedy M, Knight C (eds) Approaches to the Evolution of Language. Cambridge University Press, Cambridge, pp. 68–91
Knight C, Studdert-Kennedy M, Hurford J (eds) (2000) The Evolutionary Emergence of Language: Social Function and the Origins of Linguistic Form. Cambridge University Press, Cambridge
Krubitzer L, Kaas J (2005) The evolution of the neocortex in mammals: how is phenotypic diversity generated? Curr Op Neurobiol 15:444–453
Lachmann M et al. (2001) Cost and constraints in animals and in human language. Proc Natl Acad Sci USA 28:13189–13194
Lai CSL, Gerelli D, Monaco AP, Fisher SE, Copp AJ (2003) FOXP2 expression during brain development coincides with adult sites of pathology in a severe speech and language disorder. Brain 126:2455–2462
Levi O, Jongen-Relo AL, Feldon J, Roses AD, Michaelson DM (2003) ApoE4 impairs hippocampal plasticity isoform-specifically and blocks the environmental stimulation of synaptogenesis and memory. Neurobiol Disease 13:273–282
Li W-H, Saunders MA (2005) The chimpanzee and us. Nature 437:50–51
Lieberman P (2002) On the nature and evolution of the neural bases of human language. Am J Phys Anthropol 35:36–62
Lieberman P (2007) The evolution of human speech. Its anatomical and neural bases. Curr Anthropol 48:39–66
Marcus GF, Fisher SE (2003) FOXP2 in focus: what can genes tell us about speech and language. Trends Cogn Sci 7:257–262
Marcus GF Rethinking eliminative connectionism. Cogn Psychol 37:243–282
Maynard Smith J (1991) Honest signalling: the Philip Sidney game. Anim Behav 42:1034–1035
Maynard Smith J, Harper D (2003) Animal Signals. Oxford University Press, Oxford
Maynard Smith J, Szathmáry E (1995) Major Transitions in Evolution, WH Freeman, New York
Maynard Smith J, Szathmáry E (1999) The Origins of Life, Oxford University Press, Oxford
Maynard Smith J (1998) Evolutionary Genetics. Oxford University Press, Oxford
Miller G (2001) The Mating Mind. Anchor Books, New York
Müller R-A, Rothermel RD, Behen ME, Muzik O, Chakraborty PK, Chugani HT (1999) Language organization in patients with early and late left-hemisphere lesion: a PET study. Neuropsychol 37:545–557
Musso M, Weiller C, Kiebel S, Müller SP, Bülau P, Rijntjes M (1999) Training-induced brain plasticity in aphasia. Brain 122:1781–1790
Neville HJ, Bavelier D (1998) Neural organization and plasticity of language. Curr Op Neurobiol 8:254–258
Noble J (2000) Cooperation, competition and the evolution of prelinguistic communication. In Knight C, Studdert-Kenedy M, Hurford J (eds) Evolutionary Emergence of Language. Cambridge University Press, Cambridge, pp. 40–61
Nobre AC, Plunkett K (1997) The neural system of language: structure and development. Curr Op Neurobiol 7:262–268
Nolfi S, Floreano D (2002) Synthesis of autonomous robots through evolution. Trends Cogn Sci 6:31–37

Oldham MC, Horvath S, Geschwind DH (2006) Conservation and evolution of gene coexpression networks in human an chimpanzee brains. Proc Natl Acad Sci USA 103:17973–17978

Paterson SJ, Brown JH, Gsödl MK, Johnson MH, Karmiloff-Smith A (1999) Cognitive modularity and genetic disorders. Science 286:2355–2357

Pica P, Lemer C, Izard V, Dehaene S (2004) Exact and approximate arithmetics in an Amazonian indigene group. Science 306:499–503

Pinker S (1994) The Language Instinct. Penguin Books, London

Pinker S, Bloom (1990) Natural language and natural selection. Behavior Brain Sci 13:707–786

Pinker S, Jackendoff R (2005) The faculty of language: what's special about it? Cognition 95:201–236

Pollard KS et al. (2006) An RNA gene expressed during cortical development evolved rapidly in humans. Nature 443:167–172

Power C (1998) Old wives' tales: the gossip hypothesis and the reliability of cheap signals. In: Hurford J, Studdert-Kenedy M, Knight, C (eds) Approaches to the Evolution of Language. Cambridge University Press, Cambridge, pp. 111–129

Premack D (2004) Is language the key to human intelligence? Science 303:318–320

Preuss TM (2000) Taking the measure of diversity: comparative alternatives to the model-animal paradigm in cortical neuroscience. Brain Behav Evol 55:287–299

Raber J, Wong D, Yu G, Buttini M, Mahley RW, Pitas RE, Mucke L (2000) Apolipoprotein E and cognitive performance. Nature 404:353–353

Rapoport SI (1990) How did the human brain evolve? A proposal based on new evidence from in vivo brain imaging during attention and ideation. Brain Res Bull 50:149–165

Rizzolatti G, Arbib MA (1998) Language within our grasp. Trends Cog Sci 21:188–194

Rolls ET, Stringer SM (2000) On the design of neural networks in the brain by genetic evolution. Prog Neurobiol 61:557–579

Rose Y, Royle P (1999) Uninflected structure in familial language impairment: evidence from French. Folia Phoniatr. Logop. 51:70–90

Sampson G (1997) Educating Eve: The 'Language Instinct'. Debate. Cassell, London

Scharff C, Haesler S (2005) An evolutionary perspective on FoxP2: strictly for the birds? Curr Op Neurobiol 15:694–703

Schmitt AM, Shi J, Wolf AM, Lu C-C, King LA, Zou Y (2006) Wnt-Ryk signalling mediates medial-lateral retinotectal topographic mapping. Nature 439:31–37

Senghas A, Kita S, Özyürek A (2004) Children creating properties of language: Evidence from an emerging sign language in Nicaragua. Science 305:1779–1782

Shu W, Cho JY, Jiang Y, Zhang M, Weisz D, Elder GA, Schmeidler J, De Gasperi R, Sosa MA, Rabidou D, Santucci AC, Perl D, Morrisey E, Buxbaum JD (2005) Altered ultrasonic vocalization in mice with a disruption in the Foxp2 gene. Proc Natl Acad Sci USA 102:9643–9648

SLI Consortium (2002) A genomewide scan identifies two novel loci involved in specific language impairment. Am J Hum Genet 70:384–398

Smith K (2003) The transmission of language: models of biological and cultural evolution. PhD thesis, The University of Edinburgh, Edinburgh

Sporns O, Chialvo DR, Kaiser M, Hilgetag CC (2004) Organization, development and function of complex brain networks. Trends Cogn Sci 8:418–425

Steels L (2003) Evolving grounded communication for robots. Trends Cogn Sci 7:308–312

Steels L (2004) Constructivist development of grounded construction grammars. In: Scott D, Daelemans W, Walker M. (eds) Proceedings Annual Meeting Association for Computational Linguistic Conference, Barcelona, pp. 9–19

Steels L, Beule JD (2006) Unify and merge in fluid construction grammar. In: Vogt P et al. (ed.) Symbol Grounding and Beyond: Proceedings of the Third International Workshop on the Emergence and Evolution of Linguistic Communication. Springer-Verlag, Berlin, pp. 197–223

Stout D, Chaminade T (2007) The evolutionary neuroscience of toolmaking. Neuropsychologia 45:1091–1100

Stromswold K (2001) The heritability of language: a review and metaanalysis of twin, adoption, and linkage studies. Language 77:647–723

Sur M, Learney CA (2001) Development and plasticity of cortical areas and networks. Nat Rev Neurosci 2:251–262

Számadó S, Szathmáry E (2006) Language evolution: competing selective scenarios. Trends Ecol Evol 21:555–561

Számadó Sz (1999) The validity of the handicap principle in discrete action-response games. J Theor Biol 198:593–602

Szathmáry E (2001) Origin of the human language faculty: the language amoeba hypothesis. In: Trabant J, Ward S (eds) New Essays on the Origin of Language. Mouton/de Gruyter, pp. 41–51

Szathmáry E (2003) Cultural processes: the latest major transition in evolution. In: L. Nadel (ed.) Encyclopedia of Cognitive Science. Nature Publishing Group, Macmillan. DOI: 10.1002/0470018860.s00716

Szathmáry E et al. (2007) In silico evolutionary developmental neurobiology and the origin of natural language. In: Lyon C, Nehaniv C, Cangelosi A (eds) Emergence Of Communication And Language. Springer, London, pp. 151–187

Teramitsu I, Kudo LC, London SE, Geschwind DH, White SA (2004) Parallel FoxP1 and FoxP2 expression in songbird and human brain predicts functional interaction. J Neurosci 24:3152–3163

Teter B, Xu P, Gilbert JR, Roses AD, Galasko D, Cole MD (2002) Defective neuronal sprouting by human apolipoprotein E4 is a gain-of-negative function. J Neurosci Res 68:331–336

The Chimpanzee Sequencing and Analysis Consortium (2005) Initial sequence of the chimpanzee genome and comparison with the human genome. Nature 437:69–87

Toma DT et al. (2002) Identification of genes involved in Drosophila melanogaster geotaxis, a complex behavioral trait. Nat Genet 31:349–353

Tomblin JB, Pandich J (1999) Lessons from children with specific language impairment. Trends Cog Sci 3:283–285

Van der Lely HJK, Rosen S, McClelland A (1998) Evidence for a grammar-specific deficit in children. Curr. Biol. 8 (1998) 1253–1125

Vaneechoutte M, Skoyles JR (1998) The memetic origin of language: modern humans as musical primates. J Memetics - Evolution Models Info Trans. Available at: http://www.cpm.mmu.ac.uk/jom-emit/1998/vol2/vaneechoutte mand skoyles jr.html

Vargha-Kadem F, Watkins KE, Price CJ, Ashburner J, Alcock KJ, Connelly A, Frackowiak RSJ, Friston KJ, Pembrey ME, Mishkin M, Gadian DG Passingham RE (1998) Neural basis of an inherited speech and language disorder. Proc Natl Acad Sci USA 95:12695–12700

Washburn SL, Lancaster C (1968) The evolution of hunting. In: Lee RB, DeVore I (eds) Man the Hunter. Aldine, New York, pp. 293–303

White SA, Fisher SE, Geschwind DH, Scharff C, Holy TE (2006) Singing mice, songbirds, and more: Models for FOXP2 function and dysfunction in human speech and language. J Neurosci 26:10376–10379

Wolpert DH, Macready WG (1997) No free lunch theorems for optimization, IEEE Trans Evolution Comput 1:67–82

Wyles JS, Kunkel JG, Wilson AC (1983) Birds, behaviour, and anatomical evolution. Proc Natl Acad Sci USA 80:4394–4397

Zahavi A (1975) Mate selection - A selection for handicap. J Theor Biol 53:205–214

Zhang J, Webb DM, Podlaha O (2002) Accelerated evolution and origins of human-specific features: FOXP2 as an example. Genetics 162:1825–1835

Chapter 14

The Codes of Language: Turtles All the Way Up?

Stephen J. Cowley

Abstract Linguistic signalling is compared with using artificial and organic codes. Based on Barbieri's (2003) work, I begin by showing parallels between organic processes and how language prompts conscious attitudes and micro-semantics. Hypothetically, organic coding may shape the neural and interactional dynamics that subtend language. Turning to development, I then compare the organic process of DNA transcription with Trevarthen and Aitken's (2001) intrinsic motive formation (IMF). This shows that the organic process model can throw light on the emergence of self. As in protein manufacture, embodied adaptors use the closure of a world to promote functional change. Rather as cells synthesise proteins, IMF prompts neural reorganization. By constraining how action and perception impact on neural activation, proto-artefacts (expressions, emotions, and attitudes) gradually insinuate themselves into how we act, feel, and speak. Human customs connect intrinsic motivation and displays of affect that, over time, prompt infants to *believe* in words. Parallels between organic coding and language dynamics are thus consistent with a distributed view of language. As artefacts and organic codes coevolved, our bodies became dependent on an ability to take the *language stance*.

1 The Language Stance

> It is to be sure an immensely powerful folk hunch that complex structures are made of little 'things' and that processes decompose into the bringing together of these little things. However, it is not science. (Ross and Spurrett, 2004, p. 643)

We habitually take a language stance. Provided that people speak the same tongue – whatever that means – they use the stance to repeat, analyse, discuss, and challenge 'what is said'. In repeating things and talking about talk, we draw on what Taylor (1997) terms *linguistic reflexivity*. Further, this canny trick not only characterises human signalling but is also pivotal in learning to talk. Babies rely heavily on what adults say (and think) about their utterances. The reflexivity of language is, however, just as important to adults. Without the trick, for example, my

---

University of Hertfordshire, UK and University of KwaZulu-Natal, South Africa

M. Barbieri (ed.), *The Codes of Life: The Rules of Macroevolution.*
© Springer 2008

nonce title would remain entirely opaque. With it, even readers who are deaf to sources and resonances can use imagination in guessing what may follow. Drawing on connotations, they may surmise that codes use biology to ground language, that turtle morphology has a role to play, or that Mother Nature is a myth. Each construal – and many, many besides – uses the language stance.

One metaphor that shapes linguistic reflexivity is that of *coding*. This applies to practices like reading, transcribing, paraphrasing and translating and, influentially, working with computers. In general terms, it merely hints that cultural artefacts help us come up with meanings. Our signals partition the world. The paper uses Barbieri's notion of organic coding to clarify how such partitioning arises. Building on work by Love (2004, 2007), Ross (2007), Kravchenko (2007), and others, I contrast sense-making based on language with the use of constructed codes (like Morse). Second, emphasising semantics, I apply the organic coding model to language. By so doing, I endorse Sebeok's (2001) claim that language extends the human sensorium. The organic coding model suggests that brains are remodelled by coordinating linguistic events. Exposure to speech motivates an individual to integrate cultural artefacts with social use of vocal and visible expression.

2   Coding

We discuss language in terms of coding and its cognates. While we sometimes invoke form-based processes, we often hint vaguely at intrinsic sense-making. To hunt down contrasts, therefore, I distinguish organic from constructed codes. Next I ask whether digits reliably associate signals with *abstracta* or, conversely, if signals compress semantic categories that later connect with digital forms. If digits are basic, we expect invertability; if categories come first, coding is unidirectional. This is a live issue. First, since as languages *can be* defined as sets of utterances, there is a resemblance to codes like Morse. Second, as in machine use of constructed codes, we speak and understand automatically. For such reasons, we can ask if we use conditioned action and perception or if, like physical symbol systems, we rely on constructed and invertable processes. Accordingly, I will consider the extent to which sense-making uses wordings alone and that to which it is integrated with expression. Further, if integration matters, we need to ask whether it exploits presuppositions or favours the compression of pre-existent categories. In addressing this, I begin by distinguishing constructed coding from organic processes.

A constructed code exploits input/output rules.[1] The design has formal simplicity. Where rules operate on internal conventions, expressions can be machine coded. Like a computer program, Morse features this kind of artificiality. The property, however, is not constitutive of codes. As Barbieri stresses, it applies to coding that uses *external* agents. While bodies, artefacts, and experience make programs and Morse-operators impressive, their input/output models limit complexity to processing.

---

[1] These resemble Barbieri's (2007) 'mental codes' because silent rehearsal is 'heard' in sentence fragments.

While rules are complex, semantics are simple. Token representations 'stand in' for external invariants that also 'stand in' for the same representations. Codes of this type conform to a *constructed process model*:

Adaptor → | Process | → Output
(e.g. Morse) | P→Q; X→Y; Z→A; etc. | (e.g. English)

Morse input consists of dots and dashes which, on decoding, are re-rendered in another script (e.g. SOS). All processes (symbolised by →) are causal. In practice, both monolingual English speakers and machines can render dots or dashes into English (or Chinese). Such constructed codes bear the hallmark of invertability:

Adaptor → | Process | → Output
(e.g. English) | Q→P; Y→X; A→Z; etc. | (e.g. Morse)

English can be re-rendered as sequences of dots and dashes (e.g. ...—...). While Barbieri shows that DNA, like other organic codes, is irreducible to such models, this has rarely worried linguists. Faced with human language, they often assume input/output signalling. Since we *say* that forms map on to meaning (and vice versa), many posit that brains draw on a synchronic linguistic *system*.[2]

Barbieri's organic coding defies input/output logic. Rather, he focuses on processes that bring about synthesis. In protein manufacture, this arises in linking the 'worlds' of genotype, ribotype, and phenotype. Since proteins add to information carried by genetic structure, the process bears the organic mark of context sensitivity. It is semantic. Thus, it resembles the primitive triangle that Davidson (1997) saw as the one possible basis for language skills. These, he thought, arose in creatures that 'react in concert to both features of the world and each other's reactions' (1997, p. 27). In filling out the sense of public events, these concerted reactions constitute instances of what Wheeler (2005) has labelled *nontrivial causal spread*. As in Davidson's creatures, dual recognition synthesises perceived features of the world with human reactions to permit the rise of complex semantics. This parallels protein manufacture. Using another kind of dual recognition, the internal constraints of DNA set off context-sensitive processes. In contrast to Morse, organic triangulation gives genes sensitivity to events beyond the cell: RNA fills out information. The figure below sketches an organic process model:

---

[2] For Saussure (1916), the point of view defines the linguistic object: the language stance identifies something *real*. Thus linguists focus on language systems and Steven Pinker believes that his text about octopus mating can "cause precise new combinations of ideas to arise in ... minds" (1994, p. 15). Students have pointed out that this assumes an ignorant person who reads rather literally (and with precision).

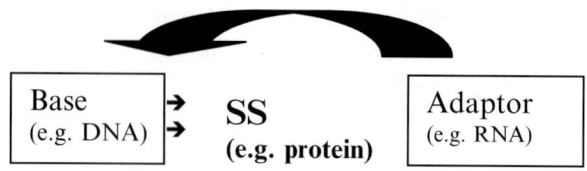

Base (e.g. DNA)	→ →	**SS** (e.g. protein)	Adaptor (e.g. RNA)

Double recognition bridges worlds (shown by boxes) and, using specifics, synthesises semantic artefacts. As Barbieri suggests, the cell draws on artefact-based *manufacture* where the regulated processes stand in for an external designer. While lacking invertability, messengers prompt adaptors to derive many outcomes from a base. Coding is unidirectional, open, and expresses more than is in organic memory. Manufacturing uses cell-internal (and, indeed, external) circumstances. Elsewhere, Barbieri (2007) shows that other codes use similar adaptors. Below, instead of focusing on adaptors or the molecular domain, I apply the organic process model to neural, fringe-conscious, and interactional aspects of language.

2.1   Language-Behaviour versus Morse Code

Subjects can *deliberately* use language, C++ or what we call *Albanian*. We express ourselves, write programs and use verbal patterns.[3] While allowing formal analysis, strings like *s'ka zot* invite semantic construal. To Albanian judgement the wording may be seen, among other things, as counterfactual. Indeed, in the Rilindja (1981) dictionary, *s'ka zot* exemplifies *zot;* in translation, *god* is illustrated by *there is no god*. Such patterns prompt both dissent and agreement. In ground-breaking work Ross (2007) argues that, uniquely, human signals prompt such judgements. The ability to identify counterfactuals is part of an evolutionary phase shift.[4] New forms of culture developed and, in historical time, changed cognition. Eschewing the language stance, Ross pursues Dennett's (1991a) view that human agency was transformed by the evolution of our signalling. In the final sections of the paper, I endorse this by arguing that language transforms us into fully human subjects. Eventually, using real-time events, we create the selves needed to adopt the language stance. Once this is done, when it suits us, we can choose how to act and speak.

Morse can spell out there is no god (or the contrary). Since, in this respect, it resembles speech or the phrase from the dictionary, some trace digital signalling to formal patterns. Instead of regarding digitality as compressing information, it is opposed to analogue signalling. The automatic nature of uttering *s'ka zot* (and other *Albanian* wordings) is ascribed to the formal properties of strings. There are, however, reasons to challenge the computational analogy. First, our judgements are prompted – not by strings – but *acts* of

---

[3] Generativists used physical symbol systems to model (inner) grammars. With the demise of computational models of mind, constructions are increasingly traced to cultural evolution (e.g. Deacon, 1997; Tomasello, 1999).

[4] In evolutionary game theory, a phase shift is a dimension increase in the adaptive complexity of selection dynamics. In evolution, a phase shift occurred with the coming of organisms that use photosynthesis and, in animals, with the rise of a central nervous systems or the emergence of vertebrates.

reading. Second, script processing is not a good model for listening which is, indisputably, more primitive. Third, as Barbieri's notes, while constructed codes rely on external agents, we hear and understand what we ourselves say. We face no equivalent to the symbol grounding problem (Harnad, 1991; Taddeo and Floridi, 2005). The reason, as MacDorman (2007) argues, is that neither humans nor brains are physical symbol systems. While modellers use ungrounded output that must be interpreted by agents, we grasp our thoughts. Language is already grounded. Far from reducing to formal strings, language depends on agents who integrate wordings with cognitive dynamics. We connect wordings, past experience and features of the world. Although we *say* meaning that depends on *the words that are actually spoken* (sic), we may be fooling ourselves. After all, this diminishes dialogue, bodies, circumstances, and people. Even lay views of language may be guilty of what Linell (2005) terms *written language bias*. These reduce language to *little things* that, it can be imagined, act as output from a naked brain. We abstract utterances from the dynamics of neural and expressive activity.

When identified with wordings, language is taken to link physical invariants to meanings. For some, the miracle depends on *minds* (e.g. Steinberg et al., 2001, p. 1). Allegedly, (inner) intentional states solve the symbol grounding problem by picking out real-world invariants. We rely on central processing to solve the *linking problem* (Tomasello, 2005). While usually attributed to a grammar, cognition, or intention-reading, this assumes a centralised process.[5] Language is input/output driven by a system that composes/decomposes structures. The brain is ascribed such a role in (most) functionalist and formalist theories of language. A neural system is taken to construct formal arrangements akin to those found in analysing dialogue into chains of utterances based on verbal and other rule-following units. Centralism is thus a cognitive counterpart of analysing speech into sets of decontextualised, standardised, and regularised *forms* (Lyons, 1977). These idealizations can be challenged. For example, Cowley and Love (2006) attack the assumption that language is separate from perception and action. While theorists avoid the grounding issue, they pay a high price. First, in separating language from bodies, physics, and biology, they reduce coordinated activity to monologue. Second, to justify appeal to central processing, they make unsupported assumptions like the following.

- Any piece of language is language only insofar as it instantiates part of a language system without which it could not be interpreted.
- A language system is a set of linguistic forms; these forms are systematically related to meanings.
- Linguistic communication takes place by virtue of form-meaning correlations that are guaranteed by shared information carried by the underlying language system.

This inauspicious beginning is Cartesian. If language systems are independent of physics, emotion, and behaviour, they cannot be present in animals. Indeed, to operate, they need a 'mental' realm. Once such a domain is posited, it will make

---

[5] Given the work of Tomasello (1999, 2003), belief that language depends on an intention-reading device has become popular. For critique, see Cowley, 2004 (also, in press b).

metaphorical space for a central system that chains verbal units together. Indeed, reduced to arrangements of little things, talking can be pictured as machine-like input/output. It will seem to follow that, like an operator of Morse, the central system manipulates sequences of forms.

Utterance →	Process	→ Meaning
(e.g. English)	P→Q; X→Y; Z→A; etc.	(e.g. English)

While we *say* that meanings depend on wordings, this may be fictional. In Section 2.2 I will show that this picture of language obscures its organic basis. Rather it presents language as an input/output process on the dubious grounds that we *say* that meanings depend on wordings (and vice versa).

2.2   Challenges to Constructed Process Models

In challenging form-based models of language, some emphasise human poetry and creativity (Vico, 1991; Croce, 1960). Many share Wittgenstein's (1958) view that language moulds the human world. If natural selection also has a part, its basis may be 'reacting in concert to features of the world and to each other's reactions' (Davidson, 1997, p. 27). Such claims parallel linguistic challenges to structural analyses. Thus, rejecting the *reality* of word-forms, Reddy (1979) denies that these *have* or *carry* meanings. The orthodox, he says, are fooled by the *conduit* metaphor. For Harris (1981) too, linguists depend on the *language myth*. Without argument, they ascribe reality to words, rules, and language systems. By reifying 'structures', linguistic communication becomes non-behavioural ('telementation'). In fact, sense-making integrates wordings with experience (Harris, 1998). Endorsing this, Cowley (2007 b) argues that language is distributed. Spread as it is across bodies, space, and time, language is analogue *and* digital (Love, 1998). Indeed, only this hybridity can unite culture, brains, and first-person experience. Given a flow of analogue expression, artefacts have become integral to a process that unites what we do feel and think. While based in coordinated *behaviour*, language is inseparable from meshwork of cultural and institutional resources. Given the importance of linguistic reflexivity – and analysis – we overlook its roots in coordinated expression. Oddly, learning to talk is pictured as acquiring (or constructing or learning) second-order cultural constructs (wordings and meanings). We overlook experience. Further, in spite of Cartesian models, brains need no more represent wordings than colours, elephants or bacteria. Language-behaviour may arise without connecting 'little things' to bursts of sound or identifying 'words' in the behavioural flow. Indeed, phonology may lack neurobiological reality (Docherty and Foulkes, 2000; Port and Leary, 2005). By challenging Saussure's (1916) point of view, we can reconceptualise language. Below, I seek the origins of dialogue in organic processes.

Working independently, Kravchenko (2007) builds on Maturana (1978; Maturana and Varela, 1987) to reach similar conclusions. Language is behaviour that connects subjects to an environment (or consensual domain). It serves in constructing a world where, using experience, we modify each other's perception and action. As structurally determined systems, signals show aspects of the world. Speaking and understanding depend – not on wordings – but physical and cultural experience. Real-time thinking depends on, above all, connotations (Maturana, 1978). In rejecting appeal to structures, Kravchenko emphasises linguistic dynamics. Playing down denotations, he stresses how linguistic flux prompts us to monitor and modify what we do. Below, I reject this epistemological approach because it fails to explain how personhood arises in creatures without digital semantics (viz. human babies). Given Kravchenko's focus on verbal patterns, he fails to ask how behavioural dynamics make us semiosis-ready. Having scrutinised interactional and neural dynamics, I claim that social events prompt babies to become human subjects. Indeed, infants use digital semantics long before *hearing* wordings or discovering (what we call) 'signs'. First, I trace organic coding to the utterance dynamics that index changing states on both sides of the skin. As self-reports confirm, language changes first-person experience. In this respect, it contrasts with constructed coding. Later, real-time dynamics prompt us to *hear* wordings.

Ross (2007) also challenges reduction of language to constructed codes. Taking an evolutionary view, he applauds Chomsky (1957) for tracing many computational powers to language. Rather than reify the objects of linguistic analysis, he stresses that human judgements depend on more than wordings. We construct meaning

spaces as we *actively* partition the world. In Wittgenstein's terms, we learn to see aspects. We can choose to *see* a shape as black or as a triangle. Indeed, we can even use its dimensions to *imagine* a red square.

Avoiding the assumption that digital semantics depend on wordings or shared meanings, Ross frames his argument in terms of evolutionary game theory. He asks how humans come to *choose* perspectives. While conceding that other primates identify objects and events, he doubts that the rise of digital semantics can be explained by natural selection. Accordingly, he uses Bickerton's (1990) work to suggest that cultural selection is also needed if we are to grasp logical relationships between hierarchially arranged objects and events. Formal representations could not, in themselves, be reduced to hard-wired synaptic dispositions. On its own, Ross thinks, natural selection could not give rise to grammatical categories. Indeed, no single, selective process could lead to the abandonment of a metaphysics. There is no reason to think that the primate world of objects and events was replaced by one based on hard-wired syntactic categories. Indeed, unlike the dots and dashes of Morse, wordings exploit complex semantics. Thinking is irreducible to operations on syntactically defined arrangements of *denotata*.

While generating infinite sets of sentences, Bickerton says nothing about how these map onto the world. Ignoring the symbol grounding problem, Ross objects that no selective advantage would accrue to an individual who grew a grammar. First, the organ would give no strategic benefit. Second, it would not prompt the identification of counterfactuals.[6] Indeed, this way of perceiving the world would fail to separate fantasies from facts. While fun, that is not a basis for selection. Given individualism, linguistic nativists overplay formal constraints while underestimating strategic signals. Rather than begin with the brain, we can ask how coordinated signalling – linguistic or metalinguistic – gives rise to compressed possibility spaces. How, in short, do human agents come to see aspects and choose perspectives.[7] To address this, we must turn to how nontrivial causal spread affects real-time decisions. Indeed, unless we explain how cultural selection favoured digital semantics, the evolutionary phase shift will defy explanations. Accordingly, Ross asks how human signalling turns infants into human subjects. Brains, he stresses can code or transduce information without any need for either little things or form-based processes. Digital semantics use interactional history and, as also Trevarthen suggests (see below), the relevant brain structures evolved as adaptations for coordinated signalling. To build bridges between words (in this case, agents), we need new ways of compressing information. Far from reducing to input/output, Ross thinks, human signalling transcends the duality of genotype and phenotype. Humans are *ecologically special*. Among primates, as Dennett (1991a) argues, we stand out because our strategic signals (and digital semantics) give us virtual *selves*.

In sketching an alternative to viewing language as a constructed code that maps forms onto meanings, I show that organic coding came to animate both subjects and their thoughts. Coaction, I will argue, gave rise to ways of life that use (strategic) forms of creativity. Language, on this view, is a distributed meshwork of resources that prompt, coordination, hearing, and conscious decision making. We depend on integrating skilled use of signalling with second-order cultural constructs (including *words, rules, and languages*). Given this hybridity, semiosis unites value-based connotations with wordings and cognitive dynamics. We make strategic moves in an ecology where brains and bodies connect emotionally and across practices. With this distributed view in place, I turn to another issue. Do these heterogeneous resources use organic codes? Do we fill out missing information by using public events in ways that enable us to function as conscious human beings?

3   From Wordings to Dynamic Language

The language stance has hypnotic power. It induces us to replace dynamics and history with systems that allegedly 'process' linguistic forms. Having challenged input/output modellings, I show how we depend on integrating wordings with more primitive language

---

[6] The model commits the mereological fallacy (Bennett and Hacker, 2003).

[7] Ross (2004), following tradition, opposes the verbal patterns of 'language' to the dynamics which he calls metalinguistic signalling. For Cowley and Love (2006), together, these constitute first-order language.

(attitudes and affect). Specifically, I suggest how integration uses our grasp of connotations. We thus make sense of what we hear and, strangely perhaps, use sensing for learning. Attention to strategic signals thus highlights how human signalling draws on dynamics. Later, I show that related processes influence how babies learn to talk. Our brains turn us into agents whose motivations connect experience with normative signals. We become persons who use what we *hear* to harness the digital power of language.

While the language stance deems expression non-cognitive, this jars with experience. While heeding wordings, we also use perceived dynamics. Multimodal expression is intrinsic to human agency (Thibault, 2004a,b). Alongside wordings, we exploit vocal, facial, and gestural expression. Cognitive dynamics allow us to *hear* voices and use the language stance to talk about individual differences, affect, mood, attitude, and circumstances (which elude constructed process models). Quite clearly, expression prompts us to anticipate vocal and other signals. While communicative, these also exert causal or cognitive effects. What we say is influenced by human bodies because, among other things, we respond to their dynamics. While wordings pick out objects and actions, interpersonal, and textual meaning draw on events. We integrate segments (speech sounds) with pitch, pace, cadence, rhythm, tempo, duration, loudness, etc. For Darwin (1871) language and song had the same roots. Far from being dependent on formal representations, we also use expressive dynamics. Given history, cultural patterns can facilitate sense making.

The temporal dimension of language is often neglected. First, Marková, Foppa, Linell and others pioneered emphasis on real-time dialogue (see, Marková and Foppa, 1990, 1991). Linking conversational analysis to phenomenology, they abandoned the monological approach to language. By focusing on the effects of dialogical patterns, they rejected centralised systems. Avoiding input/output models, Linell et al. (1988) showed that utterances are prospective and retrospective. Even turn taking may owe less to sequencing than cognitive dynamics, politeness, and a language stance. Second, we exploit asymmetries of status, power, knowledge, etc (Linell and Luckmann, 1991). In social life, strategic interaction is dominated – not by wordings – but presentations of self. Third, as Linell (2007) shows, dialogical principles also apply to the brain. Related views are increasingly corroborated. For example, in influential work, Pickering and Garrod (2004) show that semantic priming matters to conversation. We align utterances in meaningful ways. These, moreover, characterise individuals, groups, and cultures (see, Thibault, 2005). Further, artificial agents can use coordination to simulate the rise of grammar (Lyon et al., 2007; Cangelosi et al., 2006). Strikingly, robots that align behaviour can warp feature detection to simulate discovery of shared colour categories (Steels and Belpaeme, 2005).

Cognitive dynamics spread between people. In seminal work, Hutchins (1995a,b) shows that to land planes, or navigate ships, we depend on integrating wordings with artefacts, routines, knowledge, and myths. People act as *distributed cognitive systems*. Recently, Alač (2005) has extended this view by emphasising how important bodily coordination is to humans. In so doing, she documents how, without instruction, an assistant's body transforms her into an expert on fMRI displays. Although wordings have a part, she also depends on subtle coordination of words, postures, gestures, and video images. More generally, of course, language varies in space, between groups (e.g. Hudson, 1996) and across individuals (e.g.

Johnstone, 1996). Further, wordings and dynamics drive the ubiquitous affiliation and distancing that enacts accommodation (see, Giles et al., 1991). When human beings are co-present, we exploit psychobiological functions or 'emotion, mood, cognition, bodily orientation and muscular effort' (Goffman, 1983, p. 3).

In important work Erickson and Shultz (1982) showed how a counsellor's advice (i.e. what was said) correlated with shared rhythmic patterns. Worryingly, in 1970s America, coordination often echoed ethnic origins. For perceptual and social reasons, we treat 'communication' as the norm. In spite of this comfortable view (based on the language stance), human experience is influenced by both visible expression and what Abercrombie (1967) called *voice dynamics*. Indeed, the feel of social life may depend heavily on the flow of events in a micro-dimension. In Kravchenko's (2007) terms, cognitive dynamics enable us to *experience* semiosis. Strangely, we hear what people mean – not just words that are actually spoken. Documenting this, Goodwin (2002) shows how meaning arises as gestures are coordinated with vocal activity. Cowley (1994) shows that human relationships use, how voices are orchestrated. More recently, resulting duets have been traced to Haken's (1993) enslavement principle (Cowley, in press a). Finally, in bold synthesis, Thibault (2004a,b) provides theoretical elaboration of how bodily coordination shapes multi-scalar human interaction.

Dynamic aspects of language connect bodies across time and space. Language, therefore, is irreducible to centralist processing. Thus not only do bodies affect what we say but, generally, they do so in accordance with the claims of Love (2007), Kravchenko (2007), and Ross (2007). Dynamics evoke connotations as lived events are coordinated around wordings, biomechanics, and circumstances. Second, since signalling warps even robot 'perception', social strategies may well help us partition the world. Something like Davidson's primitive triangle may turn out to be the basis for digital semantics. Were this the case, human perspective taking would change as other peoples' reactions compressed information based on direct experience. We might differ from wild primates because of how cultural ecology enabled us to modify what we perceive, identify, and value.

4   External Adaptors in Language?

Can expression bridge between cultural processes and a neural world? Since real-time processes elude today's technology, I begin with how expressive dynamics affect behaviour. To separate neurally mediated evidence from reports, I trace nontrivial causal spread to interactional and, later, neural time-scales. Using examples, I show that sense making uses patterns to stabilise dynamics around semantic values. Then, turning to human symbol grounding, I argue that infants act in parallel ways. In this case, however, they use what Trevarthen (1998) calls IMF. Processes within the brain change infant motivations by attuning their activity to rewards that are likely in their social environment.

Kravchenko (2007) sketches a useful view of micro-scale language activity. In his biological theory, construals use wordings, together with reciprocal causation and self-organization ('intentionality'). Continuously changing dynamics mesh each person's initiatives with the other's responding. Cognition fuses with communication as events gain an experiential feel. In real-time, an actor's orientation affects the observer who, using experience-based sensitivity, may *modify* (or inhibit) expression. Interaction thus cascades as perceived orientations (and non-orientations) create a domain that is animated by vocal and visible patterns. As this environment changes, we *hear* how signs are meant. Thus, fused activity promotes coordination between embrained bodies. Orientations and reorientations mesh with both what we will come to call *words* and attitudes that echo past events. Over a history of interactions, a shared history accumulates. Using wordings and expression, connotations shape first-person phenomenology.

To apply the organic process model to language, I begin with a classic study by John Gumperz (1982; see also, Eerdmans et al., 2003). In this work he reports on success in improving customer response to South Asian staff. Initially, customers were often negatively affected by simple offers of items like 'gravy'. While out-group members heard the utterances as *unfriendly*, South Asians regarded them as *neutral*. Miscommunication, Gumperz suggested, drew on *contextualization cues* (CCs). Thus, a falling cadence on gravy evoked 'presuppositions' that differed between groups. In short, different representational content was triggered by a single physical event. While used to change behaviour, the theory was not successful (Eerdmans et al., 2003).[8] Offering an explanation Cowley (2006) argues that we

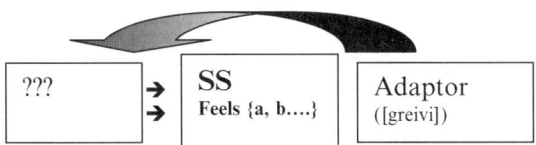

report – not hearings of objective patterns – but our own *felt responses*. These are part of the feeling of what happens (Damasio, 1999) or core consciousness. To report talk as *neutral* or *unfriendly* is to use connotations to *say* how dynamics are for us.[9] Like an adaptor, *gravy* manufactures a felt relation (semantic synthesis).

In Barbieri's terms, we hear a prosodic artefact. While nothing is known about the neural processes, this links the behavioural observations to the practical work. Staff used the CC label in a trick that works to their advantage. Having been given a reason to offer *gravy* (and other items) on a rising cadence, they changed their behaviour. Holding the idea of CC in mind, they spoke offers on a rise which, to English ears, sounded *friendly*. Using the language stance (or higher level analysis), they altered how they spoke. They changed the dynamics (Maturana's (1978) 'intentionality') that manufacture semantic syntheses or felt responses. Contextualization does not,

---

[8] In Eerdemans et al. (2003) collection, it is rejected by Prevignano, Levinson, Thibault (see Cowley, 2006). While many views are possible, contextualization lacks any specifiable physical parameters.

[9] This argument is influenced by Dennett's (1991a) discussion of the phi phenomenon (see Cowley and Love, 2006). In reporting that we see a green light turn red, he argues, we report a judgement.

therefore, depend on cues. Rather it is strategic coordination and, quite clearly, voice dynamics have *cognitive* powers.

Even if we report felt responses, this throws no light on neural processes. To consider whether these use organic coding, I use how Rączaszek et al. (1999) apply dynamical systems theory to meaning construal. In experimental work, they test the hypothesis that, in construing utterances, decision making tends to stabilise on single judgements. Thus, in one study, prosodic variables (the pause and vowel duration of rhythmic feet) are manipulated to test judgements of an ambiguous sentence. Specifically, the focus falls on construals of the wording:

(1) Pat or Kate and Bob will come.

Subjects were presented with utterances whose foot length varied such that, at one extreme they would be heard in meaning 1 and, at the other in meaning 2. In one experimental condition sentences were presented in either ascending/descending fashion and, in the other, in random order. Using a reaction-time technique, they assigned one of two senses to what they hear (Either *[Pat or Kate] and Bob will come OR Pat or [Kate and Bob] will come.*) Overall, subjects tended to show hysteresis. All things being equal, construals of (phonetically) identical utterances will be influenced by recent events. Having just synthesised a percept, we are more likely to hear the same percept again. The figure shows conditional probabilities of assigning the same construals in a subject who shows this pattern (Fig. 1).

Construal is affected by history as well as the physical stimulus. It draws on wordings, the physical event and what are termed the *dynamics of meaning*. Judgements fall into attractor basins depending on, among other things, a previous event. In further work, reaction times to the same stimuli were found to be longer

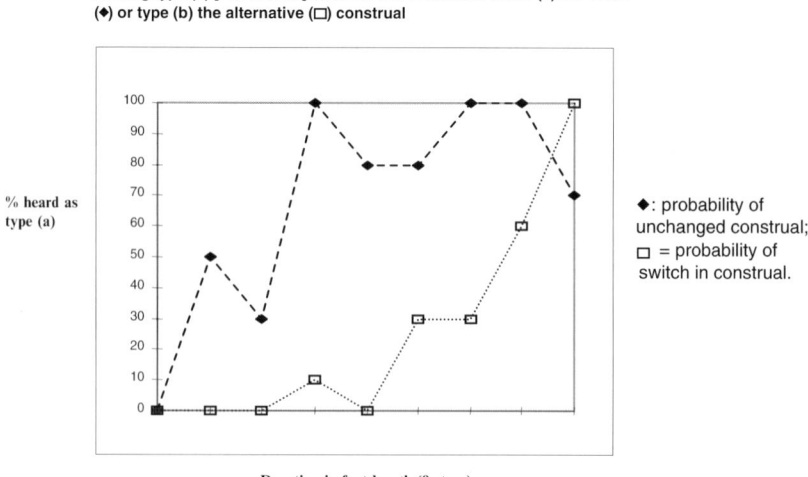

Results for subject Q:
Hearing type (a) *[Pat or Kate] and Bob will come* after either (a) the same
(♦) or type (b) the alternative (□) construal

Fig. 1 Hysteresis in meaning construal

when participants were about to switch their construals than in cases where this did not occur. Even in experimental conditions, percepts integrate verbal constraints, echoes in working memory and voice dynamics. Brains, it seems, seek out stability. Sentence construal integrates speech timing with what has been heard. If this happens in experimental conditions, internal conventions are likely to be even less powerful *in the wild*. Further, effects vary between people. While subjects show internal consistency, there are huge individual differences. Finally, Rączaszek et al. (1999) show that, when we use visual context, construal settles in a 350–400 ms time-window. Strikingly, this is long enough to be *conscious* of what is seen.

As for the customers at the canteen, utterances set off events that stabilise on a pattern. In settling on a sense, participants exploit dual recognition. While hearing wordings, orientation (or reorientation), they also use prosody. As Rączaszek et al. (1999) emphasise, symbols constrain dynamical events (c.f. Pattee, 2000). Using dual recognition, dynamics are integrated with a heard pattern. Organic coding thus offers a model for how we use connotations. Neural 'openness' uses recent experience and timing to integrate recognition that manufactures cognitive–communicative fusion. While akin to Cowley's (2006) felt response, in the work of Rączaszek et al. (1999), the parallels with organic coding apply to neural functioning. The percepts that are identified become manufactured syntheses that arise as artefacts (utterances) prompt brains to fill out incomplete information.

Wordings are construed together with accompanying prosodic and other information. Far from using constructed codes, human sense making integrates wordings with what we hear, expectations, physical events and, indeed, contents of working memory. As described by the organic process model, established routines fill out incomplete information in real-time. Thus in both construing interactional attitudes and reacting to sentence meaning, we find a parallel with protein synthesis. Next, therefore, I extend the comparison to the developmental time-scale. I ask how coaction transforms neural systems which control infant motive formation. The focus falls, first, on how caregiver expression influences baby's percepts. Second, I ask how this change contributes to cultural learning.

5   Human Symbol Grounding

Trevarthen (1997, 1998; Trevarthen and Aitken, 2001) provides behavioural, functional, and neural evidence that expressive dynamics give infants environment expectant powers. Denying that attunement-based dynamics are sufficient to explain early infant development (Kaye, 1982; Rochat, 2002; Fogel et al., 2006),

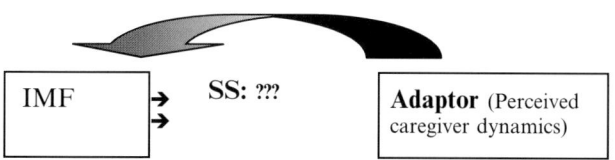

he posits a motivated process of neural development. What he terms IMF integrates caregiver expression with an infant's changing evaluative standards. Before using the organic code model to compare IMF to a coding base, I show how semantic syntheses arise from caregiver dynamics:

In empirical work, Cowley et al. (2004) argue that, at 14 weeks, babies exploit *signs of culture*. To extend this claim, I focus on an incident described in that work (also, Cowley, 2006). Specifically, a 14-week-old Zulu girl interacts as described:

> Initially, the baby shows signs of incipient distress. Like her mother, we do not know why she is (mildly) upset. Interestingly, however, frame-by-frame analysis shows that this occurs shortly after she has failed to make contact with her mother's hand. Following mutual gaze, the baby emulates her movement and, failing to achieve touch, breaks gaze and begins to cry. About 1 second later, the mother shows full direct response. Using complicated African hand movements that a Westerner finds difficult to copy, she gets the child to return gaze. As the mother produces complex waving movements, without return- ing gaze, she says or sings *nge ka* (no!) with sound spreading over the syllables. As she does so the child relaxes – she is not afraid – and the mother picks up on the change by lowering her hands and smiling. This part of the event takes about 3 seconds.

While not even a trained observer can explain *why* she falls silent, the baby acts on cue. Using Zulu expression, she acts in a socially coherent way at the *right* time. She meets her mother's wish that she *thula* (or fall silent). Since other dyads feature similar coordination, Cowley et al. (2004) regard this as part of local *culture*. Further, given the smile, the baby may experience her reward. For Trevarthen (1979, 1998), this *intersubjective behaviour* uses IMF. Subcortical structures change interactions in ways that shape a relationship. Given IMF, dyadic events enable the baby to learn what a caregiver is likely to want (and hope for). Even now the baby sensitises to (some) cultural norms. Since these are historically specific, they fill out more information than internal constraints (imitation or intention read- ing) could express. Unless it is innate (which is *very* unlikely), this complementary behaviour must draw on a shared history. Infant experience gives rise to behaviour timed to fit ongoing displays. Controlled as this is by the brain (not the whole baby), the dyad uses nontrivial causal spread.

Mother and infant engage in circular interaction. Each reorients in increasingly sophisticated activity that Cowley (2007) calls mutual *gearing*. Without (inner) understanding, their interpersonal dance serves to calibrate and recalibrate what each is likely to offer (including obedience and its rewards). As with adult construals, physical patterns index cultural categories. Indeed, it is especially striking that the 14-week-old orients to *Zulu* gestures. (On viewing the video, other South Africans have commented that *their* babies would cry more.) *Metaphorically,* she reads intentions or, simply, does what the caregiver wants. Further, her inhibition is both coherent and, as shown by a beaming smile, rewarded.[10] Infant prospects are shaped

---

[10] It is hard for adults to modify infant behaviour. Leaving aside what may be universal responses to picking up, feeding, rocking, etc., modified reorientations probably begin after about the 6th week. Once the behaviour becomes established, it is both anticipatory and culturally specific (Cowley et al., 2004).

by interactional history. By using contingencies to *relate* to patterns, the baby exploits a history of interactions to evaluate circumstances. Given this kind of active perception, her mother's dynamics ground changing routines. Without intrinsic goals (or intentional states) her repertoire grows by coming to anticipate rewards. Coregulated events thus develop by using what Cowley and MacDorman (2006) call *epistemic actions*. Babies who still lack any capacity to set goals are nonetheless prompted to exploit cultural norms. This is possible because IMF triggers motivations based on experience. Already, the Zulu girl senses something of what *her mother wants*. In organic coding terms, semantic synthesis shapes social behaviour reveals the baby's rudimentary awareness.

The baby uses indexical representations. Although unaware of *why* the caregiver acts, value laden movements enable her to discover normative patterns. She develops evaluative standards (or representations) that, as in other mammals, allow what Sterelny (2003) calls *robust tracking*. Kravchenko (2007) calls these *first-order descriptions* because their value pertains to the dyad – not the baby. In this respect, the infants resemble the airport workers. As in pre-training utterances of *gravy*, they use a *feeling of what happens* (Damasio, 2003). However, while trained workers bring the vocalization (or epistemic action) under *individual* control, the baby can rely on the caregiver's consistent use of semantic markers. Like the rewarding smile, these enact cultural display. While like the airport workers in using real-time events that represent cultural patterns, the baby is not taught to inhibit. Rather, given self-organization, action is reshaped by norms that, in time, do become relevant for the (whole) baby. Given intrinsic motivation, the baby draws on what Di Paolo (2005) calls *adaptivity*. As motivation develops, contingency recognition uses micro-responses that are taken to show anticipation. Cascading changes occur as interaction gives the baby preferences. The effect is redoubled by how infant promptings make caregivers highlight what conforms and contrasts with expectations. The baby's motivations thus draw on cultural motivations. Consistent reorientations make infant expression more subtle and, as a result, lead to marked changes in caregiver actions, feelings, and beliefs. The infant scaffolds the caregiver as both parties attribute similar values to interactional moves. Nontrivial causal spread enables the baby to develop motivations based on experience of the caregiver's doings.

Much depends on conflict and cooperation. As babies are not *automata*, caregivers can gauge distress or hunger. They gear to babies by ensuring that displays of 'when baby should thula' match the setting. In this way, infants use Zulu semiotics ('thula now') in learning about circumstances. As reactions compress features of the world, their increasing sensitivity will bring rewards. This will be integrated with the synaptic pruning and autoptosis used as neural systems discriminate *kinds* of events. Expressive artefacts thus manufacture semantic synthesis using processes that resemble those found in the adults who report attitudes and identify meanings. In identifying percepts, infant communication drives cortical differentiation. Natural artefacts trigger activity that adults can understand. Infant strategic behaviour attunes to adult expectations as changing motivations shape the baby's sense of persons. Events connect IMF with mutual gearing to invoke rewards. The result is felt response where intrinsic systems fill out incomplete information. Sensibility to contingencies enables babies to incorporate ways of picking up on what *cultural* circumstances offer.

5.1   Below the Skin

Caregiver dynamics add dimensions to infant experience. Gradually, other orientation enables the child to discover what it is to be an actor (Cowley, 2007 b). Strikingly, as IMF changes human agency – what Ross sees as the mark of our ecology – we fill in the blind spot in Kravchenko's model. Far from being semiosis-ready, *hearing* what is said depends on a history where expression links with culturally derived semantic patterns. Sophisticated brain-based response uses categories that, while only sensed, shape motivated behaviour (there is no inner process of understanding). Thus, in line with the arguments adduced by Ross (2007), the brain structures relevant to digital semantics arose as adaptations for coordinated signalling.[11] This indeed may be why activation patterns associated with IMF become a biological underpinning of language. Next, therefore, I ask how cultural artefacts, contingencies and affect allow a baby to scaffold how caregivers enact wants and desires.

The organic process model shows that attitudes, meaning-selection, and intersubjective behaviour conform to a pattern. In each case, the brain uses human expression to set off a dynamic process that manufactures a semantic synthesis (or felt response). Do the Zulu baby's syntheses drive motive formation? Might IMF be a coding *base* whose neurobiology constitutes an organic *world?*

A base is both foundational and able to trigger open, unidirectional change. Using nontrivial causal spread, we express more than is in organic memory. Since IMF develops, it contrasts with DNA. That aside, semantic syntheses provoke the 'sensitive two way mirroring of the emotional values of expressions' (Trevathen and Aitken, 2001, p. 8). As IMF develops, infant motivations engender specific expression. Conversely, changing adult behaviour alters infant expression. Of course, babies rely less on imitation than how caregivers introduce 'new or arbitrary social habits or conventions' (2001, p. 8). The results are not invertable. Functionally, higher-level systems override basic motivation (see, Trevarthen, 1998). Low-level mimesis (m) exploits caregiver displays to ensure that, gradually, a higher motivation (M) system develops. This contributes to a relationship based on interactions that

---

[11] Ross suggests (personal communication) that signal coordination uses common brain structures as 'a crucial wedge' in the 'compression of possibility spaces'. Thus, 'almost all signalling between, e.g. dolphins and humans is either analogue or requires pointing'. In this sense, he does 'begin with the brain' – even if the relevant structures evolved as adaptations for signalling coordination. Whereas Trevarthen focuses on individual brains and *motive* formation, Ross thinks in population terms. Thus, he stresses how (unspecified) brain-internal changes arose with digital signals as cultural selection began to leave 'the genetic/raw neural basis far behind'. I see these as complementary (not competing) accounts.

produce 'efficient anticipatory cognitive systems' (Trevathen and Aitken, 2001, p. 21). In the example, the M-system produces inhibition by overriding the lower level m-system. Caregiver vocalization and hand-waving prompt events based in value-marked experience. The M-system exploits a history of semantic syntheses. Not only does self-organization present just such layering, but, as Davidson (1997) suggests, it uses a triangular process. In the example, the baby links other persons' reactions (expression) to features of the world (values). Events outside the body manufacture an M-system as organic memory and normative displays become integral to the baby's movements.

IMF uses events beyond the skin to set off behaviour. In real-time, infants prompt what caregivers think and do. A developmental trajectory emerges as nontrivial causal spread organises the IMF. Trevarthen, of course, focuses on neural reorganization that uses 'core regulatory functions that emerge in the embryo period' (Trevathen and Aitken, 2001, p. 25). IMF is traced to 'additions at the end of the human brain and around the parietotemporal junction' (p. 25). In embryo, the relevant neurons perform morphogenic functions. After birth, they are co-opted to regulating what happens beyond the skin. The infant's m-system sets off, among other things, mimetic moves. These, like much the baby does, spark value laden reorientation. So, for Trevarthen, imitation is part of a 'motivation used for purposeful negotiation of the social domain' (p. 8). It is a marker whose results depend mainly on cultural wants and desires. Thus the baby's microdynamics elicit, anticipate, and frame value-laden response. Infant distress, for example, leads to norm-based reactions that strengthen the baby's action tendencies. This is consistent with Bateson's (1979) work on interactions with 4-month-olds in Iran. These *protoconversations* had a cultural tone because, as Ross suggests, coordination draws on culturally selected signals. Like an organic base, IMF uses these artefacts to make the baby 'environment expectant' (p. 25).

The semantic syntheses that result from IMF give rise to culture-sensitive motivations. Unidirectional change makes infant actions more sophisticated while altering caregiver behaviour. As the baby acts, the caregiver gears to her manifest expectations. Regardless of whether Trevarthen is right that the child has rudimentary awareness, this contrasts with constructed coding. First, no external invariants are required: experience enables the baby to track events where action (or inhibition) will precede reward. Indeed, far from finding a special form of centralised process, humans show marked sensitivity to cultural displays. As in other domains, understanding is integrated with real-time affect and action. Trevarthen is therefore correct to describe the rise of expectations as *motive formation*. In coherent action, the baby uses equivalents to a cell's external functors and multiple, transduced messages. Far from needing inner intentions, motive formation links evaluations to dynamics. Small babies use gearing based on how two brains connect perception, felt response, and action tendencies. As posited on the organic process model, the process uses two recognition functions. Physical coupling (or attunement) is integrated with (asymmetrical) evaluation of likely situation-relevant rewards.

Barbieri presents an organic coding base as a *world*. While lacking the closure of genes, IMF is biologically channelled. In spite of its scaled-up complexity, motive formation is constrained by – not just perceived values – but also an evolutionary history. The basis, it seems, lies in phylogenetically ancient interneurons in the brain stem (Panskepp, 1998; cited Trevathen and Aitken, 2001, p. 21). While humans use these to bring expression under higher control, homologous cells regulate the oral and respiratory tracts of other mammals too. Trevarthen thus traces the phylogenesis of the M-system to a history of nontrivial causal spread based on signalling that elicits care (c.f. Falk, 2005). This, of course, is consistent with using vigorous expression to calm the child. Caring behaviour sets off somatic cues used by the visceral system which links them to processes that regulate the baby's states.

Given its epigenetic and evolutionary history, IMF is open to cross-specific comparison. In primates, humans alone bring vocalizations under higher-level control. Otherwise, parallels dominate. As in chimps and bonobos, motive formation sensitises to interindividual and social norms. Content aside, IMF functions as in other mammals by using somatic feedback to define 'perceptual affordances prospectively' (Lee, 1993; cited, Trevathen and Aitken, 2001, p. 17). Human babies have special powers to discover opportunities associated with cultural expression. Trophotropic networks, which maintain the organism's functions, integrate activity with anticipatory, experience-seeking actions. These draw on ergotropic systems that, as in Christensen and Hooker's (2002) work, use memories to identify (social) affordances. Self-directed anticipatory learning exploits attention to regular polarity, symmetry' and spatial patterning of bodily form while, more specifically, babies display interest in the rhythmic dynamics of joint action. Given an IMF world, coaction gives bodies an experience of musicality (Trevarthen and Aitken, 2001).

While functionally closed, social experience drives motive formation. Gradually, the baby's amiable and tuneable muscles gain control over the uptake and censoring of (potential) stimuli for eyes, ears, lips, tongue (and other systems). Nontrivial causal spread gradually recruits all the sensory-accessory motor systems into the M-system. As a result human expression comes under the control of a single (distributed) neural system. Infants thus develop fine control over vocal and expressive gestures. Long before using 'executive' power (or the prefrontal cortex), the baby exploits value-laden caregiver response. Infant expression thus combines mimetic and instinctive expression with aspects of normative display. While scaled up from cellular processes, this resembles artefact-based manufacture. Cultures, it seems, reorganise infant brains by stabilizing norms. As Goffman puts it, 'Social structures don't "determine" culturally standard displays' but allow us to 'select them from the available repertoire' (1983, p. 11). To this broad view, Trevarthen adds neurobiological details about how value-laden expression shapes human agency. Growing motivations prompt co-management of pleasure, ease, tension, wariness, relief, and so on. Routines draw on culturally selected ways of honing an infant's expectations. Given IMF, circumstances become central to our lives. Contingencies make 'hard-wired instinctive motivating inputs from subcortical emotional systems' socially sensitive. Emotional displays become 'the essential regulatory feature in human life' (Trevathen and Aitken, 2001, p. 24). As IMF develops subcortical structures regulate the functions of a growing neocortex. It is this which guarantees increasing functional specificity. The power of such change is especially clear in the finding that, for example, blind babies learn to coordinate their hands to vocal rhythm (Tønsberg and Hauge, 1996).

Using external activity, the baby's brain meshes an increasingly differentiated social world with variable expression. Increasing functional specificity heightens sensitivity to circumstances. Interactional events become more regular as newborns 'integrate their state regulation with maternal care' (Trevathen and Aitken, 2001, p. 21). Without understanding, they scaffold caregivers who rely on integrating displays of affect with wordings. Experience reorganises events under dual control. Babies thus exploit bodies whose dynamics, unbeknown to them, are verbally constrained. Social learning draws on certainties, other people's attitudes, and feelings. At the same time, IMF ensures that affect-ridden expression regulates neurobiological change. Motive formation links semantic synthesis to value-marked experience (microdynamics) and norm-based expression. This warps and compresses unmediated (or grounded) information. Sensitivity to circumstances co-develops with a sense of persons. IMF connects expressive dynamics and the brain as organic memory adjusts to cultural values.

6 Artefactual Selves?

We find commonalities in judging attitudes, construing ambiguous sentences, and IMF. While adults use incomplete information to impose construals, infants use expression to manufacture both sensitivity to circumstances and motivated action. Value-laden dynamics augment the infant's cognitive powers as IMF channels experience towards future action. While IMF gives infants increasing self control, adults use semantic syntheses to partition the world. In using dynamics, adults settle on a judgement (e.g. unfriendly) and thus mimic users of artificial codes. Far from using invertable rule-based processes (or invariant cues), it is much more likely that the events use the brain's constructed standards. As Spurrett and Cowley (2004) argue, verbal understanding emerges as *abstraction amenable* behaviour. Dynamics, just as in adults, prompt norm-based preferences that motivate action. Eventually babies come to feel that they author actions. So, returning to Ross' (2007) view of human uniqueness, IMF helps constitute the neural control systems that are needed by the self. Motive formation brings coaction under control of cultural display that is partly regulated by what the infant expects.

The baby relies on its brain, caregiver dynamics and a group's normative patterns. As the infant becomes increasingly mindful, she gradually learns to manage coordinated display that, later, will include wordings. Eventually, human expression is found to be *interpretatively terminal* (Love, 2007). In contrast to a constructed code, certain segmental patterns (e.g. ba-by) are, in the child's world, already semantic. Given extensive experience of uttering related sounds, she may come to hear 'baby' as *baby*. In thus coming to hear a wording, more direct perceptions may be reorganised as a result of motivated action. Using compression based on cultural contingencies, aspects of the perceived world (e.g. of a sister) may become linked with a segmental percept. Repeated exposure thus compresses non-conceptual content.[12] In coming to *hear* ba-by, therefore, she may draw on a non-conceptual category (certain aspects of what we call 'sister'). As this becomes a rudimentary concept, intrinsic motivation will prompt the baby to test out its uses. Many new forms of signalling become possible. For current purposes, it is of interest to look briefly at the crib-monologues (self-talk) sometimes produced by 3–4 year olds. Here is an example from a child known as Emily:

> We bought a baby, cause, the well because, when she, well, we thought it was for Christmas, but when we went to the s-s-s store we didn't have our jacket on, but I saw some dolly, and I yelled at my mother and said I want one of those dolly. So after we finished with the store, we went over to the dolly and she bought me one. So I have one. (Nelson, 1996, p. 163)

Emily *remembers* what she said and how she acted. By conforming to English patterns, the language stance gives a perspective on 'what happened'. Once again, the necessary processes can change the baby's world. Since Emily now hears her own speaking (as shown by self-correction), she *can* learn from her utterances. In this passage, for example, she uses rudiments of logic ('So after we finished with the store … we bought one.') While this could be deliberate, it is likely that mindless patterns make such patterns familiar long before they are either heard or put to strategic ends. Given working memory (also shown by repair), she not only hears verbal patterns but uses these to react to what she is saying ('well, we thought it was for Christmas'). Having gone beyond reliance on socially appropriate vocalizations (hearing segmental patterns) Emily may be motivated to vary her wordings to fit with her telling (and her feelings about self). While lacking space to pursue such issues, this suffices to provide a sketch of the developmental context.

Even one-year-olds exploit caregiver hearings of 'more', 'milk' or 'ta' in coaction. Hypothetically, analysis amenable behaviour arises from caregiver attributions that fill out incomplete information. Indeed, not only do they often react to benefit the child but, just as strikingly, adults may repeat what *should have been said*. This surely retrains working memory to hone in on discrete, distinct (segmental)

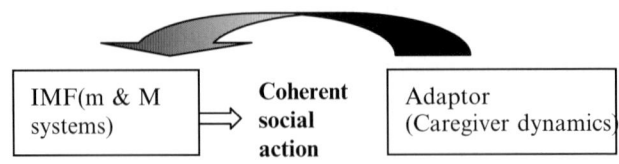

| IMF(m & M systems) | $\Rightarrow$ | **Coherent social action** | Adaptor (Caregiver dynamics) |

---

[12] Broadly, this parallels Steels and Belpaeme's (2005) work on how robots detect colours. In the imitation game, 'correct' use of patterns is rewarded. Where linked with selection, robots come to compress non-conceptual content in more similar ways.

patterns. If so, we can sketch logical steps towards gaining control over speaking. First, a baby (1) makes use of vocalizations. Later, she (2) hears distinct segmental patterns. Once achieved, she can (3) begin to align them to the caregiver's view of discrete patterns. Given reactions, she will (4) discover (roughly) what they afford. Later, given tellings about the past, she will (5) learn that they can evoke (what we call) memories. Given the power of semantic synthesis, we might present steps 1–5 as consolidating *coherent social action.*

Given that caregiver dynamics shape infant motive formation Emily's behaviour may take on layered complexity (as Kravchenko (2007) suggests).

1. IMF (m) $\rightarrow$ M (the child develops a 2nd motivation system)
2. IMF (mM) $\rightarrow$ V (IMF prompts vocalization)
3. IMF (mMV) $\rightarrow$ V_e (IMF prompts vocalizations to become effective)
4. IMF (mMV$_e$) $\rightarrow$ V_t (IMF prompts effective vocalizations to become types)
5. IMF (mMV$_e$V$_t$) $\rightarrow$ XX (IMF prompts vocalization about vocalization)

Caregiver dynamics prompt motive formation and, thus, increasingly context-sensitive behaviour. Expressive acts control the body in ways that, initially, sensitise to circumstances and later increases the child's control over events. In coming to narrate what happens, semantic syntheses can begin to tell self into existence. Given a history of sense-making, she can compress (non-conceptual) categories. In Ross' (2007) terms, she begins *define her own meaning spaces.* Using cultural norms, she partitions the world by among other things, drawing, on hearings of ba-by. Digitisation of semantics thus leads back to hearing aspects. Signalling alters motivation in ways that, hypothetically, prompt us to be self-narrative. The history of linguistic and metalinguistic expression contributes to how, as noted by Ross (2007), Dennett (1991a), and Bruner (2002), we tell selves into existence. In organic process terms, these are manufactured by artefacts that feature Emily's use of polyphony ('I yelled at my mother'). Surprisingly, perhaps, this supports Trevarthen's view that motive formation enables the rise of skills that shape awareness. Thus as narrative begins to influence what a child says, thinks and does, first-person impressions may change both working memory and how I hear who I am. Rather as in reviving a half-forgotten language, utterances gradually mutate into heard wordings that shape meaningful events. Given this sketch, we begin to extend Kravchenko's (2007) claim that language dynamics prompt us to *hear* linguistic signs. In development, the feeling of what happens may use digital signalling to consolidate itself into what we *call* linguistic experience (viz. hearing wordings).

As motivated expression shapes a consensual domain, we perceive in new ways. Above all, language assumes controlling functions. The process can be seen as one where motive formation prompts hearing in a particular sense. This suggests that, as babies come to believe in wordings, the consensual domain is increasingly influenced by expression. As wordings and meanings become familiar, social events are more routinely defined, evaluated, and interpreted around descriptions. As certain terms become familiar, we sensitise to counterfactuals. Even without a metalanguage for talking about talk, wordings come to seem word-based. Indeed, this kind of hearing readies us for actions – repeating, discussing, and challenging – that presuppose some kind of analysis. The expanded IMF (or self) thus uses circumstances to warp categories based in real-time expression. Simultaneously, cultural constraints serve to mark the (changing) limits of what can be said and done. By attending to the dynamics associated with organic coding, our vocal powers can emerge from using repetition to regularise both motivations and hearings. We take on trust that *zot, red,* or *Albanian* identify kinds. Even if the marks index normative patterns that shape a set of construals, we tend to reify their putative causes. We do this because, as children, we learned to act as if using words and, later, to use hearings in controlling and explaining social events (Cowley, 2007 b). Especially when we learned literacy skills, we picture language in terms of languages. Later, metalinguistic descriptions reinforce the language stance. We believe in *imaginary* codes (e.g. *Albanian*): we speak of words and languages as *real*. Even if we resist theories where brains *represent* such codes, such views influence our lives.

7 Turtles All the Way Up?

With the collapse of the computational model of mind, organic coding offers an alternative. First, it traces language to *turtles* or the natural artefacts that manufacture human judgements. More specifically, human brains prompt us to make strategic use of semantic syntheses (and felt relations). Organic coding unites many phenomena. Above, I have applied it to interactional attitudes, sentence construal, motivation systems, and the rise of self. Even if these do not use *adaptors*, the model throws new light on the analogue-digital process of first-order language. Indeed, it is this hybridity which allows sense making to exploit language dynamics. This is an alternative to centralism. If language is distributed, we need not picture utterances as sets of forms whose rules allow them to run through input/output devices.

In spite of parallels with Morse, language is not a constructed code. Precisely because it is not centralised, its semantics has no need for external agents. Rather, dialogical capacities set off strategic signalling that, as in all organic codes, shows context-sensitivity. Lacking invertability, digital aspects of language result from both semantic compression and a history of cultural selection. While based in Dennett's (1991b) work, the link with organic coding enables the claim to be pursued with respect to both neural and behavioural evidence. First, a simple signal ('gravy') can synthesise variable felt responses. Second, in experimental conditions, construing a sentence can depend on integrating wordings, expression and a previous hearing. Brains settle on – not established meanings – but percepts that fill in cultural display. Third, IMF can be seen as underpining interaction, relationships, and the dynamics that eventually come under the control of verbal constraints. Babies are thus transformed into human subjects by using processes that, at very least, resemble organic coding. Far from relying on invariants associated with external agents, human beings are transformed outside-in. Since we are biological creatures – and in spite of the language stance – language diverges from constructed codes. As part of strategic and motivated behaviour, it uses the same organic processes as cells, brains, and interacting human bodies.

Organic coding speaks for giving cognitive dynamics a central role in language. This, of course, is consistent with the distributed view of Love, Kravchenko, Ross, and others. Indeed, by switching one's focus from forms to expression, sense making is seen to depend on how brains motivate response to coaction. Far from drawing on *a priori* symbols (or signs), we rely on bodies that become language-ready. Given digital semantics, hybrid language comes to play the modelling role emphasised by Sebeok (2001). Far from depending on dichotomies of internal system and external use, language exploits a meshwork of heterogeneous processes and practices. Both dialogically and in monologue, we rely on integrating expression with wordings. Language is analogue and digital because adaptive signals regulate an individual's world. Indeed, it is precisely this hybridity that impacts on the IMF in ways that give us shared descriptions and an ability to identify (what we take to be) counterfactuals. Language thus has a major role in making us human subjects. Semantic syntheses transform human agency as we come to define what we believe (and want). While mammals use interaction to discover the world, syntheses are parasitic on other people's norms and expressions (of feelings, thoughts, and actions). Thus, by the age of 4 or 5, children can use autobiographical memory, story telling, and other cultural practices to become living subjects. Instead of adopting wordings, we become people who publicly present who we are.

Trevarthen traces human agency to sensitivity to expression. As motivation formation occurs, babies adjust to the caregiver's adjustments. Brain-side, IMF gives control over expression. Given how interaction changes, world-side, we slowly develop into persons. Using norms that motivate behaviour, the child develops uses for second-order constructs (e.g. *red, words, minds*). This allows linguistic reflexivity which is, I think, culture's best trick. Language used to talk about language can shape activity even among those who know nothing about what is said. By extending behaviour, in this way, we become self-conscious actors. Indeed, once 'selves' are taken for granted, we can use second-order constructs to exploit (or develop) constructed codes. Not only do individuals gain strategic advantages but, historically, social change runs in parallel to such processes. Over time, it has given us visual art, scripts, printing, mass communication and, recently, robots. Each change, of course, forces new ways of deploying language. Today, we can ask if machines could use the dynamics of organic coding. Indeed, if we can define how expression uses *adaptors*, robots can become a new theoretical test-bed. In principle, they could be used to test the hypothesis of *turtles (or artefacts) all the way up*.

Applied to language, the organic process model suggests that first-order linguistic events – human biomechanics – tidy our brains. By building human motivation systems, we exploit second-order models that exploit cultural practices. As in protein manufacture, IMF (like DNA) uses syntheses based in nontrivial causal spread. In language, neural events use values and patterns based in a history of cultural display. Humans show sensitivity to norms, attitudes, emotions, and expressions. Gradually, these shape motivations that drive interaction and, eventually, make us able to hear ourselves speaking. While based in affect, coaction conspires with culture to give children *belief* in words. Given coevolutionary processes, developmental events exploit processes that, at least, resemble organic coding. Children adopt ways of acting based in the *language stance*. Echoing what used to be called a hocus pocus view of linguistics (see Joos, 1957), I regard wordings as second-order constructs. The novel proposal, therefore, is that the organic process model be used to consider how these cultural constructs are used by brains, expression, and biology. As strategic signals make us into subjects who believe in wordings, we stand much to gain from taking the language stance.

Since biosemiosis drives expressive activity, brains synthesise meaning. They use coaction to integrate values with symbolic, normative, and other constraints. By using expression in motive formation, talk takes on modelling roles. Given semantic compression, we manufacture a fictional entity that construes events. This agent – self or soul – serves, among other things, to apply metaphors. Appeal to *coding*, I have shown, can be used to identify organic, constructed, and even imaginary processes. Finally, therefore, let me end by sounding a note of caution. Language dynamics may not, in some literal sense, function as adaptors. Semantic biology may provide no more than a model of how language influences bodies and, in development, shapes the emergence of persons. However, even on a weak reading, we can say that, as we integrate expression with segmental patterns we come up with semantic syntheses that echo experience. As is illustrated by entry into literacy, such events can transform a child's world. Historically too, the rise of grammars, dictionaries, and schools led to extensions of the language stance and, as a result, how we lead our lives. That is, I believe, beyond dispute. However Barbieri's model is used, it shows that the resulting events may depend on how we link imaginary and organic codes. By integrating wordings with multi-scalar dynamics, semantic syntheses prompt us to new forms of action. Strikingly, they push thoughts beyond the information given and, at the same time, prompt thinkers to pick out counterfactuals.

Acknowledgements This paper was made possible by colleagues in and around the *Distributed Language Group*. I owe special thanks to Marcello Barbieri who asked me to pursue parallels between distributed language and organic coding. Many thanks too to Joanna Rączaszek-Leonardi who reawakened interest in the experimental psycholinguistics. Last, I wish to express gratitude to Alex Kravchenko and Don Ross who have driven my interest in linguistic biomechanics, shown due scepticism, and provided invaluable feedback on early drafts of *Turtles*.

References

Abercrombie D (1967) Elements of general phonetics. Edinburgh University Press, Edinburgh

Alač M (2005) From trash to treasure: Learning about brain images through multimodality. Semiotica 156(1/4):177–202

Barbieri M (2003) The Organic Codes: An Introduction to Semantic Biology. Cambridge University Press, Cambridge

Barbieri M (2007) Is the cell a semiotic system? In: Barbieri M (ed.) Introduction to Biosemiotics. Springer Verlag, Berlin, pp. 179–207

Bateson MC (1979) The epigenesis of conversational interaction: a personal account of research development. In: Bullowa M (ed.) Before Speech: The Beginning of Interpersonal Communication. Cambridge University Press, Cambridge, pp. 63–77

Bennett MR, Hacker PMS (2003) Philosophical Foundations of Neuroscience. Blackwell, Oxford

Bickerton D (1990) Language and Species. University of Chicago Press, Chicago

Bruner J (2002) Making Stories: Law, Literature, Life. Farrar, Strauss, and Giroux, New York

Cangelosi A, Smith A, Smith K (eds) (2006) The Evolution of Language. World Scientific, Singapore

Christensen WD, Hooker CA (2000), An interactivist-constructivist approach to intelligence: self-directed anticipative learning. Philos Psychol 13(1):5–45

Chomsky N (1957) Syntactic Structures. Mouton & Co, The Hague

Cowley SJ (1994) The role of rhythm in conversations: a behavioural perspective. Language Comm 14:353–376

Cowley SJ (2004) Simulating others: the basis of human cognition? Language Sci 26(3):273–299

Cowley SJ (2006) Language and biosemiosis: a necessary unity? Semiotica 162(1/4):417–444

Cowley SJ (2007a) The cradle of language: making sense of bodily connexions. In: Moyal-Sharrock D (ed.) Perspicuous Presentations: Essays on Wittgenstein's Philosophy of Psychology. Palgrave MacMillan, London, pp. 278–298

Cowley SJ (2007b). How human infants deal with symbol grounding. Interaction Studies, 8(1):81–104

Cowley SJ (2006a) Human symbol grounding. Inter Studies (in press)

Cowley SJ (2006b) Beyond symbols: how interaction enslaves distributed cognition. In: Thibault PJ, Prevignano C (eds) Inte Anal Language: Discuss State-of-the-Art (in press)

Cowley SJ (2006c) How do leaky minds shape learning to talk? Language Comm (in press)

Cowley SJ, Moodley S, Fiori-Cowley A (2004) Grounding signs of culture: primary intersubjectivity in social semiosis. Mind, Culture Activity 11(2):109–132

Cowley SJ, Love N (2006) Language and cognition, or, How to avoid the conduit metaphor. In: Duszak, A, Okulska U (ed.) Bridges and Walls in Metalinguistic Discourse, Peter Lang, Frankfurt, pp. 135–154

Cowley SJ, MacDorman K (2006) What baboons, babies and Tetris players tell us about interaction: a biosocial view of norm-based social learning. Connection Sci 18(4):363–378

Croce B (1960) History: Its Theory and Practice (transl. Ainslie D). Russell & Russell, New York

Damasio A (1999) The Feeling of What Happens: Body, Emotion and the Nature of Consiousness. Heinemann, London

Darwin C (1871) Descent of Man. Murray, London

Davidson D (1997) Seeing through language. R Inst Philoso, Suppl, 42:15–28

Deacon T (1997) The Symbolic Species: Co-Evolution of Language and the Brain. W W Norton, London

Dennett D (1991a) Consciousness Explained. Little Brown, Boston, Massachusetts

Dennett D (1991b) Real patterns, J Philoso 88:27–51

Di Paolo EA (2005) Autopoiesis, adaptivity, teleology, agency. Phenomenol Cognit Sci 4(4):429–452

Docherty G, Foulkes P (2000) Speaker, speech and knowledge of sounds. In: Burton-Roberts N, Carr P, Docherty G (eds) Phonological Knowledge: Conceptual And Empirical Issues. Oxford University Press, Oxford, pp. 105–129

Eerdmans S, Prevignano C, Thibault P (2003) Language and Interaction John Benjamins, Amsterdam, The Netherlands

Erickson F, Shultz J (1982) The Counselor as Gatekeeper: Social Interaction in Interviews. Academic Press, New York

Falk D (2004) Early homonins, utterance-activity and niche construction. Behavior Brain Sci 27:509–510

Fogel A, Garvey A, Hsu HC, West-Stromming D (2006) Change Processes in Relationships: A Relational-Historical Approach. Cambridge University Press, Cambridge

Giles H, Nikolas C, Coupland J (1991) Accommodation theory: communication, context, and consequences. In: Giles H (ed.) Contexts of Accommodation: Developments in Applied Sociolinguistics. Cambridge University Press, Cambridge, pp. 1–68

Goffman E (1983) The interaction order. Am Sociol Rev 48:1–17

Goodwin C (2002) Time in action. Current Anthropol 43(Suppl Aug–Oct):S19–S35

Gumperz JJ (1982) Discourse Strategies. Cambridge University Press, Cambridge

Haken H (1993). Are synergistic systems (including brains) machines? In: Haken H, Karlqvist H, Svevin U (eds) The Machine as Metaphor and Tool. Springer-Verlag, London, pp. 123–138

Harris R (1981) The Language Myth. Duckworth, London
Harris R (1998) Introduction to Integrational Linguistics. Pergamon, Oxford
Hutchins E (1995a) How a cockpit remembers its speeds. Cognitive Sci 19:265–288
Hutchins E (1995b) Cognition in the Wild. MIT Press, Cambridge Massachusetts.
Hudson R (1996) Sociolinguistics, 2nd edn. Cambridge University Press, Cambridge
Johnstone B (1996) The Linguistic Individual: Self-Expression in Language and Linguistics. Oxford University Press, Oxford
Joos M (1957) Readings in Linguistics. American Council of Learned Societies, Washington, DC
Kaye K (1982) The Mental and Social Life of Babies: How Parents create Persons. Methuen, London
Kravchenko A (2007) Essential Properties of language or why language is not a code. Language Sciences 29(1)
Lee DN (1993) Body-environment coupling. In: Neisser U (ed.) The Perceived Self: Ecological and Interpersonal Sources of Self Knowledge. Cambridge University Press, Cambridge
Legerstee M (2005) Infants' Sense of People: Precursors to a Theory of Mind. Cambridge University Press, Cambridge
Linell P, Luckmann T (1991) Asymmetries in dialogue: some conceptual preliminaries. In: Marková I, Foppa K (ed.) Asymmetries in Dialogue. Harvester Wheatsheaf, Hemel Hempstead, pp. 1–20
Linell P, Gustavsson L, Juvonen P (1988). Interactional dominance in dyadic communication. a presentation of the initiative-response analysis. Linguistics 26(3)
Linell P (2005) The Written Language Bias in Linguistics. Routledge, Oxford
Linell P (2007) Dialogicality in languages, minds and brains: is there a convergence between dialogism and neuro-biology? Language Sci (in press)
Love N (2004) Cognition and the language myth. Language Sci 26:525–544
Love N (2007) Are languages digital codes? Language Sci (in press)
Lyon C, Nehaniv CL, Cangelosi A (2007) Emergence of Communication and Language. Springer, Berlin
Lyons J (1977) Semantics, Vol. II. Cambridge University Press, Cambridge
Marková I, Foppa K (eds) (1990). The Dynamics of Dialogue. Harvester Wheatsheaf, Hemel Hemstead
Marková I, Foppa K (eds) (1991) Asymmetries in Dialogue. Harvester Wheatsheaf, Hemel Hempstead
Maturana HR (1978) Biology of language: The epistemology of reality. In: Miller G, Lenneberg E (eds) Psychology and Biology of Language and Thought. Academic Press, New York, pp. 28–62
Maturana H, Varela F (1987) The Tree of Knowledge: The Biological Roots of Human Understanding. Shambhala, Boston
Nelson K (1996) Language in Cognitive Development: The Emergence of the Mediated Mind. Cambridge University Press, Cambridge
Panskepp J (1998) Affective Neuroscience: The Foundations of Human and Animal Emotions. Oxford University Press, New York
Pattee HH (2000) Causation, control and the evolution of complexity. In: Anderson CB, Emmeche C, Finnemman NO, Christiansen PV (eds) Downward Causation. Aarhus University Press, Aaarhus, pp. 63–77
Pickering MJ, Garrod S (2004) Toward a mechanistic psychology of dialogue. Behavior Brain Sci 27:169–225
Pinker S (1994) The Langauge Instinct. Penguin, Harmondsworth
Port RF, Leary A (2005) Against formal phonology. Language 85:927–964
Rączaszek J, Tuller B, Shapiro LP, Case P, Kelso S (1999) categorization of ambiguous sentences as a function of a changing prosodic parameter: a dynamical approach. J Psycholing Res 28(4):367–393
Reddy MJ (1979) The conduit metaphor. In: Ortony A (ed.) Metaphor and Thought. Cambridge University Press, Cambridge, pp. 284–324

Rilindja (1981) Fjalor i Gjuhës së Sotme Shqipe, Vol 2. In: Rexha, A, Pashku A, Bujuri F, Kukay R, Rrahmani N (eds) Rilindja, Prishtinë

Rochat P (2002) Dialogical nature of cognition. In: Jaffe J, Beebe B, Feldstein S, Crown C, Jasnow M (eds) Rhythms of Dialogue in Infancy; Co-ordinated Timing in Development. Blackwell, Oxford, pp. 133–144

Ross D (2004) Meta-linguistic signaling for coordination amongst social agents. Language Sci 26:621–642

Ross D (2007) H sapiens as ecologically special: what does language contribute? Language Sci 16(1)

Ross D, Spurrett S (2004) What to say to a sceptical metaphysician: a defence manual for cognitive and behavioural scientists. Behavior Brain Sci 27:603–647

Saussure F de (1916) Cours de linguistique générale. (English transl. R. Harris. Duckworth, London, 1983.)

Sebeok TA (2001) Biosemiotics: its roots, proliferation and prospects. Semiotica 134(1/4):61–78

Spurrett D, Cowley SJ (2004) How to do things without words. Language Sci 26(5):443–466

Sterelny K (2003) Thought in a Hostile World: The Evolution of Human Cognition. Blackwell, Oxford

Steels L, Belpaeme T (2005).Coordinating perceptually grounded categories through language. A case study for colour. Behavior Brain Sci 28(4):469–489

Steinberg DD, Hiroshi N, Aline DP (ed.) (2001) Psycholinguistics: Language, Mind and World, 2nd edn. Longman, London

Taylor TJ (1997) Theorizing Language: Analysis, Normativity, Rhetoric, History. Pergamon Press, Oxford

Thibault PJ (2004a) Brain, mind and the signifying body: an ecosocial and semiotic theory. Continuum, London

Thibault PJ (2004b) Agency and consciousness in discourse: self-other dynamics in a complex system. Continuum, London

Thibault PJ (2005) The interpersonal gateway to the meaning of mind: unifying the inter and intaorganism perspectives on language. In: Hasan, R, Matthiessen C, Webster J (eds) Continuing Discourse on Language: A Functional Perspective. Equinox, London, pp. 117–156

Thibault PJ and Prevignano C (eds.) *Interaction Analysis and Language: Discussing the State-of-the-Art.*

Tomasello M (1999) The Cultural Origins of Human Cognition. Harvard University Press, Cambridge, Massachusetts

Tomasello M (2003) Constructing a Language: A Usage-based Theory of Language Acquisition. Harvard University Press, Cambridge, Massachusetts

Tomasello M (2005). Beyond formalities: the case of language acquisition. *The Linguistic Review*, 22: 193–197.

Tønsberg GH, Hauge TS (1996) The musical nature of prelinguistic interaction. The temporal structure and organization in co-created interaction with congenital deaf-blinds. Nordic J Music Therapy 5(2):63–75

Trevarthen C (1979) Communication and Co-operation in early infancy: a description of primary intersubjectivity. In: Bullowa M (ed.) Before Speech. Cambridge University Press, Cambridge, pp. 321–347

Trevarthen C (1998) The concept and foundations of infant intersubjectivity. In: Bråten S (ed.) Intersubjective Communication in Early Ontogeny. Cambridge University Press, Cambridge, pp. 15–46

Trevathen C, Aitken KJ (2001) Infant intersubjectivity; Research, theory and clinical applications. J Child Psychol Psychiatr 42(1):3–48

Vico G (1991) The New Science of Giambattista Vico (transl. Bergin TG, Fisch MH). Cornell University Press, Ithaca. (Unabridged translation of Scienza Nuova, 3rd edn., 1744.)

Wheeler M (2005) Reconstructing the Cognitive World: The Next Step. MIT Press, Cambridge, Massachusetts.

Wittgenstein LW (1958) Philosophical Investigations, 2nd edn. Blackwell, Oxford

Chapter 15

Code and Context in Gene Expression, Cognition, and Consciousness

Seán Ó Nualláin

Abstract 'Code' and 'context' are two of the most ambiguous words in English, as there cognates no doubt are in other languages. In this paper, we are concerned to bring these notions down to earth, and use them as structuring schemas to understand various types of biological and cognitive activities. En route, we consider work such as Barbieri's and Smith's that impinge on our concerns in various ways. The notion of code, of course, has been discussed with respect to what constitutes the original genetic code, and indeed what the original autocatalytic sets were to facilitate the emergence of a biological code in the first place, and where in organisms they could be located. These issues are not our central concern here.

We first revisit the main themes of previous work that provides a new framework for considering gene expression vis-à-vis language production and comprehension. It is decided that the gene expression/language production analogy is a worthwhile one, and that both evince a useful dichotomy of code and context. Yet language itself, as exemplified in the case of verbally able autistic people, is not sufficient for communication. We then consider what the code for intersubjectivity is, if not language. It is argued that selfhood, which is the medium in which communication takes place, may be an artefact of data-compression occurring in the brain. Yet we have managed to extend this epiphenomenal attribute into a vehicle for "code" by narrating to ourselves. These narratives, which on a personal level maintain what is often the illusion of personal potency, likeability, and logical consistency, have on the social level the capacity to produce consensually validatable identifications like national sentiment, those parts of gender which are constructed, and other affilia tions. We finish with an extended framework in which qualia – generically viewed, subjective states – are naturalised in a way that does justice both to ourselves and the world that we inhabit.

Lecturer, Symbolic Systems, Stanford Visiting scholar, Molecular and Cell Biology, Room 230, Donner lab, UC Berkeley, CA 94704, USA, e-mail: sonual@stanford.edu

M. Barbieri (ed.), *The Codes of Life: The Rules of Macroevolution.*
© Springer 2008

1 Introduction

Barbieri (2003) insists on the existence of a multiplicity of codes in Nature. A code is defined as a set of rules relating entities between independent worlds. He argues that along with obvious genetic and linguistic codes, we must consider splicing codes, signalling codes, compartment codes, and apoptosis codes. These points granted, many apparently counter-intuitive conclusions follow. 'Meaning' arises when any given object is related to another through a code. Therefore, meaning can be said to inhere at the molecular level when the code is between organic molecules. At the genetic level, we speak of copy making; at the protein level, we can speak of code making and 'meaning'. He would agree with the 'language of thought' proposals from Fodor and his followers (see O Nualláin, 2003, pp. 91–92, and below) that we can – and, according to Fodor, *must* in order to explain many cognitive phenomena – speak about a mental code or 'language of thought' as emphatically as we speak about a genetic code at the origin of life. Finally, the notion of adaptor, defined as 'catalyst plus code', is introduced. Peirce's celebrated trio of sign, meaning, and interpretant can be biologised as sign, meaning, and adaptor. Barbieri (2006) extends the previous argument to insist that sequences and codes are 'nominable' entities.

An analogous project, arising from an impulse to extend the notion of 'computation', complements that of Barbieri. Brian Smith (1996) is unwilling, a priori, to assert even the existence of objects, let alone that of subjects. His project is an attempt to construct a theory of 'computation in the wild'; a notion of computation so encompassing that it can encompass word processing at one extreme and the most athletic feats of Lisp hackers at the other. After finding all previous accounts of computation *qua* effective computability, Turing machines, and so on, inadequate, Smith comes to a radical conclusion; a theory of computation is also a general metaphysics. To explain computation as a concept requires a thorough ontology of the world, and metaphysical preparation.

In the beginning, according to Smith, was the raw material for objects; proportionally extensive fields of particularity (1996, p. 191). A metaphysical clearing has to occur before we even speak of the existence of different objects, let alone their relation to each other through a code. He argues that the cognitive acts of sensation, perception, and syntactic analysis can be compiled together in a concept called 'registration' which results in the emergence of s-objects (roughly speaking, subjects). 'Registration' is of course broader than cognition and can occur at 'lower' biological levels than the cognitive; therefore, Smith's is a hospitable formalism, with respect to this principle at least, for biosemiotics. Objects emerge when the registration is altered to cater for the possibility of independently registered 'worlds'; to use cognitive science terminology, allocentric versus egocentric representation. Smith's conclusion is revolutionary. He argues that computation is quite as intentional, in the classical Brentano sense of relating to objects, as any human cognitive process.

Combining aspects of the two proposals in ways that do violence to neither, we may indeed speak about 'meaning' in a way that does not confine it to human cognitive process. In this paper, we are attempting to extend analyses such as those of Barbieri and Smith in a way that psychologises these analyses in the case of human cognition, and en route transforms our notion of subjectivity. So the goal here resembles that of Smith; to deconstruct and then return our notion of the 'subject' or its related term 'self' in the manner that he does for 'object', 'true', 'formal', and so on. The path is as follows: We first revisit the main themes of previous work that provides a new framework for considering gene expression vis-à-vis language production and comprehension. In many ways, this programme is consistent with and extends Piagetian genetic epistemology (Piaget, 1972). We finish with an extended framework in which qualia – generically viewed, subjective states – are naturalised in a way that does justice both to ourselves and the wonderful, terrible world that we inhabit.

2   Gene Expression and Linguistic Behaviour

In our previous work it was noted that the analogy between gene expression and language production is useful, both as a fruitful research paradigm and also, given the relative lack of success of natural language processing (nlp) by computer, as a cautionary tale for molecular biology. In particular, given our concern with the Human Genome Project (HGP) and human health, it is noticeable that only 2% of diseases can be traced back to a straightforward genetic cause. As a consequence, we argue that the HGP will have to be redone for a variety of metabolic contexts in order to found a sound technology of genetic engineering (Ó Nualláin and Strohman, 2007).

In essence, the analogy works as follows: first of all, at the orthographic or phonological level, depending on whether the language is written or spoken, we can map from phonetic elements to nucleotide sequence. The claim is made that Nature has designed highly ambiguous codes in both cases, and left disambiguation to the context. At the next level, that of syntax, we note phenomena, like alternative splicing in gene expression, that indicate a real syntactic level in the genome analogous to that in Natural Language (NL). It is argued, in turn, that the semantic level in language corresponds to protein production, and that, in apparent paradox, proteins do not in themselves specify 'meaning'. That is reserved for the concept of function in the environment. For example, while rock pocket mice in Arizona acquire an adaptive darkening melanin through expression of genes related to the MC1R protein, their fellow species members 750 km. away in New Mexico use an entirely different mechanism for the same end. Sibling species are of quite different genetic origin; the swallowtail butterfly mimics its poisonous counterparts using entirely different genetic mechanisms. Conversely, Watson (1977, p. 7) postulated 30 years ago that ' a multiplicity of proteins will be generated from "single" genes'. The suspension of disbelief required for the HGP's hype to take was considerable, given the scientific sophistication of many of the boosters of the project.

There has been an enormous amount of work on the nature and complexity of the genetic code. Di Giulio (2005) reviews stereochemical and physicochemical hypotheses about the origin of the genetic code before coming to the conclusion that the co-evolution hypothesis about this origin, with its reference to the biosynthetic relationship between amino acids and corroboration by analysis of metabolic pathways, is most nearly correct. Gene assembly in ciliates has spawned a massive amount of work (e.g. Daley et al., 2003); the existence of two different nuclei, micronucleus and macronucleus, in a single cell, and the transformation of the latter into the former has provided much grist to the mill of formalists anxious to apply the artillery of formal language theory to genetics. Yet this will not provide a context (general calculus) relating nucleotide to phenotype any more than generative grammar facilitated the advent of computers that could translate between arbitrary texts of arbitrary languages at will.

Similarly, in NL, correct interpretation of language strings often requires the processing of 'pragmatic' factors. Conversely, 'semantics' often does yield meaning; indeed there are situations where a single word or phoneme can give us all the meaning we want, without any explicit syntactic or semantic processing (Fire!!). Since Austin (1962) distinguished the 'perlocutionary' import of phrases from the other levels of messages conveyed, it has been clear that sentences like 'Would you like to help me?' are not necessarily amenable to pure semantic interpretation to arrive at their meaning. Humans who are non-opaque to such indirection quickly realise that understanding of such sentences initially requires an act of consciousness, an act of considering themselves as objects in the environment of the speaker. Of course, as we shall see, many humans never quite get it; at one extreme, sufferers from severe autism, and at the other, the milder versions of self-centredness that we classify as that of Asperger.

Are there counterparts in Nature? If the nexus of organism and environment is considered over time, perhaps we can suggest that there are. While it is absurd to suggest that the swallowtail butterfly is 'trying' to look like others, natural selection in the specific environment pushes it in a direction that makes it seem as though it had conscious knowledge of its own appearance. In the schema considered here, the HGP did little more than elicit some context-independent relations between keywords and semantic primitives. The cautionary lesson is that the HGP has to be redone in countless metabolic contexts, just as NLP needs at least another generation of work before the Turing test can be passed outside wholly trivial micro-domains.

Some simple, yet very significant, points need to be made, particularly in the context of the emerging field of evolutionary developmental biology (evo-devo). First of all, considered with respect to the Chomsky Hierarchy, the system of gene expression with its myriad feedback mechanisms is context sensitive. Therefore, to take but one example, the stance in Salzberg et al. (1998) that there are no long-distance dependencies in the genome is absurd. The issue of formal complexity in the case of the expression of certain ciliar genomes, as we saw, has inspired an impressive corpus of papers. Secondly, the notion of the 'field' is as current in biology, thanks to painstaking observation of the development of the fruitfly embryo through the expression of Hox genes, as it has been in Physics post-Faraday. In order to clarify matters, we will refer to these field effects as due to 'domain' for the rest of this paper.

The field effects are considerable. Hox genes, which determine many aspects of morphogenesis, operate in a context-sensitive manner with a recursive, phrase-structure type complexity. A context-sensitive rule for language might specify that in context A, a sentence is to be described as a noun phrase followed by a verb phrase:

$$S - A/NP\ VP$$

Similarly, Hox genes function with genetic switches at two levels. One set belongs to the Hox genes themselves, and specify their expression in the different longitudinal segments of the animal. The other set of switches involve recognition by Hox proteins in order to vary the expression of different genes. Thus, to take one example, the BCMP gene will express proteins belonging to the outer ear, or rib cartilage, depending on the field. One gene might generate different proteins, according to Watson (ibid.); conversely, one gene might be controlled by a variety of other genes, giving rise to recursivity and ambiguity fully as complex as anything in NL.

Let us continue with our analysis of the effects of external dynamics on formal symbolic systems with reference to another innovation from physics, Einstein's work on gravitation. Einstein famously inspired many excursions to view solar eclipses by predicting that a massive body like the sun would distort space time sufficiently for stars to appear displaced from their normal positions. Moreover, this distortion of space time would be more and more marked as one approached the surface of the massive body in question. The argument in the previous papers leading to this one, similarly, is that restriction of domain causes the layers of language – phonetic/orthographic, syntactic, semantic, and pragmatic – to get compressed until, ultimately, a single phoneme can have massive meaning effects. This restriction of domain is, of course, nothing other than the process of living itself.

3   Cognition

Of course, language is no more the only symbolic cognitive system than proteins manifest the only biological code. In Ó Nualláin (2003) (Chapter 7), the following attributes are predicated of human symbol systems: A hierarchical organization, formal complexity of a certain degree, a recursive structure, processing within micro-domains called contexts, metaphor, emotional impact, ambiguity, systematicity, duality of structure, the notion of a native language, and creativity. In western classical music, for example, harmonic contexts are defined so specifically that a dominant seventh chord will always demand resolution to the tonic. In Jazz, that is of course not the case; songs like 'Yesterdays', for example, involve circles of dominant seventh chords following each other like a snake swallowing its tail. In visual art, there is an enormous cultural and historical component as well. Piet Mondrian can draw a jumble of boxes in a minimalist context and label the result

'Broadway boogie-woogie', neatly inviting musical experience to enter also. Braque and Picasso, some decades earlier, arrived at a system of depiction that considers the depicted object from a variety of different perspectives, thus violating classical canons; the result, cubism, can be considered an innovation in formal complexity. Likewise, Paolo Uccello's rediscovery of the laws of perspective in early renaissance Italy adds another layer of formal complexity. Finally, in their choice of art, the public often chooses the less formally complex of two forms, when given a choice; the monodic melodies of Monteverdi gained prominence over Palestrina's dense harmonies.

Fodor (1975) has vehemently argued for the existence of an innate, language-independent, language of thought. He insists that computation over representations is the essence of cognition, and that therefore there must be a language of thought in which representations are couched for cognition to occur. He exploits the fact that absolute innatism, like its cognate absolute idealism, is irrefutable; we must, he argues, possess the complete set of representations from birth. Essentially, the argument boils down to the fact that, if concept Y is logically distinct from concept X, it cannot possibly have been learned as an extension of concept X; if it is not logically so distinct, it essentially is concept X. It is therefore, either a distinct innate concept, or an extension of an innate concept. To his opponents like Talmy who insist on compositional semantics with new meanings emergent from combinations of old such, Fodor offers ridicule; it is like saying that 'New York' is simply an up-to-date version of the Yorkshire town. Talmy remains unconverted.

A coda to Fodor's 'language of thought' argument would insist on its manifestation in various symbolic modalities like pictorial representation and music as well as language itself. There is a lot of value in this argument. We might like to explore whether folk music around the world contain equal levels of formal complexity, in the same way that languages do, to take one example. Therefore, we might conclude that the relative harmonic and melodic simplicity of Balkan music is counterbalanced by its rhythmic sophistication, with use of time signatures like 11:8 and 7:8 that are rare indeed in the 'wild'. Conversely, contemporary 'Irish' music (much of which is Scottish) is rhythmically and harmonically simple, but extremely intricate melodically. In the absence of this research, we do have some critical evidence in musical scores for how music is processed in the real world by a sophisticated audience. For example, Beethoven's Eroica symphony features the use of diminished chords repeatedly to change key in its first movement and moves imperceptibly (once the diminished innovation is accepted) away from its 'home key' before returning. In this case, the key changes can be regarded as shifts in context. Mozart, in his last G minor symphony, uses dissonance a great deal in a way that foreshadows many much later innovations; this is an elaboration in 'code'.

Gombrich (1959) underlines a history of elaborations in code in the history of pictorial representation. Constable's landscapes are, according to Gombrich, explicitly scientific. They are a Popperian attempt to set up refutable hypotheses about the constraints on visual experiences. We have spent some time on music, vision, and language at this level because we do not know what the neural code really is. If there is a unified such code, it must exist at a very high level of abstraction if implementations in media as disparate as dance, music, architecture, and so on are to be achieved. One hypothesis, featured in the contributions by Kime, Aizawa, and Hoffman to Ó Nualláin et al. (1997) is that the neural code is describable by mathematical formalisms akin to tensors and Lie groups. Specifically, these formalisms allow entities to transfer between systems of different dimensionalities while maintaining their quantitative identity. The principles of invariance we can possibly identify as Kantian quantities.

Therefore, at an explicitly symbolic level, we can say a great deal about code and context in the symbolic modalities of language, vision, and music. Of course, the analysis has been extended to other aspects of human endeavour like architecture and dance. Undoubtedly, code processing is contextual, and skilled artists play on this to provide us highly ambiguous structures for aesthetic affect. When it comes to specifying the code at a neural level, we know little.

4 Code and Context in Consciousness and Intersubjectivity

With respect to this final section, we know even less at the neural level about consciousness than we know about the production of symbolic expression. However, we have a plethora of data, available to the most unfunded observer: introspective experience. We will conclude that, whereas Smith's (1996) caveats about the necessity for the construction of subjects and objects is correct, the path to a resolution of this issue may feature the type of non-linear dynamics that he does not refer to.

The argument is that our experience of self reflects a process of context specification, as well as generating the illusion of agency, in the hope it will become a reality. We narrate to ourselves continually, assigning to ourselves the authorship of acts that in fact happened automatically; we identify ourselves with respect to our nation, gender, football team, or choice in food from second to second. In actual fact, EEG work by Walter Freeman, inter alia, indicates that, at a gross level, our gross brain state is changing much faster than that; perhaps 3–7 times a second (Freeman, 2003, 2004; Ó Nualláin, 2007). The 'conscious moment' at a perceptual level is on the order of 100 ms; stimuli presented to each eye at a gap greater than this will not be subject to binocular fusion, and will be perceived as two different objects. However, the cortex can be destabilised by a saccadic eye movement in 3 ms (Freeman, 2004). So we are getting a sample in consciousness of events that, whatever we imagine about them, are happening at a speed that is well beyond our conscious control by far most of the time. It makes sense to talk of the conscious 'perceptual' moment as being in the order of a tenth of a second, and the 'subjective' self-related moment as being perhaps an order of magnitude longer.

However, we have educational and training systems that use the accumulated experience of humanity to facilitate our getting partial control of our mental processes in restricted contexts. The lavish funds required to establish and run universities, schools, and research institutes bears witness to just how distractible we are, and just how much support we need to keep up a train of thought. One estimate is that we consciously process a few hundred bits out of the hundreds of billions that impinge on us each second. The major task for the brain, then, is filtering; context specification.

Therefore, our phenomenal experience, with its ever-changing self that yet claims permanent ownership of the whole organism, is mainly detritus from a process of context specification that occurred automatically, whole tenths of a second before. The self is fundamentally the part of this process accessible to the slow rhythms of consciousness. The pathology of the self-mechanism is manifest in many forms of asomatognosia, where the patient will insist that the left side of her body does not belong to her, or will with equal determination consistently claim that she has three hands.

Now for code, Amartya Sen has recently stressed how we are all multiple; just as his native India, he stresses, is both Hindu and Muslim, so are we all different identities at different moments. Indeed, to continue, these identifications are themselves a code, a snapshot of the Other that allows us to compress much information. They function at a level above that of language and other symbol systems. They become more elaborate, as when the nineteenth century notion of the nation-state compiled feelings of place and race into one highly morally ambiguous entity. At their best, they allow a shorthand for dealing with others, and indeed with ourselves.

To summarise up to this point, then, at the phenomenal level, as at the organic molecular level and levels in between, there are clear processes corresponding to context and code. Perhaps we shall eventually learn to ontologise the world such that our identifications, and the physiological processes on which they supervene, seem clearly to reflect the structure of both the objective and intersubjective world. With this ontology complete, we can identify whatever remains outside its grasp as our authentic and true selves, inviolable by neural contingency.

It may not be clear that there is actually a problem to be solved with respect to the code used in communication between people. Surely the existence and proper use of language is sufficient? Unfortunately, as exemplified in autism, this is not the case. Williams (1992) at times gives a harrowing account of the transition from autism to the real social world. It is all the more powerful as its author was at all times verbally fluent:

'My progress at school was not too bad. I loved letters and learned them quickly. Fascinated by the way they fitted together into words, I learned those, too. My reading was very good, but I had merely found a more socially acceptable way of listening to the sound of my own voice. Though I could read a story without difficulty, it was always the pictures from which I understood the content' (ibid., p. 25)

Freeman (2004) anticipates this, and argues that we effectively need a new type of code to understand intersubjective communication as mere symbols are not enough. Meaning is the result of synchrony at a neural level between two individuals, and need not necessarily refer to anything in the 'outside' world. Even with a full semantic analysis of each sentence, we have not necessarily achieved meaning.

He exploits the work of Barham (1996) who argues that biological systems, like the brain, are often kept at far from equilibrium states and need high energy oscillators in the environment to stabilise them. Freeman argues that the neural part of this story has been confirmed with EEG work by himself and others.

Therefore, the 'code' for social life is at a higher level than mere symbolic expression. As a consequence, the way we mediate our experience to ourselves, which in general is an introjection of the tropes of social life into phenomenal space, is going to exploit this code. At first analysis, the code involves coherent 'selves', as described above which yet can change from second to second in phenomenal experience; dialogue with others stabilises them. Obviously, much further work of the type that Freeman and his colleagues have done is necessary; however, the case for existence of a meta-level verbal conscious code, exemplified by the case of Williams (1992), is one that has many strengths.

5   Conclusion

Therefore, we can conclude that the duopoly of code and context recreate themselves at each level we have considered in this paper. To ignore context is to risk wasting further billions on doomed extensions of the HGP, of simplistic nlp projects, and of programmes for curing autism that fail to reflect the subtlety of the disorder in many cases. It is also to risk the generation of a new type of cognitive science, as remote from the reality of the grace and grit of real human existence as its intellectual forebears.

References

Austin J (1962) How to do things with words. Oxford University Press, Oxford
Barbieri M (2003) The Organic Codes: An Introduction to Semantic Biology. Oxford University Press, Oxford
Barbieri M (2006) Life and Semiosis: The real nature of information and meaning. Semiotica 158:233–254
Barham J (1996) A dynamical model of the meaning of information. Biosystems 38:235–241
Daley M, Ibarra O, Kari L (2003) Closure and decidability of some language classes with respect to ciliate bio-operations. Theoret Comp Sci 306(1–3):19–38
Di Giulio M (2005) The origin of the genetic code: theories and their relationship, a review. Biosystems 80:175–184
Fodor J (1975) The Language of Thought. Crowell, New York
Freeman W (2003) A Neurobiological theory of meaning in Perception, part 1. Inter J Bifur Chaos 14(2):515–530
Freeman W (2004) How and why brains create meaning from sensory information. Inter J Bifur Chaos 13(9):2493–2511
Gombrich E (1959) Art and Illusion. Phaidon, London
Ó Nualláin S, Mckevitt P, Mac Aogain M (eds) (1997) Two Sciences of Mind. Benjamins, Amsterdam, The Netherlands

Ó Nualláin S (2003) The Search for Mind, 3rd edn. Exeter, England
Ó Nualláin S (forthcoming) Neural correlates of consciousness of what? Target article, New ideas in psychology.
Ó Nualláin, S, Strohman R (2007) Genome and natural language: how far can the analogy be extended. In: Witzany (ed.) Proceedings of Biosemiotics. Tartu University Press, Umweb, Finland
Piaget J (1972) Principles of genetic epistemology. Routledge, London
Salzberg S, Searls D, Kasif S (1998) Computational Methods in Molecular Biology. Elsevier Science, New York
Smith B (1996) The Origin of Objects. MIT Press, New York
Watson J (1977) Cold Spring Harbour annual report
Williams D (1992) Nobody Nowhere. Random House, New York

Chapter 16

Neural Coding in the Neuroheuristic Perspective

Alessandro E.P. Villa

The diversity of neural coding properties of the units makes the cortex a difficult region to study and makes it especially unattractive to those who like their science in neat packages. Let us hope that new studies, new techniques, and new findings will move us out of what will someday be called the early phases (or even the dark ages) of neuroscientific study of the cortex.
(M.H. Goldstein, Jr. & M. Abeles in: Handbook of Sensory Physiology, Vol. V/2, Springer Verlag, Berlin-Heidelberg, 1975)

Abstract In the study of the cerebral functions the concept of cognition could not be conceived independently of its neurobiological bases and its relationship to the mental representation, the logic, and the computational theories of the animal and human performances. The psychic entities which serve as elements of thought are symbolic images, more or less decipherable, which can be reproduced or combined willingly. This process evolves necessarily from the rupture of the temporal constraint, and appears similar to the aesthetic approach as a method of recognition. The term 'coding' usually refers to a substitution scheme where the message to be encoded is replaced by a special set of symbols, a far weaker metaphor, for representation of information in the nervous system, because substitution codes are essentially static as defined by fixed rules. A single quantitative measurement becomes unable to determine an axis of congruity and paradigms other than the classical ones should be considered for studying cerebral functions. In the neuroheuristic framework, the 'result' cannot be simply positive or negative because the process itself cannot be reduced to proficiency as such. The search for a neural coding *per se* is temporarily removed and replaced by the study of neural activity in the dynamical perspective. Some hints of the 'neural catastrophe' and challenges to come are discussed as open issues for future paths of investigation.

Université Joseph Fourier Grenoble 1, INSERM U318 Neurobiophysique,
CHUG Michallon (La Tronche), Pavillon B, BP 217, F 38043, Grenoble Cedex 9,
e-mail: Alessandro.Villa@neuroheuristic.org

M. Barbieri (ed.), *The Codes of Life: The Rules of Macroevolution.*
© Springer 2008

1   Prolegomenon

The possible strategies that we could use in trying to comprehend cerebral functioning hinges on the subsequent problems arising from interdisciplinary studies of
molecular, cellular, individual, and social behaviour. Many disciplines have an
interest, and an important contribution to make, in obtaining an acceptable solution:
philosophy, psychology, neuroscience, pharmacology, physics, artificial intelligence,
engineering, computer science, and mathematics. Whilst such interdisciplinarity
makes the problem more exciting it also makes it more difficult. The languages of
various scientific disciplines have to be used, and appeal to the knowledge bases in
those disciplines also made (Segundo, 1983). Scientific thought, as we know today,
is based upon the assumption of an objective, external world. This conviction is supported by a rationale which calls upon mechanical laws of causal efficacy and
determinism. Fundamentally, it is the correspondence between the hypotheses and
their predictions through experimental research that builds the empirical success of
Science, as we know from the cultural heritage of Galileo Galilei.

One of the most striking characteristics of the organization of the nervous system is its direction towards temporal information processing, illustrated by the
phenomena of memory and future projection which are necessary components of
consciousness and of the so-called higher nervous activities. It is interesting in this
respect to note that the cultural enlightenment of the Arabic philosopher and physician Ibn Sina, or Avicenna (980–1037), influenced the prevailing brain theory
throughout the Middle Age until the end of sixteenth century. In one of his master-
pieces, *The Canon of Medicine*, he made a synthesis of Greek and Roman medical
achievements that suggested the existence of a hierarchical chain of cells (from the
Latin term *cellae,* i.e. *chambers,* and not referred to the modern meaning of cell)
inside the brain. The information arriving to the first cell was then relayed to the
second and third set of cells for progressively higher nervous activities (Fig. 1).

The first cell, the most anterior one, contained the common sense (*sensus comunis*)
recipient of all sensory modalities and fantasy (*phantasia*). The second cell was
lying between the first and the third cells and contained the imagination (*imaginativa*)
and the cognition (*cognitiva*). The third cell was located in the most posterior part
of the brain, but it was functionally intended as the most distant cell with respect to
the sensory inputs, and contained memory (*memorativa*) and according to some
authors a part of the cognitive faculty associated to the estimation (*estimativa*).
Besides the dissection of corpses very little, if any, invasive investigation of the
brain was available until the nineteenth century, and those concepts remained alive
among the scientific community until nowadays.

2   The Neuroheuristic Paradigm

The biological laws of neuroscience developed in the nineteenth and twentieth
century, as well as the construction of mathematical axioms, were derived from
observations and pertaining to a priori knowledge. If, however, the observation is

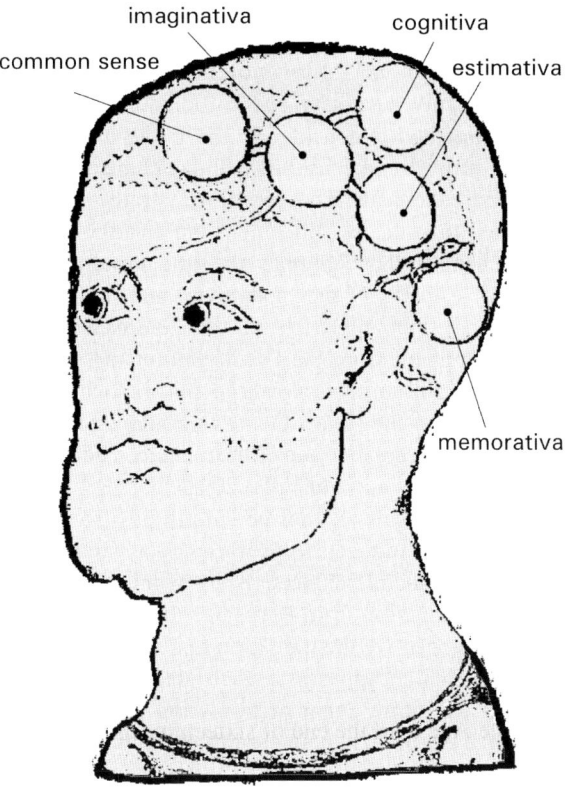

Fig. 1 Redrawing that partially reproduces a diagram of the brain (fourteenth century) illustrated on parchment illumination (University Library, Cambridge)

exclusively and stringently applied to the description, the observation becomes reductive. Since the application of scientific protocols to the investigation of how the brain works, it appeared that the dynamic relations between memory and sensorimotor activity could reveal abrupt reorganizations of information, characterised by a temporal dimension other than the one attributed by the slant of classical mechanisms. As queried by the mathematician Hadamard about his own mental experience, Einstein replied that words and language, written or spoken, did not seem to play the least role in the mechanism of his thoughts (Hadamard, 1952). The psychic entities which serve as elements of thought are symbolic images, more or less decipherable, which can be reproduced or combined at will. This process evolves necessarily from the rupture of the temporal constraint, and appears similar to the aesthetic approach as a method of recognition. A single quantitative measurement becomes unable to determine an axis of congruity and paradigms other than the classical ones should be considered for a scientific interpretation of the results, following a process likewise the introduction of the superstring theory with respect to the standard model.

The information processing effected by the brain appears then as a result of accordance between Nature ('bottom-up') and Nurture ('top-down'). Research strategy based on the 'bottom-up' information flow, the preferred view by neurobiologists, seems potentially necessary and sufficient; however, it is not wholly viable to actual experimentation considering the impossibility of simultaneously examining, even in a primitive species, all cellular elements of the brain and all variables that affect those elements. The 'top-down' strategy, with the assistance of 'dark boxes', is easier to bring to fulfillment but insufficient and irrelevant in understanding the mechanisms that coordinate the local networks of cellular elements. It seems, therefore, that a fusion of the 'bottom-up' and 'top-down' mechanisms is needed, leading to a distinct approach to the Neurosciences. Let us call it Neuroheuristics, or Neuristics, from the Greek *neuron* (nerve) and *heuriskein* (to find, to discover). Its definition corresponds to that branch of Science aimed at exploring the assumptions of the Neurosciences through an ongoing process, continuously renewed at each successive step of the advancement, towards understanding the brain in its entirety (Taylor and Villa, 2001).

In this framework, the 'result' cannot be simply positive or negative because the process itself cannot be reduced to proficiency as such. The accent here is on the dynamic and non-reducible characteristics of this approach. It is important at this point to make a distinction from Bergson's (1917) psychophysical interactionism. In Bergson's perspective the transition to a successive stage is dependent upon the vital impulse which appears at each stage. Therefore, it is the vital impulse which is the activating agent of transition between the stages. In our perspective, the change occurs when an essentially new and unexpected combination develops from preexisting properties. At the dawn of the twenty-first century, such an approach can reap benefits from the new sciences and technologies which promote the emergence of new concepts; molecular biological studies and computer science can be an integral and crucial extension to the field of Neuroscience.

The emergent process of Neuroeuristics is accompanied by a perceptive jump (*Gestalt* switch). The analogy between the abstract levels of organisation in the computer and in the brain encompasses the fundamental observation that computer programming represents a deliberate mock-up, or artificial imitation, of human intellectual activity. In reviewing hypotheses, which are destined to be out of date, the neuroheuristic perspective differs from the greater portion of cognitive studies. The Neurosciences have made only a minor contribution to the knowledge of the biological substrates of creativeness, despite progress made in comprehension of the neurobiological basis of perception, training, and memory by animal experimentation. The cognitive sciences, however, faced the problem originated in the brain to separate declarative knowledge ('know what') from procedural knowledge ('know-how'). In this respect, it is undeniable that the cognitive sciences have benefited from its interchange with the field of artificial intelligence. In our view, intellectual activity cannot be reduced to its computational dimension. We subscribe to the synergy of information processing and

Neurosciences. Such kind of synergistic process is not peculiar of our times and it follows several fundamental historical facts.

In 1753–1755, the Bernese physiologist Albrecht von Haller (1708–1777) published an essay, the 'Dissertation on the Irritable and Sensitive Parts of Animals' (original title: *De partium corporis humani sensibilius et irritabilus*). This work was based on numerous experiments of vivisection and on stimulation of organs using the new knowledge offered to physiology by physics, chemistry, and natural history. With a rudimentary technique of stimulation, Von Haller classified the parts in irritable, sensible, or elastic and noted that the reactions varied between different parts of the brain. The historical importance of the work by Von Haller is not so much related to the results obtained, but rather in systematically applying the new technologies with a scientific protocol (Brazier, 1984). This approach resulted in a turnabout in the university environment of the eighteenth century. With the introduction of currents of Galvanic fluids into the brain, a powerful new tool of investigation developed at the end of the eighteenth and beginning of the nineteenth century. The use of electricity was not strictly limited to its instrumental character, which was at the basis of electrophysiology, but rather the proper characteristics of propagation and generation of this type of energy became the basis of fertile hypotheses. Inspired by Von Haller and by the works of his compatriot Malacarne, Luigi Rolando, the Piedmontese physician, added a fundamental contribution to the succession of the naturalistic descriptive paradigm which was adopted thus far. Supported by the work of Alessandro Volta, Rolando was struck by the analogy between electric devices and the structure of the cerebellum to which he assigned a role in locomotion. In addition, Rolando was able to discern regularities in the morphology of the cerebral cortex and could establish relations between its parts while tracing the map and assigning them a name. Rolando's research was based on the metaphysical assumption that brain organisation had necessarily to be submissive to constant and recognizable laws (Rolando, 1809). His criticisms directed against the organology concepts of Gall, then diffused throughout the Occidental world, were not at all dictated by a priori concepts. Rolando did not underestimate these anatomical studies, but he denounced on several occasions the absence of objective evidence for distinct organs to the tens of mental functions identified by the phrenologists.

Plotinus (204–270), the great philosopher and founder of the Neo-Platonism, developed a unique theory of sense perception and knowledge, based on the idea that the mind plays an active role in shaping or ordering the objects of its perception, rather than passively receiving the data of sense experience. In this sense, Plotinus may be said to have anticipated the medieval insights and even the phenomenological theories of nineteenth and twentieth century (Gerson, 2005). However, the heuristic paradigm of Von Haller-Rolando did not develop further in the neuroscientifics of the nineteenth century, which possibly explains the conceptual delay taken by biomedical research in Neurosciences with regard to mathematics and to physics and chemistry. In the 1940s, the confrontation between the Neurosciences and the Theory of Communication by Shannon and Weaver led to the emergence of the cybernetic movement (Shannon, 1948). There is an analogy between what occurred 200 years ago with the initiation of electricity and technology and the situation of today in which the Neurosciences are confronted with Computer and Information Sciences. In the Enneads, Plotinus states that the knowledge of the One is achieved through the experience of its 'power' (*dynamis*) and its nature, which is to provide a 'foundation' (*arkhe*) and 'location' (*topos*) for all existents. In the neuroheuristic paradigm the term One could be replaced by complexity and experience replaced by scientifically testable hypotheses.

3   The Coding Paradox

In the last quarter of the nineteenth century Eduard Hitzig and Gustav Fritsch discovered the localisation of cortical motor areas in dogs using electrical stimulation and Richard Caton was the first to record electrical activity from the brain. Electrophysiology started to develop rapidly and Edgar D. Adrian published a seminal study suggesting the all-or-none principle in nerve (1912). In the late Twenties Hans Berger in Germany demonstrated the first human electroencephalogram and opened the way to clinical applications of electrophysiology. Nevertheless, the English School was leading the investigations in electrophysiology in the first part of the century, and for his specific research on the function of neurons, Adrian shared the 1932 Nobel Prize for Medicine with Sir Charles Sherrington. Although most remembered for his scientific contributions to neurophysiology, Sherrington's research focused on spinal reflexes as well as on the physiology of perception, reaction, and behaviour. His transdisciplinary approach, in the sense that disciplines were not only used one next to the other but were really intermingled in his protocols, was an extraordinary example of non-reductionist view of neurophysiology.

A series of experimental achievements obtained in the 1950s delineated the theory that the essential carriers of neural information are perturbations of the membrane potential of the neurons, the most dramatic of which is called *action potential* or *spike* (see Shepherd, 1994, for all references to basic neuroscience). The term spike refers to the waveform of the potential when recorded by means of a high impedence microelectrode in the vicinity of the neuron (extracellular recording) or in its interior (intracellular recording). The waveform is characterised by a peak corresponding to an initial decay of the membrane potential immediately followed by a sudden reversal of the potential and its return to the initial level. Such process lasts about 1 ms for the overall majority of nerve cells for species ranging from invertebrates to mammals.

The spike is propagated through a specific neuronal appendix – the axon – in a way that it is 'regenerated' at the branching points. At the tip of the axonal branches the cell membrane is specialised and forms the so-called pre-synaptic membrane. The electrical current carried by the spike produces changes in the pre-synaptic membrane properties that affect the membrane of another nerve cell (the post-synaptic neuron). These changes may last up to a few tenths of a millisecond and may be mediated by electrical currents or chemical reactions. As a consequence of the synaptic transmission a small post-synaptic current is generated in the post-synaptic neuron and it is propagated through specialised appendices – the dendrites – towards the cell region containing the nucleus (where the genetic material [the nucleic acids] is located in association with regulatory scaffolding proteins). Whenever the perturbation of the cell membrane of a sensitive region near the nucleus is strong enough to modify its ionic currents a spike is generated and propagated through the axon. Then, the process goes on to the next post-synaptic cells. There are two main exceptions in this system: the entry and the output points. The sensory cells – the entry points – are specialised nerve cells able to transduce the energy of the surrounding world (light, heat, mechanical pressure, chemical ligand-receptor bindings) into spikes that will be transmitted along the nervous system. The output points of the system are the muscle and the glands that allow expression of the neural response to a stimulus – the behaviour – via motor and humoral responses. The cells of the effector organs have the possibility to transduce the afferent electrical currents into another physical energy (e.g. mechanical movement for muscles, release of hormones for the humoral system).

Other electrical processes, so-called after-potentials, may last hundredths of millisecond and other spike-triggered events involving intracellular biochemical reactions and modulation of genetic expression may last orders of magnitude longer (in the order of minutes to hours). This indicates that several time scales coexist and are intermingled in the process of neural information. The discussion of these phenomena goes beyond the scope of the present article and we will restrict to the discussion of the information transmitted by the action potentials.

The generation of a spike is the consequence of ionic currents that drive ions in and out the nerve cell. A neuron spends most of its energy to keep its membrane potential to a certain level as a function of its biophysical membrane properties. Despite some small fluctuations the membrane potential *at rest* is very stable. This stability holds for many biophysical perturbations but it has often been observed that many neurons may be characterised by a bistability, with two levels of resting potentials as a function of major changes in the state of the cell membrane. Notice that in a widespread jargon used by the neurophysiologists a neuron is said to *fire* whenever it generates an action potential. This jargon produces expressions like *firing rate* referred to the rate of spikes per second, *firing pattern*, etc.

All the spikes of a sequence produced by one and the same neuron look very similar given the same resting potential. Spikes are also similar for different neurons and such all-or-none feature has often inspired the digital analogy for brain processing. Far from being real, this is an extremely rough approach because spikes tend to occur at an average frequency of five spikes or less per second.

If one assumed a binary code at the millisecond scale for the spiking (say digit '1') and not-spiking (digit '0') state of a neuron this would lead to consider that a

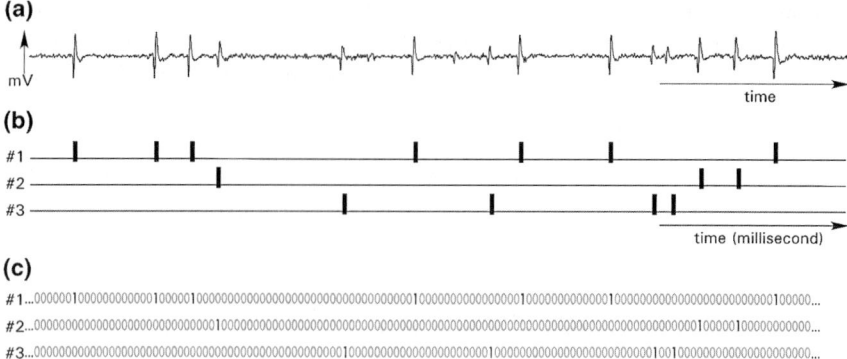

Fig. 2 (a) Extracellular recording of action potentials by a microelectrode. (b) The action potentials waveforms have been sorted at the millisecond scale according to the their shape and three spike trains are obtained. (c) The three spike trains might be coded in a binary stream where '1' means spike

neuron would transmit '0' most of the time. Action potentials occur either singly or in bursts (a rapid succession of spikes) where spikes step on each other's tails. The sequence of the spikes of a neuron is referred to as *spike train* (Fig. 2).

If one assumed long-lasting chunks of time this would lead to elementary *neural bits* that would last tens or hundredths of millisecond, a duration that can hardly be accepted when considering the precision and the time scale of movement execution, which is the major observable output of neural processing.

The problem of the neural time scale reappears despite the attempt to remove all spike-triggered effects other than the spikes themselves. In fact, random variation, noise, and reliability arise almost universally in the nervous system and the questions of the definition and meaning of neural coding are far from being trivial to pose (Segundo, 1985). The term 'coding' has a strict definition in cryptology as it refers to a substitution scheme where the message to be encoded is replaced by a special set of symbols. The above definition seems to be a far weaker metaphor for representation of information in the nervous system. Firstly, substitution codes are essentially static because they are defined by fixed rules. If the rules change over time the message cannot be deciphered and will be misinterpreted (actually this problem appears clearly with human aging but it will not be discussed here). Secondly, it seems unlikely the existence of a small fixed set of symbols to be encoded or decoded in the nervous system. In the nervous system, time sequences, delays, relatively precise coincidence relationships seem to be critically important aspects of information processing and the possibility to fit them into substitution codes appears rather remote. The search for dynamical coding schemes at variable time scales is the main goal of the Neuroheuristic paradigm. To this respect the present knowledge of the neurosciences is not advanced enough to let us formulate a testable theory. Instead, the emergence of collective properties – the properties not contained by the sum of the parties – is becoming a meaningful question thanks to the possibility that appeared in the past decades to investigate the activity of a network of neurons both theoretically and experimentally.

4   Spatio-Temporal Patterns of Neural Activity

The nervous system includes several subsystems with peculiar features that add up to the complexity. In addition to the input (sensory channels) and output (motor and humoral channels) subsystems, there is an extensive network of interconnected ganglia and nuclei that are mainly located in the brain, the most prominent of which is the cerebral cortex (for review on the quantitative analysis, see Braitenberg and Schüz, 1991). The cerebral cortex (the cortex) is formed by two hemispheres and represents the major part of the nervous system in humans, although the vital centres are located elsewhere in the so-called primordial brain. The cerebral cortex, in particular, focused the attention of neuroscientists since the early times of the scientific investigation, as discussed in the prologomenon of this chapter. It includes the overwhelming majority of all brain tissues and contains several tens billion neurons. These numbers are already impressive but the connectivity of the cortex is even more impressive.

The slow integration time of nervous cells, operating in the milliseconds range, somewhat a million times slower than presently available supercomputers and the huge number of connections established by a single neuron (in the order of tens of thousands) has suggested that information in the nervous system might be transmitted by simultaneous discharge of a large set of neurons. As much as 99% of all input neural fibres to the cerebral cortex originated from another cortical area (mostly within the same hemisphere). The reflexive nature of corticocortical connections is such that using a metaphor we could say that the cerebral cortex essentially talks to itself. Multiple dimensions of sensory and behaviorally relevant stimuli are processed by thousands of neurons distributed over the many areas of the cortex. The hypothesis that neurons process information along time both individually and jointly following precise time relationships pervaded the Neurosciences since the nervous system was conceptualised as dynamic networks of interacting neurons (McCulloch and Pitts, 1943).

The activity of each cell is necessarily related to the combined activity in the neurons that are afferent to it. Due to the presence of reciprocal connections between cortical areas, re-entrant activity through chains of neurons is likely to occur in all brains. Developmental and/or learning processes are likely to potentiate or weaken certain pathways through the network by affecting the number or efficacy of synaptic interactions between the neurons. Despite the plasticity of these phenomena it is rational to suppose that whenever the same information is presented in the network the same pattern of activity is evoked in a circuit of functionally interconnected neurons, referred to as a *cell assembly*.

In cell assemblies interconnected in this way some ordered sequences of inter-spike intervals will recur. Such recurring, ordered, and precise (in the order of few milliseconds) interspike interval relationships are referred to as *spatio-temporal patterns of discharges* or *preferred firing sequences*. (Fig. 3).

The term *spatio-temporal pattern of discharges* encompasses both their precision in time, and the fact that they can occur across different spike trains, even recorded from separate electrodes. For this to be true, temporal firing patterns must occur to a significant level above chance.

Several evidence exist of spatio-temporal firing patterns in behaving animals, from rats to primates (Villa et al., 1999; Shmiel et al., 2005), where preferred firing sequences can be associated to specific types of stimuli or behaviours. Furthermore, recent studies on active propagation of action potentials in the dendrites have provided biophysical models supporting the existence of precise neuronal timing and its modulatory role in determining the strengthening/weakening of the synaptic coupling between pre- and post-synaptic neurons, *spike timing dependent plasticity* (STDP) (Bell et al., 1997). The synaptic response increased if the pre-synaptic spike preceded the post-synaptic spike, but in the reverse order the synaptic response decreased. The time window for synaptic plasticity to occur is in the order of tens of milliseconds, and a difference in spike timing of only few milliseconds near coincidence may switch plasticity from potentiation to depression.

These findings provide an important support to the general view that a stronger synaptic influence is exerted by multiple converging neurons firing in coincidence, thus making synchrony of firing ideally suited to highlighting responses and to expressing relations among neurons with high temporal precision. An influential and remarkable model based on the assumptions of high temporal precision in brain processing is the *synfire chain* hypothesis (Abeles, 1982, 1991). This model suggests how precise timing can be sustained in the central nervous system by means of feed-forward chains of convergent/divergent links and re-entry loops between interacting neurons forming a cell assembly (Fig. 4).

A fundamental prediction of synfire chains is that simultaneous recording of activity of cells belonging to the same cell assembly involved repeatedly in the same process should be able to reveal repeated occurrences of such spatio-temporal firing patterns like those observed in the experimental studies cited above. Structures like synfire chains may exhibit patterns of activity where a group of neurons excite themselves and maintain elevated firing rates for a long period and allowing the same neuron to participate in many different synfire chains.

If one assumes that a synfire chain is associated to a basic primitive (i.e. an instance of a certain type of information), then dynamic bindings of synfire chains are able to give rise to compositionality of brain processes. Cognitive processes are likely to be compositional, that is, they must start from primitives and be capable of extracting and building up structured representations by means of progressively complex hierarchical processes (Bienenstock, 1995). Thus, the specific and dynamics relations that bind the primitives, and not only the structure of the primitives themselves, are emphasised in the role of forming the composite expressions (Abeles et al., 2004).

(a) Simultaneous recording of spike trains

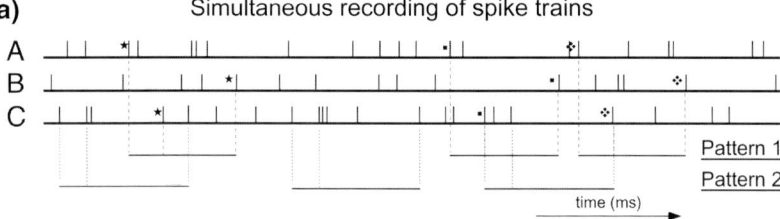

(b) Detection of statistically significant spatiotemporal firing patterns

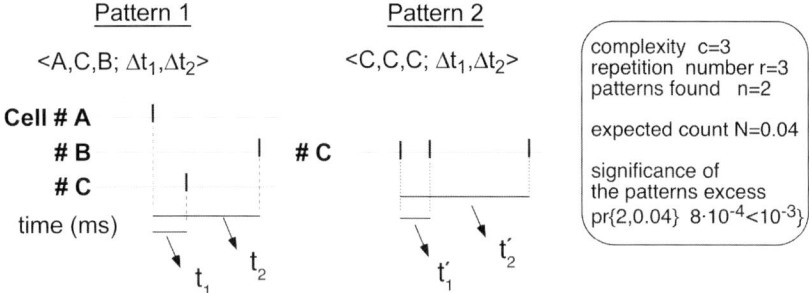

(c) Rasters of spikes aligned on pattern start

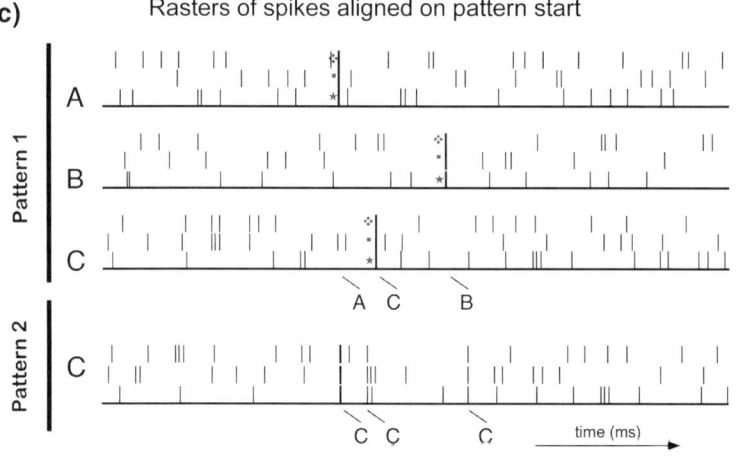

Fig. 3 Outline of the general procedure followed by pattern detection algorithms. (**a**) Analysis of a set of simultaneously recorded spike trains. Three cells, labelled A, B, and C, participate to a patterned activity. Three occurrences of two precise patterns are detected. Each occurrence of the first pattern has been labelled by a specific marker in order to help the reader to identify the corresponding spikes. The spikes belonging to the second pattern are indicated by arrows. (**b**) Estimation of the statistical significance of the detected patterns. Two patterns, n = 2, <A,C,B> and <C,C,C> were found. Each pattern was formed by three neurons, c = 3, and was repeated three times, r = 3, in the analysed record. The expected number of patterns of this complexity and repetition number was N = 0.04. The probability to observe two or more patterns when 0.04 patterns are expected is noted as pr{0.02, 4}. (**c**) Display of the pattern occurrences as a raster plot aligned on the patterns start. (Adapted from Tetko and Villa, 2001.)

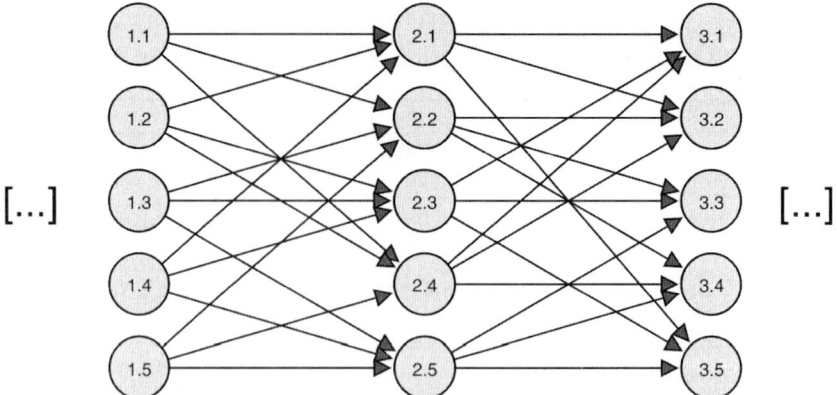

Fig. 4 A schematic synfire chain characterised by the number (w) of neurons in each pool of the chain, and its multiplicity (m), defined as the number of projections from a neuron in the pool n to a neuron in pool $n + 1$. The synfire chain is said to be incomplete if $m < w$, like in this example: $w = 5$ and $m = 3$. Note that individual neurons can appear in multiple pools of the same or different synfire chains. (Reproduced from Iglesias, 2005. With permission.)

The unlimited complexity offered by the unbound number of possible temporal combinations of preferred firing sequences permits to the semantics of composite objects to reach a level of complexity far enough to represent the highest brain activities that characterise human thoughts. This model is undoubtedly appealing and is a source of extensive investigation in computational neuroscience, experimental psychology, and neurophysiology. The theoretical framework that has been delineated can address effective elements associated to the problem of binding but the initial question of coding, as it was defined in the previous section, is not yet solved. The fact that a custom-made statistical analysis can detect the significant firing patterns that are associated to cognitive processes does not tell us much about the read-out mechanisms that should be embedded in the neural networks for decoding the transmitted information (Hopfield and Brody, 2001). Following the neuroheuristic approach we prefer to hold on provisional goals so that the question of neural coding *per se* is temporarily removed and replaced by the study of neural activity in the dynamical perspective.

5   The Neural Catastrophe

The whole time series of spike occurrences is assumed to be an expression of some fundamental process governing the activity of the neurons being recorded. When a specific input pattern activates a cell assembly, the neurons are activated following a certain mode. Then, a *mode of activity* defines how an information is processed within a neural network and how it is associated to the output pattern of activity that is generated. In this framework the *state* of the neural network is defined by a set of parameters characterizing the neural network at a certain time. Then, the state of the network at any given time is represented by the values of these parameters and a network state is fully determined if all parameters are known for each neuron. If we were able *ab absurdo* to set the same initial conditions for all elements of the neural networks we would obtain the same spike trains.

For sake of simplicity, it is rational to describe the activity of the network with the spike trains of all its elements. Spike trains are statistically expressed by point-like processes with the meaning that point process system are systems whose input and output are point processes. Let us consider a simple example of point process system, whose dynamics is characterised by discrete steps in time. Let $\{x_i\}$, $i = 1, \dots K$, be a time series with K points, where x represents the state of the system. In a *dynamical system* the subsequent state of the system is determined by its present state. The simplest expression would be to consider a map defined by $x_{i+1} = ax_i$, where a is a control parameter.

The biological systems, and the brain in particular, are often characterised by feedback mechanisms. The expression $x_{i+1} = ax_i(1 - x_i)$, known as the logistic map, illustrates a simple dynamical system with a negative non-linear feedback, defined for $x \in [0,1]$. It is clear from this expression that the time arrow is non-reversible, because it is always possible for each x_i to obtain a value x_{i+1} but there are two possible x_i for each x_{i+1}.

A dynamical system as a whole is said to be *deterministic* if it is possible to predict precisely the evolution of the system in time if one knows exactly the initial conditions. However, a slight change or incorrect measurement in the initial conditions results in a seemingly unpredictable evolution of the system. A passage in time of a state defines a *process*. Whenever a process is completely deterministic at each step of its temporal evolution but unpredictable over the long term, it is called *chaotic process*, or simply *chaos*.

An equivalent definition of a process is a path over time, or *trajectory*, in the *space of states*. The points approached by the trajectory as the time increases to infinity are called *fixed points* and the set of these points forms an *attractor*. If the evolution in time of the system is described by a trajectory forming a closed loop – also referred to as a *periodic orbit* – then the system is said to have a *limit cycle*. A closer look at the logistic equation shows that for all values $0 \le a \le 1$ the iterated series decay towards zero. For some other values of a, e.g. 1.7 or 2.1, the series converges to a fixed point, equal to 0.52381 and 0.41176, respectively. Conversely, for $a = 3.2$ the series converges to two alternating fixed points, i.e. $x_{i+2} = x_i$, and for $a = 3.52$ it converges to four alternating fixed points, i.e. $x_{i+4} = x_i$. The trajectories of the three last examples are periodic orbits, with periods equal to 1, 2, and 4, respectively.

A further analysis of the logistic equation shows that with a control parameter equal to 4.0 and an initial condition $x_0 = 0.6$ the system tends to decay to zero, but with an initial condition $x_0 = 0.4$ the dynamics never produces a repeating sequence of states. This aperiodic behaviour is different from randomness, or white noise, because an iterated value x_i can only occur once in the series, otherwise due to the deterministic dynamics of the system the next value must be also a repetition, and so on for all subsequent values. In general, brief initial perturbations applied to any combination of the governing set of parameters move a dynamical system characterised by fixed points away from the periodic orbits but with the passing of time the trajectory collapses asymptotically to the same attractor. If the system is deterministic, yet sensitive to small initial perturbations, the trajectory defining its dynamics is an aperiodic orbit, then the system is said to have a *chaotic attractor*, often referred to as a *strange attractor*.

Spike trains are treated as point process systems and a crucial requirement for a theoretical framework is to identify these point process systems without any assumption as to whether or not they are linear. Point process systems are said to be identified when an acceptable model is found. The first step of the identification is to estimate certain conditional rate functions, called kernels, of the spike trains. The one of zero order, i.e. a constant, simply measures the mean firing rate – the average rate of action potentials per unit time. The one of first order, a function of a single time argument, relates to the average effect of a single trigger spike (pre-synaptic) on the spike train. The one of second order, a function of two time arguments, relates to the interactions between pairs of spikes. And so forth for higher-order functions. Then, successive models can be constructed recursively and based on the kernel of zero order, on the kernels of zero and first order, on the kernels of zero, first, and second order, and so on.

By extending this approach to the spike trains recorded from all elements of the neural network it is theoretically possible to develop an acceptable model for the identification of the system. Notice that the goodness of fit of a certain kernel estimate as plausible is evaluated by means of a function f describing its mode of activity – the mode of activity being defined by how information is processed within a neural network and how it is associated to the output pattern of activity that is generated. In formal terms, let us define a *probability function f* which describes how a state x is mapped into the space of states. If the function is set by a control parameter μ we can write $f\mu(x) = f(x,\mu)$. A *dynamical system x'* is a subset of the space of states and can be obtained by taking the gradient of the probability function with respect to the state variable, that is $x' = \text{grad } f\mu(x)$. Mathematically speaking, the space of states is a finite dimensional smooth manifold assuming that f is continuously differentiable and the system has a finite number of degrees of freedom (Smale, 1967).

For periodic activity the set of all possible perturbations define the inset of the attractor or its *basin of attraction*. In case of the logistic map, for $a = 3.2$ all initial conditions in the interval $(0 < x_0 < 0.6875; 0.6875 < x_0 < 1)$ end up approaching the period 2 attractor. This interval is known as the basin of attraction for the period 2 attractor, whereas the value $x_0 = 0.6875$ is a fixed point. If the activity is generated by chaotic attractors, whose trajectories are not represented by a limit set either before or after the perturbations, the attracting set may be viewed through the geometry of the topological manifold in which the trajectories mix. The function f corresponding to the logistic map, $f(x) = ax(1 - x)$, is the parabolic curve containing all the possible solutions for x. This function belongs to the single humped map functions which are smooth curves with single maxima (in here the single maximum is at $x = 0.5$).

Let us consider again the case of large neural networks where the complexity of the system is such that several attractors may appear, moving in space and time across different areas of the network. Such complex spatio-temporal activity may be viewed more generally as an *attracting state*, instead of simply an attractor. In particular, simulation studies demonstrated that a neural circuit activated by the same initial pattern tends to stabilise into a timely organised mode or in an asynchronous mode if the excitability of the circuit elements is adjusted to the first-order kinetics of the post-synaptic potentials (Villa and Tetko, 1995; Hill and Villa, 1997).

Let us assume that the dynamical system is structurally stable. In terms of topology structural stability means that for a dynamical system x' it exists a neighborhood $N(x')$ in the space of states with the property that every $Y \in N(x')$ is topologically equivalent to x'. This assumption is extremely important because a *structurally stable dynamical system* cannot degenerate. As a consequence, there is no need to know the exact equations of the dynamical system because qualitative, approximate equations – i.e. in the neighbourhood – show the same qualitative behaviour (Andronov and Pontryagin, 1937) (Fig. 5).

In the case of two control parameters, $x \in IR$, $\mu \in IR^2$, the probability function f is defined as the points μ of IR^2 with a structurally stable dynamics of $x' = $ grad $f\mu(x)$ (Peixoto, 1962). That means the qualitative dynamics x' is defined in a neighbourhood of a pair (x_0, μ_0) at which f is in equilibrium (e.g. minima, maxima, saddle point). With these assumptions, the equilibrium surface is geometrically equivalent to the Riemann-Hugoniot or *cusp catastrophe* described by Thom (1975). The cusp catastrophe is the universal unfolding of the singularity $f(x) = x^4$ and the equilibrium surface is described by the equation $V(x,u,v) = x^4 + ax^2 + bx$, where a and b are the

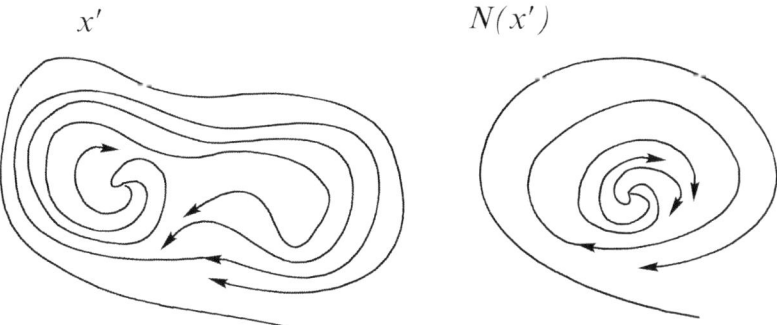

Fig. 5 Two-dimensional representation of dynamical systems that are topologically equivalent (see text for definition of equivalency)

control parameters. According to this model the equilibrium surface could represent stable modes of activity with post-synaptic potential kinetics and the membrane excitability as control parameters (Fig. 6).

The paths drawn on the cusp illustrate several types of transitions between network states. Point (*a*) on the equilibrium surface of Fig. 6 corresponds to a high level of excitability and a relatively long decay time of the post-synaptic potentials, e.g. 12 ms. This may be associated to the tonic mode of firing described in the thalamo-cortical circuit, where bi-stability of firing activity has been well established. This is in agreement with the assumption that the same cell would respond in a different mode to other conditions.

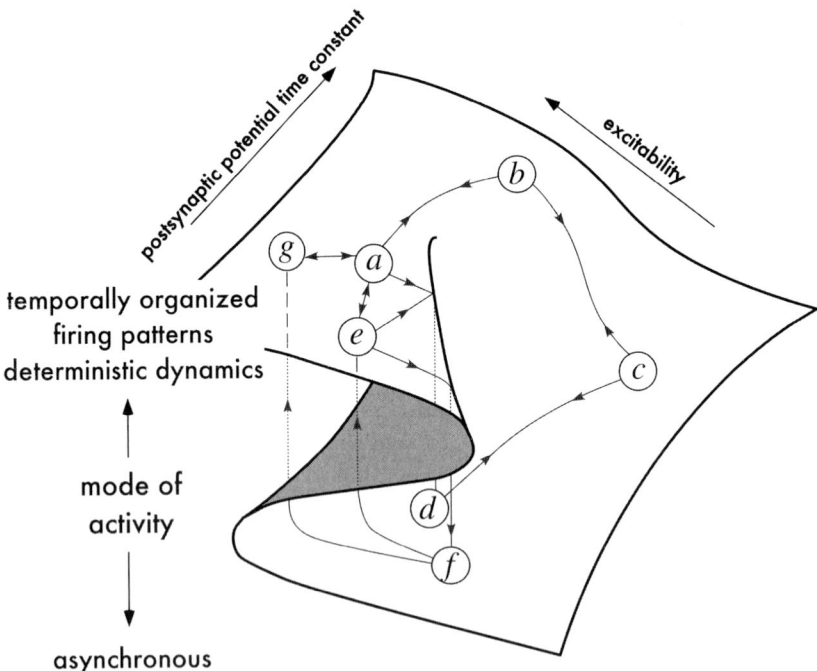

Fig. 6 Topological interpretation of neural dynamics as a function of two control parameters, the cell excitability and the kinetics of the post-synaptic potentials. The equilibrium surface is represented by a cusp catastrophe where transitions can occur either suddenly or continuously between temporally organised firing patterns and asynchronous activity. This equilibrium surface refers to the activity of a specific neural network and it is possible the same cell belongs to more than one cell assembly. If the cell assemblies are controlled only by one parameter in common, then temporal and rate code are not mutually exclusive

In general, the same neural network may subserve several modes of activity through modulation of its connectivity, e.g. according to learning or pathological processes, or by modulation of its excitability, e.g. by modulation of the resting potential or of the synaptic time constants. Remember that the state of the neural network is defined by a set of characteristic control parameters at a certain time. Then, at any given time, the state of the network is represented by the values of control parameters and a network state is fully determined if all parameters are known for each neuron. It is not possible to know all variables determining brain dynamics, yet the progresses made by computational and statistical physics have brought a number of methods allowing to differentiate between random, i.e. unpredictable, and chaotic, i.e. seemingly unpredictable, spike trains.

In this theoretical framework at point (a) in Fig. 6 the network state is such that an input pattern will evoke precisely time-structured activity detectable by preferred firing sequences. These sequences should not be interpreted as 'Morse' code because different firing patterns might be evoked by the same input if the synaptic dynamics is changed within a certain range of cellular excitability, as suggested for neuromodulatory mediators.

Also, different input patterns of activity may produce similar modes of activity, somewhat like attractors. The transitions between these states are represented by paths (a-b-a), (a-e-a) and (a-g-a). Indeed it has been observed in the cortex and in the thalamus that several types of neurons tend to switch towards a rhythmic or bursty type of firing if the excitability is decreased. This effect may be provoked by a hyperpolarization of the cell membrane or by modifying the spike threshold level (Foote and Morrison, 1987). In the former case a smooth passage between timely structured activity and asynchronous firing is likely to occur, as suggested by path (b-c-b), especially if the synaptic decay is long. Conversely, if the synaptic decay is fast and a modulatory input modifies the threshold potential a sudden switch from temporal patterns of firing to desynchronised activity will occur, as indicated by paths (a-d) and (e-f).

Complex spatio-temporal firing patterns may also occur with low levels of excitability, i.e. point (e) in Fig. 6, as suggested by cholinergic switching within neocortical networks (Villa et al., 1996; Xiang et al., 1998). Point (e) on the equilibrium surface can be particularly unstable because a further decrease in excitability, path (e-f), but also an increase in synaptic decay, path (e-d), may provoke a sudden change in the mode of activity, as observed in simulation studies (Villa, 1992; Hill and Villa, 1997).

It is important to notice that if the excitability is low, e.g. during long lasting hyperpolarisation, the kinetics of the post-synaptic potential is often irrelevant regarding the input pattern so that the output activity will always tend to be organised in rhythmic bursts. Conversely, if the excitability is increased from a starting point (f) and the time constant of the synaptic decay is fast, say 4–5 ms, the input patterns could turn on either stable, path (f-g) or unstable temporally organised modes of activity only through sudden transitions, path (f-e).

6   Postlude

Defining a coherent workplan for an innovative research project is considerably dependent on the ongoing difficulties that can be foreseen but also subject to unforeseeable factors. Artificial neurons have been modelled in the last years trying to characterise the most important features of real neurons and to achieve further aspects associated to their computational characteristics and learning. Biologically inspired neuron models, so-called neuromimes, have become an important topic of research aimed at studying selected mathematical models of single neurons with statistical properties that could fit real data and to study the sensitivity of these models to noise. Noise is unavoidable in any living system and the experimental results obtained from very different neuronal structures may show the important role of noise in the transfer of information in neuronal networks. It plays indeed an important role in neural transmission and a fundamental question can be raised about the meaning of neural code, as previously defined in cryptology, if noise matters so much (García-Ojalvo and Sancho, 1999).

A workplan flexible enough to allow for the emergence of new concepts that are not bound by a restrictive definition of coding should include the investigation of the significance of synchronised spike patterns in the brain and the mechanism of their generation and propagation. In Section 5 the neuronal network appears as a complex system characterised by coupled non-linear dynamical systems. Synchronization of chaotic systems has been discovered (e.g., Fujisaka and Yamada, 1983, Pecora and Carroll, 1990) and since then it has become an important research topic in mathematics (e.g. Ashwin et al., 1994), physics (e.g. Ott and Sommerer, 1994), and engineering (Chen, 1999). In a broader context, this workplan is related to the open question of how structured information is represented in the brain. As engineers increasingly are inclined to look for inspiration from biology, this question is also of general relevance for information technologies. The confrontation between Neurosciences and Computer and Information Sciences will allow the emergence of new provisional hypotheses, i.e. functional hypotheses limited to the *topoï* where these hypotheses evolved, for an ongoing 'stepwise' understanding of the higher cognitive functions.

This last section cannot be an epilogue or a conclusion because we are still far from designing a fully intellectually satisfactory framework defining the neural coding. The postlude represents a closing phase and a new opening at the same time. To use a metaphor, we could state that the neuroheuristic approach observes the experimental results beyond the surrounding wall of the hypothesis by coupled conjecture and testing, similarly to a child playing in a garden while observing what happens beyond whatever enclosure surrounds him, which could be a hedge, a gate or a lattice. All of these are closures but they are all different and belong to distinct *topoï*. This metaphor also serves to indicate the character, sometime with passion, sometimes with pleasure, that must cloth the act of the scientific discovery. In order to retract the former maxim *ludendo docere*, research directed strictly by a registered finality according to a settled perspective is dangerous for scientific creativity. Remarked upon by the neurobiologist, Alan Hodgkin (1977):

> The stated object of a piece of research often agrees more closely with the reason for continuing or finishing the work than it does with the idea which led to the original experiments. In writing papers authors are encouraged to be logical, and, even if wished to admit that some experiment that turned out in a logical way was done for a perfectly dotty reason, they would not be encouraged to "clutter up" the literature with irrelevant personal reminiscences.

Since Neolithic times, when the first flint was produced, the tools have changed in form, but the interaction that they have with the pathway of discovery has remained essentially constant. Due to modern computer technologies, investigation methods without precedent have emerged in research of the Neurosciences. However, the complexity of problems presented to the researcher of today is of such a broad span that the database approach is unable of reducing its computational performance to a disciplinary dimension. The neuroheuristic approach to brain sciences attempts to promote a paradigm based upon synergy between modelling and experiments which parallels the synergy between Computer and Information Sciences with the Neurosciences.

The hard sciences of engineering, physics and mathematics are now making important contributions to a theoretical framework for the various micro and ever more global information strategies used in the brain. Non-invasive brain imaging, lesion studies and single and multi-electrode studies on animals and humans are now producing results of deep import as to the networks of neural modules supporting a broad range of psychological tasks. The increasing information about these networks is raising optimism as to the possibility of constructing a global model of the brain as a multi-modular system. The philosophy behind such an approach is to start with simple nerve cells – the neurons – and modules, and develop increasing complexity in due course. The task itself requires transdisciplinarity and the neuroheuristic approach may lead to a framework from which to tackle the study of higher nervous activities from a global viewpoint.

References

Abeles M (1982) Local Cortical Circuits: An Electrophysiological Study. Springer Verlag, Berlin

Abeles M (1991) Corticotronics: Neural Circuits of the Cerebral Cortex. Cambridge University Press, New York

Abeles M, Hayon G, Lehmann D (2004) Modeling Compositionality by Dynamic Binding of Synfire Chains. J Comput Neurosci 17:179–201

Andronov A, Pontryagin L (1937) Systèmes grossiers. Dokl Akad Nauk SSSR 14:247–251

Ashwin P, Buescu J, Stewart I (1994) Bubbling of attractors and synchronisation of oscillators. Phys Lett A 193:126–139

Bell CC, Han VZ, Sugawara Y, Grant K (1997) Synaptic plasticity in a cerebellum-like structure depends on temporal order. Nature 387:278–281

Bergson H (1917) L'évolution créatrice. Presses Universitaires de France, Paris

Bienenstock E (1995) A model of neocortex. Network: Comput Neural Syst 6:179–224

Braitenberg V, Schüz A (1991) Anatomy of the Cortex: Statistics and Geometry. Springer Verlag, Berlin

Brazier MAB (1984) A History of Neurophysiology in the 17th And 18th Centuries: From Concept To Experiment. Raven Press, New York

Chen G (1999) Controlling Chaos and Bifurcations in Engineering Systems. CRC Press LLC, Florida

Foote SL, Morrison JH (1987) Extrathalamic modulation of cortical function. Ann Rev Neurosci 10:67–96

Fujisaka H, Yamada T (1983) Stability theory of synchronized motion in coupled-oscillator systems. Prog Theor Phys 69:32–47

García-Ojalvo J, Sancho JM (1999) Noise in Spatially Extended Systems. Springer-Verlag, New York

Hadamard J (1952) La psychologie de l'invention en mathématiques. Gauthier-Villars/Gauthier-Villars, Paris

Hill SL, Villa AEP (1997) Dynamic transitions in global network activity influenced by the balance of excitation and inhibition. Network 8:165–184

Hodgkin AL (1977) Chance and design in electrophysiology: an informal account of certain experiments on nerve carried out between 1934 and 1952. In: The Pursuit of Nature. Informal essays on the history of physiology, Cambridge University Press, Cambridge

Hopfield JJ, Brody CD (2001) What is a moment? Transient synchrony as a collective mechanism for spatiotemporal integration. Proc Natl Acad Sci USA 98:1282–1287

Gerson LP (2005) The Cambridge Companion to Plotinus, Series: Cambridge Companions to Philosophy. Cambridge University Press, Cambridge

McCulloch WS, Pitts W (1943) A logical calculus of the ideas imminent in nervous activity. Bull Math Biophys 5:115–133

Ott E, Sommerer JC (1994) Blowout bifurcations: the occurrence of riddled basins and on-off intermittency. Phys Letts A 188:39–47

Pecora LM, Carroll TL (1990) Synchronization in chaotic systems. Phys Rev Lett 64:821–824

Peixoto M (1962) Structural stability on 2-dimensional manifolds. Topol 1:101–120

Rolando L. (1809, 1974) Saggio sopra la vera struttura del cervello dell'uomo e degli animali e sopra le funzioni del sistema nervoso, Biblioteca di storia della medicina, Vol. 6. A. Forni ed., Bologna

Segundo JP (1983) Rationalism in an age of reason. J Theoret Neurobiol 2:161–165

Segundo JP (1985) Mind and matter, matter and mind? J Theoret Neurobiol 4:47–58

Shannon CE (1948) A mathematical theory of communication. Bell Syst Tech J 27:379–423

Shepherd GM (1994) Neurobiology, 3rd edn. Oxford University Press, New York/Oxford

Shmiel T, Drori R, Shmiel O, Ben-Shaul Y, Nadasdy Z, Shemesh M, Teicher M., Abeles M (2005) Neurons of the cerebral cortex exhibit precise interspike timing in correspondence to behavior. Proc Natl Acad Sci USA 102:18655–18657

Smale S (1967) Differential dynamical systems. Bull Am Math Soc 73:747–817

Taylor JG, Villa AEP (2001) The 'Conscious I': A neuroheuristic approach to the mind. In: Baltimore D, Dulbecco R, Jacob F, Montalcini RL (eds) Frontiers of Life, Vol III. Academic Press, New York, pp. 249–270

Thom R (1975) Structural stability and morphogenesis (Eng transl.). W.A. Benjamin, Reading, Massachusetts

Villa AEP (1992) Les catastrophes cachées du cerveau. Le Nouveau Golem 1:33–63

Villa AEP, Tetko IV (1995) Spatio-temporal patterns of activity controlled by system parameters in a simulated thalamo-cortical neural network. In: Hermann H (ed.) Supercomputing in Brain Research: from Tomography To Neural Networks. World Scientific Publishing, Singapore, pp. 379–88

Villa AEP, Bajo VM, Vantini G (1996) Nerve growth factor (NGF) modulates information processing in the auditory thalamus. Brain Res Bull 39:139–147

Villa AEP, Tetko IV, Hyland B, Najem A (1999) Significance of spatiotemporal activity patterns among rat cortex neurons in performance of a conditioned task. Proc Natl Acad Sci USA 96:1006–1011

Xiang Z, Huguenard JR, Prince DA (1998) Cholinergic switching within neocortical inhibitory networks. Science 281:985–988

Chapter 17

Error Detection and Correction Codes

Diego L. Gonzalez

Abstract This article presents an elementary introduction to the encoding methods used for error detection and correction in communication processes. These methods have been developed following the pioneering work of Claude Shannon in the 1940s that founded *Information Theory*. Information theory studies in mathematical terms how to transmit messages in a reliable way using communication channels that necessarily introduce "noise" or errors in the messages. The key point for the implementation of error-free communication is the encoding of the information to be transmitted in such a way that: (a) some extent of redundancy is included in the encoded data, and (b) a method for efficient decoding at the receiver is available. These two requirements together usually imply that the data to be transmitted need to be mathematically organized, often following principles borrowed from discrete group theory. In this article a review of encoding methods so far developed for this end is given.

Just as it is clear that error-correcting coding methods represent a key feature for the development of successful practical communication technologies, it is also becoming ever more clear that living organisms need to resort to analogous strategies for optimizing the flux and integrity of biological information. In addition to the general theoretical constraints which every communication system needs to obey, the possible role of such error detection and correction codes in biological (genetic and neural) systems is briefly discussed here, and a new possibility for implementing error detection and correction based on generic properties of non-linear dynamical systems and their associated symbolic dynamics is presented.

1 Introduction

From the origin of mankind, humans have tried to represent thoughts, sensations, and ideas through symbols. From the first pictorial signs featuring natural forms, animals, and also numbers, the level of representation has become ever more

St. George School Foundation and National Research Council of Italy,
e-mail: diego.gonzalez@cini.ve.cnr.it

M. Barbieri (ed.), *The Codes of Life: The Rules of Macroevolution.*
© Springer 2008

abstract. At the end of this first encoding process primitive spoken and written languages were born. In the majority of known languages, an abstract correspondence between a represented concept (a thing, an idea) and a symbol (not necessarily related in any way to the concept that it represents) is established. This abstract process of representation is indeed a coding: we accept that the word *house* represents any kind of construction made for the protection and habitation of human beings but this particular sequence of five letters of the alphabet is indeed completely arbitrary. It is universally accepted that language has developed in order to establish a common set of symbols allowing for communication between different individuals. With the passage of time this initial objective has been enlarged, for example, to ensuring the secrecy of important written messages, or, more recently, to allowing the transmission of information at great distances, as in the case of first the telegraph, and successively of different kinds of radio communications. Despite this ancient cultural origin of spoken and written codes, their systematic study on a rigorous scientific basis is very recent; indeed it was born in parallel with the development of technologies for communication at a distance. The main problem that has been faced in the development of such technologies is how to ensure the accuracy of the transmitted information using transmission media that necessarily introduce unpredictable errors. The answer to this simple problem in mathematical terms was not solved until the fundamental contributions of Claude Shannon in the 1940s that gave rise to "Information Theory" (Shannon and Weaver 1949a, Shannon 1949b, Pierce 1961, Yockey 1992).

2   Number Representation Systems

At a digital level, for the processing and transmission of information, we need first to convert information into numbers. A digital photograph observed on a computer screen or in a color print give us high-level information that our brain interprets at a holistic level: for example, as a red poppy in a corn field. For the computer, instead, this image is nothing but a long sequence of numbers encoding the intensity and the color of any image point called a *pixel*. Moreover, the numbers representing pixels are in a particular number representation called the *binary system* (Ore 1988). Binary digits are the only numbers that our computers are able to manage at the hardware level. This technological choice is due to the fact that binary numbers are represented using only two digits, i.e. 0 and 1. The binary representation system pertains to the class of *positional power representation systems*. In positional representation systems, any ordered digit in the representation represents the number of times that a given power of an integer number called the *base* (b) is included in the additive decomposition of the number. The position of the digit determines the power of the base by which the digit is multiplied: the nth digit is multiplied by b^n. The digit position is computed starting with the rightmost one, because the rightmost digit is the least significant, corresponding to a positional weight of $1 = b^0$. In a particular power representation, the digits should take values from 0 to b − 1 in order to assure that the representation is univocal: given a number, its representation is unique (there is only one sequence of signs), and given a sequence of signs, these specify through additive decomposition only this same number. Of course it is necessary also that all numbers may be represented, and this property is assured by the fact that the sum up to order n of all the powers of the base times the greatest possible digit, i.e. (b − 1), equals the following power of the base, (n + 1), minus one,

$$\sum_{n=0}^{k-1}(b-1)\cdot b^{n}=b^{k}-1$$

Thus, the only thing that is needed in order to completely specify a positional power representation system is the value of the base b. Our usual number representation system, the decimal one, pertains also to this class of representation systems with the value of the base equal to 10. The same is true for other common number representation systems, for example, b = 2 for the binary system, b = 8 for the octal one, and so on.

As an example we show the representation of the number 421 in the usual decimal system (base 10). As stated above, the digits can take only the values from 0 to b − 1 = 9. From right to left (see Table 1), any digit is multiplied by the corresponding power of 10 and the result sums to 421.

For the case of the binary system, as remarked above, the digits can take only the values 0 and 1. For this reason it is the system preferred in computing applications: the two digits can be represented by two complementary states of an electronic system, for example, an electronic switch with the "on"–"off" states representing the 1 and 0 values of the binary digits. Really, there is no limitation to the number of stable states that an electronic system can have, but historical reasons have determined the use of the binary system as a privileged one for computing applications, for example, thanks to the development of a very consistent theoretical corpus known as *Boolean Algebra* (Kohavi, 1970) that allows a great simplification in the design of logical functions.

As an example we can see in Table 2 the representation of the decimal number 21 in the binary system (1,0,1,0,1)

Table 1 Representation in base 10 of the number 421. The decimal representation system is one example of positional power representation systems, which include also the binary system

Base powers	$10^2 = 100$		$10^1 = 10$		$10^0 = 1$		Total
	x		X		X		
Multiplicative digits	4		2		1		
	=		=		=		
Partial results	400	+	20	+	1	=	421

Table 2 Representation of the decimal number 21 in the binary system. As in the decimal case the digits (0 or 1), are multiplied by the powers of the base (2) from right to left. The sum of these products equals 21

Base powers	$2^4=16$	$2^3=8$	$2^2=4$	$2^1=2$	$2^0=1$	Total
	x	x	x	x	X	
Multiplicative digits	1	0	1	0	1	
	=	=	=	=	=	
Partial result	16	0	4 +	0 +	1 =	21

The univocal property of positional power representations precludes the existence of redundancy in the represented numbers. As we will see later, redundancy is a fundamental ingredient for detecting and correcting eventual errors in a communication process. For this reason, if we want to implement error correction codes, redundancy has to be artificially introduced. It may be remarked that, although redundancy is absent in power representation systems, other number representation systems exist that introduce redundancy in a natural fashion. This is the case for the *non-power positional representation systems*. An example of the utility of this particular kind of number representation for the description of biological information is given in this same volume (Chapter 6, this volume).

3   Information Theory, Redundancy, and Error Correction

Having defined the primary number representation system on which we intend to implement the coding methods for error correction, that is the binary system, we may study in more general terms the problem of information transmission. This is the main scope of *Information Theory*. As previously remarked, the main question is to understand the information transmission limitations in speed and accuracy due to the nonideal character of the transmission channels, i.e. due to the introduction of undesirable but unavoidable errors in the transmitted information.

First of all we need to quantify the information. For our purposes, information is equivalent to the quantity of different states that can be transmitted. If we think of ordinary language, if we transmit Italian words, for example, of only two letters, there being 21 different letters in the Italian alphabet, we have $21 \times 21 = 441$ possible words (of course, some of these do not have a defined meaning in the language). This means that we can transmit 441 different messages, and the quantity of information will be proportional to this number. If we take the base-2 logarithm of that number as a definition of the quantity of information, such a quantity coincides with the number of bits in the binary system needed for transmitting all the possible messages. As the digits in the binary system can take only two values, a binary word of n bits encodes a maximum number of 2^n different messages. In our case, the number of bits is 9, which is the minimum integer number ensuring the transmission of all the possible messages (in excess, because $2^9 = 512 > 441$ but $2^8 = 256$ which does not suffice for representing all the possible words).

Now it is possible to define a correspondence (which may be arbitrary or, alternatively, defined by the coding of individual letters) between the messages; that is, the words of two letters, and the binary coding; that is, binary words of length 9. For example, using an arbitrary correspondence, the first words can be described as follows (observe that any word of two Italian letters is assigned a length 9 binary string),

aa » 000000000
ab » 000000001
ac » 000000010
ad » 000000011

As stated above, the length of the binary strings will be at least 9 in order to be able to transmit all the possible messages. This number is independent of the particular choice of the correspondence between the words (information) and the binary strings. For this reason it defines the quantity of information, which is independent of the particular representation we choose. In the present example, because the number of binary strings is greater than the number of two-letter words, some binary strings may be empty, that is, with no assigned word (strictly speaking the quantity of information can take non integer values, indeed, is the base-2 logarithm of the number of possible messages; for practical reasons, however, the quantity of information as a number of bits is always an integer).

If now we aim to utilize this binary code for transmitting the words through a nonideal communication channel, a significant problem appears: a minimal error of 1 bit in the received information lead us to interpreting the message in a completely wrong manner. If, for example, we intend to transmit the word "aa" by the string "000000000" and this string suffers only one error in the last – least significant – binary bit, becoming "000000001," the transmitted message will be interpreted as "ab" as can be seen in Table 1. From the point of view of the accuracy of the transmitted information this is a disaster, and in order to cope with it we need to introduce some means for detecting and correcting errors. As we will see later this ability, in turn, relies necessarily on the introduction of redundancy at the level of the transmitted information.

The fact that error detection and correction is possible in a communication system is a consequence of the Shannon theorem, with which information theory was born as a scientific discipline. The Shannon theorem ensures that reliable communication is possible simply by augmenting the length of the strings used in the transmission process. For example, by using 10 bits instead of 9 for encoding the two-letter words of the former example, we can assign to every word two different binary strings because the total number of strings is greater than twice the total number of words, i.e. $2^{10} = 1024 > 2 \times 441 = 882$.

This nonunivocal representation of the transmitted information (any word is represented by more than one binary string) is a redundant one. It is through the use of this redundancy that the reliability of communication can be improved. It is clear, however, that there is a practical limit for increasing redundancy: very long strings can assure an almost error-free communication but the efficiency of the communication process decreases. For transmitting the redundant information in the same time we need more channels and more complex and faster means for coding and decoding operations. In technical terms we say that we need more bandwidth in the communication process, which is necessary for transmitting information at higher rates.

3.1 The Shannon Theorem

Suppose that one needs to transmit a message of M bits. This message is a word among the 2^M possible words that can be transmitted with a string of M bits. At this point we can ask how much redundancy (how many extra bits) needs to be added in order to assure that the message can be completely recovered at the receiver. The answer to this question is given by the second Shannon theorem:

If the probability to have an error in a single bit is p, and this probability is the same for all the bits of the message, in order to be sure of recovering the original message we need to add at least $N.H(p)$ redundancy bits. N is the total number of bits of the transmitted message (original bits plus redundancy or checking bits), and $H(p)$ is a quantity that depends only on the probability of introducing an error:

$$H(p) = p.\log_2(p) - (1-p).\log_2(1-p)$$

This means that given the original number of bits, M, and the error probability, p, the total number of bits N will be:

$$N = M/(1-H(p))$$

However, this is a theoretical result and it is licit to ask how a practical code ensuring the complete recovery of the message can be implemented.

Human languages furnish a good example of how redundancy can contribute to message reliability. If a scribe has introduced an error in the transcription of an ancient text, for example, in the word *house*, the word may become for example *houte*, which has no meaning; it is not a word in the English language. By a simple reading of the text it is easy to understand that an error has been introduced. In this case it is also relatively simple to retrieve the right word, taking into account the context in which the word is placed; the meaning of the paragraph in which the word is inserted. The main problem is that other possible errors in a single letter of the word may lead to another valid word, for example *mouse*. It is clear that longer

Table 3 Hamming distance between binary strings. This is defined as the number of differing bits between two strings of the same length. In (a) the Hamming distance is 1; in (b) the Hamming distance is 3 (the differing bits are shown in grey).

(a)

String 1	0	0	0	0	0	0	0	0	0
String 2	0	0	0	0	0	0	0	0	1

(b)

String 1	0	0	0	0	1	0	0	0	1
String 2	0	0	0	0	0	0	0	1	0

words have lesser probability that a single error leads to another valid word; for example, the word *communication* does not lead to another valid word of the English dictionary from an error in only a single letter. In this case the recovery of the original word is ensured. This is due to the fact that the number of possible words with 13 letters is very high; 21^{13} or approximately 10^{17}, while the total number of words in the English language is at least 11 orders of magnitude less.

Thus this error-correction property of common languages is based on the fact that not all of the possible combinations of letters of the alphabet represent allowed words. In this way, if an accidental change leads to an inexistent word, we are sure that an error has been introduced. This property of languages introduces us to the concept of the distance between words, or equivalently for the transmission of coded messages, of the distance between binary strings.

The easiest way to define such a distance between two strings is to count the number of bits that differ (comparing bits in the same digit position). This is known as the Hamming distance (Sweeney, 2002); For example, the strings 000000000 and 000000001 are at a Hamming distance of 1 because they differ only in the last bit, while the strings 0000100001 and 000000010 (Table 3a,b) are at a Hamming distance of 3 because bits 1, 2, and 5 are different.

3.2 Parity Based Error Detection/Correction Methods

A simple method for detecting errors consists in encoding the information (two-letter words in our previous example) by a subset of the redundant binary strings. Doing this, the words can be encoded in such a way that the Hamming distance between different words is maximized. If the least distance between strings pertaining to different words is 2, the arrival at the receiver of a string at a Hamming distance of 1 from one of the strings of the coding table (the allowable strings) is a sure sign of the introduction of an error in the transmission of the information. In fact, a single error in a word leads to a Hamming distance of 1 with respect to some other string of the allowed subset (accessible, for example, through a table). However, if the typical Hamming distance is two, we are not able to correct the error (the incorrect string will be at equal distance, i.e. 1, from two different valid strings). More sophisticated techniques allow for error detection and correction in an arbitrary number of bits. In order to give an example of such a possibility, we consider binary strings of length 6. One of the first methods employed for error detection was the control of the parity of the string. Suppose we want to transmit the string 1,0,1,0,1,1, and to be able at the receiver to understand if a one-bit error has been introduced. In order to achieve this, as mentioned above, we need to introduce redundancy in the information. For example, we can add an additional bit, represented by the new grey cell in Table 4.

In order to define the value of this additional bit we can use the concept of the parity of the string: the parity of a binary string coincides with the parity of the number of ones that the string contains. An odd number of ones defines an odd string, and an even number of ones, an even string. The succeeding step is to impose the condition that all the transmitted strings must be odd (equally we could choose that they be even). As the original string 1,0,1,0,1,1 is even (it contains an even number, i.e. 4, of 1's,), in order that the enlarged string (the original string plus the additional parity bit) be odd, the additional bit must be a 1, as shown in Table 5.

In this way the enlarged (redundant) string will be odd (it contains an odd number of 1's, i.e. 5). If at this point an error is introduced in the transmission of the string (including the additional bit) as shown in Table 6, at the receiver there will arrive an even string, signaling in this way that an error has been introduced. This method does not allow for the detection of an even number of errors, because in such a case the parity of the received string returns to its correct – odd – value. Moreover, the method does not provide the knowledge of in which bit the error has been introduced, and consequently the error cannot be corrected. The method can be improved by adding more parity bits, that is, by adding further redundancy and thus transmitting longer strings as required by the Shannon theorem. On the basis of the theoretical predictions of Shannon, as early as the beginning of the 1950s,

Table 4 Additional redundancy bit converting a length-6 binary string into a length-7 one

1	0	1	0	1	1
1	0	1	0	1	

Table 5 In order for the string to be of total odd parity, the additional bit must be a 1 (grey cell); the total number of 1's is now odd (5)

1	0	1	0	1	1	1

Table 6 An error is introduced in the fifth digit, changing it from 1 to 0 (shown in grey)

1	0	0	0	1	1	1

Richard Hamming (1950) succeeded in developing a coding method based on the parity of the strings that provides for the localization of the errors (the position of the corrupted bits) and thus allows their correction, for errors of only one bit. The method also permits the detection of errors of two bits but not their correction. The quantity of additional bits required for implementing the Hamming method is $\log_2(N) + 1$. This implies the curious result that doubling the data string length requires only one additional parity bit for maintaining the 1-bit accuracy in the receiver.

Suppose that we have strings 8 bits in length. The former result indicates that we must add $\log_2(N) + 1 = \log_2(8) + 1 = 3 + 1 = 4$ parity check bits. We can distribute the additional parity bits, P_1, P_2, P_3, and P_4, in the following way (shown in grey),

P_1	P_2	1	P_3	2	3	4	P_4	5	6	7	8
		1		1	0	1		0	1	1	0

For calculating the values of the parity bits it suffices to consider that the position of any data bit is encoded by the binary number defined by the parity bits; for example, the position of the 7th bit of the data string (binary value 1), is related to the parity bits P_1, P_2, and P_3, because 7 is represented in binary digits as 1,1,1,0 (the last 0 means that the 7th bit does not participate in the coding of P_4). Table 7 displays all the string data positions as a binary function of the parity check bits.

Through simple inspection of this table we can deduce that, for example, the parity bit P_1 is involved in the coding of the data in the positions 1, 3, 5, and 7 (the positions marked by a 1 in the column corresponding to the bit P_1). If we want to implement an even parity coding, it suffices to define the parity bit P_1 in such a way that the group of bits 1, 3, 5, 7 and P_1, be of even parity. As the subgroup 1, 3, 5, 7

Table 7 To determine the value of the parity check bits in the Hamming method, the groups defined by data corresponding to a 1 in a particular parity column and the parity check bits are forced to be of the same parity. For example, the group formed by the column corresponding to P_1 is determined by data in the positions 1, 3, 5, and 7 (1, 0, 0, 1), and this same number (P_1). In order to ensure that this group be of even parity we are forced to choose P_1 as a 0; in this way, the total parity of the group is even: we have an even number of ones in the group (2)

P_1	P_2	P_3	P_4	Data position	Data values
1	0	0	0	1	1
0	1	0	0	2	1
1	1	0	0	3	0
0	0	1	0	4	1
1	0	1	0	5	0
0	1	1	0	6	1
1	1	1	0	7	1
0	0	0	1	8	0

(1,0,0,1) is even, P_1 should be a 0; in this way, the complete group 1, 3, 5, 7 and P_1, (1,0,0,1,0) will be of even parity.

The other parity bits, P_2, P_3, and P_4, can be determined in an analogous way. Observing Table 7 we deduce that the bit P_2 is determined by the data bits 2, 3, 6, and 7 (1,0,1,1); as this subgroup is odd, we need to assign a 1 to P_2 so that the complete group, 2, 3, 6, 7, and P_2, becomes of even parity. The other complete groups are: 4, 5, 6, 7, and P_3, and, 8 and P_4. The final result is: P_1, P_2, P_3, P_4 = (0,1,1,0). Adding these parity bits to the original string we obtain,

P_1	P_2	1	P_3	2	3	4	P_4	5	6	7	8
0	1	1	1	1	0	1	0	0	1	1	0

This string is the redundant one – the original data plus 4 additional parity bits – that will be transmitted. If at the receiver this string arrives corrupted by an error in one of its bits, thanks to the parity coding, detection of the position of the error and thus its correction will be possible. Suppose that the 7th bit is corrupted:

P_1	P_2	1	P_3	2	3	4	P_4	5	6	7	8
0	1	1	1	1	0	1	0	0	1	0	0

Such an error will cause the groups, (1,3,5,7, and P_1), (2,3,6,7, and P_2), and (4,5,6,7, and P_3), to be of odd (wrong) parity. Returning to Table 7, we observe that the only bit which changes the parity of the groups containing P_1, P_2, and P_3 is the bit in the 7th position, as shown in grey in Table 8. Thus the position of the corrupted bit should be the 7th one, which can be immediately corrected by changing its value from 0 to 1. With this change all the parity groups return to their original even parity coding.

If in the transmission process two errors (instead of one as in the previous example) are introduced, for example, at the 3rd and 7th positions,

P_1	P_2	1	P_3	2	3	4	P_4	5	6	7	8
0	1	1	1	1	1	1	0	0	1	0	0

only one parity group becomes of odd parity: the parity group, 4,5,6,7, and P_3, (1,0,1,0,1), is odd, as shown in Table 9.

But this information suggests that the wrong data bit is in the 4th position because changing it from 1 to 0 makes all the parity bits even. Nevertheless, we know that the wrong bits are those in the 3rd and 7th positions. This example shows that this coding scheme does not allows for adequate correction of more that one simultaneous error.

The previous example, which we developed in an intuitive approach, is part of the error detection and correction methods based on the parity-coding scheme called "block parity coding" (Sweeney, 2002). These coding schemes are characterized by

Table 8 We observe that groups involving P_1, P_2, and P_3, parity check bits are of the wrong parity (odd). The only data bit involved in these three groups is the 7th one; thus this must be the corrupted binary digit. Correction of the error can be immediately performed by changing its value from 0 to 1. It can be checked that by doing this all the groups recover their correct, even, parity

P_1	P_2	P_3	P_4	Data position	Data values
1	0	0	0	1	1
0	1	0	0	2	1
1	1	0	0	3	0
0	0	1	0	4	1
1	0	1	0	5	0
0	1	1	0	6	1
1	1	1	0	7	0
0	0	0	1	8	0

Table 9 With the introduction of two errors at digit positions 3 and 7 only the group involving the parity check P_3 becomes of wrong (odd) parity. In this case we know that an error has been introduced but we cannot correct it (the change of data digit in the 4th position does not allow the recovery of the correct original data)

P_1	P_2	P_3	P_4	Data position	Data values
1	0	0	0	1	1
0	1	0	0	2	1
1	1	0	0	3	1
0	0	1	0	4	1
1	0	1	0	5	0
0	1	1	0	6	1
1	1	1	0	7	0
0	0	0	1	8	0

the fact that parity information is processed in blocks of finite length (8 bits in the previous example). In a more formal approach, the mathematical structure that allows the implementation of this kind of coding scheme is based on the theory of finite groups and the related concept of fields. In general terms, a field is defined by a set of elements between which can be established two kinds of mathematical operations, usually called the product and sum, with well-defined properties. The action of these operations produces elements inside the field set. Moreover, if such an original set is composed of a finite number of elements, the field is a finite field and can have only a number of elements equal to a power of two (2^n). Such finite fields are also called Galois fields (Schroeder, 1986) in honor of their discoverer, Pierre Evariste Galois.

4 Other Error Detection/Correction Methods, Genetic and Neural Systems, and a Nonlinear Dynamics Approach for Biological Information Processing

Other coding methods exist that are based on iterative processes, that is, the output is fed to the input by some feedback mechanism. These methods are called convolutives, in analogy with the mathematical operation of convolution that involves an integral calculus with delayed inputs. Convolutive methods are usually of the soft type. This means that they do not give a deterministic value of data at the decoder output, as in the case of the former parity block methods, but they assign a different probability to the different possible output values. In this way, a decision system may choose the one with the greater probability or follow other suitable criteria.

Among the more interesting convolutive methods we find the recently discovered ones called turbo codes (Battail, 1998; Berrou et al., 1993) which use also recursive decoding. The main advantage of these methods is that they approach the theoretical limits foreseen by the Shannon theory regarding their performance in error detection and correction (they approach the *channel capacity* (Sweeney, 2002) which represents a supremum of the code rate achievable). It should be also mentioned that other competing methods exist that reach similar performances: the so-called low-density sparse parity check methods (indeed these are classified inside the class of linear block coding methods that use iterative algorithms for decoding) (Gallager, 1962).

Because of the great optimization of most of the vital mechanisms of life, the emergence of errors in the transmission of the related information through the appropriate communication channels can be a source of serious or fatal problems for the organism involved. Researches at the level of genetic and neural information demonstrate that sophisticated mechanisms for error detection/correction indeed exist at the two above-mentioned levels (see for example, Fuss and Cooper, 2006; Farabaugh and Björk, 1999; Stiber, 2005; Stiber and Pottorf, 2004). Leaving aside the identity of the actual biological mechanisms implementing such functions, the flux of biological information cannot avoid the general laws governing information transmission, such as, for example, the results of the second Shannon theorem. For this reason is licit to ask if life does not use systems similar to those described above, developed by humans for digital communications, for implementing at a practical level error detection and correction of relevant biological information. The main lesson that can be learned from the Shannon theorem is that for transmitting error-free information through unreliable transmission channels it is necessary to introduce redundancy, that is, additional information or "check bits" to the original information. In the case of the genetic code, this redundancy is evident: different codons may code for a single amino acid. The information to be transmitted is represented by a finite number of words, in this case 21, i.e. the 20 amino acids plus the stop signal. But for encoding these 21 words we have 64 elements, i.e. the 64 codons. Necessarily the representation of 21 words by means of 64 elements is redundant: more than one element encodes the same word (amino acid). Recalling the Shannon theorem, the scope of this redundancy seems to be clear: to ensure the fidelity of the transmitted message, i.e. the chain of amino acids defining a particular protein.

At this point it seems opportune to try to understand a second question: how can a coding and decoding system be implemented ensuring fidelity of the information transmission by means of the natural redundancy of the genetic system? The most immediate answer should be that the system is based on a random coding, that is, there does not exist a pre-programmed order facilitating the coding and decoding operations as in the case of man-made digital communication systems.

Although a random coding can show a strong error correction capability, the main drawback at a practical level is that the decoding will be limited by an efficiency constraint. A random coding can give results as good as a structured one, but the efficiency of the former can be dramatically lower; this efficiency is made apparent in practical terms in excessively long times for retrieving the error-free information. However, in the genetic machinery, the ribosome is a highly efficient machine, assembling amino acids with incredibly low error rates and operating in a reactive medium that contains a very mixed soup of the components to be assembled. Is it possible to achieve this performance without resorting to sophisticated methods of coding and decoding of the genetic information? This is a question of fundamental importance that needs a cooperative effort from different fundamental and experimental branches of science to be definitively resolved. Its solution may contribute to a qualitative jump in different fields, and not least, to create a new basis for genetic therapies.

In the rest of this paper, instead of developing more technically complicated topics related to error detection and correction, such as, for example, those related to the above-mentioned turbo codes, we develop an interesting conceptual alternative: the use of nonlinear dynamics for implementing error control and related signal processing in biological systems. The theory of dynamical systems has produced the last scientific revolution in the physico-mathematical sciences (Strogatz, 1994). This new research discipline has led in a few years to important theoretical and experimental results, such as those related to chaos theory, chaos control, synchronization in all its variants, stochastic resonance, emergent behavior, spatiotemporal auto-organization (Bak, 1996), etc. These important areas related to nonlinear dynamics have found key applications in many different disciplines, such as for example, chemical dynamics, meteorology, astronomy, electronics, etc., but have also penetrated into the biological and social sciences, with examples such as ecology and population dynamics, neural and cardiac dynamics, metabolic pathways, psychology, and economics. This corpus of knowledge has contributed also to create a new science of complexity based on nonlinear dynamics and the emergence of spatiotemporal structures. The power of nonlinear modeling has been demonstrated through the properties of complexity, universality, and dimensionality reduction, which are associated with the generic behavior of dynamical nonlinear systems. Among others, these properties allow for qualitative modeling of complex spatial and temporal behavior, and the implementation of logical functions including the generation of universal computation machines such as Turing machines (Prusha and Lindner, 1999; Sinha and Ditto, 1998). Moreover, two important properties of dynamical systems have also been demonstrated regarding error correction, i.e. stochastic resonance (Moss and Wiesenfeld, 1995), and chaos communication (Bollt, 2003). The first corresponds to a paradoxical effect in which the addition of noise to a system can improve the signal to noise ratio. Many different examples and variants of this phenomenon have been studied, corresponding more or less to a statistical filtering of the signal – the meaningful information – that needs to be improved. Outstanding examples are found in the neural processing of sensorial information (Mitaim and Kosko, 1998; Longtin, 1993).

The second case, instead, is more similar to that of standard deterministic systems for error correction, but maintains, however, interesting peculiarities. In the following we try to show how a communication system can be implemented on the basis of the second alternative, leading to general properties of usual communication systems including error detection and correction capabilities.

A typical configuration for chaos synchronization represented as a communication system is shown in Fig. 1 (Corron and Pethel, 2003; Pethel et al., 2003). Chaos synchronization can be observed by comparing the output of the master oscillator with the output of the slave. When, to a certain precision, the two signals are equal we say that the system is synchronized. Usually the oscillators are continuous systems producing continuous signals, thus we need to convert their outputs into a discrete set of values (pertaining to some finite alphabet of signs).

The encoder is in charge of this operation, first, by converting the continuous trajectory of the master oscillator into a discrete iteration, for example using a Poincaré map, and secondly, by posterior discretization of the continuous variables of the discrete iteration, converting them into discrete variables which pertain to an alphabet of signs. This last operation is performed through symbolic dynamics and consists in partitioning the space state into a discrete number of non-overlapping regions (Hao and Zheng, 1998). Through successive iterations, the state of the system moves between these regions, defining a symbolic trajectory; to every region is assigned a discrete symbol. In order to attain synchronization we do not need complete information about the state of the master oscillator. Master and slave oscillators, beside their chaotic behavior, are also identical to a certain precision, which can be measured at the level of the parameters defining them. Supposing that initially both oscillators are in the same state, at the following iteration the output of the slave oscillator becomes different to that of the master oscillator to a certain extent. This is because of the chaotic character of both oscillators, which produces divergence in their evolution arising from minimal differences in their initial conditions. It is clear at this point that in order to maintain synchronization we do not

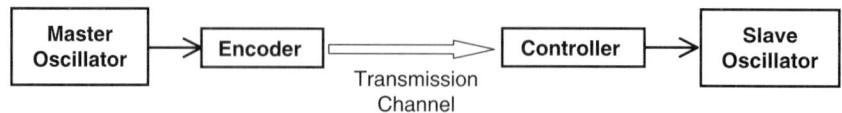

Fig. 1 Block diagram of a synchronized system of two identical chaotic oscillators with unidirectional coupling represented as a communication system

need to send through the communication channel all the information characterizing the state of the master oscillator; the information specifying the difference between the two states is sufficient. In technical terms, we may say that this condition requires that the Kolmogorov-Sinai entropy of the master system be lower than the channel capacity (the supremum of the coding bit rate) (Stojanovski et al., 1997). It must be remarked that to ensure the synchronization to a certain degree of precision automatically implies an error correction capability: we recover to the desired precision the information sent through the transmission channel. Moreover, more oscillators can be coupled in series or in parallel to the same master oscillator. This possibility allows for practical implementations, which implies communication between spatially separated points. As examples of possible biological applications we can think of the ribosome complex in protein synthesis (see Chapter 6, this volume), or of neural communication of sensorial information. For this last case, it has been demonstrated that high-level sensorial parameters can be described using dynamical attractors (Cartwright et al., 1999, 2001). Neurons are highly nonlinear systems and thus the behavior of neuronal populations can describe this kind of phenomenon in a natural way (Tonnelier et al., 1999). Furthermore it has been demonstrated that on the same basis the global dynamics of neural systems can show error-correction capabilities (Stiber, 2004).

Error detection/correction codes are unavoidable in man-made digital information transmission systems because of the necessity to ensure integrity of the transmitted information. Living organisms are subjected to similar demands regarding the integrity of crucial biological information. Because any kind of communication system is ruled by very basic principles of communication theory, the possible biological solutions to the problem of information protection need to resort to the same kind of mechanisms that have been sketched in this chapter. Because the most sophisticated artificial error-correction methods may be put in terms of dynamical systems theory (Richardson, 2000; Agrawal, 2001) and because many biological processes are highly nonlinear in nature, the hypothesis of using dynamical systems for error detection and correction in biological information control and management seems to be a very natural one, and should be explored in depth.

Acknowledgements I wish to thank Professor Marcello Barbieri for his invitation to write this contribution. I am also profoundly indebted to Dr. Julyan Cartwright for his useful suggestions and careful reading of the manuscript.

References

Bak P (1996) How Nature Works. Copernicus, New York

Battail G (1998) A conceptual framework for understanding Turbo Codes. IEEE J. Select Areas Comm 16(2):245–254

Berrou C, Glavieux A, Thitimajshima P (1993) Near Shannon limit error-correcting coding and decoding: Turbo-codes. Proc. IEEE ICC'93, Geneva, Switzerland, pp. 1064–1070

Bollt E (2003) Review of chaos communication by feedback control of symbolic dynamics. Int J Bifur Chaos 13(2):269–283

Cartwright J, Gonzalez DL, Piro O (1999) Nonlinear dynamics of the perceived pitch of complex sounds. Phy Rev Lett 82(26):5389–5392

Cartwright J, Gonzalez DL, Piro O (2001) Pitch perception: a dynamical-systems perspective. PNAS 98(9):4855–4859

Corron NJ, Pethel SD (2003) Phys Lett A 313:192

Farabaugh PJ, Björk GR (1999) How translational accuracy influences reading frame maintenance. EMBO J 18(6):1427–1434

Fuss JO, Cooper PK (2006) DNA repair: dynamic defenders against cancer and aging. PloS Biol 4(5):899–903

Gallager R (1962) Low-density parity check codes. IRE Trans Info Theory 7:21–28

Hamming RW (1950) Error detecting and error correcting codes. Bell Syst Tech J 26

Hao B, Zheng W (1998) Applied Symbolic Dynamics and Chaos. World Scientific, Singapore

Kohavi Z (1970) Switching and finite automata theory. McGraw-Hill, New York

Longtin A, (1993) Stochastic resonance in neuron models. J Stat Phy 70(1/2):309327

Mitaim S, Kosko B (1998) Adaptive stochastic resonance, Proceedings of the IEEE, 86(11)

Moss F, Wiesenfeld K (1995) The benefits of background noise. Sci Am 273:66–69

Ore O (1988) Number Theory and its History. Dover Publications, New York

Pethel SD, Corron NJ, Underwood QR, Myneni K (2003) Phys Rev Lett 90:254101

Pierce JR (1961) Symbols, Signals and Noise. Harper Modern Science Series, New York

Prusha BS, Lindner JF (1999) Nonlinearity and computation: implementing logic as a nonlinear dynamical system. Phy Lett A 263:105–111

Shannon CE, Weaver W (1949a) The Mathematical Theory of Communication. The University of Illinois Press, Illionis.

Shannon CE (1949b) Communication theory of secrecy systems. Bell Syst Tech J 28:656–715

Schroeder M (1986) Number Theory in Science and Communication, 2nd edn. Springer Verlag, Berlin

Sinha S, Ditto WL (1998) Phys Rev Lett 81:2156–2159

Stiber M (2005) Spike timing precision and neural error-correction: local behavior. Neural Comp 17:1577–1601

Stiber M, Pottorf M (2004) Response space construction for neural error-correction. Proc. IJCNN'04, Budapest, Hungary

Stojanovski T, Kocarev L, Harris R (1997) IEEE Trans Circuit Syst 44:1014

Strogatz SH (1994) Non-linear Dynamics and Chaos with Applications in Physics, Biology, Chemistry, and Engineering. Addison-Wesley, Reading, Massachusetts

Sweeney P (2002) Error Control Coding: From Theory to Practice. Wiley, New York

Yockey HP (1992) Information Theory and Molecular Biology. Cambridge University Press, New York

Tonnelier A, Meignen S, Bosch H, Demongeot J (1999) Synchronization and desynchronization of neural oscillators. Neural Net 12:1213–1228

Chapter 18

The Musical Code between Nature and Nurture

Ecosemiotic and Neurobiological Claims

Mark Reybrouck

Abstract This contribution is about sense-making in music. In an attempt to bring together such diverging fields as semiotics and neurobiology, it argues for a processual approach to music, which conceives of 'music users' as organisms that 'cope' with their environment. It is a position which calls forth ecological and epistemological assumptions and which stresses the importance of a conception of music as dealt with rather than a static conception of music as structure or artefact. As such, it considers music as a sounding and temporal art which appeals to lower-level mechanisms of reactivity, as well as to the acquired mechanisms of sense-making which are the outcome of a learning history. It is argued, further, that there is a continuum between lower level sensory processing and higher-order cognitive elaboration. The musical code, accordingly, holds a hybrid position between innate and wired-in dispositions and higher-level cognitive processing mechanisms. The very concept of code, further, is given some theoretical grounding as well as empirical evidence from the domains of psychophysics, psychobiology, and neurobiology.

1 Introduction

This contribution is about musical sense-making. It argues for an interdisciplinary approach to the process of 'dealing with music' in an attempt to bring together such diverging fields as semiotics and neurobiology. Its major aim is not to present a theoretical and empirical state of the art (see Reybrouck, 2001a,b, 2003b, 2004, 2005a,b) but to provide an operational approach that has both descriptive and explanatory power. As such, it argues for a broadening of the scope of traditional music research in stressing the role of the 'music user' besides the music. Dealing with music, in this view, is a generic term that encompasses traditional musical behaviours – such as listening, performing, improvising, and composing – as well as more general perceptual and behavioural categories (Reybrouck, 1999, 2006a).

Catholic University of Leuven, Section of Musicology Blijde-Inkomststraat 21, PO Box 03313, B-3000 Leuven, Belgium, e-mail: Mark.Reybrouck@arts.kuleuven.be

M. Barbieri (ed.), *The Codes of Life: The Rules of Macroevolution.*   395
© Springer 2008

All these categories can be defined in a broader sense as a continuous process of sense-making that is grounded in our biology and our possibilities for adaptive control, and which can be typically described in ecological terms of coping and interacting with the sonic world.

The claims are challenging: they bring about a whole research program that argues for a kind of continuity between lower-level sensory processing and higher-order cognitive elaboration. Major topics are the biological foundations of musical epistemology, the concept of adaptation and adaptive control, the role of interaction with the environment, and the concept of experience.

2 Dealing with Music: Towards an Operational Approach

Musicology as a discipline has many topics of research. There are, in fact, prevailing paradigms such as historical research, music analysis, and performance studies. On the other hand, new paradigms are evolving which challenge these traditional approaches by focussing on four major claims: (i) music as a sounding art; (ii) the process of dealing with music; (iii) the role of the musical experience; and (iv) the process of sense-making while dealing with the sounding music. As such, there is a major turn in contemporary musicological research which argues for the broadening of the field from a 'structural approach' to music – with a major emphasis on the delimitation of musical units and their interrelations – to a 'processual approach' that also takes into account the 'musical experience' (see Blacking, 1955; Määttänen, 1993; Westerlund, 2002; Reybrouck, 2004). Dealing with music, in this view, is a processual experience which is related to the definition of music as a 'sounding' and 'temporal' art. Hence the importance of the *listening experience* which is both sensory driven and time consuming. Listening, moreover, can be defined in ecological terms as 'coping with the sounds', encompassing several levels of processing such as the perceptual, the computational/representational, and the behavioural one (for an elaboration of the terms, see below). All these levels can be treated in isolation but they can be integrated in a more encompassing framework as well. As such, it is appealing to argue for an 'operational' description of the major moments of dealing with music with at least two major claims: (i) the definition of music should be broadened from a rather restricted and limited category – the classical Western art music – to a more encompassing definition of music as a subset of the sonic environment; and (ii) the process of dealing with music should be approached from the positions of cybernetics and systems theory (Reybrouck, 2005a, 2006b,c). The latter, especially, provide a useful 'operational' terminology for describing the multiple interactions of an organism with its environment, relying on some very basic functions such as perceptual input, internal processing, effector output, and feedback.

The major moments of this interaction are exemplified in the 'cybernetic' concept of a *control system* (see Fig. 1). Cybernetics, as a whole, is a unifying discipline

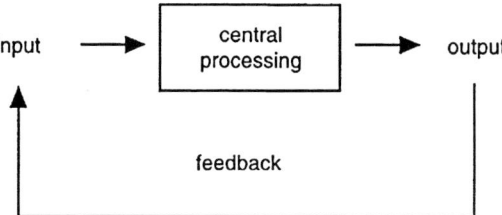

Fig. 1 The basic schema of a control system

that brings together concepts as different as the flow of information, control by feedback, adaptation, learning. and self-organisation (see Bateson 1973, 1978, 1985; Brier, 1999; Cariani, 2003). It has reintroduced the role of the observer into science, and has stressed the major role of subjectivity as well. This has been emphasised particularly in 'second order cybernetics', which typically conceives the observer as a participant and as part of the observed system (see Luhmann, 1990, 1995; Maturana and Varela, 1980; Pask, 1961a,b, 1992; von Foerster, 1974, 1984). There is, in fact, a lot of freedom in the way observers construct their knowledge as the outcome of their interactions with the environmental world.

Dealing with music, accordingly, can be considered as a kind of 'knowledge acquisition'. It allows us to conceive of music in *epistemological* terms and to conceive of listeners as observers who construct and organise their knowledge and bring with them, their observational tools (Maturana, 1978, pp. 28–29). The central point in this approach is the role of *subjectivity* and the way it influences our reactions to the environment. Living organisms, in fact, behave as subjects that respond to 'signs' and not to 'causal stimuli'. This is a major claim of semiotic functioning that stresses the emancipation from mere causality and time-bound reactivity.

The latter has been advocated most typically by the pioneers of Russian objective experimental psychology in what they presented as *reflex theories* (Bechterev, 1917, 1933; Kornilov, 1930; Pavlov, 1926). Both Kornilov's *reactology* – which held the view that an organism is nothing but a bundle of responses to the environment – and Bechterew's *reflexology*, which studied reflex responses, especially as they affect behaviour, were major exponents of a naturalistic and mechanistic view on the relation between subjective experience and the material world. Both approaches also tried to offer a needed alternative for mere idealistic psychology. They have proven, however, to be somewhat inadequate in being too behavioural and too reductionistic in their methodology. Reactive activity, for example, involves a direct coupling between sensory input and resulting effects, based on a 'historyless transfer function' of a particular automatism (Meystel, 1998). Such a mechanism, however, seems to be inadequate for the explanation of goal-directed behaviour with deliberate planning.

Arguing on these lines it is obvious to question also the role of *direct perception*. This conception of the perceptual process – as elaborated in ecological psychology

(Gibson, 1966, 1979, 1982) – holds that perception is possible without the mind intervening in the process. It involves *presentational immediacy* of the sensory stimuli and launches immediate reactions to the solicitations of the environment, proceeding mainly in real time. There is a lot of empirical support in favour of this claim, yet it is possible to go beyond the constraints of mere reactivity and to put *intermediate variables* between the sensory stimuli and the reactions which are triggered by these stimuli (Paillard, 1994; Reybrouck, 2001b).

The position I hold is a *biosemiotic* and *ecosemiotic* one. Rather than relying on wired-in circuitry with well-defined reactions to well-defined stimuli, I argue for the introduction of mechanisms of sense-making and epistemic autonomy, as music users do not merely react to sounds, but to *sounds as signs* which are the outcome of current and previous epistemic interactions with the sounding world (Reybrouck, 1999, 2005a). Music knowledge, in this view, is to be 'constructed' and 'acquired' rather than being merely 'innate' and 'wired-in'. The mechanisms of knowledge construction, however, are not totally autonomous: there are perceptual, ecological, and psychophysiological constraints which function as filters with respect to the information that listeners can actually pick up and process.

As such, we should consider the *biological bases* of musical epistemology (Reybrouck, 2001a) with a corresponding tension between 'nature' and 'nurture' – referring respectively to the neurobiological claims of wired-in circuitry (innate and natural mechanisms) for perceptual information pickup – as against the learned mechanisms for information processing and sense-making. The claims may seem to be diverging at first glance, but they are complementary to a great extent. To quote Damasio:

> as we develop from infancy to adulthood, the design of brain circuitries that represent our evolving body and its interaction with the world seems to depend on the activities in which the organism engages, and on the action of innate bioregulatory circuitries, as the latter react to such activities. This account underscores the inadequacy of conceiving brain, behavior, and mind in terms of nature versus nurture, or genes versus experience. Neither our brains nor our minds are tabulae rasae when we are born. Yet neither are they fully determined genetically. The genetic shadow looms large but is not complete. (1994, p. 111)

3 Musical Sense-making and the Concept of Code

Understanding music is a multifaceted problem. There is no unambiguous answer to 'what' listeners hear and 'how' they hear. Much depends on attentional strategies and strategies of sense-making which may differ considerably between listeners. There are, however, basic mechanisms of sense-making which are general across ages and cultures and which rely on levels of processing that are grounded in our biology (see Drake and Bertrand, 2003, for an overview). This is basically the 'nature' side of the nature/nurture debate, which has coloured musical discussions for decades. It is arguable that at least some mechanisms of musical sense-making rely on genetics rather than on culture. Yet there is the related question whether some reactions to music can spread across the Earth and become universals without genetic basis. As such, it may be interesting to look for mechanisms of 'coding' and 'decoding' which are so general that they are accessible to common music users all over the world.

These 'universal' claims challenge that music research which has been rather narrowly focused on the canon of classical Western music – the common practice tradition – with as paradigm a very particular way of concert culture where listeners are taught to listen 'silently and respectfully' with a minimum of bodily movement or emotional expression (Cook, 1998; Frith, 1996; Small, 1999). This listening habitus is typical of certain classical music traditions – such as the Western and the Hindustani 'classical' music – with an emphasis on stillness, focused attention, and inner withdrawal. Anthropological studies (Becker, 2001), however, reveal other patterns of dealing with music, to mention only the role of social interactions and dialogical and situational forms of musical expression and emotion.

What I argue for, therefore, is a broadening of the definition of music from a restricted body of musical works to an approach which is both transcultural and transhistorical – from 'music' to 'musics' of the world – and which relies on psychophysical and neurobiological commonalities that are typical hallmarks of Homo sapiens as one of the major developments of hominid phylogenetic evolution. Dealing with music, in this view, can be considered in terms of basic coding and decoding mechanisms which can be approached from different perspectives.

3.1   Universals of Perception, Cognition, and Emotion

And are there universals of music perception and cognition? Are there universals of musical emotion as well? Can we conceive of perception and cognition as separate faculties or should we rely on combined cognitive–perceptual activity of the brain in creating the experienced real world (Jerison, 2000)?

The major contributions in cognitive musicology have focused mainly on cognitive aspects of dealing with music. Musical emotions have been introduced only recently as a research topic with sufficient scientific relevance (Juslin and Sloboda, 2001b; Peretz, 2001). Emotions, in fact, are difficult to study adequately, but the same holds true also for the idiosyncrasies of individual sense-making that rely on perceptual processing and cognitive functioning. A major question, therefore, is related to the 'universal' claims of our reactions to music: can we rely on cross-cultural generalities or must we conceive of particulars that are embedded in cultural and historical contextual settings? Should we conceive of the primacy of the self-contained 'individual' or must we take into account also the 'supra-individual', which extends beyond the body and the mind of the single individual (Becker, 2001)?

The answer is not clear. There is empirical evidence for both *universals* and *specifics* in the process of dealing with music. The problem, further, is exemplary of the tension between two positions in science: those that stress the 'general' laws and instances as against those that stress the 'cases' and 'interpretations' (Becker, 1983; Geertz, 1983). The search for *universals* in human functioning, however, has been controversial up to now. There are universals of language (Chomsky, 1965; Greenberg, 1963; Pinker, 1994), human universals in general (Brown, 1991; Murdock, 1945; Pinker, 2002) and even universals which are related to the mechanisms of arousal and primary emotions (Becker, 2001).

Conceiving of universals in music, however, is less obvious. The strongest criticisms have been provided by ethnomusicologists who have stressed again and again the importance of cultural relativism (Becker, 2001). It is necessary, therefore, to broaden the scope of research and to relate the process of musical sense-making to the search for universals in general.

A starting point is the experimental principle from *developmental research* that allows us to tease apart the processes that appear to be innate or hard-wired from those that develop with maturation and acculturation (Drake, 2003). The procedure is quite simple: if young infants, children, and adult musicians display similar modes of functioning on particular task, then it can be concluded that the process is innate or at least functional at an early age (Drake and Bertrand, 2003, p. 22). The same principle can be applied to the study of children (developmental research) and intercultural research (comparative research) with the aim to answer the question of the true 'universal' nature of psychological processes in general.

A major contribution to these problems has already been made in the field of *evolutionary linguistics*. In a continuous attempt to motivate the emergence of 'linguistic universals', there has been a first reduction to three major categories: phonological, syntactic, and semantic universals. The major efforts of the endeavour, however, have focused traditionally on 'phonetic' and 'syntactic' rules and principles. Most languages, in fact, have vocabularies of words whose articulatory and acoustic definition is mediated by a *phonological* system (Liberman, 1975, for a critical discussion see Port and Leary, 2005). The majority of the contributions, however, have dealt with *syntax* rather than with phonetics. The Chomskian tradition in linguistics is illustrative of this approach: it tries to find the set of abstract grammatical principles that underlie all human languages, a kind of 'universal grammar' that relies on the assumption of universals as innate features of a 'language faculty'. According to Chomsky, humans are born with an innate knowledge of language – the *language faculty* – as a biologically autonomous system in the brain that has an initial state which is genetically determined (Chomsky, 1965, p. 13, see also Hauser et al., 2002). The basic tenet in this approach is the innateness of linguistic ability and the belief that the genetic base for language came about as the result of Darwinian evolution. The question, however, is still subject of discussion.

The search for linguistics universals has played a less important role in *semantics* (van Rooij, 2005). The only exception is the discussion of 'semantic universals' which are related to colour and kinship terms (Berlin and Kay, 1969; Leech, 1974). The prototypical example of colour terms seems to argue for semantic universals that are induced by perceptual universals with pan-human perceptual processes constraining lexical semantics. There exist eleven basic colour terms – black, white, red, green, blue, yellow, brown, purple, pink, orange, and grey – which may correspond to a universal pattern of perceptual salience that determines an apparent lexical salience of certain focal colours. Other examples are the kinship relations (father-of, mother-of, husband-of, wife-of, son-of, daughter-of, brother-of, sister-of, identical-to) and the case of pronouns that are used in talking to someone about ourselves (singular pronouns – 'I' and 'you'; plural pronouns – 'we').

Besides this search for semantic 'universals', there is another approach, which has focused on semantic 'primes'. If languages have irreducible semantic cores, it must be possible to reduce them to the smallest and most versatile set of words, which are, in a way, indefinable. These are *semantic primes* which are linguistic expressions whose meaning cannot be paraphrased in any simpler terms and which have a lexical equivalent in all existing languages (Goddard, 2002; see also Wierzbicka, 1996).

The question, now, is how to relate these universals to the realm of music? Here, also, there has been a major focus on 'syntax' and 'structure' with a corresponding neglect of 'semantics'. The question, therefore, remains as to the desirability and viability of musical semantic universals with as major questions: what do listeners hear? What kind of distinctions do they make? What kind of sense-making do they rely on? Are there any causal relations between the music as a stimulus and the listener's response?

The answer is complicated by the twofold nature of music as a *temporal* and *sounding* stimulus. Music, in this view, is not to be conceived as a symbolic stimulus with semantic meaning which can be dealt with out-of-time, but as a time-consuming and sounding processual event which impinges upon our sensory and perceptual apparatus. As such, its primary meaning is not *lexical* but *experiential* (Reybrouck, 2005a,b). Musical semantics, therefore, is in search not only of the 'lexicosemantic' – conceived almost exclusively in terms of referential semantics which is couched in the form of abstract and emotional neutral cognitive representations – but also of the 'experiential' dimension of meaning, which, in turn, is somewhat related to the 'affective' dimension of meaning. *Affective semantics*, further, carries the mark of the ties that connect humans with their environment. It has been neglected still more than the lexicosemantic dimension of language (Molino, 2000).

The claims have already received some empirical support. There is, first, evidence from evolutionary perspective with as typical example a class of primate vocalisations – called *referential emotive vocalisations* – which embrace a type of call that serves an online, emotive response to some object in the environment. These vocalisations have a dual acoustic nature in having both a referential and emotive meaning (Frayer and Nicolay, 2000). Additional empirical grounding comes from 'physiological reactions' that have shown to be correlates of listening to music (Krumhansl, 1997; Scherer and Zentner, 2001; Peretz, 2001). Some of them are physiological constants, such as the induction or modification of *arousal*, which can be defined in its most narrow sense as a stimulation of the autonomic nervous system. They play a major role in numerous studies of music and emotion, but many questions are still open with regard to their affective and interpretative components.

3.2   Universals in Music: Do they Exist?

The mere existence of music is universal as every culture, which is known today, has music in some form. There is, however, no agreement as to the nature of this universal claim: is music a cultural–social universal or is it grounded in our biology? The stance of a famous ethnomusicologist as Blacking is exemplary of the case. He commonly admitted that music-making is a fundamental and universal attribute of the human species, but he warned against 'universals' in musical meaning, stating that the conventions of musical systems and people's emotional response to them are cultural constructs. Even within well-defined cultures, there are so many idiosyncrasies and subjectivities that the mere claim of universal meaning in music seems problematic. Yet, he accepted the existence of *psychophysical commonalities* between disparate tonal systems, as well as the major role of relatively unchanging biological processes of aural perception in discovering patterns of sound:

> it is clear that the creation and performance of most music is generated first and foremost by the human capacity to discover patterns of sound and to identify them on subsequent occasions. Without biological processes of aural perception, and without cultural agreement among at least some human beings on what is perceived, there can be neither music, nor musical communication. (Blacking, 1973, p. 9; see also Balkwill, 1997)

This quote foreshadows the current alignment between *music* and *biology*. It is exemplary of a paradigm shift in thinking of music in terms of universals, which was initiated by the 1970s – two journals published a special issue on the topic: *Ethnomusicology* (1971, 15(3)) and *The World of Music* (1977, 19(1/2); see Nettl, 2000). More recently, the topic of universals has received renewed attention in 'musicological' research (see Brunner, 1998; Kon, 1998; Marconi, 1998; Miereanu and Hascher, 1998; Normet, 1998; Padilla, 1998). But, above all, there is the new field of *biomusicology* (see Wallin, 1991; Wallin et al., 2000) which aims at resuscitating the concept of musical universals, in taking full advantage of current developments in Darwinian anthropology (Durham, 1991), evolutionary psychology (Barkow et al., 1992) and gene-culture co-evolutionary theory (Feldman and Laland 1996, Lumsden and Wilson, 1981). It addresses the basic question of the origins of music, and comprises three main branches: evolutionary musicology, neuromusicology, and comparative musicology. To quote Brown et al.:

> Evolutionary musicology deals with the evolutionary origins of music, both in terms of a comparative approach to vocal communication in animals and in terms of an evolutionary psychological approach to the emergence of music in the hominid line. Neuromusicology deals with the nature and evolution of the neural and cognitive mechanisms involved in musical production and perception, as well as with ontogenetic

development of musical capacity and musical behavior from the fetal stage through to old age. Comparative musicology deals with the diverse functional roles and uses of music in all human culture, including the contexts and contents of musical rituals, the advantages and costs of music making, and the comparative features of musical systems, forms, and performance style throughout the world. (2000, p. 6)

There is need of additional foundational and methodological research in order to give this new research domain its full academic flavour, but the interdisciplinary approach is likely to be fruitful as a starting point. The combined contributions from ethnomusicology, biomusicology, evolutionary musicology, neuromusicology, cognitive musicology, comparative musicology, and zoomusicology are likely to provide a lot of insights that go beyond approaches that have been rather intuitive up to now. Some questions, however, remain unresolved. One of them is the tension between culture and biology.

The *biological* position argues for wired-in mechanisms of reactivity to sounds or music, the *cultural* position argues for variables that are the outcome of immersion in a culture. Rather than thinking in terms of a dichotomy, however, it is fruitful to consider the dependency between both variables (see Wallin, 1991, p. 6). Listeners, in fact, are biological organisms, which are immersed in a culture. But even culture did not evolve in a vacuum. Both culture and music are born out of man's 'animal characteristics', which are rooted in the biology of perception and cognition, and this may be universal to a great extent (Goldschmidt, 1959). Even at the level of acquired habits and learned responses, there seems to be a lot of commonality. It allows us to conceive of *universals* in music, which exist beneath the surface of cultural variation and mere subjectivity (Balkwill, 1997).

It may be fruitful, in this context, to consider the contributions from neurobiology and ethnomusicological research. As Nettl puts it: 'If ethnomusicological involvement has some justification, it concerns the interface among three areas of concern: cultural universals, musical universals specifically, and the origins of music.' (Nettl, 2000, p. 464). Musical universals, especially, can be helpful in theorising about the origins of music: to the extent that surface structures – musical features – are showing similarities which correspond to concepts provided by evolutionary theories, it can be stated that the universal and the biological coincide (Imberty, 2004, p. 450).

There are, further, three major topics of concern: (i) the fact that music is a widespread phenomenon all over the world; (ii) the neurophysiological substrates that underlie common responses to sounding music; and (iii) the observation that there are musical traits which are common among humans regardless of origin and culture. Additional issues include the question whether music is to be dealt with at a 'structural' or an 'experiential level' (see Reybrouck, 2005a,b, 2006a) and the role the body plays in this experience. It really makes a distinction if we conceive of music at a level with surface features that are describable in an objective way – structural level – as against music as a process, which is enacted in a social setting.

As to the first topic – music as a widespread phenomenon all over the world – it is possible to conceive of music as a universal phenomenon with *adaptive power*.

According to Miller (2000, see also Trehub, 2003), music exemplifies many of the classic criteria for complex human adaptation: (i) no culture in any period of recorded history has been without music (universality); (ii) the development of musical abilities is orderly; (iii) the ability is widespread in the sense that almost every normal adult can appreciate music and carry a tune (basic perceptual and performing skills); (iv) adults can recognise thousands of melodies, implying specialised memory; (v) special-purpose cortical mechanisms are involved; (vi) there are analogues in the signals of other species such as songbirds, gibbons, and whales, raising the possibility of convergent evolution; and (vii) music can evoke strong emotions which implies receptive as well as productive adaptation.

As to the second topic, the *neurophysiological substrates* of music perception, there are common biological characteristics which all human beings share. Some perceptual activity is even common to various species beyond the human species and there is good reason to assume that in most vital features the world as experienced is stable among species (Jerison, 2000, p. 187). The same probably holds true for dealing with music. There is, however, great scepticism among ethnomusicologists who conceive of universals of perception as 'smacking too much of biological determinism, and ... of denying the importance of historical forces and cultural traditions in explaining the properties of musical systems and musical behavior' (Brown et al., 2000, p. 14). A possible solution to this reluctance could be the *biocultural view*, which argues for a balance between genetic or biological constraints on the one hand, and historical contingencies on the other. Musical universals, in this view, place all humankind on equal ground and focus on the unity that underlies the great diversity present in the world's musical systems, attributing this unity to neural constraints underlying musical processing (Brown et al., 2000).

As such, much can be learned from the study of perceptual processing dispositions, which are likely to be genetically pre-wired. A technique for studying them stems from *human infant studies*: if substantial adult–infant similarities are evident in the perception of music, one could argue that at least some aspects of music processing have a biological basis. It means further that perception of music is inherently biased rather than unbiased. The musics of the world, in this view, have capitalised on these biases or universals of auditory pattern processing, which should mean that music from different cultures could be expected to share some fundamental properties (Trehub, 2000, p. 428).

A somewhat related field of research is the topic of *infant-directed speech* as used by caregivers over the world. Most of them enhance their vocal messages to pre-linguistic infants by making them more musical than usual. This is done by several techniques: using simple but distinctive pitch contours but articulating words poorly; raising or lowering the pitch level and expanding or reducing pitch contours; slowing tempo and making the utterances more rhythmic and repetitive (Trehub, 2000, p. 437).

The observation, thirdly, that there are musical traits, which are common among humans, regardless of origin and culture, is generally accepted by musicologists though there is no absolute agreement on what they are. We should remember, however, that universals need not necessarily apply to all existing music (Nettl, 2000).

In an attempt to generalise as much as possible, it should be recommended to focus not exclusively on 'great' music but also to embrace all kinds of primitive and popular music. As Molino puts it: 'our conception of music, based on the production, perception, and theory of "great" European classical music, distances ourselves irremediably from the anthropological foundations of human music in general' (2000, p. 170). It surely makes sense to study some well-established musical notions with quasi-universal character, such as scales, key notes, bars, melody, harmony, and tones as opposed to noises (Mâche, 2000), but much of their relevance is questionable when applied to non-Western musical cultures.

As such, it is arguable to list up common features of music all over the world, or, as Arom puts it, to conceive of 'anthropomusicology' as 'the scientific discipline that would deal with the suite of human musical properties as they are manifested in the ensemble of known musics'(2000, p. 27). In an attempt to determine such a minimal set of criteria, he provides a provisional list of universals or quasi-universals which are specific to music: (i) intentionality or an act of intentional construction; (ii) a formal process which detaches the music from the sound environment and which gives the musical substance an internal articulation in terms of proportions (formalisation of time through temporal ratios, periodicity and the principle of symmetry); (iii) a set of contrastive pitches (musical scales), and if the music involves more than one individual; (iv) modes of coordination between them (ordered and simultaneous interactions).

Another contribution considers two universal features to account for a large part of what music is at the structural level: 'pitch blending' and 'isometric rhythms'. The first refers to the simultaneity of pitched sounds. In opposition to speech, which proceeds by a successive alternation of parts, music has the intrinsic capacity for promoting cooperative group performance and interpersonal harmonisation with pitches sounding together at the same moment (Frayer and Nicolay, 2000). The feature of isometric rhythm, on the other hand, involves the level of temporal organisation that is regular and periodical and which corresponds to the regular points in the music to which one could tap one's foot (Cross, 2003).

Other universal features could be listed, but even if the list should be exhaustive, the question remains as to what is innate and what is acquired. As Huron puts is:

> Music is now deeply embedded in a cultural/historical context where human musical memories span centuries, and the fashion cycle is a significant engine of change. Music is now part of a Lamarckian system where acquired characteristics are transmitted in Dawkinsean 'meme pool' rather than in Mendelian 'gene pool'. (2003, p. 73; see also Dawkins, 1976)

3.3 *Primary and Secondary Code*

The problem of musical universals is directly related to the origins of music. It calls forth evolutionary claims and the related question whether music has some adaptive function both at the phylogenetic and ontogenetic level of human beings as biological organisms. A central issue in this discussion is the tension between human's innate reactivity to stimuli in general and those reactions, which are acquired through ontogenetical development. Or put in other terms: is musical sense-making coded in our genetic programmes with reactions which rely on pre-wired and innate programmes or should we conceive of music as something which calls forth higher functions of the brain? And can we conceive of musical sense-making in terms of a dichotomy (innate/acquired) or should we think in terms of complementarity between distinct levels of processing? There is, in fact, a lot of evidence that points towards a hybrid position, with music acting as a stimulus that induces primary reactions, which can be filtered by the cognitive apparatus as well.

The problem is further complicated by the critical distinction between perceptual 'development' – which is merely the outcome of maturation and aging – and perceptual 'learning'. It is necessary, therefore, to tease apart those changes that are the outcome of experience from those that are the outcome of the normal growth of individuals.

An interesting starting point here is the theoretical work on *codes* and *code acquisition* in general. There is, in fact, a distinction between quasi-automated wired-in reactivity and those reactions that are the outcome of at least some learning history. As such, it is instructive to observe human behaviour in general and to examine the way it can be classified in different cultures all over the world. An important representative of this approach was Murdock (1945) who introduced a classification of *cultural-social universals* in anthropology with three major categories: (i) universals which correspond to primary, genetically coded (instinctive) impulsion; (ii) universals which correspond to acquired habits and which are rooted in fundamental bio-psychical demands; and (iii) universals which correspond to cultural habits with only very thin links to the conditions of the secondary level. Murdock did not totally deny the existence of an 'animal level' of culture and society, but these animal characteristics can serve only the most general conditions for culture and society. All attempts, which go further in explaining, should be considered as mere reductionism (see also Goldschmidt, 1959).

Murdock's view is somewhat related to Bystrina's concept of 'codes', which, in his view, are defined as relatively invariant systems of rules that regulate the processing of information. A distinction can be drawn between three layers of processing: the phylogenetically earlier *primary* (or hypolinguistic) *codes* – the genetic code, the intraorganic and perception codes – are of an innate nature; superimposed on them are *secondary codes* (linguistic or sign codes) which are the result of a learning process; they provide the basis for *tertiary* (hyperlinguistic or text codes) – also called cultural codes – which operate at a level above the secondary codes. It is assumed, further, that the invariants of tertiary codes have developed as imprintings in especially sensitive phases of the evolution of man (see also Jiranek, 1998).

There is much of value in these ideas, even if some of them are too general to be operational. To quote Wallin:

> There are certainly genetically coded impulses …, but the most forceful of them are not reserved for man alone: to satisfy sexual needs, hunger, thirst, etc. Language and music, two very human abilities, do not correspond to impulses or instincts. They are, as such, not coded into the genetic program. Into the genetic code is inserted information or constraints concerning how, in situations of vital emergency, to mobilise the most effective displays in order to attract or detract other organisms; among these constraints are included optimal functional pathways between mechanisms, through which signals and signs can be produced. (1991, pp. 8–9)

The genetically programmed codes are fast and efficient. They are, however, restricted in their scope and functions. As Barbieri puts it:

> The interpretation process that we observe in many animals can easily be understood as an evolution of their signal processing systems. It is likely that the most primitive reactions were heavily determined by genes, but the number of hard-wired responses could not increase indefinitely, and animals became more and more dependent on processes of learning in order to increase their behavioral repertoire. (2005, p. 124)

Innate reactions, further, are perceptually primitive in a radical sense. Most animals and men, according to Cariani, have neural coding strategies that are used in representing and processing sensory information. They encompass basic body plans, sensory organs, and neural representations that are roughly similar (Cariani, 1998). There is, in other words, a whole domain of 'sensory coding', which can be studied in a straightforward way.

The domain of sensory or perceptual systems is prototypical. It is one of the rare domains, which allow a kind of low-level automatic and elementalist approach to the processing of information with mapping functions between perceptual input and behavioural reactions that suggest some kind of causal relationships. But even at this level of processing, there should be made an effort to go beyond the computation of correlation coefficients between a single independent variable and a single dependent variable as perceptual phenomena have multidimensional rather than unidimensional determinants (Uttal, 1998, p. 200). Put in other terms: the relation between perceptual sense-making and sensory stimulation is not as causal and linear as would be assumed, even at the level of sensory processing.

3.4 The Concept of Coding

The concepts of code and coding entail a lot of ambiguity. They have been used in semiotics and communication theory, in cryptographics, in neurophysiology and sensory physiology as well as in genetics. The word coding, however, has a relatively constant meaning, which is related to the nature of 'representation'. To quote Uttal:

> how do signals or symbols from one universe of discourse represent patterns of information from another. The major theoretical notion that underlies coding theory is that there are

invariants of organization, patterns, or meaning, which can be conveyed from sources to destination even though represented by different symbols and in different kinds of physical energies in the communication process. (1973, p. 207)

And further:

What, then, is a formal definition of a code? We may define a code as a set of symbols that can be used to represent message concepts and meanings (patterns of organization) and the set of rules that governs the selection and use of these symbols. (1973, p. 208)

Coding theory resolves some problems of psychobiological equivalence. It states that any dimension is capable of representing any other dimension. Rather than starting from dimensional isomorphism between stimulus and neural representation, it involves a system of codes or cues which are able to symbolically represent, rather than geometrically replicate, external stimulus patterns (Uttal, 1978, p. 26). The representation of the amplitude of a stimulus by a train of nerve impulses is a typical example.

The key problem of coding theory, however, is the relation between a stimulus with a certain physical intensity, its neural coding through neural fluctuation and some perceived sensory dimension (the subjective experience) (Uttal, 1978, p. 416). Two questions should be answered here: which are the relevant *neural parameters* – properties of neurophysiological signals – that vary as a function of the stimulus variables and which are the common *sensory dimensions*?

Many properties of signals can function as symbols for the dimensions of perceptual experience: place of activation, number of activated units, neural event amplitude, and temporal pattern of the activated units. All of them are *candidate neural codes*, which can be directly established by neurophysiological research. They, however, do not necessarily bear any resemblance to 'perceptual significance'. In order to be *true codes* some additional criteria must be met: they must vary as a function of some stimulus variable, they must be interpreted by some subsequent mechanism, they must be both necessary and sufficient for the concomitant variation of some behavioural experience and they must actually become the equivalent, at some stage, of a mental process (Uttal, 1973, p. 212). As such, there is a distinction between what is merely concomitant and what is truly relevant.

The common sensory dimensions, on the other hand, can be considered from their perceptual aspects rather than from the dimensions of the physical stimulus itself. Such a set of *discriminable dimensions* is modality independent which means that it can be used to describe patterns in any sensory modality whatever (Uttal, 1978, p. 416). Examples of such coding factors are 'perceived quantity' or sensed intensity or magnitude of a physical stimulus, 'perceived quality' or the kind or quality of a sensory experience, and 'temporal discriminations' such as relative temporal order – which of two different stimuli came first? – temporal acuity (ability to distinguish two identical stimulus events which are sequential in time as separate events) and duration or interval (ability to replicate the sustained duration of an event or interval between two events (Uttal, 1978, p. 418)).

3.5   Coding and Representation

Coding theory is related to two major problems of *psychobiological* equivalence: the problem of the mind-body or–in a more reduced sense–the mind-brain relationship and the particular ways in which mental processes are encoded and represented. Its major claims are that the nervous system provides the immediate, necessary and sufficient mechanisms for the embodiment of all mental processes and that mental processes are reducible to the function, arrangement, and interaction of neurones as the constituent building blocks of the nervous system (Uttal, 1978, pp. 26, 355).

Coding theory, further, argues against *isomorphically oriented theories* that claim a kind of dimensional isomorphism or similarity between stimulus and representation. It also rejects the need of linear relationships between neural coding dimensions and psychophysical dimensions. It only claims an equivalence of maintained information from the neural states to the psychological state (Uttal, 1978, p. 68).

Real coding, in this view, means that any dimension is capable of *representing* any other dimension. It involves a system of codes or cues which are able to symbolically represent external stimulus patterns (Uttal, 1978, p. 26). As such, it is not sufficient to discuss merely the physical stimulus dimensions. The latter can be very quickly re-encoded into entirely different stimulus dimensions, even within the sensory receptor itself. It is more fruitful, therefore, to consider the codes that are situated at least at the level of conversion of physical energy into neural energy. This is the 'transductive process' with the concept of 'neural code' as a prototypical example. It has been proposed as the outcome of the observation that single neurone responses vary concomitantly with some variant in a physical stimulus dimension (Uttal, 1973, p. 212).

The fundamental problem of a theory of *sensory coding* and *neural representation*, further, is the association between percepts, experience and thought on the one hand, and the candidate neural dimensions on the other hand (Uttal, 1978, p. 430). It entails a kind of *neural reductionism* in focusing rather narrowly on the relationship between single cell responses and psychological functions. Yet this has proven to be valuable in the domain of sensory or perceptual systems–especially in the visual domain (dark adaptation, the Mach band, the blind spot, colour sensitivity, and a number of other phenomena (Uttal, 1998, p. XI), where perceptual phenomena seem to reflect neuronal transformations produced by the early or peripheral portions of the sensory pathways.

Sensory coding is characterised by its relative simplicity: there is an initial isomorphism of stimulus and receptor response, and a primarily unidirectional flow of information (Uttal, 1978, p. 356). But even here, the problem is much more complex and multifaceted than 'the' sensory coding problem. The latter has emphasised traditionally the single aspect of singular determination of the functional relationships between stimulus intensity and nerve impulse frequency. As is obvious from psychophysical research, however, there is no simple psychophysical law that determines the relationship between perceived magnitude and stimulus dimensions. Many other dimensions, such as quality, temporal and spatial parameters, must be encoded by neural signals as well (Uttal, 1978, p. 415).

A central question, therefore, is how this 'representation' maps in some direct way onto the domain of 'psychological responses'. It represents the general process by which neural signals or responses come to be the physical equivalent of mental processes (Uttal, 1978, p. 357). As such, it is one of the major research issues of modern psychobiology: 'The fundamental problem of representation is to determine how discrete neural processes can serve as the transactional equivalents of molar psychological processes.' (Uttal, 1978, p. 24)

'Sensory coding', in this view, is a rather poor model of cognitive coding theory, in the sense that the central cognitive or symbolic coding moves away from straightforward mappable representations. To quote Uttal once more:

> The central symbolic processes ... allow considerable "patching" to be done to the relatively poor sensory information that does make its way to the intrinsic areas. Holes may be filled in, clues interpreted, past experience introduced, and judgments made about relationships that differ considerably from the simple geometric or temporal characteristics of the original stimulus. The criteria of excellence of integrative mechanisms are, therefore, based on the richness of the mixture of information and the deviation from, rather than fidelity to, simple reproduction of the input stimulus patterns. (Uttal, 1978, pp. 414–415)

As such, one can question the claims of psychophysics as a major attempt to provide reliable correlations between the sensory levels of stimuli and their perceptual processing. It is interesting, therefore, to consider additional mechanisms of processing and sense-making that go beyond mere linear response functions of given receptors.

4 Principles of Perceptual Organisation: Steps and Levels of Processing

Dealing with music is a psychological process of 'coping' with the sounds. There are, however, different levels of sense-making with a major distinction between low level, quasi-automatic reactivity to typical stimuli, as against cognitive processing that entails some intermediate variables between the stimuli and the reactions to these stimuli. The distinction is critical, both in terms of processing efforts and cognitive economy–information 'pickup' as against information 'processing'–, and in terms of sense-making. There is, in fact, a long-lasting debate on the causal relation between stimulus and response, with the related issue of direct perception and naturalistic approaches to semiotics and sense-making in general (see Leman, 1999). There are surely advantages in recognising the consistency of cause and effect. It allows us to use invariant and consistent sensory responses as guides to objects or events, which, in turn, provides adaptive power for coping with our environment.

4.1 Levels of Processing

A central question in the perceptual processing of music is the transition from *sensation* to *perception*. It is an old topic in the psychology of perception in general. 'Sensation' is usually defined as the conscious response to the stimulation of a sense organ or nerve receptor, but this definition does not account for the totality of the contents of consciousness in an act of perception, which involves selection among sensations, combination, organisation and sometimes even supplementations from imagination (Lee, 1938, pp. 24–25). Perception, further, is a rather complex phenomenon. According to Uttal, it is the '… relatively immediate, intrapersonal, mental [experience] evoked as a partial, but not exclusive, result of impinging multidimensional stimuli, as modulated by the transforms imposed by the neural communication system, previous experience, and contextual reasonableness. Each percept is the conscious end product of both simple computational transformations and more complex constructionist interpretations. However, the underlying neural and symbolic processes are not part of the observer's awareness.' (Uttal, 1981a, p. 14) Perception, in this view, always takes place against the background of previous experiences and its study should be conducted in terms of processing of information.

Major efforts have been undertaken in order to find linear and causal relations between stimuli and responses with the field of 'psychophysics' as a prototypical example. In a somewhat broader sense, this was also the aim of early 'psychobiological research': to find out the relation between mental processes and their underlying physiological processing mechanisms. This 'body-mind relationship', however, has proven to be less predictable than was hoped for. The biological and physiological roots of perceptual processing may be interesting as a starting point, but there are so many other mechanisms –cognitive, affective, attentional, motivational–that interfere in the processing, that the early hope for unidimensional correlations has proven to been illusory. There is, further, a whole machinery of 'transactional' processes between the listener and the music (see Mowbray and Rodger 1963 for perception as transaction between perceiver and environment, see Berne 1964 for the topic of transaction). As Deschênes puts it:

> If we appreciate a particular type of music, it is due neither to the music itself or to the musicians, but rather to the interaction going on between our mind content and body, the music, the musicians, the values carried by the music (social, moral, or otherwise), and the sensations we are looking for in the music, and to the correlation that we can create between this music and these sensations and perceptions. Music surely activates in a listener quantifiable percepts, but what is retained after any listening are the qualities that these awaken in the listener's consciousness during the listening process. (1991, p. 199)

As such, there seems to be a lot of subjectivity. Yet, there are universal processing mechanisms that go beyond the idiosyncrasies of individual and subjective reactions to the sound. As a rule this applies to the lower levels of processing–with the risk of psychobiological reductionism–, and the ecological levels of coping with the world. There are, however, lawful commonalities at higher levels of perceptual processing as well. This seems to suggest that perception is not merely a primitive data-driven process which works independently of top-down influences. The *Gestalt laws* of perception are illustrative of this point (see Köhler 1947, Koffka 1935, Wertheimer 1923 for the seminal publications, and Bregman 1981, 1990, Leman 1999, Reybrouck 1997, for musical applications). According to Pomerantz (1986) there are no less than 114 Gestalt laws for object perception, suggesting that they can hardly be built into a primitive data-driven process. This sheer number simply calls forth higher levels of processing and information organisation, although some rules for recognising the natural world's regularities seem to be embodied in the hardware of our brains. It is interesting, therefore, to consider some of the basic mechanisms of perceptual organisation in general and, more specifically, the levels of perceptual processing.

According to Neisser (1967), there are at least three stages of processing: the *preperceptual, perceptual* and *cognitive* stage. Applied to music, this means that the level of conscious and active listening is preceded by a passive stage of filtering and feature-detection (Neisser, 1967, p. 199, see also Seifert, 1993, pp. 180–182): the passive mode is fast and less demanding as to processing efforts– it relies on wired-in programs and mechanisms–, the higher levels of processing, on the contrary, are more demanding and are not reducible to causal stimulus-reaction chains.

It has been shown, moreover, that there are distinctive and successive stages in perception (Levine and Shefer, 1981, see also Rosenberg and Adelstein, 1993, Sekuler and Blake, 1985). At first, there is a *distal stimulus* which corresponds to the actual object in the environment. It is the distant source of sensory information which corresponds to the energy emitted by or reflected from some object. Secondly, there is a *proximal stimulus* which is commonly defined as the pattern of energy that impinges on the sensory receptors. It is to this kind of stimuli that human beings have access when they interact with an environment rather than to the stimuli per se. Or put in other terms: observers depend most directly on proximal and not on distal stimuli in perceiving their environment. The act of perception, further, involves the *transduction, transmission*, and *processing* of the stimuli. Transduction means the conversion of energy from one form to another as a result from the sensory action. It involves a number of steps, the first of which is the alteration of the chemical equilibrium of the receptor by the physical energy of the stimulus (Uttal, 1973, p. 12). The transmission step is the transfer of information-bearing electropotentials propagated from peripheral portions of the nervous system to other more central portions. The processing process, further, is hypothesised to involve a judgement of which distal stimuli most likely caused the sensation, a kind of compromise between what is presented and what is selected to denote a systematic response tendency. It involves an act of 'inference' which is commonly known as the *perceptual hypothesis* (see Krechevsky, 1932a,b,c) and which results in the generation of an internal representation of the outside world: the *percept*. In order, however, to make a correct perceptual hypothesis the perceptual system must be presented with enough salient sensory information in a proximal stimulus. If this is the case, an appropriate internal model of the actual distal stimulus will be created. The last stage of perception, finally, is *action* or *reaction*, which may consist of cognitive or motor responses – both manifest and internalised – to the sensory input.

All this has implications for a genuine perceptual theory. In the main, it is not sufficient to explain perception in terms of distal stimuli or of separate senses. What is needed is a theory that does justice to the relation between 'distal' and 'proximal stimuli'. In an unpublished paper Gibson (1961, unpublished data) provided an interesting attempt to classify the senses and the sensory inputs accordingly, together with some criticisms of the traditional approach to perception. The latter embrace the following remarks: the classical sensations are only poor guides to the description of the receptive process in general; the typical variables of stimulation, borrowed from physics, are oversimplified; and the relation between stimuli and their sources in the environment – the 'proximal' and 'distal' stimuli – has never been worked out in detail. The channels of sense-perception, further, cannot be represented by any fixed list of mutually exclusive 'senses'.

What he argues for, on the contrary, is an *overall picture* of the receptive process as a whole with several categories and subcategories at different levels of the process: (i) the *receptive system* which can be excited and which can be split up into *receptive elements* that are connected to an afferent neurone, *sensory surfaces* that group together bundles of adjoining neurones, and *sensory systems*, which consist of a sensory surface with its neural connections and motor equipment for adjusting and exploring stimulation; (ii) the types of *stimulus energy* that can excite them; (iii) the *environmental fields of potential stimulation* which do not relate primarily to substantial and material things but rather to potential stimuli for the sense organs; (iv) the *environmental sources* of these fields of stimulation which comprise the environment that can be registered by the senses; and (v) the *general sources of stimulation*, which are inner and outer sources of variation of proximal stimulation, involving changes in the environment, the muscular actions of the body, sensory adjustment, and exploration by the perceiver and moving in the world.

The overall picture is challenging: it goes beyond the classical distinction between mutually exclusive modalities of senses and sense data as well as beyond the basic variables of stimulation such as intensity, frequency, duration, location, and extension. Even the distinction between 'proximal' and 'distal' stimuli is insufficient as there are sources of stimulation at more than one level of perceptual analysis.

4.2   Nativism and the Wired-in Circuitry

Dealing with music relies on low levels of reactive behaviour and higher levels of cognitive processing. As such, there are several related questions: (i) what is the role of innate and wired-in circuitry? (ii) Can music be processed automatically and at a preattentive level? (iii) Can we conceive of musical competence as an innate faculty somewhat analogous to linguistic competence? (iv) Are there brain substrates for music processing? (v) What is the role of emotion in music processing?

There are, first, information-processing mechanisms that are basic and innate and that rely on low-level neural architecture which functions as *wired-in reactive circuitry*. According to Damasio, they are designed to solve automatically the basic problems of life, such as finding sources of energy, incorporating and transforming energy, maintaining a chemical balance of the interior of the body, maintaining the organism's structure by repairing possible impairment and fending off external agents of threat. Most of these mechanisms are part of the regulatory *homeostatic machinery* that provides the innate and automatic equipment of life governance. At the bottom of this organisation are very simple responses such as approaching and withdrawing, or increases (arousal) and decreases in activity (calm or quiescence) (Damasio, 2004, p. 31).

Most of this reactive behaviour points into the direction of automatic processing beyond conscious and deliberate control. It involves a lot of biological regulation that engages evolutionary older and less developed structures of the brain such as the hypothalamus, brain stem, and limbic system. Such a 'primitive' processing mechanism, however, has considerable adaptive value: it provides the organisms with levels of elementary forms of decision making which rely on sets of neural circuits which do the deciding (Damasio, 1994, pp. 123–127).

According to Damasio, there are at least four levels which precede the stage of conscious control: the level of metabolic processes, basic reflexes and the immune system; the level of behaviours which are associated with the notion of pleasure or pain (reward or punishment); the level of drives and motivations – hunger, thirst, curiosity and exploration, play, and sex – and the level of emotions-proper (Damasio, 2004, p. 34). The whole machinery, however, ensures an organism's survival. Based on innate strategies, it permeates the human brain from the start of life with knowledge regarding how to manage the organism, including both the running of life processes and the handling of external events. The mechanisms, however, are rather general. It is interesting, therefore, to look for similar ones, which are related to the processing of music.

4.3   *Arousal, Emotion, and Feeling*

There are two major ways of dealing with music: music can be considered as something which catches us and induces several reactions which are beyond conscious control, or music can be processed in a conscious way that appeals to several higher functions of the brain. As such, it is possible to respond to musical sound patterns with two distinctive strategies of sense-making: the *acoustic* and *vehicle mode* of perceiving and responding (Frayer and Nicolay, 2000, p. 278) – see also the distinction between rhythmo-affective and representational semantics (Molino, 2000). The first involves emotive meaning with particular sound patterns being able to convey emotional meanings; the second involves referential meaning, somewhat analogous to the lexicosemantic dimension of language with arbitrary sound patterns as vehicles to convey symbolic meaning.

The acoustic mode of responding refers to the immediate 'emotive aspect' of sound perception and production. It relies on the process of *sentic modulation* (see Clynes, 1977), which is a general modulatory system that is involved in conveying and perceiving the intensity of emotive expression by means of three graded spectra: tempo modulation (slow-fast), amplitude modulation (soft-loud), and register selection (low pitch-high pitch), somewhat analogous to the well-known rules of prosody which encompass our most basic idea about intonation. It refers to the local risings and fallings, quickening and slowing, and loudenings and softenings that are involved in expressively communicating meanings. The sentic modulation process appears to be invariant across modalities of expression – this is Clynes' 'equivalence principle', which states that a sentic state may be expressed by any number of different output modalities in humans, such as speech, music, and gesture (Clynes, 1977, p. 18, see Frayer and Nicolay, 2000).

This *sentic modulation*, further, hints at an important aspect, namely how expressive qualities vary and change in a dynamic way. Emotional expressions are not homogeneous over time, and many of music's most expressive qualities relate to structural changes over time, somewhat analogous to the concept of prosodic contours (Frick, 1985) – dynamic patterns of voice cues over time – which are found in vocal expressions (Juslin, 2001a, p. 317). The same mechanism is to be found in the practices of caregivers all over the world who sing to young infants in an 'infant-directed' singing style – using both lullaby and playsong style – which is probably used in order to express emotional information and to regulate their infant's state (Trainor and Schmidt, 2003).

This brings us to the domain of *feelings* and *emotions*. Music, as a rule, has an emotive meaning which calls forth mechanisms of 'affective semantics' with corresponding reactions which can vary between mild and strong emotional experiences (see Gabrielsson and Lindström, 2001). A core assumption in this context is the role of the body and its physiological responses as a 'theatre for emotional processing' and the mind–brain–body relationship in particular. It reminds us of James' view on the mechanisms of feelings and emotions (see Damasio, 1994, p. 129 for an extensive review) which strip the emotions down to a process that involves the body or the body sense that he considered to be the essence of the emotional response. To quote his famous words: 'If we fancy some strong emotion and then try to abstract from our consciousness of it all the feelings of its bodily symptoms, we find we have nothing left behind, no "mind-stuff" out of which the emotion can be constituted, and that a cold and neutral state of intellectual perception is all that remains' (James 1901, 1890, p. 451).

There has been a lot of controversy about this claim, which, for short, can be stated as the postulation of a mechanism in which particular stimuli in the environment excite – by means of an innate set of inflexible mechanisms – a specific pattern of body reactions. In this view, the body is the main stage for enactment of emotions, either direct or via its representation in somatosensory structures in the brain. Feelings, accordingly, are considered to be reflections of body-state changes.

One of the criticisms of this theory concerns the claim that we 'always' use the body as a theatre for emotions. It is possible, in fact, to re-enact emotions in emotional bodily states as well as to bypass this bodily enactment in a kind of 'as if' devices. This latter mechanism is made possible through neural devices that help us feel 'as if' we were having an emotional state, as if the body were being activated and modified. It means, further, that the brain learns to concoct the fainter image of an 'emotional' body state without having to re-enact it in the body proper (Damasio, 1994, p. 155). These 'as if' devices, however, are acquired only as the outcome of repeatedly associating the images of given entities or situations with the image of freshly enacted body states. In order, therefore, for the image to trigger the bypass device, the process must have been run or looped through the body theatre first (Damasio, 1994, p. 156). It should be noticed, further, that emotional responses target both the body proper and the brain. The latter, especially, is able to produce major changes in neural processing that constitute a substantial part of what is perceived as a feeling (Damasio, 2000, p. 288).

Another controversial point, according to Damasio (1994, p. 131), is James' claim that we do no need to 'evaluate' the significance of the stimuli in order for the reaction to occur. This holds true for some *primary* or *basic emotions* that human beings experience early in life, and for which a Jamesian pre-organised mechanism would suffice. They are easy to define – their listing embraces fear, anger, disgust, surprise, sadness, and happiness – and reflect the basically innate neural machinery that is required to generate somatic states in response to certain classes of stimuli. As such, they are inherently biased to process signals and to pair them with adaptive somatic responses with the processing proceeding in a preorganised fashion, which relies on wired-in dispositions to respond with an emotional reaction when certain features of stimuli in the world or in our bodies are perceived.

There are, however, other emotions that people experience in adult life and whose scaffolding has been built gradually on the foundations of these early emotions. These *secondary emotions* embrace reactions to a broad range of stimuli, which are filtered by an interposed voluntary and mindful evaluation. As such, they allow room for variation in the extent and intensity of preset emotional patterns (Damasio, 1994, p. 131).

The distinction is critical. According to Oatley (1992), it stresses the hybrid position of emotional processing, where 'nature meets nurture': emotive meaning can rely both on pre-programmed reactivity and on culturally established patterns or conventions of coping with the sound. It is commonly agreed that human beings are wired-in to respond with an emotion to certain features of stimuli in the world, but emotional reactions cannot be exhaustively explained by pre-programmed reaction patterns alone.

This same evolutionary perspective conceives of 'musical' emotions as *adaptive responses*, which according to Peretz, can be aroused silently in every human being as the product of neural structures that are specialised for their computation.

They can even be seen as *reflexes* in their operation, occurring with rapid onset, through automatic appraisal, and with involuntary changes in physiological and behavioural responses (Peretz, 2001, p. 115). This explains also that we often experience emotions as 'happening' to us, rather than chosen by us.

Further, there is evidence suggesting that emotional reactions to music do activate the same cortical, sub-cortical, and autonomic circuits which are considered to be the essential survival circuits of biological organisms in general (Trainor and Schmidt, 2003, p. 320). The cortical processing involves the higher functions of the brain; the sub-cortical processing affects the remainder of the body through the basic mechanisms of chemical release in the blood and the spread of neural activation. The latter, especially, incites people to react bodily to music with a whole bunch of *autonomic reactions* – changes in heart rate, respiration rate, blood flow, skin conductance, brain activation patterns, and even hormone release (oxytocin, testosterone) – which are induced by music or sound (see Becker, 2001, p. 145, for an overview) and which are driven by the phylogenetically older parts of the nervous system. They can be considered to be the 'physiological correlates' of listening to music and demonstrate convincingly that music does produce autonomic changes, which are associated with emotion processing (Trainor and Schmidt, 2003, p. 312).

Music, thus, 'induces' emotions; it is not simply 'about' emotions. It activates the autonomic system and physiological reactions can be demonstrated. There is, however, a distinction between mere physiological arousal and the interpretations or feelings that are the outcome of this arousal. As Damasio (2000, p. 79) puts it: we should distinguish between emotions and their resultant feelings, with the former referring to the autonomic nervous system arousal and the latter to the complex cognitive, culturally inflected interpretation of emotion. As such, it is possible to separate emotional *arousal*, which is clearly a universal response to musical listening, from the more cognitive concept of *feelings* which relate directly to fundaments and beliefs that are learned within a cultural context and rely upon linguistics categories (Becker, 2001, p. 145).

The question, then, arises which musical emotions can readily be distinguished as proper and distinct categories. Certain musical emotions can be categorised as happiness, sadness, anger, and fear, with the happy and sad emotional tones tending to be among the easiest ones to communicate in music (Gabrielsson and Juslin, 1996; Krumhansl, 1997). They might be expressed by similar structural features across musical styles and cultures, somewhat like facial expressions. The claim, nonetheless, is still somewhat tentative as the search of universals in the expression of musical emotions has just begun (see Balkwill and Thompson, 1999; Juslin, 2001). There is, however, increasing evidence that musical emotions are quickly and easily perceived by members of the same culture. Emotional judgements, as a rule, exhibit a high degree of consistence, suggesting that perception of emotions in music is natural and effortless for the large majority of listeners (Peretz, 2001, p. 114).

It can be questioned, further, whether it is possible to relate specific emotional reactions to specific *structural features* of the music. In an attempt to tease out some of them, Sloboda (1991) collected a list of features that produced particular physical emotional responses. A harmonic descending cycle of fifths to tonic, a melodic appoggiatura, and a melodic or harmonic sequence all tend to produce a 'tears' response. New or unprepared harmony seems to produce a 'shivers' response. Another example is the finding by Schubert (1999) that the best predictor of arousal is loudness.

There is, finally, a difference between *experiencing* and *recognising* emotions, somewhat related to paradigms of emotion, which rely on two broad systems for classifying them: *categories* and *dimensions* (Schubert, 2001, p. 401). The 'categorical' classification assumes that emotions, which carry different meanings – happiness, sadness, anger, fear, disgust – are distinct and independent entities which can be labelled as one single category. The 'dimensional' classification holds that all emotions are in some way related within an n-dimensional semantic space, or, more correctly, emotion space, with *valence*, *activity*, and *arousal* as dimensions. Happy and sad, for example, are opposite emotions along the valence dimension of emotion and sad and angry emotions express low and high arousal.

As such, it is clear that there is a distinction between tapping only into the arousal dimension as a default response to music, and having an articulated emotional response. The latter, in fact, represents complex combinations of perceptual, cognitive, and emotional aspects, which brings us to the role of cognitive penetration in the process of dealing with music.

4.4   The Role of Cognitive Penetration

Listening to music can be considered as a process that impinges upon innate mechanisms of coping with the sounds. Some of them act as lock-and-key and are likely to be induced by the music in a direct way. The human brain, however, is able to transcend its wired-in and reactive circuitry in order to build up new epistemic relations with its environment (Cariani, 2001, 2003; Reybrouck, 2005a). Understanding musical structures with at least some complexity, for example, is not to be explained merely in terms of mere reactive machinery, but relies on higher levels of nervous activity that can determine the perceptual outcome and cognitive functioning as well. To quote Damasio:

> The evidence on biological regulation demonstrates that response selections of which organisms are not conscious and are thus not deliberated take place continuously in evolutionarily old brain structures. Organisms whose brains only include those archaic structures and are avoid of evolutionarily modern ones—reptiles, for instance—operate such response selections without difficulty. One might conceptualise the response selections as an elementary form of decision making, provided it is clear that it is not an aware self but a set of neural circuits that is doing the deciding. ... Yet it is also well accepted that when social organisms are confronted by complex situations and are asked to decide in the face of uncertainty, they must engage systems in the neocortex, the evolutionarily modern sector of the brain. (1994, p. 127)

As soon as environments are getting more unpredictable and complex, however, the organism must not only rely on highly evolved genetically based biological mechanisms but also on suprainstinctual survival strategies that have developed in society and are transmitted by culture. The latter, as a rule, require consciousness, reasoned deliberation and willpower with response selections that are no longer reducible to the functioning of a set of neural circuits in the older structures of the brain.

Many perceptual phenomena, in this view, are *cognitively penetrated*, which means that the determination of the perceptual response depends on factors beyond the raw physical attributes of the stimulus (Uttal, 1998, p. 3). This holds true for perception in general, which is not to be explained merely in reductionist terms, which try to peripheralise the explanation of perception by locating the critical locus of perceptual experience in the earliest or lowest level of the sensory pathways. As Uttal puts it, there is simply no causal relation between sensory input and cognitive processing and resultant reactive behaviour: 'stimuli do not lead solely and inexorably to responses by simple switching circuit-like behaviour'. And further: 'The main point is that "cognitive penetration" plays a very much more important role in perception than is often acknowledged, in particularly by those who seek to show the neural correlates of some psychophysical response. The ubiquity of cognitive penetration argues strongly that neuroreductionist strategies attributing perceptual experience to a few peripheral neurones will be hard to justify' (1998, p. 213).

The only exception to these claims has been the centuries old trichotomy of the *sensory-transformational-response* mechanism with a grouping of cognitive processes into three classes: input, output and central transformation. There is, however, no convergence on a more detailed taxonomy (Uttal, 2001, p. 145–147).

5 Psychobiology and the Mind–Brain Relationship

There is a lot of freedom in the process of sense-making, which is induced by the impingement of physical stimuli upon the sensory system. Yet there is a whole body of empirical research about sensory coding and wired-in reactivity to environmental stimulation, which claims consistency in sensory responses. This has been the major claim of *psychophysics*, which should be seen as the exact science of the functional relations between the body and the mind (Fechner, 1860). Being concerned basically with the relations between the dimensions of 'physical stimuli' and the dimensions of 'sensory experiences', it has made enormous progress towards an understanding of the information processing carried out by sensory systems. *Sensory psychophysics*, especially, has progressed considerably towards this understanding. It is one of the most mature of the psychobiological sciences (Uttal, 1978, p. 16).

5.1 Psychophysics and Psychophysical Elements

Psychophysics, in its most narrow and traditional aspect, has concentrated on the psychological problem of judged magnitudes or intensities, with a major focus on the relationships between perceived magnitude and stimulus dimensions (Uttal, 1978, p. 415). Its experimental paradigm has been the *psychophysical experiment* in which a single stimulus dimension (S) is manipulated and a single behavioural response dimension (R) measured (Uttal, 1998, p. 199). The trouble, however, with this narrow conception is that it mistakes procedure for problems and precision for goals. Psychophysics, in this narrow sense, becomes synonymous with a few methods for the determination of thresholds (Stevens, 1951, p. 31).

In its broader aspects, psychophysics can be seen as the science of the response of organisms to stimulating configurations. As such, it sees the responses as indicators of an attribute of the individual which vary with the stimulus and which are relatively invariant from person to person. As a science it tries to contribute to the 'invariance' of the human nature in relying on seven major categories of *psychophysical problems*: (i) absolute thresholds (what are the stimulus values that mark the transition between response and no response?); (ii) differential thresholds (what is the resolving power of the organism, what is the smallest detectable change in a stimulus?); (iii) equality (what values of two different stimuli produce the same response or appear equal on the scale of some attribute?); (iv) order (what different stimuli produce a set of responses or psychological impression that can be set in serial order?); (v) equality of intervals (what stimuli produce a set of responses successively equidistant on the scale of some attribute?); (vi) equality of ratios (what stimuli produce a set of responses bearing constant ratio?); and (vii) stimulus rating (with what accuracy (validity) and precision (reliability) can a person estimate the 'physical' value of a stimulus?) (Stevens, 1951, p. 33).

It is easy to translate these psychophysical claims to the realm of music and to argue for some reliable correlation between *acoustic signals* and their *perceptual processing*. This correlation, further, embraces the conversion between the acoustic level of musical stimuli and the level of meaning, as well as the lawfulness or arbitrariness of this transformation.

The use of the term 'psychophysical' with respect to music, however, should be distinguished from its use in the broader field of psychophysics. Traditionally, the term refers to the relationship between a physical variable, like 'decibels, and a perceptual variable, like 'perceived loudness' (see Boring, 1942, for a historical perspective). The relationship may be a logarithmic function as described in Fechner's Law, or an exponential relationship as described in Stevens' Power Law (Stevens, 1975). In the literature of music psychology, however, the term *psychophysical* has been used to refer to physical properties of music (e.g. tempo, pitch range, melodic complexity, rhythmic complexity) that may be defined and assessed independently of the musical conventions of any particular culture. It represents those qualities, which are not restricted to a particular musical style or culture, and which do not require detailed musical knowledge in order to interpret them or respond to them at an emotional level (Balkwill, 1997, p. 3–4). They mainly correspond to those characteristics of music to which basic auditory processes naturally respond (e.g. tempo/pulse speed). Tempo is an example: it can be measured in beats per minute (bpm), which is a quantifiable measure of occurrence over time, which could apply to any type of stimulus.

As such, there have been attempts to list up *psychophysical elements of music*. For the assessment of emotional content of Western music, e.g., the following elements have been summed up: tempo, modality, melodic contour, harmonic complexity, melodic complexity, rhythmic complexity, articulation, dynamics, consonance/dissonance, pitch register, and timbre (Holbrook and Anand, 1990; Gabrielsson and Juslin, 1996; Gerardi and Gerken, 1995; Kratus, 1993; Nielzén and Cesarec, 1982; Scherer and Oshinsky, 1977; see Balkwill, 1997, p. 13 for an overview).

5.2 Psychobiology and its Major Claims

The transition from physical or acoustical stimuli to mental experiences is a *psychobiological* problem. It revolves around the single axiom that the external world is represented within the organisms by patterns of neural activity. A main goal of psychobiology, therefore, is to integrate psychology and neurophysiology into one unified 'monistic' framework, which reduces sensory and psychological activities to patterns of neural activity (Uttal, 1973, p. 6). This *neural reductionism* aims at discovering how the brain makes neural patterns in its nerve-cell circuits and how it manages to turn them into mental patterns (Damasio, 2000, p. 9). As such, it tries to establish the psychoneural equivalents of perceptual responses and mental experiences.

Rather than being concerned with the relationship between the stimulus and the neural responses – the neurophysiological search for neural codes – or with the relationship between the stimulus and the total systems behaviour of the organism – the psychological claim – psychobiology is concerned with the classic *mind-body problem* itself, namely the relationship between the neural response dimension and the dimensions of experience (Uttal, 1973, p. 213). This is, in fact, the basic reductionism of the psychobiological axiom that psychological properties arise out of physiological processes, and that it is possible to seek explanations of psychological functions by observing the underlying physiological processes (Uttal, 1978, p. 10).

The three main themes of psychobiology, therefore, address three fundamental questions: (i) the localisation of function in the brain (a problem at the macroanatomical level: where in the brain are particular processes mediated?); (ii) the representation or coding (a problem at the microanatomic level: how do neural networks represent, encode, or instantiate cognitive processes?); and (iii) the dynamic change or learning (how does our brain adapt to experience, what changes occur in its neural networks as a result of experience, and how do these changes correspond to externally observed behaviour?) (Uttal, 1978, p. XV, 2001, p. 2). The questions point into the direction of neurobiological research.

6 The Neurobiological Approach

The last decades have brought a depth of understanding about the mind–brain relationship that goes beyond the more intuitive approaches of previous research. Scholars can now rely on empirical evidence from both neuroimaging and morphometric studies in order to elucidate brain areas, which mediate both the production and perception of music. As such, it is possible to refine the findings in three major domains: (i) localisation studies which try to find anatomical markers for musical skills; (ii) studies on the role of neural plasticity; and (iii) the search for structural and functional adaptation of the brain as the result of musical experience, both at a macrostructural and microstructural level.

6.1 Brain and Mind: Towards a New Phrenology

Localisation studies are typical of physiological psychology and modern cognitive neuroscience. They aim at relating mental functions to particular structures or localised regions of the brain and give, in a way, new impetus to the former science of *phrenology* – a theory which claimed to be able to determine the character, personality traits, and even criminology on the basis of the shape of the head. This science is mainly considered to be a pseudoscience with many historical aberrations, but nevertheless, it has received some credit as a protoscience for the finding that some brain areas can be related to specific mental functions.

Neuromusicology, similarly, is in continuous search of *anatomical markers* of typical or special skills (see Altenmüller, 2003, p. 347). It has been profoundly influenced by the idea of the *modularity of mind* (Fodor, 1983, 1985; Gardner, 1983; see Peretz and Morais, 1989 for musical applications) which conceives of 'modules' as specialised computational devices that are devoted to the execution of some biologically important functions. According to Fodor, it is possible to conceive of *input systems* as modular systems, relying on a whole bunch of arguments which all define the scope of modular functions: an input system is defined by a separate sensory modality; our awareness of sensory messages is obligatory; we are able to perceive only the input or output system without having knowledge of the codes or interpretative processes that led to a particular perception; the speed of perception precludes processing analyses; the system is relatively impermeable to cognitive, topdown influences; an input system is clearly associated with particular receptors, structures, and regions of the brain; when it breaks down, it produces highly specific behavioural distortions; and the ontogeny of the input system seems to be very similar from one person to another (Fodor, 1983, see also Uttal, 2001, p. 115).

It is possible also to conceive of a kind of modularity of *musical functions* with as controlling assumption that mental processes are analysable into distinct and separable modules. The assumption is twofold. It is related, first, to the topic of *musical capacities* or *competencies,* which should constitute a set of aptitudes or innate capacities with proper functions that depend very little on particular conditions of concrete training during childhood and adulthood (Imberty, 2004, p. 450). An additional postulate, secondly, is that these competencies have their equivalence in the internal functioning of the brain, which should mean that they correspond to defined and independent *modular neuronal systems.* We should be careful, however, not to generalise too quickly about modular functions with respect to music. According to Imberty (2004), they pertain only to those aspects of music, which are probably the least developed and least musical.

There have been several attempts, further, to solve the localisation problem: mapping mental functions and their localisation in the brain (functional brain mapping studies), investigating the acquisition of new skills and the corresponding neural changes which are associated with their mastering, and studying the effects of brain lesion in patients. The most important sources of information, however, have been provided by the ever-expanding array of studies which use both 'structural' and 'functional' brain-imaging techniques. They have been able to assign specific brain areas that mediate both the production and perception of music – including tonal, rhythmic, and emotive aspects of music processing (see Brown, et al., 2000, p. 17, for an overview). Musical skills or competencies, further, are multifaceted, they involve motor, auditory, and visual regions of the brain. We should be careful, however, not to confuse perceptual and performing skills in dealing with music, as both rely on different areas of the brain.

Some musical abilities, finally, have neural correlates and the asymmetry or dominance of certain homologue structures in the brain points in the direction of functional implications (Schlaug, 2001, p. 296). More in general, there are well-known hypotheses which assign analytical, step-by-step processing to the left frontotemporal brain regions and the more global way of processing and visuo-spatial associations to the right frontal and bilateral parieto-occipital lobes (Altenmüller, 2003, p. 349). As to the musical skills, the findings suggest that there are distinct neural subsystems – separate modules so to speak – for the processing of *pitch discrimination* and *time discrimination.* Some key brain areas in the distributed neural circuits underlie the perceptual and cognitive representation of central aspects of musical rhythm such as musical metre, tempo, pattern, and duration (Parsons, 2003, p. 260), while other areas underlie other aspect of music processing. Generally speaking, time structure seems to be processed to a greater extent in the left temporal lobe whereas pitch structure may be processed primarily in right temporal and supratemporal lobe networks (Altenmüller, 2003, p. 347).

More in particular, there is the prominent role of the *planum temporale* in the left superior temporal lobe of the human brain. Its surface area has long been taken as a structural marker for left hemisphere language dominance in right-handers – it is found to be asymmetric in normal right-handed samples with a greater leftward bias – but this same structure is also involved in auditory processing of musical stimuli. As such, it is of great interest for studies investigating the laterality of auditory processing with as major finding that musicians with *absolute pitch* – the ability to identify the pitch of any tone in the absence of a musical context or reference tone – show an increased left-sided asymmetry of the planum temporale. Its functional significance probably must be seen in the context of their ability to assign any pitch to a verbally labelled pitch class, this is categorical perception, which, in turn, influences the recognition memory for pitch (Keenan et al., 2001, p. 1402; Schlaug, 2001). Other cortical areas play a particular role in the temporal control of sequential motor tasks and their integration in bilateral motor behaviour – as in key and string players. Areas that have been established involve motor related regions such as the premotor and cerebellar cortex and the supplementary motor area of the cortex (Schlaug, 1995, p. 1048).

A somewhat related finding is the observation that there is a pattern of differences in the grey matter distribution between professional musicians, amateur musicians and non-musicians (Gaser and Schlaug, 2003, p. 9244).

6.2   Neural Plasticity and the Role of Adaptation

Localisation studies may seem both seductive and promising. Much of the 'localisation research' effort, however, is seriously misleading as well. According to Uttal, it has been based on incorrect assumptions that cannot be validated either in principle or in practice (Uttal, 2001, pp. 205, 214). It is argued, further, that localisation of high-level cognitive functions in specific regions of the brain is not achievable. In spite of some major accomplishments in the field of perception and motor behaviour, the picture is changing gradually from a rather static conception of particular brain *modules* to a dynamic conception of *reorganisational plasticity* of the adult cortex (see Pantev et al., 2003, p. 383).

Adaptability, however, is not working in a vacuum. It is triggered by challenging environments, which are able to modify the design of our brain circuitries. There are, further, two mechanisms that shape our interactions with the environmental world: the recruitment of more and more evolutionary younger structures of the brain as environments get more unpredictable and complex, and structural and functional adaptations of the brain tissue as the outcome of interactions with the world (Damasio, 1994, pp. 111–112).

Experiences, according to Damasio, cause synaptic strengths to vary within and across neural systems. They shape our neural design, not only as the outcome of our first experiences, but throughout the whole life span. Connections between neurones are built primarily during cerebral maturation processes in childhood, but even later in their life humans respond with considerable flexibility to new challenges. This means that the connectivity of the adult brain is only partially determined by genetics and early development, with a major role for modification through sensory experiences. The neural circuits, in other words, are not totally pre-given and wired-in but are repeatedly pliable and modifiable by continued experience (Damasio, 1994, p. 112; Pantev et al., 2003, p. 381).

Neural plasticity, in this view, may lead to use-dependent regional growth and structural adaptation in cerebral grey matter in response to intense environmental demands (Gaser and Schlaug, 2003, p. 9244). The basic principles are based on Hebb's ideas that learning may alter synaptic connectivity. In what is commonly known as 'Hebb's rule', it is asserted that effective connections between neurones are formed depending on synchronous activation: 'Cells that fire together, wire together' (Hebb, 1949; see also Pantev et al., 2003, p. 383). What really matters, are changes of synaptic efficacy with the conjunction of pre- and post-synaptic activity contributing to a strengthening of synaptic connections (Rauschecker, 2003, p. 357). The organisation of the brain thus seems capable of significant change to adapt to the changing demands of the environment, mainly as the result of an augmented simultaneous stimulation of neurones, which entrains a re-organisation of the functional neuronal networks in the brain. Our cerebral cortex, in other words, has the ability to self-organise in response to extrinsic stimuli.

All this holds also true for the process of dealing with music, with performing musicians as a typical example. Their unique training and practice involves both co-ordination and synchronisation of somatosensory and motor control on the one hand and audition on the other hand, along with the ability to memorise long and complex bimanual finger sequences, to translate musical symbols into motor sequences (sight-reading), and to perceive and identify tones absolutely in the absence of a reference tone (absolute pitch) (Pantev et al., 2003, p. 38; Schlaug, 2001). Musicians, therefore, can be used as an ideal subject pool to investigate the plastic nature of the human brain and to study functional and structural adaptation of the motor and auditory system in response to extraordinary challenges, which involve the unique requirements of skilled performance.

6.3   *Structural and Functional Adaptations*

The plastic changes in the brain that are the outcome of enhanced musical experience can be found at two levels: the gross-anatomical differences between professional musicians with absolute pitch and amateurs or laymen, and the subtle functional differences after musical training, which have to be sought in ever finer modifications of synaptic strength in distributed cortical networks (Rauschecker, 2003, p. 357). As such, it is possible to distinguish between *macrostructural* and *microstructural* adaptations.

It makes a difference, further, whether musical experience is meant to be 'active performance' or only 'listening' to music. The plastic changes that are triggered by playing a musical instrument are numerous and well-documented. They include changes such as the rapid unmasking of existing connections and the establishment of new ones. As such, both functional and structural changes take place in the brain in an attempt to cope with the demands of the activity of skilful playing (see Pascual-Leone, 2003, p. 396). Some of these changes, however, are triggered by listening as well.

As far as the *macrostructural* changes are concerned, there are changes, which are the outcome of instrumental playing, such as a difference in a measure of primary motor cortex size (Amunts et al., 1997, p. 210). Other findings are related to the planum temporale, the corpus callosum, and some representation areas.

The *planum temporale*, first, shows an increased leftward symmetry in musicians with absolute pitch recognition. There is, however, no agreement as to the development of this skill. Two critical notions relate to early exposure to music: almost all musicians with absolute pitch started musical training before the age of 7 and is it unlikely that an individual will develop absolute pitch if he/she commences musical training after the age of 11 (Keenan et al., 2001, p. 1406).

The *corpus callosum*, secondly, is the main inter hemispheric fibre tract that plays an important role in interhemispheric integration and communication. Its midsagittal size correlates with the number of fibres passing through this structure. In spite of its late development – it is one of the last main fibre tracts to mature in humans – increases in corpus callosum size have been observed until at least the third decade of human life. The maximal growth, however, is in the first decade, which is the period of presumed callosal maturation, which coincides with childhood increases in synaptic density and fine-tuning of the neural organisation. It has been proposed, further, that environmental stimuli, especially early in life, might affect callosal development (Lee et al., 2003, p. 205).

The corpus callosum, further, has been found to be larger in performing musicians who started their musical training before the age of 7. This adaptation can be interpreted as a morphological substrate of increased inter-hemispheric communication between frontal cortices (such as the pre-motor and supplementary motor cortex) subserving complex bimanual motor sequences (Schlaug, 1995, p. 1050). Somewhat generalising, it could be stated that environmental factors, such as intense bimanual motor training of musicians, could play an important role in the determination of callosal fibre composition and size, which, in turn, can be considered as an adaptive structural-functional process. It remains to be determined, however, whether the large corpus callosum of musicians with early commencement of training contains a greater total number of fibres, thicker axons, more axon collaterals, stronger myelinated axons, or a higher percentage of myelinated axons (Schlaug, 1995, pp. 1050–1051).

The *microstructural* adaptations, finally, must be located at the level of individual neurones and synapses. They reflect the functional plasticity of the brain which can lead to microstructural changes which have been found both as the result of brain lesion (de-afferentiation) and motor skill learning, and which aim at changing the efficacy of the neural connectivity.

As to the first, it has been shown that there is some cortical remodelling which is induced by de-afferentiation. This includes microstructural changes such as the strengthening of existing synapses, the formation of new synapses (synaptogenesis), axonal sprouting and dendrite growth (Pantev, et al., 2003, p. 385). As to the latter, there are similar significant microstructural changes, which are induced in motor-related brain regions as a consequence of intense and prolonged motor activity. The adaptations include an increased number of synapses per neurone as well as changes in the number of microglia, and capillaries, which can lead to volumetric changes detectable at a macrostructural level (Hutchinson et al., 2003, p. 943; Schlaug, 2001, p. 283).

It is clear that the brain may show some form of adaptation to extraordinary challenges. This is the case with the requirements for musical performance, which may cause some brain regions to adapt. The adaptation, however, does not always involve plastic structural changes. Musical expertise, in fact, influences auditory brain activation patterns, and changes in these activation patterns depend on the teaching strategies applied. To quote Altenmüller:

> brain substrates of music processing reflect the auditory learning 'biography', the personal experiences accumulated over time. Listening to music, learning to play an instrument, formal instruction, and professional training result in multiple, in many instances multisensory, representations of music, which seem to be partly interchangeable and rapidly adaptive. (Altenmüller, 2003, p. 349)

7 Conclusion

In this contribution I have argued for a broadening of the scope of traditional music research. I have stressed the role of the 'music user' besides the music, relying heavily on a 'processual approach' to music which conceives of music users as organisms that cope with their environment. Rather than thinking of music as a structure or artefact, I have emphasised the music user's mechanisms of sense-making and knowledge construction. This is basically an ecological approach to cognition, which considers the organism–environment interactions in a continuous attempt to make sense out of the perceptual flux.

Musical sense-making, accordingly, is shaped by previous and current interactions with the sonic world. There is a whole dispositional machinery which is coded in our genetic programs with reactions relying on pre-wired and innate neural circuitry. As such, it is possible to consider the biological grounding of musical epistemology. The musical code, in this view, holds a hybrid position between nature and nurture with as major questions: what is wired-in and what is acquired?

In order to answer these questions, I have elaborated on the contributions from psychophysics and psychobiology, which provide a lot of empirical evidence as to the topic of sensory coding. Their major claims evolve around the central axiom of 'psychobiological equivalence' between percepts, experience, and thought. They address the central question whether there is some lawfulness in the co-ordinations between sounding stimuli and the responses of music users in general. The answer is not obvious: there are psychophysical commonalities, which can be considered to be universal, but there is a lot of subjectivity as well. It makes sense, therefore, to conceive of biological, perceptual, and ecological constraints which act as 'biases', rather than to think in terms of 'causal relationships' between sounding stimuli and reactions to these stimuli. It allows us, further, to consider the idiosyncrasies of the individual music user in his/her attempts to make sense out of the sounding flux, and to conceive of the process of sense-making in terms of epistemic autonomy.

As such, there should be a dynamic tension between the 'nature' and the 'nurture' side of music processing, stressing the role of the musical 'experience proper'. Music, in fact, is a sounding and temporal art which has inductive power. The latter involves ongoing epistemic interactions, which rely on low-level sensory processing as well as on principles of cognitive economy, with the former referring to the nature side, and the latter to the nurture side of music processing. Cognitive processing, however, should be 'coperceptual' in order to take into account the full richness of the sensory experience. What I argue for, therefore, is the reliance on the nature side again, which ends up, finally, in what I call the 'nature-nurture-nature sequence' of music processing.

References

Altenmüller E (2003) How many music centres are in the brain? In: Peretz I, Zatorre R (eds) The Cognitive Neuroscience of Music. Oxford University Press, Oxford/New York, pp. 346–353

Amunts K, Schlaug G, Jäncke L, Steinmetz H, Schleicher A, Dabringhaus A, Zilles K (1997) Motor cortex and hand motor skills: structural compliance in the human brain. Human Brain Map 5:206–215

Arom S (2000) Prolegomena to a biomusicology. In: Wallin N, Merker B, Brown S (eds) The Origins of Music. The MIT Press, Cambridge, Massachusetts/London, pp. 27–29

Balkwill L-L (1997) Perception of emotion in music: a cross-cultural investigation. Unpublished PhD dissertation. York University, North York

Balkwill L-L., Thompson W (1999) A cross-cultural investigation of the perception of emotion in music: psychophysical and cultural cues. Music Percep 17:43–64

Barbieri M (2005) Life is "artifact-making". J Biosemiotics 1:113–142

Barkow JH, Cosmides L Tooby J (1992) The Adapted Mind: Evolutionary Psychology and the Generation of Culture.Oxford University Press, Oxford

Bateson G (1985) Mind and Nature. Fontana Paperbacks, London

Bateson G (1978/1973) Steps to an Ecology of Mind. Paladin: London/Toronto/Sydney/New York/Granada

Bechterew VM (1933/1917) General principles of human reflexology: An Introduction to the Objective Study of Personality.International Publishers, New York (Transl. by Murphy, E., Murphy, W)

Becker J (2001) Anthropological perspectives on music and emotion. In: Juslin PN Sloboda J (eds) Music and Emotion: Theory and Research.Oxford University Press, Oxford, pp. 135–160

Berlin B Kay P (1969) Basic Color Terms: Their Universality and Evolution.University of California, Berkeley/Los Angeles

Berne E (1964) Games People Play. Grove Press, New York

Blacking J (1973) How Musical is Man? Universtiy of Washington Press, Seattle.

Blacking J (1995) Music, Culture, and Experience: Selected Papers of John Blacking. Byron R (ed) University of Chicago Press, Chicago.

Boring EG (1942) Sensation and Perception in the History of Experimental Psychology. Appleton-Century-Crofts, New York

Bregman A (1981) Asking the "what for" question in auditory perception. In: Kubovy M, Pomerantz J (eds) Perceptual Organization, Lawrence Erlbaum Associates, Hillsdale, New Jersey, pp. 99–118

Bregman A (1990) Auditory Scene Analysis: The Perceptual Organization of Sound. The MIT Press, Cambridge, Massachusetts

Brier S (1999) Biosemiotics and the foundations of cybersemiotics. Semiotica, 127(1/4): 169–198

Brown D (1991) Human Universals. McGraw-Hill, New York

Brown S, Merker B, Wallin N (2000) An introduction to evolutionary musicology. In: Wallin N, Merker B, Brown S (eds) The Origins of Music. The MIT Press, Cambridge, Massachusetts/London, pp. 3–24

Brunner R 1998 'Univers de genèse, univers de réception'. In Miereanu C, Hascher X (eds.), Les Universaux en musique, Actes du 4e Congrès international sur la signification musicale, Publications de la Sorbonne, Paris, pp. 637–648.

Bystrina I (1983) Kodes und Kodewandel. Zeitschrift für Semiotik 5:1–22

Cariani P (1998) Life's journey through the semiosphere. Semiotica 120(3/4):243–257

Cariani P (2003) Cybernetic systems and the semiotics of translation. In: Petrilli S (ed) Translation Translation. Rodopi, Amsterdam, pp. 349–367

Cariani P (2001) Symbols and dynamics in the brain. Biosystems 60(1–3):59–83. Special issue on "Physics and evolution of symbols and codes"

Chomsky N (1965) Aspects of the Theory of Syntax. MIT Press, Cambridge, Massachusetts

Clynes M (1977) Sentics: The Touch of Emotion. Souvenir Press, London

Cook N 1998 Music: A very short introduction. Oxford, UK: Oxford University Press

Cross I (2003) Music, cognition, culture, and evolution. In: Peretz I, Zatorre R (eds) The Cognitive Neuroscience of Music. Oxford University Press, Oxford/New York, pp. 42–56

Damasio A (1994) Descartes' Error: Emotion, Reason, and the Human Brain. Harper Collins, New York

Damasio A (2000) The Feeling of What Happens. Body and Emotion in the Making of Consciousness. Vintage, London

Damasio A (2004) Looking for Spinoza. Joy, Sorrow and the Feeling Brain. Vintage, London

Dawkins R (1976) The Selfish Gene. Oxford University Press, Oxford

Deschênes B (1991) Music and the new scientific theories. International Review of the Aesthetics and Sociology of Music, 22(2):193–202

Drake C, Bertrand D (2003) The quest for universals in temporal processing in music. In: Peretz I, Zatorre R (eds) The Cognitive Neuroscience of Music. Oxford University Press, Oxford/New York, pp. 21–31

Durham W (1991) Coevolution: Genes, Culture and Human Diversity. Stanford University Press, Stanford, CA

Ekman P (1972) Universals and cultural differences in facial expressions of emotion. In: Cole J (ed) Nebraska Symposium on Motivation. University of Nebraska Press, Lincoln, pp. 207–283

Ekman P (1992) An argument for basic emotions. Cognition Emotion. 6:169–200

Ekman P (1992) Facial expressions of emotions: new findings, new questions. Psychological Science 3:34–38

Fechner T (1860) Elemente der Psychophysik. Breitkopf & Härtel, Leipzig

Feldman MW, Laland KN (1996) Gene-culture coevolutionary theory. Trends in Ecology and Evolution, 11:453–457.

Fodor J (1983) The Modularity of Mind. MIT Press, Cambridge, Massachusetts

Fodor JA (1985) Précis of the modularity of mind. Behavioral and Brain Sciences 8:1–42

Frayer DW, Nicolay C (2000) Fossil evidence for the origin of speech sounds. In: Wallin N, Merker B, Brown S (eds) The Origins of Music. The MIT Press, Cambridge, Massachusetts/London, pp. 271–300

Frick R (1985) Communicating emotion: the role of prosodic features. Psychological Bulletin 97:412–429

Frith S (1996) *Performing rites: on the value of popular music*. Oxford, UK: Oxford University Press.

Gabrielsson A, Juslin PN (1996) Emotional expression in music performance: between the performer's intention and the listener's experience. Psychol Music 24:68–91

Gabrielsson A, Lindström E (2001) the influence of musical structure on emotional expression. In: Juslin PN, Sloboda J (eds) Music and Emotion: Theory and Research. Oxford University Press, Oxford, pp. 223–248

Gardner H (1983) Frames of Mind. The Theory of Multiple Intelligences. Basic Books, New York

Gaser Ch, Schlaug G (2003) Brain structures differ between musicians and non-musicians. J Neurosci 23(27):9240–9245

Geertz C (1983) Blurred genres: the refiguration of social thought. In: Geertz C. Local knowledge: Further essays in interpretative anthropology. Basic Books, New York, pp. 19–35

Gerardi GM, Gerken L (1995) The development of affective responses to modality and melodic contour. Music Perception 12(3):279–290

Gibson J (1961) Outline of a New Attempt to Classify the Senses and the Sensory Inputs. Unpublished manuscript. Available at: www.huwi.org/gibson/classify.php

Gibson J (1966) The Senses Considered as Perceptual Systems. Allen & Unwin, London

Gibson J (1979) The Ecological Approach to Visual Perception. Houghton Mifflin Company, Boston/Dallas/Geneva/Illinois/Hopewell/New Jersey/Palo Alto/London

Gibson J (1982) Reasons for Realism: Selected Essays of James J.Gibson. In: Reed E, Jones R (eds) Lawrence Erlbaum, Hillsdale, New Jersey

Goddard C (2002) The search for the shared semantic core of alle languages. In: Goddard C Wierzbicka A (eds) Meaning and Universal Grammar - Theory and Empirical Findings. Volume I. John Benjamins, Amsterdam, The Netherlands, pp. 5–40

Goldschmidt W (1959) Man's Way: A Preface to the Understanding of Human Society. Holt, New York

Greenberg H (ed) (1963) Universals of Language. MIT Press, Cambridge, Massachusetts/London

Hauser M, Chomsky N, Fitch WT (2002) The faculty of language: what is, who has it, and how did it evolve? Science 298:1570–1579

Hebb DO (1949) The Organization of Behavior. Wiley, New York

Hevner K (1935a) Expression in music: a discussion of experimental studies and theories. Psychol Rev 42:186–204

Hevner K (1935b) Experimental studies of the elements of expression in music. American J Psychol 48:248–268

Holbrook MB, Anand P (1990) Effects of tempo and situational arousal on the listener's perceptual and affective responses to music. Psychol Music 18(2):150–162

Huron D (2003) Is music an evolutionary adaptation? In: Peretz I, Zatorre, R (eds) The Cognitive Neuroscience of Music. Oxford University Press, Oxford/New York, pp. 57–75

Hutchinson S, Hui-Lin Lee L, Gaab N, Schlaug G (2003) Cerebellar volume in musicians. Cerebral Cortex 13:943–949

Imberty M (2004) The question of innate competencies in musical communication. In: Wallin N, Merker B, Brown S (eds) The Origins of Music. The MIT Press, Cambridge, Massachusetts/London, pp. 449–462

James W (1901, 1890) The Principles of Psychology. Vol.1, Macmillan, London.

Jerison H (2000) Paleoneurology and the biology of music. In: Wallin N, Merker B, Brown S (eds) The Origins of Music. The MIT Press, Cambridge, Massachusetts/London, pp. 177–196

Jiranek J (1998) 'Réflexions sur les constances de sémantique musicale' In Miereanu C, Hascher X (eds.), *Les Universaux en musique, Actes du 4e Congrès international sur la signification musicale*, Publications de la Sorbonne, Paris, pp. 60–65.

Juslin P (2001) Communicating emotion in music performance: a review of and theoretical frame-work. In: Juslin PN, Sloboda J (eds) Music and Emotion: Theory and Research. Oxford University Press, Oxford, pp. 309–337

Juslin P, Sloboda J (2001a) Music and emotion: introduction. In: Juslin PN, Sloboda J (eds) Music and Emotion: Theory and Research. Oxford University Press, Oxford, pp. 3–20

Juslin PN, Sloboda J (eds) (2001b) Music and Emotion: Theory and Research. Oxford University Press, Oxford

Keenan JP, Thangaraj V, Halpern A, Schlaug G (2001) Absolute pitch and planum temporale. NeuroImage 14:1402–1408

Koffka K (1935) Principles of Gestalt Psychology. Harcourt Brace, New York

Köhler W (1947) Gestalt psychology: An introduction to new concepts in modern psychology. Liveright, New York

Kon J (1998) 'Einige Bemerkungen über das Problem der musikalischen Universalien' In Miereanu C, Hascher X (eds.), Les Universaux en musique, Actes du 4e Congrès international sur la signification musicale, Publications de la Sorbonne, Paris, pp. 197–205

Kornilov KN (1930) Psychology in the light of dialectical materialism. In: Murchison C (ed) Psychologies of 1930. Clark University Press, Worcester, MA

Krechevsky I (1932a) Hypotheses in rats. Psychological Review, 39:516–532

Krechevsky I (1932b) "Hypothesis" versus "chance" in the presolution period in sensory discrimi-nation learning. University of California Publications of Psychology 6:27–44

Krechevsky I (1932c) The hereditary nature of "hypothesis". J Compar Psychol 16:99–116

Kratus J (1993) A developmental study of children's interpretation of emotion in music. Psychology of Music, 21:3–19

Krumhansl C (1997) An exploratory study of musical emotions and psychophysiology. Canadian J Exper Psychol 51:336–353

LeDoux JE (1996) The Emotional Brain: The Mysterious Underpinnings of Emotional Life. Simon & Schuster, New York

Lee D, Chen Y, Schlaug G (2003) Corpus callosum: musician and gender effect. NeuroReport, 14, 2, 205–209.

Lee H (1938) Perception and Aesthetic Value. Prentice-Hall, New York

Leech GL (1974) Semantics. Penguin Books, Middlesex, England

Leman M (1999) Naturalistic approaches to musical semiotics and the study of causal musical signification. In: Zannos I (ed) Music and Signs, Semiotic and Cognitive Studies in Music, ASCO Art and Science, Bratislava, pp. 11–38

Levine M, Shefer J (1981) Fundamentals of Sensation and Perception, Brooks-Cole Publishing, Pacific Grove, California

Liberman MY (1975) The international system of English. Indiana University Linguistics Club, Bloomington

Luhmann N (1990) Essays on Self-Reference.: Columbia University Press, New York

Luhmann N (1995) Social Systems. Stanford University Press, Stanford, California

Lumsden C, Wilson E (1981) Genes, Mind and Culture. Harvard University Press, Cambridge, UK

Mâche F-B (2000) The Necessity of and Problems with a Universal Musicology. In: Wallin N, Merker B, Brown S (eds) The Origins of Music. The MIT Press, Cambridge, Massachusetts/ London, pp. 473–479

Marconi L (1998) 'Universals in Music and Musical Experiences' In Miereanu C, Hascher X (eds.), Les Universaux en musique, Actes du 4e Congrès international sur la signification musicale, Publications de la Sorbonne, Paris, pp. 657–666

Maturana H (1978) Biology of language: the Epistemology of Reality. In: Miller G, Lenneberg E (eds) Psychology and Biology of Language and Thought. Academic Press, New York, pp. 27–46

Maturana H, Varela F (1980) Autopoiesis and Cognition: The Realization of the Living. Reidel, London

Meystel A (1998) Multiresolutional Umwelt: towards a semiotics of neurocontrol. Semiotica 120(3/4):343–380

Miereanu C, Hascher X (eds.) (1998) *Les Universaux en musique, Actes du 4e Congrès international sur la signification musicale*, Publications de la Sorbonne, Paris.

Miller G (2000) Evolution of human music through sexual selection. In: Wallin NL, Merker B., Brown S (eds) The Origins of Music. MIT Press, Cambridge, Massachusetts, pp. 329–360

Molino J (2000) Toward an evolutionary theory of music and language. In: Wallin N, Merker B, Brown S (eds) The Origins of Music. The MIT Press, Cambridge, Massachusetts/London, pp. 165–176

Mowbray R, Rodger T (1963) Psychology in Relation to Medicine. Lingstone, Edinburgh/London

Murdock GP (1945) The common denominator of cultures. In: Linton R (ed.) The Science of Man in the World Crisis. Columbia University Press, New York, pp. 123–142

Neisser U (1967) Cognitive Psychology. Appleton-Century/Crofts, New York

Nettl B (2000) An ethnomusicologist contemplates universals in musical sound and musical cultures. In: Wallin N, Merker B, Brown S (eds) The Origins of Music. The MIT Press, Cambridge, Massachusetts/London, pp. 463–479

Nielzén S, Cesarec Z (1982) Emotional experience of music as a function of musical structure. Psychology of Music 10(2):7–17

Normet L (1998) 'Universals and their subdivisions' In Miereanu C, Hascher X (eds.), *Les Universaux en musique, Actes du 4e Congrès international sur la signification musicale*, Publications de la Sorbonne, Paris, pp. 191–195.

Oatley K (1992) *Best Laid Schemes. The psychology of the Emotions.* Cambridge: Cambridge University Press.

Padilla A (1998) 'Les universaux en musique et la définition de la musique', in C. Miereanu, Hascher X (eds.), *Les Universaux en musique, Actes du 4e Congrès international sur la signification musicale*, Publications de la Sorbonne, Paris, pp. 219–230.

Paillard J (1994) L'intégration sensori-motrice et idéo-motrice. In: Richelle M, Requin J, Robert M (1994) Traite de Psychologie Experimentale.1, Presses Universitaires de France, Paris, pp. 925–996

Pantev C, Engelien A, Candia V, Elbert T (2003) Representational Cortex in Musicians. In: Peretz I, Zatorre R (eds) The Cognitive Neuroscience of Music. Oxford University Press, Oxford/New York, pp. 381–395

Parsons L (2003) Exploring the functional neuroanatomy of music performance, perception, and comprehension. In: Peretz I, Zatorre R (eds) The Cognitive Neuroscience of Music. Oxford University Press, Oxford/New York, pp. 247–268

Pascual-Leone A (2003) The brain that makes music and is changed by it. In: Peretz I, Zatorre R (eds) The Cognitive Neuroscience of Music. Oxford University Press, Oxford/New York, pp. 396–409

Pask G (1961a) An Approach to Cybernetics. Science Today Series. Harper & Brothers, New York

Pask G (1961b) The cybernetics of evolutionary processes and of self-organizing systems. Namur. Third International Conference on Cybernetics, Namur, Belgium, pp. 27–74

Pask G (1992) Different kinds of cybernetics. In: van de Vijver G (ed.) New Perspectives on Cybernetics: Self-Organization, Autonomy and Connectionism. Kluwer Academic, Dordrecht, pp. 11–31

Pavlov I (1926) Conditioned Reflexes: An Investigation of the Phyisological Activity of the Cerebral Cortex. Transl. and edited by Anrep GV, Dover, New York

Peretz I (2001) Listen to the brain: a biological perspective on musical emotions. In: Juslin PN, Sloboda J (eds) Music and Emotion: Theory and Research. Oxford University Press, Oxford, pp. 105–134

Peretz I, Morais J (1989) Music and modularity. Contemporary Music Review 4:277–291

Pinker S (2002) The Blank Slate. Viking Press, New York

Pinker S (1994) The Language Instinct. Morrow, New York

Pomerantz JR (1986) Visual form perception: an overview. In: Schwab E, Nusbaum H (eds) Pattern Recognition By Humans And Machines: Visual Perception, Vol. 2. Academic, Orlando, Florida, pp. 1–30

Port R, Leary A (2005) Against formal phonology. Language 81(4):927–964

Rauschecker JP (2003) Functional organization and plasticity of auditory cortex. In: Peretz I, Zatorre R (eds) The Cognitive Neuroscience of Music. Oxford University Press, Oxford/New York, pp. 357–365

Reybrouck M (1997) Gestalt concepts and music: limitations and possibilities. In: Leman M (ed.) Music, Gestalt and Computing. Studies in Cognitive and Systematic Musicology. Springer Verlag, Berlin-Heidelberg, pp. 57–69

Reybrouck M (1999) The musical sign between sound and meaning. In: Zannos I (ed.) Music and Signs, Semiotic and Cognitive Studies in Music. ASCO, Art and Science, Bratislava, pp. 39–58

Reybrouck M (2001a) Biological roots of musical epistemology: functional cycles, Umwelt, and enactive listening. Semiotica 134(1–4):599–633

Reybrouck M (2001b) Musical imagery between sensory processing and ideomotor simulation. In Godøy RI Jörgensen H (eds) Musical Imagery. Swets & Zeitlinger, Lisse, pp. 117–136

Reybrouck M (2003b) Musical semantics between epistemological assumptions and operational claims. In: Tarasti E (ed) Musical Semiotics Revisited. Acta Semiotica Fennica XV. International Semiotics Institute, Imatra, pp. 272–287

Reybrouck M (2004) Music cognition, semiotics and the experience of time. Ontosemantical and Epistemological Claims. Journal of New Music Research 33(4):411–428

Reybrouck M (2005a) A biosemiotic and ecological approach to music cognition: event perception between auditory listening and cognitive economy. Axiomathes. An International Journal in Ontology and Cognitive Systems 15(2):229–266

Reybrouck M (2005b) Body, mind and music: musical semantics between experiential cognition and cognitive economy. Trans. Transcultural Music Review 9. Available at: http://www.sibetrans.com/trans/ [ISSN:1697-0101], 46 pp

Reybrouck M (2006a) Music cognition and the bodily approach: musical instruments as tools for musical semantics. Contemporary Music Review 25(1/2):59–68

Reybrouck M (2006b) Musical creativity between symbolic modelling and perceptual constraints: the role of adaptive behaviour and epistemic autonomy. In: Deliège I, Wiggins G (eds) Musical Creativity: Multidisciplinary Research in Theory and Practice. Psychology Press, Oxford, pp. 42–59

Reybrouck M (2006c) The listener as an adaptive device: an ecological and biosemiotical approach to musical semantics. In: Tarasti E (ed.) Music and the Arts. Acta Semiotica Fennica XXIII - Approaches to Musical Semiotics 10. International Semiotics Institute/Semiotic Society of Finland, Imatra/Helsinki, pp. 106–116

Rosenberg L, Adelstein B (1993) Perceptual decomposition of virtual haptic surfaces. In: Proceedings IEEE 1993 Symposium on Research Frontiers in Virtual Reality, San Jose, California, October 1993, pp. 46–53

Scherer KR, Oshinsky JS (1977) Cue utiiization in emotion attribution from auditory stimuli. Motivation and Emotion 1:331–346

Scherer KR, Zentner KR (2001) Emotional effects of music: production rules. In:. Juslin PN, Sloboda JA (eds) Music and emotion: theory and research. Oxford University Press, Oxford, pp. 361–392

Schlaug G (1995) Increased corpus callosum size in musicians. Neuropsychologia 33(8):1047–1055

Schlaug G (2001) The brain of musicians. a model for functional and structural adaptation. In: Zatorre R, Peretz I (eds) The Biological Foundations of Music, Vol. 930. Annals of the New York Academy of Sciences, New York, , pp. 281–299

Schubert E (1999) Measuring emotion continuously: validity and reliability of the two dimensional emotion space. Australian Journal of Psychology 51:154–165

Schubert E (2001) Continuous measurement of self-report emotional response to music. In: Juslin PN, Sloboda J (eds) Music and Emotion: Theory and Research. Oxford University Press, Oxford, pp. 393–414

Seifert U (1993) Systematische Musiktheorie und Kognitionswissenschaft. Zur Grundlegung der kognitiven Musikwissenschaft. Verlag für systematische Musikwissenschaft, Bonn, Paris

Sekuler R, Blake R (1985) Perception. Alfred J.Knopf, New York

Sloboda J (1991) Music structure and emotional response: some empirical findings. Psychology of Music 19:110–120

Small C (1999) Musicking: the meanings of performance and listening. A lecture. *Music Education Research*, 1:9–21.

Stevens S (1951) Mathematics, measurement, and psychophysics. In: Stevens S (ed.) Handbook of Experimental Psychology.Wiley/Chapman & Hall, New York/London, pp. 1–49

Stevens SS (1975) Psychophysics: Introduction to its perceptiial, neural and social prospects. Wiley, New York

Trainor LJ, Schmidt, LA (2003) Processing emotions induced by music. In: Peretz I Zatorre R (eds) The Cognitive Neuroscience of Music. Oxford University Press, Oxford/New York, pp. 310–324

Trehub S (2000) Human processing predispositions and musical univerals. In: Wallin N, Merker B, Brown S (eds) The Origins of Music.The MIT Press, Cambridge, Massachusetts/London, pp. 427–448

Trehub S (2003) Musical predispositions in infancy: an update. In: Peretz I, Zatorre, R The Cognitive Neuroscience of Music. Oxford University Press, Oxford, pp. 3–20

Uttal W (1973) The Psychobiology of Sensory Coding. Harper & Row, New York/Evanston/San Francisco/London

Uttal W (1978) The Psychobiology of Mind. Lawrence Erlbaum, Hillsdale, New Jersey

Uttal W (1981) A Taxonomy of Visual Processes. Lawrence Erlbaum Associates, Hillsdale, New Jersey

Uttal W (1998) Toward a New Behaviorism. The Case Against Perceptual Reductionism. Lawrence Erlbaum Publishers, Mahwah, New Jersey/London

Uttal W (2001) The New Phrenology. The Limits of Localizing Cognitive Processes in the Brain. The MIT Press, Cambridge, Massachusetts/London

van Rooij R (2005) Evolutionary motivations for semantic universals. Paper presented at the Blankensee conference, Berlin

von Foerster H (ed) (1974) Cybernetics of Cybernetics. University of Illinois, Illinois

von Foerster H (1984) Observing Systems. Seaside. Intersystems Press, California

Wallin N (1991) Biomusicology. Neurophysiological and Evolutionary Perspectives on the Origins and Purposes of Music. Pendragon Press, New York

Wallin N, Merker B, Brown S (eds) (2000) The Origins of Music. The MIT Press, Cambridge, Massachusetts/London

Wertheimer M (1923) Untersuchungen zur Lehre von der Gestalt, II. Psychologische Forschung, 4:301–350. Translated as Laws of organization in perceptual forms. In: Ellis WD (ed.) A Source Book of Gestalt Psychology. Routledge & Kegan Paul, London

Wierzbicka A (1996) Semantics: Primes and Universals. Oxford University Press, Oxford

Index

Biosemiotics

1. M. Barbieri: *The Codes of Life*. The Rules of Macroevolution. 2008

ISBN 978-1-4020-6339-8